Exploring Monte Carlo Methods

Exploring Monte Carlo Methods

Second Edition

William L. Dunn

J. Kenneth Shultis

ELSEVIER

Elsevier
Radarweg 29, PO Box 211, 1000 AE Amsterdam, Netherlands
The Boulevard, Langford Lane, Kidlington, Oxford OX5 1GB, United Kingdom
50 Hampshire Street, 5th Floor, Cambridge, MA 02139, United States

Notices

Knowledge and best practice in this field are constantly changing. As new research and experience broaden our understanding, changes in research methods, professional practices, or medical treatment may become necessary.

Practitioners and researchers must always rely on their own experience and knowledge in evaluating and using any information, methods, compounds, or experiments described herein. In using such information or methods they should be mindful of their own safety and the safety of others, including parties for whom they have a professional responsibility.

To the fullest extent of the law, neither the Publisher nor the authors, contributors, or editors, assume any liability for any injury and/or damage to persons or property as a matter of products liability, negligence or otherwise, or from any use or operation of any methods, products, instructions, or ideas contained in the material herein.

ISBN: 978-0-12-819739-4

For information on all Elsevier publications
visit our website at https://www.elsevier.com/books-and-journals

Publisher: Katey Birtcher
Editorial Project Manager: Sara Valentino
Production Project Manager: Kamatchi Madhavan
Designer: Patrick Ferguson

Typeset by VTeX

Printed in the United States of America

Last digit is the print number:
9 8 7 6 5 4 3 2 1

Working together
to grow libraries in
developing countries

www.elsevier.com • www.bookaid.org

To the women in our lives

Contents

3.   Pseudorandom Number Generators

4.   Sampling, Scoring, and Precision

5.   Variance Reduction Techniques

6.   Markov Chain Monte Carlo

7. Inverse Monte Carlo

8. Linear Operator Equations

9.   The Fundamentals of Neutral Particle Transport

10.   Monte Carlo Simulation of Neutral Particle Transport

E.   Some Popular Monte Carlo Codes for Particle
       Transport

F.   Minimal Standard Pseudorandom Number Generator

About the Authors

William L. Dunn, born in Rapid City, South Dakota, in 1946, received a BS (1968) degree in Electrical Engineering from the University of Notre Dame in Notre Dame, Indiana. He then received MS (1970) and PhD (1974) degrees, both in Nuclear Engineering, from North Carolina State University (NCSU) in Raleigh, North Carolina. Dr Dunn was employed as an in-house Nuclear Engineering consultant at Carolina Power and Light Company from 1974 until 1979, when he returned to NCSU. From there, he entered into a career in contract research, first at Research Triangle Institute, then with Applied Research Associates, and finally with Quantum Research Services, a firm that he and Dr Fearghus O'Foghludha formed in 1988.

During those years in contract research, Dr Dunn was Principal Investigator on projects for several Federal and State government agencies. He also performed research for University and corporate clients. For several years he was Manager and then Director of the North Carolina bid to serve as the sight of the Superconducting Super Collider (SSC). This led to his involvement in research on the solenoidal detector collaboration. In 2002, Dr Dunn returned to academia by accepting a position in the Mechanical and Nuclear Engineering Department at Kansas State University (KSU), where he has remained. During his tenure there, he served as Department Head from 2013 to 2019. In 2015, he was recognized with the Radiation Science and Technology Award of the Isotopes and Radiation Division of the American Nuclear Society (ANS).

J. Kenneth Shultis, born in Toronto, Canada, in 1941, graduated from the University of Toronto with a BASc degree in Engineering Physics. He gained his MS (1965) and PhD (1968) degrees in Nuclear Science and Engineering from the University of Michigan. After a postdoctoral year at the Mathematics Institute of the University of Groningen, the Netherlands, he joined the Nuclear Engineering faculty at Kansas State University in 1969, where, 53 years later, he continues teaching, researching, and writing technical papers and books.

Professor Shultis teaches and conducts research in neutron and radiation transport, radiation shielding, reactor physics, numerical analysis, particle combustion, remote sensing, and utility energy and economic analyses. He has had a rich collaboration in research and scholarship. Besides being coauthor of this book, he has coauthored the books *Fundamentals of Nuclear Science and Engineering, Radiation Shielding, Radiological Assessment, Principles of Radiation Shielding*, and *Radiation Detection*. In addition, he has produced over 150 research papers and reports and served as a consultant to many private and governmental organizations. He is a Fellow of the ANS, and he has received many awards for his teaching and research, including the infrequently awarded ANS Rockwell Lifetime Achievement Award for his contributions over 50 years to the practice of radiation shielding.

Preface to the Second Edition

When the publisher approached us to write a second edition of this book, we did not hesitate because such an endeavor would provide us an opportunity to explore further the use of Monte Carlo (MC) methods. In the 10 years since the first edition, the scientific literature has seen a burgeoning of MC use in studies made in numerous quantitative disciplines. New applications, new MC methods, more robust random number generators (RNGs), and adoption of MC algorithms to the massively parallel computer systems currently being introduced have made MC an indispensable tool for very many researchers.

The vast number of new MC applications and the often sophisticated mathematical analyses that support these new approaches or better algorithms for MC calculations make a detailed review of the advances of the past two decades a Herculean task far beyond what can be covered in an introductory book on MC. A detailed review would also be a Sisyphean endeavor because of the continual advances being made in MC methods, their use in new applications, the great overlap in "new" discoveries among disciplines, and the different semantic descriptions used within different disciplines.

In this second edition, we have added new results that have broad application, that piqued our curiosity, and that can be described using only basic undergraduate mathematics or can, at least, be summarized without the need to understand the advanced, and often opaque, mathematics that underpin these results. Such is the case for the chapter describing RNGs. Over the past 30 years, the production of new RNGs has become a cottage industry with numerous "new" generators being proposed. The chapter on RNGs has been expanded substantially. In particular, generators based on linear feedback shift registers and on cellular automata have now been included.

Other new additions to the book include slightly expanded histories of MC and the Metropolis–Hastings algorithm. Likewise the discussion on Markov chain Monte Carlo (MCMC) has been expanded to emphasize the differences and similarities between MCMC and ordinary MC. The great utility of the Metropolis–Hastings algorithm, introduced in the context of MCMC, has revolutionized the power of Bayesian statistical analyses over the past quarter-century. In this second edition, the Bayesian approach has been given far greater attention. The presentation on inverse and symbolic MC has also been revised with an enhanced discussion of the applications of inverse techniques to various problems. The appendix listing many probability distributions and their properties has been rearranged and augmented by the inclusion of yet more distributions.

The chapter on the use of MC to solve mathematical equations that involve a linear operator has doubled in size. New material includes coupled ordinary differential equations, different boundary conditions for partial differential equations, and time-dependent partial differential equations. In the last category, transient heat conduction, based on the less well-known integral form of the heat conduction equation, is discussed. The MC example results are usually compared to analytical results that often are given as infinite sums. To help the interested reader understand the analytic results, a tutorial on linear operators and the importance of their eigenfunctions is given in a new appendix.

As before, the new material is accompanied with example problems. We have also tried to improve certain sections that our students found lacking. Of course, we have tried to fix the embarrassing number of known typographical errors in the first edition, but, undoubtedly, have replaced them with new errors in this edition. Because the electronic files used by the publisher were again prepared by the authors using LaTeX, we must again accept responsibility for all errors that appear in the second edition. The authors welcome comments and would like to be informed of typographical and other errors that a reader may discover.

Finally, we are forever indebted to "the women in our lives," namely, our wives and daughters, who have endured too many evenings of neglect as we labored on this treatise and our other professional activities.

William L. Dunn
J. Kenneth Shultis
Manhattan, KS
July 2021

Preface to the First Edition

Monte Carlo methods for solving various numerical problems have been used for over a century, but saw relatively limited use until the past several decades when inexpensive digital computers with immense computational power became ubiquitous. Today, Monte Carlo methods dominate many areas of science and engineering, largely because of their ability to treat complex problems that previously could be approximated only by simplified deterministic methods. Various Monte Carlo methods now are used routinely by scientists, engineers, mathematicians, statisticians, economists, and others who work in a wide range of disciplines.

This book explores the central features and tools that allow Monte Carlo methods to be useful in such a wide variety of applications. The emphasis in this book is on the practical implementation of Monte Carlo rather than on strict mathematical rigor. Our references cite such rigor, but most Monte Carlo users need not wade through the arcane proofs that underlie Monte Carlo methods in order to use those methods effectively. The primary purpose of this book is to serve as an introductory textbook on Monte Carlo, at the undergraduate or graduate level. Another purpose is to provide a text that readers can use to teach themselves how to utilize Monte Carlo methods. The book also is intended to be of use to readers who explore new subjects for the joy that can be derived from learning. Finally, it is hoped that those experienced in Monte Carlo methods may find the book useful as a general reference.

The idea for a basic book on Monte Carlo was conceived many years ago, shortly after one of us (WLD) learned how to apply Monte Carlo to radiation transport problems from reading various accounts written in the 1950s and 1960s. Events conspired to keep him from devoting the time necessary to complete the task, until he joined the faculty of the Mechanical and Nuclear Engineering Department at Kansas State University in 2002. There he found not only the opportunity but also a willing colleague (JKS) to provide valuable assistance to the endeavor. Fortunately, new ideas have emerged in the interim and the book now serves a greater purpose than originally envisioned.

The first eight chapters of this book are generic, in that they apply to Monte Carlo methods that can be used in any field of application. The next two chapters focus on applications to radiation transport, because both authors are nuclear engineers and have experience in this area. However, the authors have endeavored to make the entire book sufficiently generic and accessible so that it can be useful to practitioners in many fields. Thus, the chapters on radiation transport focus on basic issues and provide guidance as to how the generic Monte Carlo methods can be applied in a particular field.

Each of the chapters incorporates examples and is followed by a list of problems or exercises. Five appendices are included that provide useful supporting material, such as descriptions of a number of probability distributions and a summary of several of the more popular general-purpose radiation transport simulation codes.

We cannot adequately recognize and thank all of those who have contributed to the creation of this work. Dr Robin P. Gardner first exposed one of us (WLD) to Monte Carlo in the late 1960s and has continued to be an inspiration in many ways. We gratefully acknowledge the reviews of substantial portions of drafts of this book that Richard E. Faw and Clell J. Solomon provided. We also thank Richard Hugtenburg, who kindly reviewed the chapter on Markov Chain Monte Carlo. Additionally, we note that many practitioners in the field of Monte Carlo provided substantial input to Appendix D. These people are true experts in Monte Carlo who not only have produced valuable general-purpose codes but also have taken the time to summarize their codes for the benefit of the readers of this book. The many students who have waded through early drafts and who pointed out deficiencies and places where clarifications were needed have been a tremendous help to us. Also, we thank the LAB for its hospitality during the many hours spent there revising early drafts.

The electronic files used by the publisher were prepared by the authors using LATEX and, consequently, we must accept responsibility for all errors that appear in the book. The authors welcome comments and would like to be informed of typographical and other errors that readers may discover.

Finally, we acknowledge the support and inspiration that our wives and children have provided us throughout our professional careers.

William L. Dunn
J. Kenneth Shultis
Manhattan, KS
September 2010

Chapter 1

Introduction

Many things Comte Buffon did adore,
among them dropping needles galore.
He did this you see,
because to his glee,
he could find π as never before.

The Monte Carlo (MC) method for obtaining estimates of answers to well-posed problems, i.e., problems for which quantitative answers exist, has become an essential tool for analysts. Its current widespread application is due partially to the availability in the past few decades of affordable and powerful digital computers. However, the real appeal of MC is that it permits the analyst to address more complex problems than can be treated by analytical techniques, which can often solve only idealized problems such as those, for example, with simple geometries. Moreover, to obtain MC solutions, simple arithmetic is often all that is needed!

But why does MC work? Simply put, it uses a sequence of random numbers (RNs)[1] which, by definition, have no structure, no matter how subtle. Given one RN one cannot predict the next number in the sequence, or even give a probability estimate of what the next number is. And yet from these formless chains of RNs an estimate of the quantitative answer to a well-posed problem can be obtained. How is this possible?

Each RN (or subset of the sequence of RNs), when "applied" to the problem of interest, acts as a probe that elicits a response or score. Some numbers produce a relatively large score; others generate little or no score. From all the scores produced by many RNs (often thousands to trillions), an answer to the problem can be estimated. Further, the uncertainty in this estimate can itself be estimated. Furthermore, these estimates are obtained using only simple arithmetic.

The transformation of scores from random probes to an answer is a consequence of two important theorems that underpin the MC method: Bernoulli's law of large numbers (LLN) and the central limit theorem (CLT), neither of

---

[1] More often a sequence of *pseudo*random numbers are used. These numbers are generated by a structured algorithm and, while not truly random, exhibit many characteristics of being random, as is discussed in Chapter 3.

Exploring Monte Carlo Methods. https://doi.org/10.1016/B978-0-12-819739-4.00009-3

which needs to be fully understood by an analyst using the MC method, although the rules of how to apply them must be. We now begin to explore the Monte Carlo method and apply it to a wide variety of problems.

1.1 What Is Monte Carlo?

The analysis technique called *Monte Carlo* is, in essence, a methodology to use sample means to estimate population means. The term MC was coined when digital computers were first used to implement the procedure that, until then, had been called statistical sampling. In its simplest form, MC turns bits into meaningful rational numbers. Run an experiment that leads to either success (one) or failure (zero). Repeat the experiment N times. Call N the number of *histories* or *trials*. Then divide the number of successes S by the number of trials. There are $N + 1$ possible outcomes $(0, 1/N, 2/N, \ldots, N/N = 1)$. If N is large, then the quantity S/N gives a good approximation to the average or expected value of the experiment. In this form, each history is a bit (a zero or a one) and the result is a rational number. In the next section, it is shown how this simple process can be used to estimate the irrational number π.

The discrete formulation of MC becomes much more powerful in the fuller, continuous framework. MC allows one to create viable answers to complex questions by appropriate actions with repetitive histories. In this more usual form, each history yields a real number (or a floating-point number created on a digital computer) or a collection of such numbers. In other forms, histories can lead to algebraic functions. However, the result or "score" is still the sum of the histories divided by the number of trials.

More broadly, MC is a widely used numerical procedure that allows people to estimate answers to a wide variety of problems (Dunn and Shultis, 2009). In essence, MC is a highly flexible and powerful form of quadrature, or numerical integration, that can be applied to a very wide range of problems, both direct and inverse. One of the unique aspects of MC is that it allows the user a high degree of flexibility. One can introduce intuition into the solution, as long as one follows certain simple rules. This means that there are many ways to solve a given problem by MC, and the investigator can experiment with different ways. Because MC involves simple techniques to "solve" complex problems, it is powerful; because the technique is flexible, it is interesting and even fun.

Partially because of its fanciful name and partially because it seems to rely on games of chance, the first impression one may get about MC is that it is a somewhat mysterious hit-or-miss technique that really should not to be trusted. Indeed, according to Halton (1970), Trotter and Tukey (1954) claimed "the only good Monte Carlos are dead Monte Carlos." However, MC simulation is a sound technique based on two mathematical theorems. If you were to go to the city of Monte Carlo and try your hand at some games of chance offered there, you might consider yourself lucky to win more money than you lost; however, if you apply the MC technique to your problem, i.e., to your "game," and play the game correctly, you almost always win.

How can MC be used to solve "real-world" problems? What are the tools one needs to extend this technique to solve increasingly complex problems? Can MC be used to solve design and optimization problems? How can one quantify the uncertainty in one's estimate of a complex outcome? These and many other issues are explored in the remainder of this book.

1.2 A Brief History of Monte Carlo

This section is based on brief accounts of the background of MC provided by several authors. We especially wish to acknowledge the review articles prepared by Nicholas Metropolis (one of the founders of MC) (Metropolis, 1987), Chapter 1 of Hammersley and Handscomb (1964), the interesting biographical summaries available on the world-wide web provided by the School of Mathematics and Statistics at the University of St. Andrews, Scotland, Chapter 11 of the shielding book by Shultis and Faw (2000), and a recent article by Sood et al. (2021).

The history of MC can be traced back to the interesting Bernoulli family. Nicolaus Bernoulli (1623–1708) was a Belgian merchant who took over the family spice business in Basel, Switzerland. (His father, a Protestant, had left Amsterdam for religious and political reasons and moved his spice business to Basel.) Among Nicolaus' many children were two brothers gifted in mathematics, Jacob (1654–1705) (see Fig. 1.1) and Johann (1667–1748). Jacob (a.k.a. Jacques or James) earned his degrees in philosophy and theology but loved mathematics and, in

Figure 1.1 Jacob Bernoulli (1654–1705).

1687, was appointed professor of mathematics in Basel. He initially collaborated with his brother Johann, but the relationship soured as the two were fiercely competitive. Jacob produced, in 1689, the original statement of the LLN, on which MC is based (see Chapter 2). According to Maistrov (1974), this law was eventually published in Jacob's posthumous "Ars Conjectandi" (Bernoulli, 1713) or "The Art of Conjecture." Whereas Jacob must have conducted some exploratory tests, the law does not appear to have been put to extensive use at the time to estimate expectations of complex phenomena.

Also it is interesting to note that Johann tutored the French mathematician Guillaume François Antoine Marquis de l'Hôpital (1661–1704), and in fact Johann first enunciated what is now known as l'Hôpital's rule, in notes he supplied (for pay) to l'Hôpital. Both Jacob and Johann made many important contributions to mathematics. Together, they were founders of the branch of mathematics known as calculus of variations. In his history of calculus, Jacob was the first to use the term *integral* in its present sense. He was also the first to solve what is known as the *Bernoulli equation*, he gave us the *Bernoulli numbers*, and he introduced probability as a quantity and not just a concept.

Almost a century later, George-Louis Leclerc, Comte de Buffon (see Fig. 1.2), proposed (Buffon, 1777) that if needles or sticks of length L were dropped randomly onto a plane surface that had parallel lines at a spacing of $D \geq L$, then the probability P_{cut} that a needle would cut (overlap or intersect) one of the lines could be expressed as (see Example 1.1)

$$P_{cut} = \frac{2L}{\pi D}. \tag{1.1}$$

Pierre-Simon de Laplace showed (Laplace, 1786) that such a procedure could be used to estimate the value of π, because L and D are known and one can estimate P_{cut} from the number of successes (needles cutting a line, n_c) over trials (needles dropped, n_d). Thus,

$$\pi = \frac{2L}{P_{cut}D} \simeq \frac{2L}{(n_c/n_d)D}. \tag{1.2}$$

This problem is an excellent example of analog MC, in which one phenomenon (dropping sticks or needles) "simulates" another phenomenon (estimating the irrational number π). The situation is depicted in Fig. 1.3, in which 10 needles of length $L = 0.6D$ units are shown. For this case, Eq. (1.2) gives

$$\pi \simeq \frac{2(0.6)}{0.4(1)} = 3.$$

More accurate estimates of π can be obtained with larger numbers of needle drops.

The practice of estimating π by needle drops became somewhat fashionable in the nineteenth century and is recounted, for instance, by Hall (1873). Some, however, may not have played the game fairly. For example, the mathematician Lazzarini stated in 1901 that he, using needles with $L/D = 5/6$, had obtained $n_c = 1808$ intersections in $n_d = 3408$ needle drops, thereby producing an estimate of $\pi \simeq 3.1415929$, which is accurate to seven significant figures! However, as pointed out by Rao (1989) (reported by Gentle (1998)), this high accuracy is suspect. The best rational approximation for π with integers less than 16,000 is $\pi \simeq 355/113 \simeq 3.141592920....$[2] From Eq. (1.2) with $P_{cut} \simeq n_c/n_d$,

$$n_d \simeq \pi \frac{D}{2L} n_c \simeq \frac{355}{113} \frac{6}{2 \times 5} n_c = \frac{213}{113} n_c.$$

Because n_d must be an integer, n_c must be a multiple of the prime number 113. If $n_c = 16 \times 113$, then $n_d = 3408$. Assuming Lazzarini actually obtained the stated results, he could have done so by dropping 213 needles k times, until he obtained precisely $113k$ successes, which he apparently did after $k = 16$ attempts.

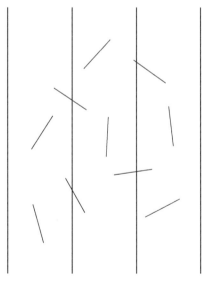

Figure 1.2 George-Louis Leclerc, Comte de Buffon (1707–1788).

Figure 1.3 A grid of parallel lines with 10 needles dropped at random.

---

[2] The next rational approximation for π is $52,163/16,604 \simeq 3.141592387...$, which has essentially the same accuracy as $\pi = 355/113$.

Example 1.1 Derivation of Eq. (1.1)

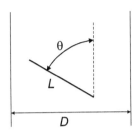

Consider the needle of length $L \leq D$ at an angle θ to the vertical shown to the left. The horizontal component of the needle is $L \sin\theta$, and, because the needle end has a uniform probability of being anywhere between the two lines, the probability a needle with angle θ cuts the left line is simply $P_{cut}(\theta) = (L \sin\theta)/D$ provided $L \leq D$. The probability a randomly dropped needle has an angle in $d\theta$ about θ is $d\theta/\pi$. Hence, the probability that a randomly dropped needle cuts a grid line is

$$P_{cut} = \int_0^\pi P_{cut}(\theta) \frac{d\theta}{\pi} = \int_0^\pi \frac{L \sin\theta}{D} \frac{d\theta}{\pi} = \frac{L}{\pi D} \int_0^\pi \sin\theta \, d\theta = \frac{2L}{\pi D}. \tag{1.3}$$

This derivation holds only for $L \leq D$. An alternative derivation, which can be generalized to needles of any length (see Problem 1.7), is given in Example 1.2.

Example 1.2 A More General Buffon's Needle Derivation

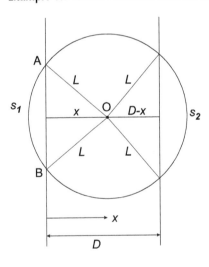

Consider the case shown in the figure to the left when one end of the needle is dropped at O, a distance x from the left line. Because all orientations of the needle are equally probable, the other end of the needle can be anywhere on the circle of radius L centered at O. The probability a needle intersects either the right or left line is just the sum of the arc lengths s_1 and s_2 outside the strip divided by the circle's circumference $2\pi L$. The arc length s_1 is L times the angle AOB, which can be expressed as

$$s_1(x) = \begin{cases} 2L\cos^{-1}(x/L), & 0 < x < L, \\ 0, & x \geq L. \end{cases} \tag{1.4}$$

Similarly

$$s_2(x) = \begin{cases} 2L\cos^{-1}((D-x)/L), & x > D - L, \\ 0, & 0 < x \leq D - L. \end{cases} \tag{1.5}$$

The end of the needle can land anywhere in the strip with equal probability, so the probability the needle end is dropped in dx about x is simply dx/D. The probability a needle with an end at x intersects one of the bounding lines is $[s_1(x) + s_2(x)]/(2\pi L)$. Hence the probability a dropped needle intersects either bounding line is

$$P_{cut} = \int_0^D \frac{dx}{D} \frac{s_1(x) + s_2(x)}{2\pi L}. \qquad (1.6)$$

Substitution for $s_1(x)$ and $s_2(x)$ and a change of variables yields

$$P_{cut} = \begin{cases} \dfrac{2}{\pi D} \displaystyle\int_0^D \cos^{-1}\left(\dfrac{x}{L}\right) dx & \text{for } D < L, \\[3ex] \dfrac{2}{\pi D} \displaystyle\int_0^L \cos^{-1}\left(\dfrac{x}{L}\right) dx & \text{for } D \geq L. \end{cases} \qquad (1.7)$$

For $D \geq L$, Eq. (1.7) simplifies to Eq. (1.1), namely, $P_{cut} = 2L/(\pi D)$. Evaluation of Eq. (1.7) for $D < L$ is slightly more complicated (see Problem 1.7).

The statistical nature of solutions to physical problems was studied in the early twentieth century by many famous scientists, including Einstein, Lord Rayleigh, Courant, and Fermi. Einstein's analysis concerned Brownian motion, a phenomenon first described in 1827 by Robert Brown, who observed with a microscope the apparent random motion of pollen grains in a droplet of oil. The explanation of this motion was slow in coming. In the 1870s, Ludwig Boltzmann introduced and developed the kinetic theory of gases which was based on the atomicity (or microscopic "graininess") of matter. The acceptance of this theory would have to wait until Einstein in 1905 published his doctoral dissertation and extended his concepts in a publication (Einstein, 1906). Einstein showed that Boltzmann's kinetic theory of atoms could explain very accurately Brownian motion. This is among many contributions that may have deserved the Nobel Prize for Einstein.

Lord Kelvin (born William Thomson, in 1824) would generate numbers at random by drawing bits of paper, from a glass jar, on which numbers had been written. He used these numbers for some of his studies of the kinetic theory of gases (Kelvin, 1901). This practice of picking RNs from jars was also used by other statisticians in Britain in the early twentieth century as a means of demonstrating statistical concepts to students. In addition, William S. Gosset, who published under the pseudonym "Student," used height and middle finger lengths from 3000 convicts to generate random samples from two correlated normal distributions. With such sampling, he studied the correlation coefficient and certain parameters of the t distribution (Student, 1908).

Lord Rayleigh commented on Bernoulli's theorem in an article before the turn of the twentieth century (Rayleigh, 1899). In this article, he showed that a random walk in one dimension could approximate the solution of a parabolic differential equation. In a seminal paper, Courant et al. (1928) applied the idea of random walks to elliptic partial differential equations with Dirichlet boundary conditions.[3] In 1931, Kolmogorov (1931) showed how stochastic processes (using what is now called Markov chains) could be used to analyze certain integro-differential equations.

The Italian physicist Enrico Fermi, in the early 1930s, used statistical sampling methods, before the name MC had been coined, in his studies of the slowing down of neutrons. According to Emilio Segrè, Fermi's student and collaborator, Fermi wowed his colleagues with his uncannily accurate predictions of their experimental results, which, unknown to them, were obtained by statistical sampling methods (Metropolis, 1987). The fact that Fermi had no computers to generate his RNs and had to do his calculations with mechanical devices or even in his head makes his accomplishments all the more remarkable. However, Fermi never published the results of these MC calculations and, as a consequence, he often is neglected in the historical record of MC.

MC became a practical analysis technique only in the mid-twentieth century, with the advent of digital computers. Once one could generate pseudorandom numbers (PRNs) and compute long summations, it became evident that the MC formalism was a great tool for estimating the behavior of neutrons and other radiation particles. Neutrons were of interest in the mid-twentieth century because of the effort to construct nuclear weapons. Just as the space program 20 years later gave impetus to computing and materials research, so too the nuclear weapons program gave impetus to the MC method as a numerical procedure for design and analysis. In fact, it was during efforts to design and test the hydrogen bomb that MC got its name.

Nicholas Metropolis (see Fig. 1.5) gives a fascinating account (Metropolis, 1987) of the contributions of Stanislaw Ulam (see Fig. 1.6), John von Neumann (see Fig. 1.7), and himself in resuscitating statistical mechanics methods using the ENIAC, one of the first electronic computers.[4] In 1945, von Neumann suggested to Metropolis that he develop a computational model to test the ENIAC. Metropolis, Stan Frankel, and Anthony Turkevich produced a model related to the hydrogen bomb, then being worked on by Edward Teller and others, and this model was run on the ENIAC.

---

[3] In the pre-1940 era, random walks in more than one dimension were usually referred as *random flights* (Haji-Sheikh and Howell, 2006).

[4] The ENIAC digital computer (see Fig. 1.4), funded by the U.S. Army Ordinance Corps, was designed and built at the University of Pennsylvania. It began operation in 1946. At the end of its life in 1956, the ENIAC contained 20,000 vacuum tubes, 70,000 resistors, 10,000 capacitors, and 1500 relays, all connected by about 5 million hand-soldered joints. It occupied about 300 square feet and consumed 150 kW of power. With special high-quality vacuum tubes, the frequency of tube failure was reduced from initially a few per day to one about every two days. The ENIAC could perform 5000 additions or subtractions, 385 multiplications, 40 divisions, or 3 square root operations per second.

Figure 1.4 The ENIAC after its relocation from the University of Pennsylvania to the Aberdeen Proving Ground, Maryland in 1947.

Figure 1.5 Nicholas Metropolis (1915–1999).

Figure 1.6 Stanislaw Ulam (1909–1984).

Figure 1.7 John von Neumann (1903–1957).

Ulam attended a session in early 1946 in which Metropolis presented the results of the ENIAC calculations. Ulam then suggested to von Neumann that the ENIAC could be used to conduct more general studies using statistical methods. All three (von Neumann, Metropolis, and Ulam) were gifted mathematicians who quickly realized the value of this suggestion. In early 1947, von Neumann proposed (in a letter to his superior Robert Richtmyer) that statistical methods be applied to solving neutron diffusion problems related to nuclear bomb research. It was Metropolis who, well aware of Ulam's interest in games of chance (for which the capital of Monaco, Monte Carlo, was well known), suggested the name *Monte Carlo*, and he and Ulam wrote a paper titled "The Monte Carlo Method" that was published in 1949 (Metropolis and Ulam, 1949). Several MC calculations were performed on the ENIAC (after it had been moved to Aberdeen, MD, in 1947) and the results were so useful that a widespread interest in MC methods was born.

In fact, the name *Monte Carlo* is most appropriate. In games of chance popular at Monte Carlo and in casinos, the "house" assures itself, by a small bias in the rules of each game, that it will win when averaged over a large number of players. However, any individual player also retains a chance to win in any particular game, hence providing the motivation and allure to play. In an MC simulation, an individual history can go anywhere in the problem space, teasing out various possibilities and their consequences; however, the investigator is assured that if enough histories are run (a large enough *sample* is chosen) and the game is played according to the rules, the procedure will converge toward the desired result.

Within a few years Herman Kahn wrote a Research Memorandum (Kahn, 1954) that formally discussed the methods of MC, including straightforward and rejection sampling, various probability distributions, evaluation of integrals, and variance reduction. Soon, several other books at least touched on MC as a means of numerical integration. One book (Cashwell and Everett, 1959) provided a practical manual on the use of MC methods and another (Hammersley and Handscomb, 1964) discussed the use of MC not only in radiation transport applications but also in statistical mechanics, polymer science, the solution of linear operator equations, and so forth. The book *The Monte Carlo Method* (Shreider, 1966) was translated from Russian by Tee and published in 1966.

A number of early monographs, chapters, and papers deal with MC applications in radiation transport. Some, written for specialists in nuclear reactor calculations, include Goertzel and Kalos (1958), Kalos et al. (1968), and Spanier and Gelbard (1969). More general treatments are available in the works of Bauer (1958), Hammersley and Handscomb (1964), Halton (1970), Carter and Cashwell (1975), Kalos and Whitlock (1986), and Lux and Koblinger (1991). The compilation edited by Jenkins et al. (1988) considers coupled photon and electron transport. Practical information can also be obtained from the manuals for well-developed MC computer codes. Especially recommended are those for the EGS5 code (Hirayama et al., 2015) and its related EGSnrc code (Kawrakow and Rogers, 2003), the GEANT4 toolkit (GEANT4, 2020), and the MCNP code (Werner et al., 2017, 2018). Other references for these and other widely used MC codes are provided in Appendix E along with brief descriptions of the codes.

Other general works on MC include Rubinstein (1981), Gentle (1998), Mooney (1997), Landau and Binder (2002), Fishman (2003), and the recent chapter by Spanier (2010). Markov chain MC (MCMC) is discussed in detail by Gilks et al. (1996). As the twenty-first century emerged, MC was ubiquitous in simulating nuclear and high-energy physics problems, weather, cosmological models, business and finance, transportation, and many other phenomena. The literature and the world-wide web are replete with articles, books, and reports on the theory and application of MC methods.

1.3   Monte Carlo as Quadrature

Our exploration of the MC method begins by considering continuous functions of a single variable. Later it is shown how the methodology developed here can be applied to discrete functions and to functions of several variables. Consider the function $z(x)$, which is dependent on a stochastic (or random) variable, x. Its mean, or expected value,[5] is

$$\langle z \rangle = \int_a^b z(x) f(x) \, dx, \tag{1.8}$$

where[6] $f(x)dx$ is the probability the random variable x has a value within dx about x. Here it is assumed that values of x are in the range $[a, b]$. The Buffon needle problem of Example 1.2 is precisely of this form. In Eq. (1.6), $f(x) = (1/D)$, $0 < x < D$, and the function $z(x) = [s_1(x) + s_2(x)]/(2\pi L)$, the expected value of which is the probability that the needle intersects a line when its end is a distance x from one line.

A vast number of physical problems involve the evaluation of integral expressions such as Eq. (1.8) (or its multivariable or discrete analogs). Ideally, the right-hand side of Eq. (1.8) can be evaluated analytically, as was done for Buf-

---

[5] The expected value $\langle z \rangle$ is sometimes denoted by $E(z)$, although the former notation is generally used in this book.

[6] The function $f(x)$, which is called a *probability density function* (PDF), is discussed more thoroughly in Chapter 2.

fon's needle problem in Example 1.2 for $D > L$. However, in many situations, the integrand is too complex to allow analytic evaluation. Even more frequently, an explicit expression for the integrand is unknown! In both of these cases, MC techniques can be used with great effectiveness to *estimate* the value of $\langle z \rangle$.

The evaluation of $\langle z \rangle$ from Eq. (1.8) is far from the most complex problem to which MC can be applied. However, it serves as a useful simple problem that allows many of the essential features of MC to be demonstrated, discussed, and understood. It is shown later how MC can be applied to problems with much higher complexity than that of Eq. (1.8).

In this section, the case is considered in which the form of the integral of Eq. (1.8) is known but the value of the integral is too difficult to evaluate analytically. In such cases, the integral must be evaluated using numerical techniques. Perhaps the most frequently used numerical technique is to apply *numerical quadrature* (or just quadrature, for short). Quadrature is nothing more than a numerical prescription for approximating integrals by finite summations. A quadrature scheme estimates the value of $\langle z \rangle$ by constructing a summation of weighted evaluations of the integrand, namely,

$$\langle z \rangle \simeq \sum_{i=1}^{N} w_i \, z(x_i) f(x_i), \tag{1.9}$$

where x_i are called the quadrature *nodes* or *abscissas* and w_i are called the *weights* of the quadrature scheme.

The very definition of a Riemann integral is expressed in this form. Break the interval $[a, b]$ into N contiguous subintervals, the ith subinterval width being denoted Δx_i. Identify x_i as the midpoint of the ith subinterval. Then the integral

$$I = \int_a^b z(x) f(x) \, dx \tag{1.10}$$

is defined (in the Riemann sense) as

$$I = \lim_{N \to \infty} \sum_{i=1}^{N} z(x_i) f(x_i) \Delta x_i. \tag{1.11}$$

Several quadrature schemes, such as Gauss–Legendre quadratures, are based on approximating functions by polynomials and their effectiveness depends on the forms of the functions $f(x)$ and $z(x)$ in Eq. (1.10). In such quadrature schemes, the abscissas and weights are prescribed and, thus, known beforehand.

MC is a technique for estimating expected values, but because these can be expressed as integrals and because almost any integral can be interpreted as an expectation, MC can be used to estimate the values of definite integrals.[7] In the MC approach to numerical quadrature, the abscissas are chosen randomly according to the probability density function (PDF) $f(x)$ and the expected value

---

[7] The simple estimation of integrals by MC has sometimes been called *deterministic* MC to distinguish it from *probabilistic* or *analog* MC (see Section 1.4).

is estimated from a sum such as in Eq. (1.9). Explicitly, the straightforward MC quadrature scheme proceeds as follows:

- generate N values x_i of the random variable x from the PDF $f(x)$,
- define the quadrature abscissas as the sampled values x_i,
- form the arithmetic average of the corresponding values of $z(x_i)$, i.e.,

$$\bar{z} \equiv \frac{1}{N} \sum_{i=1}^{N} z(x_i). \tag{1.12}$$

Note that the terms in the summation do not include the PDF. In effect, the weight factors in this basic MC scheme are of the form $w_i = 1/(Nf(x_i))$. Later it is shown that $\lim_{N \to \infty} \bar{z} = \langle z \rangle$. MC evaluation of an integral like Eq. (1.10) is shown in Example 1.3.

Example 1.3 Buffon's Needle Problem by Monte Carlo Quadrature

Consider the analysis of Buffon's needle problem given in Example 1.2. Suppose the integral of Eq. (1.6) could not be evaluated analytically. This integral is of the form of Eq. (1.8) with $a = 0$, $b = D$, $f(x) = 1/D$, and

$$z(x) = \frac{s_1(x) + s_2(x)}{2\pi L}, \tag{1.13}$$

where $s_1(x)$ and $s_2(x)$ are given in Example 1.2. The probability that a needle intersects a line is, from Eqs. (1.6) and (1.12),

$$P_{cut} \equiv \langle z \rangle \simeq \bar{z} = \frac{1}{N} \sum_{i=1}^{N} z(x_i), \tag{1.14}$$

where x_i are uniformly distributed over the interval $(0, D)$. For the special case that $D = 2L$, the probability of line intersection $P_{cut} = 1/\pi \simeq 0.318310$. The value of $\bar{z}$ and the percent error are shown in Fig. 1.8 as a function of the number of samples N used to evaluate the right-hand side of Eq. (1.14).

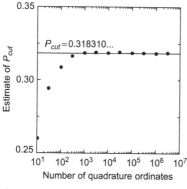 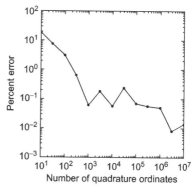

Figure 1.8 Evaluation of Buffon's needle intersection probability P_{cut} using MC quadrature as described in Example 1.3 (left figure). The corresponding absolute percentage error is shown in the right figure.

Now suppose one has to evaluate the integral

$$I = \int_a^b g(x)\,dx, \tag{1.15}$$

which is not in the form of an expected value. Such an integral can be converted into the same form as Eq. (1.8), for instance by rewriting the integral as

$$I = (b-a) \int_a^b \frac{g(x)}{b-a}\,dx = (b-a) \int_a^b g(x)f(x)\,dx, \tag{1.16}$$

where $f(x) \equiv 1/(b-a)$ is, for reasons demonstrated in the next chapter, a PDF on the interval $[a, b]$. Then an MC quadrature evaluation gives

$$I \simeq \overline{I} = \frac{b-a}{N} \sum_{i=1}^{N} g(x_i), \tag{1.17}$$

where x_i are sampled from $f(x) = 1/(b-a), a < x < b$.

Equation (1.8) is, in some sense, just a special case of a more general "expectation." In general, the expected value of a function $z(x)$ with respect to a function $h(x)$ can be expressed as follows:

$$\langle z \rangle = \int_a^b z(x)h(x)\,dx \Big/ \int_a^b h(x)\,dx. \tag{1.18}$$

For instance, the x-coordinate of the center of gravity of a suitably symmetric body can be computed as

$$\langle x \rangle = \int_a^b xm(x)\,dx \Big/ \int_a^b m(x)\,dx, \tag{1.19}$$

where $m(x)\,dx$ is the mass in the interval dx about x.

Equations such as Eq. (1.18) can also be written in the form of Eq. (1.8) by a simple normalization procedure. First let

$$M = \int_a^b h(x)\,dx. \tag{1.20}$$

Then $\langle z \rangle$ can be written as

$$\langle z \rangle = \int_a^b z(x)\frac{h(x)}{M}\,dx \equiv \int_a^b z(x)f(x)\,dx, \tag{1.21}$$

which is in the form of Eq. (1.8) and thus can be estimated by MC. This procedure may seem forced, because if the integral of Eq. (1.20) could be evaluated without MC, then the integral in Eq. (1.21) most likely could also be evaluated without MC. However, this example merely is intended to demonstrate that MC can be applied, in principle, to the evaluation of any definite integral. Further, $\langle z \rangle$ could be estimated by using MC to estimate both integrals in Eq. (1.18) separately, i.e., M could be evaluated by the basic MC estimator

$$M \simeq \overline{M} = \frac{b-a}{N} \sum_{i=1}^{N} h(x_i), \qquad (1.22)$$

where x_i are sampled from the PDF $1/(b-a)$, and then $\langle z \rangle$ could be estimated as

$$\langle z \rangle \simeq \frac{1}{N} \sum_{i=1}^{N} z(x_i), \qquad (1.23)$$

where x_i are sampled from $f(x) = h(x)/\overline{M}$. Thus, MC can be applied to a great many problems indeed.

Later it is shown that MC quadrature is not limited to estimating single integrals. In fact, MC demonstrates its superiority to many other quadrature schemes when the integrals are highly multidimensional. It is shown in Chapter 2 that, regardless of the number of dimensions m, the precision of the MC estimator varies at worst as $N^{-1/2}$, where N is the number of histories. Traditional quadratures, with prescribed abscissas and weights, have precisions that vary as $N^{-1/m}$ (Mosegaard and Sambridge, 2002). Hence, MC is superior to other quadrature schemes for integrals of dimension greater than two! Further, there are a host of ways, called *variance reduction techniques*, to reduce the MC estimation errors even more rapidly (see Chapter 5).

Thus, it is seen that the variable x really need not be stochastic in order to apply the MC quadrature procedure. As long as f has the properties of a PDF, then the integral of Eq. (1.8) is a number that can be treated as if it is an average or expected value. If the integral of Eq. (1.8) is not easily evaluated analytically, then quadrature is generally used. MC provides a powerful form of quadrature, as is demonstrated in subsequent chapters.

1.4 Monte Carlo as Simulation

It is not necessary that a problem be explicitly posed as an integral in order for MC to be applied. For example, Laplace (1786) and Hall (1873) showed that the value of π could be estimated by a "game" in which needles are dropped randomly on a striped grid. This implementation of MC is sometimes called *simulation* or *probabilistic* MC. In truth, all implementations of MC can be viewed as simulations, but here is stressed the difference between (1) problems formulated and identified explicitly as integrals and (2) problems in which the integral formulation is hidden and unspecified.

Consider a rectangular map that shows a country with an irregular boundary and whose area inside the enclosed irregular boundary is sought. One way to estimate the area of this country is to randomly throw darts at the map. The ratio of the number of darts that are within the region of interest to the total number that hit anywhere on the map multiplied by the area of the map (which is trivially calculated) is a reasonable estimate of the area of the region. Of course, in a fundamental way, calculation of an area is an estimation of an integral, but, by throwing darts, the area can be estimated without ever formulating the problem as an integral. It is also interesting to note that in this example, a regular grid of points could be overlayed on the map and the area could be estimated by counting the number of grid points within the country's border. The area estimate improves the finer the grid used. This procedure is an example of simulation without sampling and is akin to *systematic* sampling, which is discussed in Chapter 5.

There are numerous other examples of quantities that can be estimated by simulating the underlying processes without explicitly formulating an integral. For example, the number of "counts" that a radiation detector registers from cosmic radiation can be estimated by simulating the process of cosmic-ray interactions with the atmosphere, the laboratory, and the detector. In such an application, some assumption must be made about the incident distribution of cosmic rays, perhaps based on balloon and satellite measurements, but the rest of the problem can be simulated without having to go into a laboratory and make an actual measurement. In this sense, MC is often viewed as a virtual experiment conducted on a computer (or by some other means, such as dropping needles on a striped plane or throwing darts at a map).

Similarly, high-energy physicists often design their specialized detectors using MC simulation of the processes involved in accelerator interactions. Why would one want to simulate the process rather than just make the measurement? Well, high-energy physics detectors are very expensive to build. One would rather run some relatively inexpensive simulations to be assured that a detector will behave as hoped and that the detector design is optimal before investing large sums of money for its construction. Further, once the detector is built, MC simulation can be used to interpret the data that the detector collects. In these applications, one rarely sets up the problem as an expectation; rather, one simulates the various processes that the interacting particles experience. The ways this is done are discussed in later chapters.

One also can simulate various models for galaxy formation, expected swings in the stock market, traffic flow patterns on highways, or many other processes. In each case, of course, one must start with models of the processes involved (e.g., the laws of gravity) and often one must make assumptions (e.g., there is a Gaussian distribution of speeds about the posted speed limit by cars on a certain highway). The MC estimates developed are only as good as the underlying models and assumptions. Nevertheless, these estimates are often of great utility to researchers, analysts, practitioners, planners, and others. For instance, MC is now being used in treatment planning for cancer patients because the MC dose estimates are, in general, superior to those obtained in other ways.

In such MC simulations, the quantity of interest x usually has a distribution $h(x)$ of possible values as determined by the underlying processes. Of course, if one knew $h(x)$ a priori, the desired expected value could be evaluated readily by quadrature, i.e., $\langle x \rangle = \int x h(x)\,dx / \int h(x)\,dx$. However, in the vast majority of problems, $h(x)$ is unknown. But if the underlying processes are known, then an MC simulation can be used not only to estimate the expected or average value of the quantity, $\langle x \rangle$, but to also estimate the distribution $h(x)$ itself! An example follows.

Example 1.4 Buffon's Needle Problem with Monte Carlo Simulation

From Example 1.2, it was found that the probability P_{cut} a dropped needle intersects a line is given by Eq. (1.7), which has the form

$$P_{cut} = \langle z \rangle = \int_0^D z(x) f(x)\,dx. \tag{1.24}$$

The ends of the needles fall uniformly between the two lines, so that $f(x) = 1/D$. However, suppose we had no idea of the probability $z(x)$ that a needle, whose end is a distance x from a line, intersects a line. An MC simulation could then be used to estimate the function $z(x)$ and then find its average value, which is an estimate of P_{cut}.

The simulation begins by first dividing the distance between the two lines into m contiguous intervals, each with width $\Delta x = D/m$. A scoring bin is created for each interval. Then N needle drops are simulated, with each needle history calculated as follows:

- randomly select the needle's end location x_i uniformly in $(0, D)$,
- randomly select the angle θ_i between the needle and the x-axis uniformly in $(0, 2\pi)$,
- calculate the x-coordinate of the needle's other end $x_j^o = x_i + L \cos\theta_i$,
- if $x_j^o < 0$ or $x_j^o > D$, then add a count to the scoring bin for the interval containing x_i^o.

After the simulation is completed, the function $z(x)$ can be estimated from the number of counts in each scoring bin. Bin k tallies the number n_k of needles with an end that fell between x_{k-1} and x_k and that intersected a line. Because the needle ends are uniformly distributed in $(0, D)$, the number starting in each bin is $n_o = N/m = N\Delta x/D$. Thus, if $x_{k-1/2}$ is the midpoint location for the bin, $z(x_{k-1/2}) \simeq n_k/n_o$. Two such estimated $z(x)$ distributions for the case $D = 2L$ are shown in Fig. 1.9 together with the exact distribution given by Eq. (1.13). The intersection probability

$$P_{cut} = \langle z \rangle = \int_0^L z(x) \frac{dx}{D}$$

is then estimated by

$$\bar{z} = \frac{1}{D} \sum_{k=1}^m \Delta x\, z(x_{k-1/2}) \simeq \frac{1}{D} \sum_{k=1}^m \Delta x \frac{n_k}{n_o} = \frac{1}{N} \sum_{k=1}^m n_k. \tag{1.25}$$

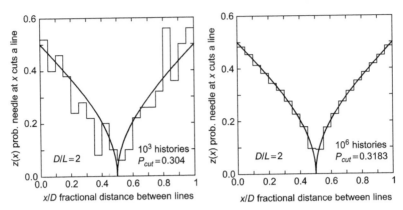

Figure 1.9 MC estimates (histograms) of the Buffon intersection probability function $z(x)$ for 10^3 (left) and 10^6 (right) simulated needle drops. The heavy line shows the exact $z(x)$ of Eq. (1.13). Results are for the case $D = 2L$. The correct value is $P_{cut} = 0.318310$.

1.5 Preview of Things to Come

The overall purpose of this book is to explore, in some detail, many of the facets and variations of the MC method. Key quantities (such as PDFs, cumulative distribution functions (CDFs), population means and variances, and sample means and variances) are defined in Chapter 2 and the two fundamental theorems (the LLN and the CLT) that form the theoretical basis of the MC method are discussed there. Chapter 3 summarizes some of the methods for generating what are called PRNs, which are used repeatedly in MC analyses. Truly RNs cannot be generated using deterministic algorithms; however, PRNs that exhibit certain useful statistical properties can. The concepts of "sampling" and "scoring" are explained for both discrete and continuous random variables in Chapter 4. MC calculations can be computationally expensive, and thus Chapter 5 reviews many of the "variance reduction" procedures that frequently are used to improve the efficiency of MC simulations. These first five chapters present the introductory material that forms the general basis for use of MC in a variety of applications.

Chapter 6 introduces MCMC, which provides an alternative method for sampling distributions; MCMC often is applied when attempting to sample multidimensional and complex distributions, such as can arise in statistical inference and decision theory. Chapter 7 presents an overview of techniques for solving inverse and optimization problems using MC. These inverse MC techniques include both iterative simulation procedures and the noniterative "symbolic" MC method. Chapter 8 reviews the use of MC methods for solving linear operator equations. Chapter 9 introduces the essential concepts of neutral particle transport, which are needed in Chapter 10, where MC simulation of neutral particle transport is discussed. The field of radiation transport was chosen to demonstrate applications of MC for two reasons. First, it is the field that sparked the greatest interest and growth in MC (and indeed radiation transport specialists gave the MC method its name, although the method had been used sparingly by others previously). Second, the authors are familiar with this particular field. However, MC methods are widely used in the physical sciences, the social sciences, many engineering fields, medicine, finance, business, transportation, and many other disciplines.

Six appendices are provided. Appendix A discusses several of the commonly used probability distributions. Appendix B discusses the subtle differences between the weak and strong forms of the LLN and, for the purists, provides a proof of the weak LLN. Appendix C provides a proof of the CLT. Appendix D discusses linear operators. Appendix E provides a synopsis of some radiation transport codes. Appendix F is included in order to identify the PRN generator that was used to solve the example problems. Readers may wish to use this PRN generator with the specific seed 73,907 in order to duplicate the results obtained in the example problems. This can be helpful in debugging codes.

1.6 Summary

MC is, in essence, a very powerful form of quadrature or numerical integration, in which finite summations are used to estimate definite integrals. Although MC is inherently involved with the concept of probability, it can be applied with much success to problems that have no apparent connection with probabilistic phenomena. The methods are based on rigorous mathematics that began with the LLN, as enunciated by Jacob Bernoulli over 300 years ago. MC is particularly valuable when considering multidimensional integrals, where it generally outperforms traditional quadrature methods. Significantly, it also can be applied to a great variety of problems for which the integral formulation is not posed explicitly. Often, the complex mathematics needed in many analytical applications can be avoided entirely by simulation. Thus, MC methods provide extremely powerful ways to address realistic problems that are not amenable to solution by analytic techniques. Today, with the widespread availability of powerful and inexpensive digital computers, MC methods are widely used in almost every discipline that requires quantitative analysis.

Problems

1.1 Find a "needle" (a toothpick, stick, or some other circular cylinder of large aspect ratio) and measure its length L. Draw parallel lines on a large piece of paper at an equal spacing of $D > L$. Drop the needle from a height of about three feet onto the paper 20 times, each time noting whether or not the stick crosses one of the lines. Use different orientations of the needle and start from different positions above the paper. Disregard any drops for which the needle does not end up on the paper. Calculate the success rate from

$$p = \frac{S}{N},$$

where S is the number of successes (times that the needle overlapped one of the lines) and $N = 20$ is the number of trials. (a) Use the data to estimate the value of π. (b) Repeat the process for another 20 trials and use the cumulative results to estimate π. Is the result now closer to the actual value of π?

1.2 Consider the function

$$u(x) = 6x - 1$$

and the PDF

$$f(x) = 1, \qquad 0 \le x \le 1.$$

(a) What is the expected value of u with respect to this PDF? (b) Use the following sequence of "pseudorandom" numbers to form the basic MC estimate $\bar{u}$ of the mean value $\langle u \rangle$: {0.985432, 0.509183, 0.054991, 0.641903, 0.390188, 0.719905, 0.876302, 0.251175, 0.160938, 0.487301}.

1.3 Random numbers $x_i \in (0, 1)$ can be used to simulate the outcome of flipping a fair coin. For example if $x_i < 0.5$, then a "head" occurs on the ith flip, otherwise a "tail" occurs. After N flips, the probability of a head is $P_h = n_h/N$, where n_h is the number of flips that lead to a head. Use the random numbers of Problem 1.2 to plot your estimate of P_h as of function of the number of coin tosses N.

1.4 Consider the definite integral

$$I = \int_{-4}^{4} e^{-x^2/2} dx. \tag{1.26}$$

Use MC to estimate the value of this integral using the 10 PRNs given in Problem 1.2.

1.5 Consider a horizontal beam 20 feet long. The mass distribution $m(x)$ along the beam varies with the distance x from the left end as

$$m(x) = 40 - 3x + 0.05x^2,$$

where x is in feet. (a) Use Eq. (1.19) to evaluate analytically the center of gravity $\langle x \rangle$ of this beam. (b) Then use the following scaled pseudorandom values of x_i to form a basic MC estimate, $\bar{x}$, of the x-coordinate of the center of gravity: {19.70864, 10.18366, 1.09982, 7.80376, 14.39810, 17.52604, 5.02350, 3.21876, 12.83806, 9.74602}.

1.6 In Buffon's needle problem, with $L \leq D$, the probability a randomly dropped needle cuts a grid line was shown in Example 1.1 to be given by

$$P_{cut} = \frac{L}{\pi D} \int_0^\pi \sin\theta \, d\theta.$$

Use the 10 PRNs given in Problem 1.2 to estimate this integral for the case $2L = D$ by MC quadrature and compare your result with the analytical result.

1.7 Consider Buffon's needle problem for the case that the needle length L is greater than the distance D between the grid lines. Following an analysis similar to that of Example 1.2, show that the probability a needle intersects a grid line is given by

$$P_{cut} = \frac{2}{\pi D} \left[D \cos^{-1} \frac{D}{L} + L - \sqrt{L^2 - D^2} \right].$$

1.8 Buffon's needle problem can use needles of any length $L/D \in (0, \infty)$. (a) What is the maximum number of grid intersections $n_{max}(L)$ a dropped needle of length L/D can have? Plot $n_{max}(L)$ versus L/D. (b) What is the probability $P_{cut}(n, L/D)$ a needle of length L/D has n grid intersections for $n = 0, 1, \ldots, n_{max}$?

1.9 Needles of lengths L_1 and L_2 with relative frequencies of f_1 and f_2 ($f_1 + f_2 = 1$), respectively, are dropped randomly on a grid of parallel lines with spacing D. (a) What is the probability that a dropped needle intersects a grid line when averaged over a large number of drops? (b) Generalize to the case when there are M different needle lengths.

1.10 Consider Buffon's needle problem in which the needles have a continuum of different lengths between L_1 and L_2. The needle lengths are defined by the function $\mathcal{L}(L)$ such that $\mathcal{L}(L) \, dL$ is the probability a needle has a length in dL about L. For $\mathcal{L}$ to be a proper probability distribution function, $\mathcal{L}$ has the property

$$\int_{L_1}^{L_2} \mathcal{L}(L) \, dL = 1.$$

What is the probability a dropped needle intersects a grid of parallel lines spaced a distance D apart? Consider cases (a) $L_1 < L_2 < D$, (b) $L_1 < D < L_2$, and (c) $D < L_1 < L_2$.

1.11 Show that the results of the previous problem reduce to those of Example 1.2 when $\mathcal{L}(L) = \delta(L - L_o)$, i.e., all needles have the same length L_o.

1.12 Consider an infinite rectangular grid of lines that form square cells with sides of length D. Onto this, grid circular disks of radius R are dropped

at random. Show that the probability P_{cut} that a disk lands on a grid line
is

$$P_{cut} = \begin{cases} 4\dfrac{R}{D}\left(1 - \dfrac{R}{D}\right), & 2R \le D, \\ 1, & 2R > D. \end{cases}$$

1.13 Consider Buffon's needle problem in which n needles of length $L < D$
are randomly dropped onto a square grid of lines with sides of length D.
If n_1 is the number of needles cutting exactly one grid line and n_2 the
number cutting two grid lines, show that for large n

$$\frac{n_1 + n_2}{n} \simeq \frac{1}{\pi}\frac{L}{D}\left(1 - \frac{L}{D}\right).$$

1.14 A Buffon needle fanatic could find only a felt-tip pen to form a grid upon
which to drop her needles. She thus drew parallel *stripes* of width t spaced
a distance D apart.
  (a) What is the probability a needle of length L falls entirely within a
  stripe?
  (b) What is the probability one end of a needle lies within a stripe and
  the other end lies between two stripes?
  (c) What is the probability a needle lies entirely between two stripes?
1.15 The Buffon needle aficionado of the previous problems decides to add
to her grid of vertical stripes horizontal stripes, also of thickness t, and
spaced a distance L apart, thereby forming a square grid of stripes.
  (a) What is the probability a needle of length L falls entirely within a
  stripe?
  (b) What is the probability one end of a needle lies within a stripe and
  the other end lies in the square bounded by four stripes?
  (c) What is the probability a needle lies entirely in the square bounded
  by four stripes?
  (d) For a needle of arbitrary length L, what is the maximum number of
  stripe edges it can intersect?
1.16 Consider a variant of Buffon's needle
problem in which needles of length L are
randomly dropped into an equilateral tri-
angular grid such as the one shown to the
right. The perpendicular from a vertex to
the opposite side has a length $D > L$.
Derive an expression for the probability,
a randomly dropped needle cuts a grid
line.

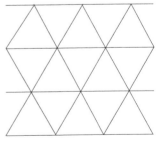

References

Bauer, W.F., 1958. The Monte Carlo method. J. Soc. Ind. Appl. Math. 6 (4), 438–451.

Bernoulli, J., 1713. Ars Conjectandi. Impensis Thurnisiorum, Fratrun, Basilea. This posthumous Latin work was subsequently published in German as Wahrscheinlichkeitsrechnung. Englemann, Leipzig, 1899, two volumes.

Buffon, G.-L.L., 1777. Essai d'Arithmètique Morale. In: Supplément à l'Histoire Naturelle, vol. 4.

Carter, L.L., Cashwell, E.D., 1975. Particle-Transport Simulation with the Monte Carlo Method. TID-26607. National Technical Information Service, U.S. Department of Commerce, Springfield, VA.

Cashwell, E.D., Everett, C.J., 1959. A Practical Manual on the Monte Carlo Method for Random Walk Problems. Pergamon Press, Inc., New York, NY.

Courant, R., Friedrichs, K., Lewy, H., 1928. Über die Partiellen Differenzengleichungen der Mathematischen Physik. Math. Ann. 100, 32–74. Reprinted (in English), IBM J. Res. Dev. 11, 215–234 (March, 1967).

Dunn, W.L., Shultis, J.K., 2009. Monte Carlo methods for design and analysis of radiation detectors. Radiat. Phys. Chem. 78, 852–858.

Einstein, A., 1906. On the theory of Brownian motion. Ann. Phys. 19, 371–381.

Fishman, G.S., 2003. Monte Carlo Concepts, Algorithms, and Applications. Springer, New York, NY. Corrected printing.

GEANT4 Collaboration, 2020. Physics reference manual. Release 10.7, Rev 5. Available at https://geant4-userdoc.web.cern.ch/UsersGuides/PhysicsReferenceMannual/fo/PhysicsReferenceManual.pdf.

Gentle, J.E., 1998. Random Number Generation and Monte Carlo Methods. Springer, New York, NY.

Gilks, W.R., Richardson, S., Spiegelhalter, D.J. (Eds.), 1996. Markov Chain Monte Carlo in Practice. Chapman & Hall, New York, NY.

Goertzel, G., Kalos, M.H., 1958. Monte Carlo methods in transport problems. In: Hughes, D.J., Sanders, J.E., Horowitz, J. (Eds.), Progress in Nuclear Energy II, Series 1, Physics and Mathematics. Pergamon Press, New York, NY, pp. 315–369.

Haji-Sheikh, A., Howell, J.R., 2006. Monte Carlo methods. Ch. 8. In: Minkowycz, W.J., Sparrow, E.M., Murthy, J.Y. (Eds.), Handbook of Numerical Heat Transfer, 2nd ed. Wiley, New York, pp. 249–295.

Hall, A., 1873. On an experimental determination of π. Messeng. Math. 2, 113–114.

Halton, J.H., 1970. A retrospective and prospective survey of the Monte Carlo method. SIAM Rev. 12 (1), 1–63.

Hammersley, J.M., Handscomb, D.C., 1964. Monte Carlo Methods. Spottiswoode, Ballantyne & Co. Ltd., London, UK.

Hirayama, H., Namito, Y., Bielajew, A.F., Wilderman, S.J., Nelson, W.R., 2015. The EGS5 Code System. SLAC-R-730. Stanford Linear Accelerator Center, Stanford University, Palo Alto, CA. Available at http://rcww.kek.jp/research/egs/egs5_manual/slac730-150228.pdf.

Jenkins, T.M., Nelson, W.R., Rindi, A., 1988. Monte Carlo Transport of Electrons and Photons. Plenum Press, New York, NY.

Kahn, H., 1954. Applications of Monte Carlo. AECU-3259. United States Atomic Energy Commission, Rand Corporation, Santa Monica, CA.

Kalos, M.H., Nakache, F.R., Celnik, J., 1968. Monte Carlo methods in reactor computations. In: Greenspan, H., Kelber, C.N., Okrent, D. (Eds.), Computing Methods in Reactor Physics. Gordon and Breach, New York, NY.

Kalos, M.H., Whitlock, P.A., 1986. Monte Carlo Methods. Wiley, New York, NY.

Kawrakow, I., Rogers, D.W.O., 2003. The EGSnrc Code System: Monte Carlo Simulation of Electron and Photon Transport. Technical Report PIRS-70 (4th printing). National Research Council of Canada, Ottawa, Canada.

Kelvin, W., 1901. Nineteenth century clouds over the dynamical theory of heat and light. Philos. Mag. 2 (6), 1–40.

Kolmogorov, A., 1931. Über die Analytischen Methoden in der Wahrscheinlichkeitsrechnung. Math. Ann. 104, 415–458.

Landau, D.P., Binder, K., 2002. A Guide to Monte Carlo Simulations in Statistical Physics. Cambridge University Press, Cambridge, UK. Originally published in 2000 and reprinted 2002.

Laplace, P.S., 1786. Théorie Analytique des Probabilités. Livre 2. Oeuvres Completes de Laplace, de L'Académie des Sciences, vol. 7, part 2, pp. 356–366.

Lux, I., Koblinger, L., 1991. Monte Carlo Particle Transport Methods: Neutron and Photon Calculations. CRC Press, Boca Raton, FL.

Maistrov, L.E., 1974. Probability Theory, A Historical Sketch. Translated and Edited by S. Kotz. Academic Press, New York, NY.

Metropolis, N., 1987. The beginning of the Monte Carlo method. Los Alamos Sci. (Special issue), 125–130.

Metropolis, N., Ulam, S., 1949. The Monte Carlo method. J. Am. Stat. Assoc. 44, 335–341.

Mooney, C.Z., 1997. Monte Carlo Simulation. Sage Publications, Thousand Oaks, CA.

Mosegaard, K., Sambridge, M., 2002. Monte Carlo analysis of inverse problems. Inverse Probl. 18, R29–R54.

Rao, C.R., 1989. Statistics and Truth. Council of Scientific & Industrial Research, New Delhi, India.

Rayleigh, L., 1899. On James Bernoulli's theorem in probabilities. Philos. Mag. 47, 246–251.

Rubinstein, R.Y., 1981. Simulation and the Monte Carlo Method. John Wiley & Sons, New York, NY.

Shreider, Y.A. (Ed.), 1966. The Monte Carlo Method. Pergamon Press, Inc., New York, NY.

Shultis, J.K., Faw, R.E., 2000. Radiation Shielding. American Nuclear Society, La Grange Park, IL.

Sood, Avneet, Forster, R. Arthur, Archer, B.J., Little, R.C., 2021. Neutronics calculation advances at Los Alamos: Manhattan Project to Monte Carlo. Nucl. Technol. 207 (sup1), S100–S133. https://doi.org/10.1080/00295450.2021.1956255.

Spanier, J., 2010. Monte Carlo methods. In: Azmy, Y., Sartari, E. (Eds.), Nuclear Computational Science. Springer, New York, NY.

Spanier, J., Gelbard, E.M., 1969. Monte Carlo Principles and Neutron Transport Problems. Addison-Wesley, Reading, MA.

Student, 1908. The probable error of a mean. Biometrika 6, 1–25;
Student Probable error of a correlation coefficient. Biometrika 6, 302–310 (1908).

Trotter, H.F., Tukey, J.W., 1954. Conditional Monte Carlo for normal samples. In: Meyer, H.A. (Ed.), Symposium on Monte Carlo Methods. University of Florida. John Wiley, New York, NY.

Werner, C.J., et al., 2017. MCNP User's Manual. Code ver. 6.2. LA-UR-17-29981. LANL, Los Alamos, NM. Available at https://mcnp.lanl.gov/pdf_files/la-ur-17-29981.pdf.

Werner, C.J., Bull, J.S., Solomon, C.J., Bown, F.B., McKinney, G.W., Rising, M.E., Dixon, D.A., Martz, R.L., Hughes, H.G., Cox, L.J., Zukaitis, A.J., Armstrong, J.C., Forster, R.A., Casswell, L., 2018. MCNP Version 6.2 Release Notes. LA-UR-18-20808. LANL, Los Alamos, NM. Available at https://doi.org/10.2172/1419730.

Chapter 2

The Basis of Monte Carlo

Jacob tossed and turned without any slumbers
'til he discovered his Law of Large Numbers.
But Laplace did yet more
with his CL encore,
which gives Monte Carlo its awesome powers.

The Monte Carlo (MC) method is firmly based on two mathematical theorems: the *law of large numbers* (LLN) and the *central limit theorem* (CLT). However, before these theorems are examined, some basic definitions and nomenclature must be established. The concepts of a random variable, a probability density function (PDF), a cumulative distribution function (CDF), marginal and conditional probabilities, population means and variances, and sample means and variances all must be understood. These are discussed in the following sections.

A variable is considered random (also called stochastic) if its value cannot be specified in advance of observing it. For instance, the maximum temperature in your town (as measured, for instance, by a certain thermometer at a certain location in town) varies from day to day and the maximum temperature for tomorrow can be predicted, based on detailed data analysis, but cannot be known exactly until tomorrow arrives. Similarly, it is assumed that a fair coin tossed fairly should yield heads 50% of the time and tails 50% of the time, but the exact outcome of the next toss cannot be predicted with certainty. In fact, most observable or measurable quantities show, *at some scale*, a random variation from one observation to the next. For example, a mint tries to make each coin of a given type have the same mass; but it is extremely unlikely that each coin has exactly the same number of atoms and, consequently, there is a very small but real random variation in mass among the coins produced by the mint. Nevertheless, as a practical matter, many variables can be treated as nonrandom. Thus, all coins from the mint can be treated as having the same mass macroscopically. Similarly, the position at which you hang a picture on a wall can be determined by measuring distance along the wall and then pounding a nail into the wall at a certain position. Once that position has been determined, it is "known," and so the variable describing that position is not random; it can be specified now and be expected to be the same tomorrow. As another example, the time the sun will rise tomorrow can be predicted with very high accuracy and hence is considered nonrandom.

Exploring Monte Carlo Methods. https://doi.org/10.1016/B978-0-12-819739-4.00010-X

Many variables in nature, however, are random at a large scale. The exact time a radioactive nuclide will disintegrate cannot be predicted with certainty. (This is an example of a continuous random variable, because time to decay varies continuously between 0 and ∞.) Also, the number of individual pieces of glass that will result if you drop a mirror on the floor cannot be specified in advance. (This is an example of a discrete random variable because the number will be a finite integer.) A random "variable" also can be a vector of several individual random variables, just like position is a vector of three spatial variables. Each random variable has associated with it a PDF, a CDF, a population mean, a population variance, and other measures of the variable's randomness.

2.1 Functions of a Single Continuous Random Variable

Let x denote a single continuous random variable defined over some interval. The interval can be finite or infinite. The value of x on any observation cannot be specified in advance because the variable is random. Nevertheless, it is possible to talk in terms of probabilities. The notation $\text{Prob}\{x_i \leq X\}$ represents the probability that an observed value x_i will be less than or equal to some specified value X. More generally, $\text{Prob}\{E\}$ is used to represent the probability of an event E. The event E can be almost anything. For instance, event E can be the temperature in Dallas exceeding 295 K on New Year's day, in which case one may write $\text{Prob}\{T_{NY} > 295 \text{ K}\} = p$, where p is the estimated probability. Alternatively, event E might be the Chicago Cubs winning the World Series next year, or event E could be obtaining a measurement of 10^6 counts in a radiation detector under certain external conditions. Later, in Chapter 6, the concept of probability is examined in more detail when statistical inference and Markov chains are discussed. For now, only the essential concepts that are needed to implement MC analyses are presented.

2.1.1 Probability Density Function

A *probability density function* (PDF) of a single stochastic variable is a function that has the following three properties: (1) it is defined on an interval $[a, b]$, where $b > a$, (2) it is nonnegative on that interval, although it can be zero for some $x \in [a, b]$, and (3) it is normalized such that $\int_a^b f(x)\,dx = 1$. Here a and b represent real numbers or infinite limits (i.e., $a \to -\infty$ and/or $b \to \infty$) and the intervals can be either closed or open.[1]

A PDF is a *density* function, i.e., it specifies the probability *per unit of x*, so that $f(x)$ has units that are the inverse of the units of x. For a continuous

---

[1] Strictly speaking, $x \in [a, b]$ implies that x can assume the values a and b and that f is defined for such values. However, the properties of PDFs also apply whether the intervals are closed $\{[a, b]\}$, half-open $\{(a, b]$ or $[a, b)\}$, or fully open $\{(a, b)\}$. From a probability standpoint, whether or not the endpoint is included or excluded is immaterial because the probability that x *exactly* equals the endpoint is zero.

random variable, $f(x)$ is not the probability of obtaining x (there are infinitely many values that x can assume and the probability of obtaining a single specific value is zero). Rather, the quantity $f(x)\,dx$ is the probability that a random sample x_i will assume a value within dx about x. Often, this is stated in the form

$$f(x)dx = \text{Prob}\{x \le x_i \le x + dx\}. \tag{2.1}$$

2.1.2 Cumulative Distribution Function

The integral defined by

$$F(x) \equiv \int_a^x f(x')\,dx', \tag{2.2}$$

where $f(x)$ is a PDF over the interval $[a, b]$, is called the *cumulative distribution function* (CDF) of f. Note that from this definition, a CDF has the following properties: (1) $F(a) = 0$, (2) $F(b) = 1$, and (3) $F(x)$ is monotone increasing, because f is always nonnegative.

The CDF is a direct measure of probability. The value $F(x_i)$ represents the probability that a random sample of the stochastic variable x will assume a value between a and x_i, i.e., $\text{Prob}\{a \le x \le x_i\} = F(x_i)$. More generally,

$$\text{Prob}\{x_1 \le x \le x_2\} = \int_{x_1}^{x_2} f(x)\,dx = F(x_2) - F(x_1). \tag{2.3}$$

In Chapter 4, what is meant by the term *random sample* is examined in more detail. For the present, it is sufficient to use the intuitive notion that a random sample is a value of x selected or measured according to the governing probability distribution.

2.1.3 Some Example Distributions

Three widely used distributions are presented below for illustration.

1. The *uniform*, or rectangular, PDF on the interval $[a, b]$ is given by

$$f(x) = \frac{1}{b-a}. \tag{2.4}$$

The CDF for this PDF is given by

$$F(x) = \int_a^x \frac{1}{b-a}\,dx' = \frac{x-a}{b-a}. \tag{2.5}$$

Note that either a or b (or both) can be negative but both must be finite. The uniform distribution is plotted in Fig. 2.1 for the particular case $a = 15$ and $b = 25$.

2. The *exponential* PDF is given by

$$f(x) = f(x|\alpha) = \alpha e^{-\alpha x}, \tag{2.6}$$

which is defined on the interval $x \geq 0$ with a single parameter α; its CDF is given by

$$F(x) = \int_0^x \alpha e^{-\alpha x'} \, dx' = 1 - e^{-\alpha x}. \tag{2.7}$$

The exponential distribution is plotted in Fig. 2.2 for the particular case $\alpha = 0.25$.

3. The *normal*, or *Gaussian*, PDF is given by

$$f(x) = f(x|\mu, \sigma) = \frac{1}{\sigma\sqrt{2\pi}} \exp\left[-\frac{(x-\mu)^2}{2\sigma^2}\right], \tag{2.8}$$

which is defined on the interval $(-\infty, \infty)$. In this case, the distribution has two parameters, the mean value μ and the standard deviation σ. Its CDF is given by

$$F(x) = \frac{1}{\sigma\sqrt{2\pi}} \int_{-\infty}^x \exp\left[-\frac{(x'-\mu)^2}{2\sigma^2}\right] dx', \tag{2.9}$$

which must be evaluated numerically or from tabulations. This distribution is plotted in Fig. 2.3 for the particular case $\mu = 20$ and $\sigma = 5$.

These and several other commonly encountered probability distributions are summarized in Appendix A.

Note that f and F are defined only in the interval $[a, b]$. For finite values of a and b, they are not explicitly defined outside the interval. However, by setting them to constants outside $[a, b]$, $f(x)$ and $F(x)$ can be defined for all values of x. For instance, the uniform distribution of Eqs. (2.4) and (2.5) can be alternatively written in the form

$$f(x) = \begin{cases} \dfrac{1}{b-a}, & a \leq x \leq b, \\ 0, & x > b \quad \text{or} \quad x < a \end{cases} \tag{2.10}$$

and

$$F(x) = \begin{cases} 0, & x < a, \\ \dfrac{x-a}{b-a}, & a \leq x \leq b, \\ 1, & x > b. \end{cases} \tag{2.11}$$

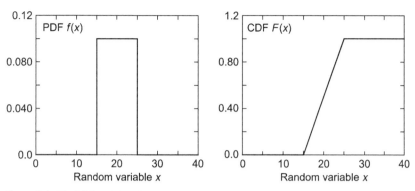

Figure 2.1 The PDF and CDF for the rectangular distribution on the interval [15, 25].

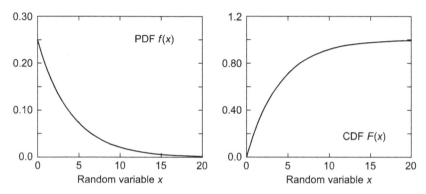

Figure 2.2 The PDF and CDF for the exponential distribution with parameter $\alpha = 0.25$.

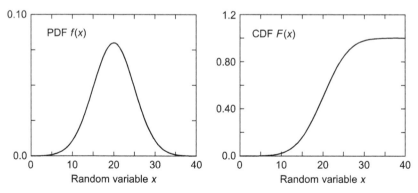

Figure 2.3 The PDF and CDF for the normal or Gaussian distribution with $\mu = 20$ and $\sigma = 5$.

This interpretation is actually used in Fig. 2.1, where the CDF is shown as zero for $x < 15$ and unity for $x > 25$. Using this convention, a CDF can be defined as $F(x) = \int_{-\infty}^{x} f(x')\,dx'$, because $f(x) = 0, x \le a$.

2.1.4 Population Mean, Variance, and Standard Deviation

Two important measures of any PDF $f(x)$ are its mean μ and variance σ^2. The mean is the expected or average value of x and is defined as

$$\langle x \rangle \equiv E(x) \equiv \mu \equiv \int_a^b x f(x)\, dx. \tag{2.12}$$

The variance describes the spread of the random variable x from the mean and is defined as

$$\sigma^2(x) \equiv \langle [x - \langle x \rangle]^2 \rangle = \int_a^b [x - \langle x \rangle]^2 f(x)\, dx. \tag{2.13}$$

This result can be written equivalently as follows:

$$\sigma^2(x) = \int_a^b \left[x^2 - 2x\langle x \rangle + \langle x \rangle^2 \right] f(x)\, dx$$
$$= \int_a^b x^2 f(x)\, dx - 2\langle x \rangle \int_a^b x f(x)\, dx + \langle x \rangle^2 \int_a^b f(x)\, dx. \tag{2.14}$$

By Eq. (2.12), the first integral is just the expected value of x^2, i.e., $\langle x^2 \rangle$, the second integral is just $\langle x \rangle$, and because $f(x)$ is a PDF, the third integral is unity. Thus, Eq. (2.14) can be written as

$$\sigma^2(x) = \langle x^2 \rangle - \langle x \rangle^2. \tag{2.15}$$

The square root of the variance is called the *standard deviation* σ. Thus, one has the trivial result that the standard deviation is $\sigma(x) = \sqrt{\sigma^2(x)}$.

Consider a function $z(x)$, where x is a random variable described by a PDF $f(x)$. The function $z(x)$ itself is a random variable. Thus the *expected* or mean value of $z(x)$ is defined as

$$\langle z \rangle \equiv \mu \equiv \int_a^b z(x) f(x)\, dx. \tag{2.16}$$

The variance of a random variable $z(x)$, by analogy to Eq. (2.13), is defined as

$$\sigma^2(z) = \langle [z(x) - \langle z \rangle]^2 \rangle = \int_a^b [z(x) - \langle z \rangle]^2 f(x)\, dx, \tag{2.17}$$

which, by the same reduction used for Eq. (2.15), is simply

$$\sigma^2(z) = \langle z^2 \rangle - \langle z \rangle^2, \tag{2.18}$$

and the standard deviation is $\sigma(z) = \sqrt{\sigma^2(z)}$.

Example 2.1 Statistics of Radioactive Decay

Radionuclides decay spontaneously to more stable nuclides. The stochastic radioactive process is well described by the exponential distribution, and the *expected* number of radionuclides at time t is given by

$$n(t) = n_0 e^{-\lambda t},$$

where n_0 is the number of radionuclides at time $t = 0$. The parameter λ is called the *decay constant* and has a fixed value for each radionuclide species. The decay constant is the probability a radionuclide decays per unit differential time interval. The probability that a radionuclide will decay within dt about t is

$$p(t)dt = \{\text{prob. it survives to time } t\}\{\text{prob. it decays in } dt\} = \{e^{-\lambda t}\}\{\lambda dt\}.$$

Thus, the PDF for radioactive decay is $p(t) = \lambda e^{-\lambda t}$. From this PDF, the mean lifetime of the radionuclides is given by

$$\langle t \rangle \equiv \int_0^\infty t\, p(t)\, dt = \lambda \int_0^\infty t e^{-\lambda t}\, dt = \lambda \left[\frac{e^{-\lambda t}}{(-\lambda)^2} (\lambda t - 1) \right]_0^\infty = \frac{1}{\lambda}$$

and the population variance of this process is given by

$$\sigma^2(t) = \int_0^\infty (t - \langle t \rangle)^2 p(t)\, dt = \langle t^2 \rangle - \langle t \rangle^2,$$

where

$$\langle t^2 \rangle = \int_0^\infty t^2 p(t)\, dt = \frac{2}{\lambda^2}.$$

Thus, the population standard deviation is

$$\sigma(t) = \sqrt{\frac{2}{\lambda^2} - \frac{1}{\lambda^2}} = \frac{1}{\lambda}.$$

2.1.5 Sample Mean, Variance, and Standard Deviation

The *sample mean* of a function z of a random variable x is an estimate of the population mean $\langle z \rangle$ obtained from a finite number of N samples or "histories" and is defined as

$$\boxed{\bar{z} = \frac{1}{N} \sum_{i=1}^{N} z(x_i).} \tag{2.19}$$

From Eq. (2.18), it is seen that the population variance $\sigma^2(z)$ depends on $\langle z \rangle$, which often is not known. Thus, one must estimate the population variance with what is called the *sample variance*. This is done as follows. Equation (2.17) is in the form of Eq. (2.16), with the function z replaced by the function $[z - \langle z \rangle]^2$. Thus, to estimate the variance, one naturally considers the quantity

$$\sigma^2(z) \cong \frac{1}{N} \sum_{i=1}^{N} [z(x_i) - \langle z \rangle]^2. \tag{2.20}$$

However, $\langle z \rangle$ is unknown; that is, after all, what is being sought. What is usually done is to approximate $\langle z \rangle$ by $\overline{z}$. This estimate of $\sigma^2(z)$ with $\langle z \rangle$ approximated by $\overline{z}$ is called the *sample variance* and is denoted by $s^2(z)$. However, in replacing $\langle z \rangle$ by $\overline{z}$, one *degree of freedom* has been used and, for reasons discussed below, N in Eq. (2.20) is, by convention, replaced by $N - 1$. Thus, the *sample variance* is defined as

$$s^2(z) \equiv \frac{1}{N - 1} \sum_{i=1}^{N} [z(x_i) - \overline{z}]^2. \tag{2.21}$$

The number of degrees of freedom v is an often used but not always well-understood parameter. In applications such as that being considered, where there are N *independent* samples, the number of degrees of freedom is simply the number of independent variables less the number of constraints c, i.e., (see, for example, Dunn (2005)),

$$v = N - c. \tag{2.22}$$

The original N samples were unconstrained (they were drawn independently) and, thus, in calculating the sample mean $c = 0$ and $v = N$, and N is used in the denominator of Eq. (2.19). However, in forming the sample variance, one of the values of $[z(x_i) - \overline{z}]^2$ is constrained ($c = 1$) by using $\overline{z}$ to estimate $\langle z \rangle$. The differences in $N - 1$ cases are independent of each other; they are random samples from which a constant has been subtracted. However, the remaining difference must assume a constrained value, that value being the value that ensures that the samples produce $\overline{z}$. In this sense, the sample mean used one degree of freedom, so $c = 1$ and $v = N - 1$.

Special Cases

Let us look at the simple cases of $N = 1, 2$, and 3. If one were to draw a single sample z_1 from an unknown distribution, then this sample would constitute the sample mean, i.e., $\overline{z} = z_1$. However, one would not have *any* information about the variance. A single sample is insufficient to indicate whether the variance is small or large. Application of Eq. (2.21) leads to the correct conclusion that $s^2(z)$ is unknown. However, putting N in the denominator would lead to the absurd conclusion that $s^2(z) = 0$ and that the population mean is known exactly. Now, draw a second sample z_2 from the same distribution. The sample average is $\bar{z} = (z_1 + z_2)/2$. However, both samples are equally removed from the sample average (i.e., $\bar{z} - z_1 = \Delta = z_2 - \bar{z}$), and hence there is only one *difference* from the mean that is available to be used in estimating the variance. Thus, $\nu = N - 1 = 1$, and the sample variance reduces to $s^2(z) = 2\Delta^2$.

For the case $N = 3$, consider three independent identically distributed (iid) samples, x_1, x_2, and x_3. The sample mean is simply

$$\bar{x} = \frac{x_1 + x_2 + x_3}{3}. \tag{2.23}$$

There are three differences, $\Delta_1 = x_1 - \bar{x}$, $\Delta_2 = x_2 - \bar{x}$, and $\Delta_3 = x_3 - \bar{x}$. However, only two of these are iid; the third is constrained by the other two. To see this, consider Δ_1 and Δ_2 to be iid (they are independent samples from which a constant has been subtracted). It is easy to see that $\bar{x} = x_1 - \Delta_1$ and $\bar{x} = x_2 - \Delta_2$. Substitute these into the expression for Δ_3 to obtain two expressions for Δ_3, namely,

$$\Delta_3 = x_3 - x_1 + \Delta_1 \tag{2.24}$$

and

$$\Delta_3 = x_3 - x_2 + \Delta_2. \tag{2.25}$$

The value of x_3 must assume the value that makes the two expressions for Δ_3 the same. Hence Δ_3 is constrained, i.e., its value is determined by $\bar{x}$, Δ_1, and Δ_2. There are only two independent random variables among Δ_1, Δ_2, and Δ_3, so the number of degrees of freedom for the sample variance is $\nu = 2$ and the sample variance is given by

$$s^2(x) = \frac{\Delta_1^2 + \Delta_2^2 + \Delta_3^2}{2}. \tag{2.26}$$

The General Case

Thus, use of $1/N$ in estimating the sample mean and of $1/(N-1)$ in the estimation of the sample variance follows from consideration of degrees of freedom. As a practical matter, this point is largely irrelevant in most MC calculations because N is usually large and the difference between N and $N-1$ is negligible. Nevertheless, Eq. (2.21) tends to avoid underestimating the sample variance and is the conventional way to express the sample variance in MC calculations.

Equation (2.21) can be converted to a form that is easier to calculate. Note that

$$s^2(z) = \frac{1}{N-1} \sum_{i=1}^{N} \left[z^2(x_i) - 2\bar{z}z(x_i) + \bar{z}^2 \right]$$

$$= \frac{N}{N-1} \left[\frac{1}{N} \sum_{i=1}^{N} z^2(x_i) - \frac{2\bar{z}}{N} \sum_{i=1}^{N} z(x_i) + \bar{z}^2 \right]$$

or

$$s^2(z) = \frac{N}{N-1}(\overline{z^2} - \overline{z}^2).$$ (2.27)

The utility of this result is that in practice it is easy to estimate the sample variance by simply collecting two summations, $\sum_{i=1}^{N} z(x_i)$ and $\sum_{i=1}^{N} z^2(x_i)$, as is demonstrated more fully in Chapter 4. Finally, the *sample standard deviation* is simply $s(z) = \sqrt{s^2(z)}$.

2.2 Discrete Random Variables

Some PDFs are defined only for discrete values of a random variable x. For example, in rolling a die, the number of dots on the upward face has only six possible outcomes, namely, $x_1 = 1, x_2 = 2, \ldots, x_6 = 6$. For a discrete random variable, $f_i \equiv f(x_i)$ represents the probability of event i, where $i = 1, 2, \ldots, I$. Then for $f_1, f_2, \ldots, f_I$ to be a PDF, it is required that (1)

$$f_i \geq 0$$

and (2)

$$\sum_{i=1}^{I} f_i = 1.$$

Further, the discrete CDF

$$F_i \equiv \sum_{j=1}^{i} f_j$$ (2.28)

can be interpreted as the probability that one of the first i events occurs. This discrete CDF has the properties that (1) $F_1 = f_1$, (2) $F_I = 1$, and (3) $F_i \geq F_j$, if $i \geq j$.

In the discrete case, the population mean and variance take the forms

$$\langle x \rangle \equiv \mu \equiv \sum_{i=1}^{I} x_i f_i$$ (2.29)

and

$$\sigma^2 \equiv \sum_{i=1}^{I} [x_i - \langle x \rangle]^2 f_i.$$ (2.30)

The discrete PDF can also be written as a continuous PDF $f(x)$ with $x \in (-\infty, \infty)$ if the Dirac delta function $\delta(x - x_o)$ is used. This function vanishes at all real x-values except at $x = x_o$, where it is infinite. Moreover, this

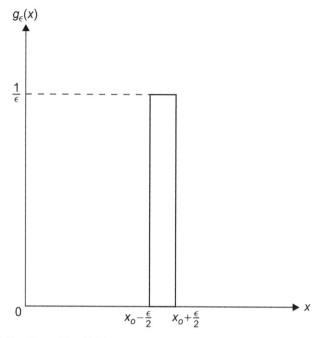

Figure 2.4 Function $g_\epsilon(x)$, which becomes a Dirac delta function as $\epsilon \to 0$.

function, with zero width and support at only one point x_o, has unit area. One can represent this function as a limiting case of the function $g_\epsilon(x)$, illustrated in Fig. 2.4 and defined as

$$g_\epsilon(x) = \begin{cases} \dfrac{1}{\epsilon}, & x_o - \dfrac{\epsilon}{2} \le x \le x_o + \dfrac{\epsilon}{2}, \\ 0, & \text{otherwise.} \end{cases} \tag{2.31}$$

The Dirac delta function is then defined as

$$\delta(x - x_o) = \lim_{\epsilon \to 0} g_\epsilon(x). \tag{2.32}$$

The Dirac delta function has the following properties:

$$\delta(x) = \delta(-x), \tag{2.33}$$

$$\delta(ax) = \frac{1}{|a|}\delta(x), \tag{2.34}$$

and

$$x\delta(x) = 0. \tag{2.35}$$

These properties really have meaning only when the delta function is inside an integral. For example, the most widely used relation involving the delta function is

$$\int_a^b g(x)\,\delta(x-x_o)\,dx = \begin{cases} g(x_o), & \text{if } a \le x_o \le b, \\ 0, & \text{otherwise}, \end{cases} \qquad (2.36)$$

where $g(x)$ is any continuous function.

With this delta function, the discrete PDF can be written as a continuous PDF using the "Dirac comb":

$$f(x) = \sum_{i=1}^I f_i \delta(x - x_i), \qquad -\infty < x < \infty. \qquad (2.37)$$

Substitution of this representation into Eqs. (2.12) and (2.13) immediately yields Eqs. (2.29) and (2.30). The CDF can also be written as a continuous function:

$$F(x) = \int_{-\infty}^x f(x)\,dx = \sum_{x_i \le x} f_i. \qquad (2.38)$$

2.2.1 Special Cases of Discrete Distributions

In most practical cases, there are a finite number of terms and I is finite. However, it is possible to construct discrete distributions where $I \to \infty$. For instance, the exponential series is of the form

$$e = 1 + \frac{1}{1!} + \frac{1}{2!} + \frac{1}{3!} + \cdots \qquad \text{or} \qquad e - 1 = \sum_{i=1}^\infty \frac{1}{i!}.$$

If $x_i = 1, 2, 3, \ldots$, then define

$$f(x_i) = f_i \equiv \frac{1}{e-1} \frac{1}{i!},$$

which has the properties of a discrete PDF with an infinite number of terms and

$$\lim_{i \to \infty} F_i = \sum_{i=1}^\infty f_i = 1.$$

As a practical matter, it is generally possible to consider some finite I for which $F_I = 1 - \epsilon$, where ϵ is as small as desired, and neglect the terms thereafter. For instance, in the above example, $F_8 = 0.999998$ and one can represent this PDF with only eight terms to within $\epsilon = 2 \times 10^{-6}$.

Another situation that deserves mention is that some discrete PDFs have states where $x_1 = 0$ and $f_1 > 0$. For instance, the binomial PDF (see Appendix A.1.2) can be written in the form

$$f(x) = \frac{n!}{x!(n-x)!} p^x (1-p)^{(n-x)}, \ x = 0, 1, 2, \ldots, n,$$

where n is a finite integer and $0 < p < 1$. In this and similar cases, there are $n+1$ probability states. In order to be consistent with the notation used earlier, one could use a new integer variable $y = x + 1$ so that $y = 1, 2, \ldots, I(= n+1)$. Then, the binomial PDF becomes

$$g(y) = f(y-1) = \frac{(I-1)!}{(y-1)!(I-y)!} p^{(y-1)} (1-p)^{(I-y)}, \ y = 1, 2, \ldots, I.$$

It is now possible to consider this a discrete PDF in the form of the former equations. Thus, the original formulation is sufficient for almost all discrete PDFs.

2.3 Multiple Random Variables

In most MC applications, multiple random variables are involved that often depend on each other. In this section, some of the important properties of such probability functions are summarized.

2.3.1 Two Random Variables

The concept of a PDF of a single random variable can be extended to PDFs of more than one random variable. For two random variables x and y, $f(x, y)$ is called the *joint* PDF if it is defined and nonnegative on the intervals $x \in [a, b]$, $y \in [c, d]$ and if

$$\int_a^b \int_c^d f(x, y) \, dy \, dx = 1. \tag{2.39}$$

The functions

$$f_x(x) = \int_c^d f(x, y) \, dy \quad \text{and} \quad f_y(y) = \int_a^b f(x, y) \, dx \tag{2.40}$$

are called the *marginal* PDFs of x and y, respectively. The joint PDF can be written in terms of these marginal PDFs as

$$f(x, y) = f(x|y) f_y(y) = f(y|x) f_x(x), \tag{2.41}$$

where $f(x|y)$ is called the *conditional* PDF of x given y and $f(y|x)$ is called the conditional PDF of y given x. Equation (2.41) can also be used to define the conditional PDFs, namely,

$$f(x|y) = \frac{f(x, y)}{f_y(y)} \quad \text{and} \quad f(y|x) = \frac{f(x, y)}{f_x(x)}. \tag{2.42}$$

The mean and variance of x are given by

$$\langle x \rangle \equiv \int_a^b \int_c^d x f(x, y) \, dy \, dx = \int_a^b x f_x(x) \, dx \tag{2.43}$$

and

$$\sigma^2(x) \equiv \int_a^b \int_c^d (x - \langle x \rangle)^2 f(x, y) \, dy \, dx = \int_a^b (x - \langle x \rangle)^2 f_x(x) \, dx. \tag{2.44}$$

Similar expressions hold for $\langle y \rangle$ and $\sigma^2(y)$. A measure of how x and y depend on each other is given by the *covariance*, defined as

$$\text{covar} \equiv \int_a^b \int_c^d (x - \langle x \rangle)(y - \langle y \rangle) f(x, y) \, dy \, dx. \tag{2.45}$$

Finally, if $f(x, y) = g(x)h(y)$, where $g(x)$ and $h(y)$ are also PDFs, then x and y are *independent random variables* and it is easily shown that $\text{covar}(x, y) = 0$.

Example 2.2 Isotropic Scattering

In simulating particle transport, one must keep track of a particle's direction. For instance, as neutrons migrate through matter, they scatter from nuclei. The probability of scattering from one direction into another direction is often expressed as a joint PDF of two angles, a polar angle θ (measured from the initial direction) and an azimuthal angle ψ. Generally, neutron scattering is *rotationally invariant*, meaning that the angle ψ is independent of the scatter angle θ and all azimuthal angles are equally likely. Thus, θ and ψ are independent and the joint PDF for neutron scattering can be expressed as

$$f(\theta, \psi) = g(\theta)h(\psi), \tag{2.46}$$

where

$$h(\psi) = \frac{1}{2\pi}, \quad \psi \in [0, 2\pi]. \tag{2.47}$$

In the special case of isotropic scattering

$$g(\theta) = \frac{\sin \theta}{2}, \quad \theta \in [0, \pi]. \tag{2.48}$$

Bayes' Theorem

Bayes' theorem, which follows from the axioms of probability, relates the conditional probabilities of two events, say x and y, with the joint PDF $f(x, y)$ just discussed. For two random variables,

$$f(x|y) = \frac{f_x(x) f(y|x)}{f_y(y)}. \tag{2.49}$$

This result is easily verified by using the definitions of Eq. (2.42). This theorem and its application to nonclassical statistical analysis are discussed at greater length in Chapter 6.

Example 2.3 Buffon's Needle Problem

The position and orientation of a "randomly" dropped needle are described by two independent variables. Here, x is the distance of the needle midpoint from the nearest grid line and θ is the angle between the needle and the grid lines. The angle θ is uniformly distributed in the interval $[0, \pi]$ and the distance x is uniformly distributed in the interval $[0, D/2]$, where D is the distance between grid lines. Because θ and x are independent, their joint distribution is the product of their PDFs, i.e.,

$$f(x, \theta) = \left(\frac{1}{\pi}\right) \left(\frac{1}{D/2}\right) = \frac{2}{\pi D}.$$

For the case $L < D$, a needle cuts a grid line if $x < (L/2) \sin \theta$. Hence the probability a dropped needle cuts a grid line is

$$P_{cut} = \frac{2}{\pi D} \int_0^\pi \int_0^{(L/2) \sin \theta} dx \, d\theta = \frac{2}{\pi D} \frac{1}{2} \int_0^\pi \sin \theta = \frac{2L}{\pi D},$$

a result found earlier in Example 1.1.

2.3.2 More Than Two Random Variables

The concept of a PDF for two random variables can be generalized to a collection of n random variables $\mathbf{x} = \{x_1, x_2, \ldots, x_n\}$. Note that each component random variable x_j, $j = 1, \ldots, x_n$, can be either discrete or continuous. The function $f(\mathbf{x})$ is the joint PDF of $\mathbf{x}$ if it obeys both $f(\mathbf{x}) \geq 0$ for all values of $\mathbf{x} \in V$ and $\int_V f(\mathbf{x}) d\mathbf{x} = 1$, where V defines the "volume" over which $\mathbf{x}$ is defined. If $\mathbf{x}$ is decomposed such that $\mathbf{x} = \{\mathbf{x}_j, \mathbf{x}_k\}$, where $\mathbf{x}_k \in V_k$ and $j + k = n$,[2] then

$$f_{\mathbf{x}_j}(\mathbf{x}_j) = \int_{V_k} f(\mathbf{x}) \, d\mathbf{x}_k \tag{2.50}$$

---

[2] It is understood that j refers to the first j elements of $\mathbf{x}$ and k refers to the remaining components, i.e., those from $j + 1$ to n.

is called the marginal PDF of $\mathbf{x}_j$ and

$$f(\mathbf{x}_k|\mathbf{x}_j) = \frac{f(\mathbf{x})}{f_{\mathbf{x}_j}(\mathbf{x}_j)} \tag{2.51}$$

is called the conditional PDF of $\mathbf{x}_k$, given $\mathbf{x}_j$. Also, in analogy to Eq. (2.44), each x_i has a variance and, in analogy to Eq. (2.45), each pair of random variables has a covariance. Finally, if

$$f(\mathbf{x}) = f_{x_1}(x_1) f_{x_2}(x_2) \ldots f_{x_n}(x_n) = \prod_{i=1}^{n} f_{x_i}(x_i), \tag{2.52}$$

then all n random variables are independent and all covariances between any two pairs are zero.

Again, suppose z represents a stochastic process that is a function of the random variable $\mathbf{x}$, where $\mathbf{x}$ is governed by the joint PDF $f(\mathbf{x})$. Then $z(\mathbf{x})$ is also a random variable and one can define its expected value, by analogy with Eq. (2.16), as

$$\langle z \rangle \equiv \int_V z(\mathbf{x}) f(\mathbf{x}) \, d\mathbf{x} \tag{2.53}$$

and its variance, by analogy with Eq. (2.17), as

$$\sigma^2(z) \equiv \langle [z(\mathbf{x}) - \langle z \rangle]^2 \rangle \equiv \int_V [z(\mathbf{x}) - \langle z \rangle]^2 \, f(\mathbf{x}) \, d\mathbf{x}. \tag{2.54}$$

This last result can be reduced to

$$\sigma^2(z) = \langle z^2 \rangle - \langle z \rangle^2. \tag{2.55}$$

The quantity $\langle z \rangle$ is properly called the *population mean*, while $\sigma^2(z)$ is the *population variance* and $\sigma(z) = \sqrt{\sigma^2(z)}$ is the *population standard deviation*. However, the formality of this nomenclature is often ignored and the terms mean, variance, and standard deviation are commonly used, respectively, for these quantities.

2.3.3   Sums of Random Variables

The purpose of an MC calculation is to estimate some expected value $\langle z \rangle$ by the sample mean or average given by Eq. (2.19). But to know how good this estimate is, the variance of $\overline{z}$, which is also a random variable, is also needed.

Begin by considering the random variable $\xi = \sum_{i=1}^{N} z_i$, where, in general, the z_i are distributed by a joint PDF $f(z_1, z_2, \ldots, z_N)$. The expected value of a sum can be shown (see Problem 2.6) to be the sum of expected values, i.e.,

$$\langle \xi \rangle = \left\langle \sum_{i=1}^{N} z_i \right\rangle = \sum_{i=1}^{N} \langle z_i \rangle = \sum_{i=1}^{N} \mu_i. \tag{2.56}$$

The variance of ξ is calculated as

$$\sigma^2(\xi) = \sigma^2\left(\sum_{i=1}^{N} z_i\right) = \left\langle \left(\sum_{i=1}^{N} z_i - \left\langle \sum_{i=1}^{N} z_i \right\rangle \right)^2 \right\rangle$$

$$= \left\langle \left(\sum_{i=1}^{N} (z_i - \langle z_i \rangle)\right)^2 \right\rangle = \left\langle \sum_{i=1}^{N} \sum_{j=1}^{N} (z_i - \langle z_i \rangle)(z_j - \langle z_j \rangle) \right\rangle$$

$$= \sum_{i=1}^{N} \langle (z_i - \langle z_i \rangle)^2 \rangle + \sum_{i=1}^{N} \sum_{\substack{j=1 \\ j \neq i}}^{N} \langle (z_i - \langle z_i \rangle)(z_j - \langle z_j \rangle) \rangle$$

$$= \sum_{i=1}^{N} \sigma^2(z_i) + 2 \sum_{i=1}^{N} \sum_{\substack{j=1 \\ j < i}}^{N} \text{covar}(z_i, z_j). \tag{2.57}$$

Now if z_i are all independent, then $f(z_1, z_2, \ldots, z_N) = \prod_{i=1}^{N} f_i(z_i)$ and it is easy to show that $\text{covar}(z_i, z_j) = 0$ (see Problem 2.7) so that

$$\sigma^2(\xi) = \sum_{i=1}^{N} \sigma^2(z_i). \tag{2.58}$$

One final property of the variance of a random number is needed, namely, for $\alpha = $ a constant,

$$\sigma^2(\alpha\xi) = \langle (\alpha\xi - \langle \alpha\xi \rangle)^2 \rangle = \alpha^2 \langle (\xi - \langle \xi \rangle)^2 \rangle = \alpha^2 \sigma^2(\xi). \tag{2.59}$$

Now apply these results to the MC estimate of the sample mean $\bar{z} = (1/N) \sum_{i=1}^{N} z(x_i)$. Here the $z(x_i) \equiv z_i$ are all identically distributed because z_i are independently sampled from the same PDF $f(z)$. Thus, all z_i have the same variance $\sigma^2(z)$ given by Eq. (2.18). Then the variance of the sample mean is, from Eqs. (2.58) and (2.59),

$$\sigma^2(\bar{z}) = \sigma^2\left(\frac{1}{N}(z_1 + z_2 + \ldots + z_N)\right)$$

$$= \frac{1}{N^2} \left[\sigma^2(z) + \sigma^2(z) + \ldots + \sigma^2(z)\right] = \frac{1}{N}\sigma^2(z). \tag{2.60}$$

Although $\sigma^2(z)$ is not known, it can be approximated by the sample variance $s^2(z)$, as given by Eq. (2.27), so that

$$\sigma(\bar{z}) \simeq \frac{s(z)}{\sqrt{N}} = \frac{1}{\sqrt{N}} \sqrt{\frac{N}{N-1}(\overline{z^2} - \bar{z}^2)} \simeq \frac{\sqrt{\overline{z^2} - \bar{z}^2}}{\sqrt{N}} \text{ for large } N. \qquad (2.61)$$

Hence the MC estimation of $\langle z \rangle \pm \sigma(z)$ is calculated as

$$\langle z \rangle \pm \sigma(z) \equiv \int_V z(\mathbf{x}) f(\mathbf{x}) d\mathbf{x} \pm \sqrt{\int_V (z - \langle z \rangle)^2 f(\mathbf{x}) d\mathbf{x}} \simeq \bar{z} \pm \sqrt{\frac{\overline{z^2} - \bar{z}^2}{N}}.$$

$$(2.62)$$

Here $\bar{z} = (1/N) \sum_{i=1}^{N} z(\mathbf{x}_i)$ and $\overline{z^2} = (1/N) \sum_{i=1}^{N} z(\mathbf{x}_i)^2$, where the $\mathbf{x}_i$ are distributed according to $f(\mathbf{x})$ over the "volume" V.

2.4 The Law of Large Numbers

MC is based on two fundamental statistical results: the *law of large numbers*[3] and the *central limit theorem* (CLT). Many MC books (e.g., Carter and Cashwell (1975), Ross (1970), Fishman (2003), or Landau and Binder (2002)) discuss the LLN and the CLT because of their importance to MC.

The heart of an MC analysis is to obtain an estimate of an expected value such as

$$\langle z \rangle = \int_a^b z(x) f(x) \, dx. \qquad (2.63)$$

If one forms the estimate

$$\bar{z} = \frac{1}{N} \sum_{i=1}^{N} z(x_i), \qquad (2.64)$$

where x_i are suitably sampled from $f(x)$, the law of large numbers states that, as long as the mean exists and the variance is bounded,

$$\lim_{N \to \infty} \bar{z} = \langle z \rangle. \qquad (2.65)$$

This law states that eventually the normalized summation of Eq. (2.64) approaches the expected value of Eq. (2.63). Here, the quadrature nodes x_i are "sampled" from the PDF $f(x)$ and the quadrature weights are equal to $1/(Nf(x_i))$. Details are given in Chapter 4 about how to obtain the samples

---

[3] The law of large numbers has strong and weak forms; this is discussed in Appendix B for those interested in the subtle details.

$\{x_i, i = 1, 2, \ldots, N\}$. A proof of the weak version of this law is given in Appendix B. The law of large numbers can be written in the alternative notation

$$\frac{1}{N}(z_1 + z_2 + \ldots + z_N) \xrightarrow[N \to \infty]{} \langle z \rangle. \tag{2.66}$$

Note that the conditions placed on Eq. (2.65) indicate that in some pathological cases, Eq. (2.65) is not valid. For instance, for certain "fat-tailed" or "heavy-tailed" distributions no population mean can be defined. Brian Hayes (Hayes, 2007) presented a discussion of the so-called "factoidal" function. The familiar factorial function of an integer n can be expressed as

$$n! = n(n-1)(n-2) \ldots (2)(1), \tag{2.67}$$

the (1) being unnecessary, because it does not change the result, but conventionally displayed. The factoidal function, identified as $n?$, can be expressed as

$$n? = (r_n)(r_n) \ldots (1), \tag{2.68}$$

where r_n is a random integer sampled between 1 and n. Thus, the factoidal function begins by sampling an integer between 1 and n and recording this as $n_1 = r_n$. If $n_1 = 1$, then sampling stops and $n? = 1$. Otherwise, a second integer is randomly sampled between 1 and n and recorded as n_2. If $n_2 = 1$, then $n? = n_1$ and sampling stops; otherwise, the procedure is repeated. Eventually, on the kth sampling, a 1 is randomly sampled ($n_k = 1$) and $n? = n_1, n_2, \ldots, n_{k-1}$, where $n_k = 1$. This function has no well-defined population mean. Hayes calculated for $n = 10$ the sample mean after 1000 samples to be about 10^{50}, whereas after 100,000 samples he estimated it to be about 10^{80}! Clearly, there is no convergence to a well-defined population mean. Hence, the law of large numbers does not apply, as the sample mean is unable to converge to a population mean that has the audacity not to exist.

The law of large numbers provides a prescription for determining the nodes and weights of an MC quadrature scheme for estimating *well-defined* definite integrals, but it does not tell us how large N must be in practice. Do 10, 10^4, 10^{20}, or $10^{10^{10}}$ samples suffice? The answer to this question is provided by the CLT, discussed next.

2.5 The Central Limit Theorem

One of the important features of MC is that one can obtain not only an estimate of an expected value (by the law of large numbers), but also an estimate of the uncertainty in the estimate (by the CLT).[4] Thus, at the end of an MC simulation, one can have an idea not only of what the answer is but also of how good the estimate of the answer is.

The CLT is a very general and powerful theorem. In one form (Kendall and Stuart, 1977) it states that for $\bar{z}$ obtained by samples from a distribution with

---

[4] Other advantageous features of MC include that it can be applied to highly multidimensional and complex problems and that the user has a high degree of flexibility in how the sampling is done and results computed.

mean $\langle z \rangle$ and standard deviation $\sigma(z)$,

$$\lim_{N \to \infty} \text{Prob}\left\{\alpha \leq \frac{\bar{z} - \langle z \rangle}{\sigma(z)/\sqrt{N}} \leq \beta\right\} = \frac{1}{\sqrt{2\pi}} \int_{\alpha}^{\beta} e^{-u^2/2}\, du. \qquad (2.69)$$

In another form, the CLT can be written as[5]

$$\lim_{N \to \infty} \text{Prob}\left\{\frac{|\bar{z} - \langle z \rangle|}{\sigma(z)/\sqrt{N}} \leq \lambda\right\} = \frac{1}{\sqrt{2\pi}} \int_{-\lambda}^{\lambda} e^{-u^2/2}\, du. \qquad (2.70)$$

Stated in words, the CLT provides the following insights.

1. The CLT tells us that the asymptotic distribution of $(\bar{z} - \langle z \rangle)/[\sigma(z)/\sqrt{N}]$ is a unit normal distribution or, equivalently, $\bar{z}$ is asymptotically distributed as a normal distribution with mean $\mu = \langle z \rangle$ and standard deviation $\sigma(z)/\sqrt{N}$.

2. The astonishing feature about the CLT is that nothing is said about the distribution function used to generate the N samples of z, from which the random variable $\bar{z}$ is formed. No matter what the distribution is, provided it has a finite variance, the sample mean $\bar{z}$ has an approximately normal distribution for large samples. The restriction to distributions with finite variance is of little practical consequence because, in almost any practical situation, the range of the random variable is finite, in which case the variance is finite.

3. As $\lambda \to 0$, the right-hand side of Eq. (2.70) approaches zero. Thus, the sample mean $\bar{z}$ approaches the true mean $\langle z \rangle$ as $N \to \infty$, a result that corroborates the law of large numbers.

4. Finally, the CLT provides a practical way to estimate the uncertainty in an MC estimate of $\langle z \rangle$, because the sample standard deviation $s(z)$ of Eq. (2.27) can be used to estimate the population standard deviation $\sigma(z)$ in Eq. (2.70).

The CLT gives teeth to the MC method because it provides an estimate of the uncertainty in the estimated expected value. Most importantly, it states that the uncertainty in the estimated expected value is proportional to $1/\sqrt{N}$, where N is the number of histories or samples of $f(x)$. If the number of histories is quadrupled, the uncertainty in the estimate of the sample mean is halved. A proof of the all important CLT is given in Appendix C.

The uncertainty in the MC estimation of the true mean can be obtained from Eq. (2.70) by approximating the unknown population standard deviation $\sigma(z)$ by the calculated sample standard deviation $s(z)$ given by Eq. (2.27). Thus, for large N, Eq. (2.70) can be written as

$$\text{Prob}\{\bar{z} - \lambda s(z)/\sqrt{N} \leq \langle z \rangle \leq \bar{z} + \lambda s(z)/\sqrt{N}\} \simeq \frac{1}{\sqrt{2\pi}} \int_{-\lambda}^{\lambda} e^{-u^2/2}\, du. \qquad (2.71)$$

---

[5] The right-hand side of Eq. (2.70) can be evaluated in terms of the error function as $\text{erf}(\lambda/\sqrt{2})$.

The right-hand side of this result can be evaluated numerically and is called the *confidence coefficient*. The confidence coefficients, expressed as percentages and rounded or truncated, are called *confidence limits*. Results for various values of λ are shown in Table 2.1. Here λ is the number of standard deviations from the mean over which the unit normal is integrated to obtain the confidence coefficient.

Table 2.1 Confidence limits for various values of λ.

λ	Confidence coefficient	Nominal confidence limit
0.25	0.1974	20%
0.50	0.3829	38%
1.00	0.6827	68%
1.50	0.8664	87%
2.00	0.9545	95%
3.00	0.9973	99%
4.00	0.9999	99.99%

Thus, for a given confidence level, the MC estimate of $\langle z \rangle$ is usually reported as $\bar{z} \pm \lambda s(z)/\sqrt{N}$. For example, the interval (for $\lambda = 1$) $[\bar{z} - s(z)/\sqrt{N}, \bar{z} + s(z)/\sqrt{N}]$ has a 68% chance of containing the true mean $\langle z \rangle$. Both the random interval $[\bar{z} - s(z)/\sqrt{N}, \bar{z} + s(z)/\sqrt{N}]$ and its specific numerical value for given values of $\bar{z}$ and $s(z)$ (e.g., [3.5, 5.2]) are called *confidence intervals*. In the latter case, what is meant is that the interval [3.5, 5.2] has a 68% chance of containing the true value of the mean. By contrast, the interpretation that $[\bar{z} - s(z)/\sqrt{N}, \bar{z} + s(z)/\sqrt{N}]$ is a random variable means that if samples of size N are repeatedly constructed and values of the uncertainty interval are calculated for each, then the relative frequency of those intervals that contain the true unknown mean would approach 68%.

In summary, the CLT guarantees that the deviation of the sample mean from the true mean approaches zero as $N \to \infty$. The quantity $\sigma/\sqrt{N}$ provides a measure of the deviation of the sample mean from the population mean after N samples. Use of the sample standard deviation $s(z)$ to approximate the population standard deviation $\sigma(z)$ allows the construction of a confidence interval about $\bar{z}$ that has a specified probability of containing the true unknown mean. As the sample size N increases, this confidence interval, whose width is proportional to $1/\sqrt{N}$, becomes progressively smaller.

Example 2.4 Evaluation of an Integral by Monte Carlo Quadrature

It can be shown that

$$Q = \int_0^{2\pi} \frac{dx}{a + b\sin x} = \frac{2\pi}{\sqrt{a^2 - b^2}}. \tag{2.72}$$

Thus, for $a = 4$ and $b = 2$, $Q = 1.813880$. Using the minimal standard generator with an initial seed of 73,907, we used MC to estimate this integral and obtained the following results:

N	$\overline{Q}$	$s(\overline{Q})$
100	1.868765	0.073536
10,000	1.817711	0.007143
1,000,000	1.813886	0.000714

Note that the sample standard deviation of the estimate varies approximately as $1/\sqrt{N}$, in accordance with the CLT. After reading the next two chapters, you will learn how to generate these results yourself.

2.6 Monte Carlo Quadrature

The results of Eq. (1.17) for a one-dimensional integral, as illustrated in Example 2.4, can be extended to a multidimensional integral over a volume V of a function $g(\mathbf{x})$. An analysis similar to that used for Eq. (2.62) yields

$$\langle g \rangle \pm \sigma(g) \equiv \int_V g(\mathbf{x})d\mathbf{x} \pm \sqrt{\int_V (g - \langle g \rangle)^2 d\mathbf{x}} \simeq V\overline{g} \pm V\sqrt{\frac{\overline{g^2} - \overline{g}^2}{N}},$$

(2.73)

where $\overline{g} = (1/N)\sum_{i=1}^N g(\mathbf{x}_i)$ and $\overline{g^2} = (1/N)\sum_{i=1}^N g(\mathbf{x}_i)^2$, with the $\mathbf{x}_i$ being uniformly distributed over V.

For V of one or two dimensions, deterministic numerical quadratures such as trapezoid, Simpson, or Gauss–Legendre quadrature are usually as efficient as MC quadrature, which requires more evaluations of $g(\mathbf{x})$ to achieve the same accuracy. However, in many practical problems, the dimension of V may be very large. For example, in computing the distributions of stars in a galaxy or the energy of electrons in condensed matter, dimensions of 10^5–10^{30} are not uncommon. For these cases, MC quadrature is the only practical way to evaluate the integrals. Even for dimensions of a few, MC quadrature may be preferred, especially if evaluation of the integrand is computationally complex. Example 2.5 illustrates the effectiveness of MC quadrature for multidimensional integrals.

Example 2.5 A Multidimensional Integral

Consider the problem of calculating a fivefold integral

$$I = \int_0^1 dx_1 \int_0^1 dx_2 \int_0^1 dx_3 \int_0^1 dx_4 \int_0^1 dx_5 \, g(x_1, x_2, x_3, x_4, x_5). \qquad (2.74)$$

One approach is to approximate each integral by a trapezoid at n equispaced ordinates between $x_1 = 0$ and $x_n = 1$, so that for a single integral

$$\int_0^1 f(x)dx \simeq \delta \left\{ \frac{1}{2}[g(x_1) + g(x_n)] + \sum_{i=2}^{n-1} g(x_i) \right\},$$

where $\delta = 1/(n-1)$ and $x_i = (i-1)\delta$. For the five-dimensional integral of Eq. (2.74), the trapezoid approximation leads to the fivefold nested sum

$$I_{trap} = \delta^5 \sum_{i=1}^{n} w_i \sum_{j=1}^{n} w_j \sum_{k=1}^{n} w_k \sum_{l=1}^{n} w_l \sum_{m=1}^{n} w_m \, g(x_i, x_j, x_k, x_l, x_m),$$

where the weight $w = 1/2$ for the endpoints and equals 1 otherwise. A typical MC approach would uniformly sample points in the five-dimensional unit hypercube, so that each coordinate is uniformly distributed between 0 and 1.

Consider Eq. (2.74) in which $g(\mathbf{x}) = \sqrt{x_1 x_2 x_3 x_4 x_5}$. This simple example can be easily integrated analytically to find $I_{exact} = (2/3)^5 \simeq 0.13169$. In Fig. 2.5, the absolute error in estimating this integral by both the trapezoid and MC approaches is plotted as a function of the number of evaluations of the integrand $g(\mathbf{x}) = \sqrt{x_1 x_2 x_3 x_4 x_5}$. It is seen from these results that the MC approach requires far fewer integrand evaluations compared to the trapezoid approach to achieve the same accuracy. This can result in considerable computational savings, especially if the integrand evaluation is time consuming. For multidimensional integrals of dimension ≥ 3, the MC approach is often by far the most efficient.

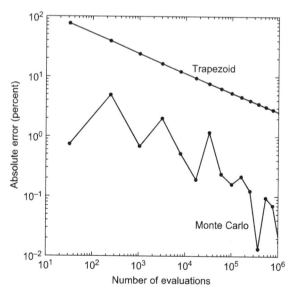

Figure 2.5 Trapezoid and MC estimates of the five-dimensional integral of Eq. (2.74).

2.7 Monte Carlo Simulation

In many complex problems, an estimate is sought of some mean value

$$\langle z \rangle = \int_a^b z(x) f(x) \, dx, \tag{2.75}$$

in which the underlying PDF $f(x)$ is not known a priori. Suppose, for example, the expected radiation dose to a thyroid, $\langle E \rangle$, arising from gamma rays emitted by some radioactive source outside the body, is sought. The probability a source photon deposits energy in dE about E in the thyroid has a PDF $f(E)$. Then the expected energy deposition per source photon is simply $\langle E \rangle = \int E f(E) \, dE$. Unfortunately, $f(E)$ is unknown. But, as shown in Example 1.3, the unknown PDF for Buffon's needle problem can be estimated numerically by performing a simulation of needle drops. Similarly, $f(E)$ can be constructed by simulating the movement of source photons as they migrate through the problem geometry and by recording the energy each simulated photon history deposits in the thyroid. With enough simulated photon histories, the PDF $f(E)$ can be approximated and the average value $\overline{E} \simeq \langle E \rangle$ can be estimated. In such simulations, one strives to mimic nature as accurately as possible and, hence, the simulation must be based on probability models describing the stochastic manner in which radiation interacts with ambient atoms as the radiation migrates through the problem geometry. MC simulations that numerically mimic the actual physical processes involved are called *analog* MC simulations.

However, in many problems an analog simulation is computationally impractical. For example, if the radiation source was far distant from the thyroid and there were many objects between the thyroid and source, very few simulated photons would ever reach the modeled person, let alone deposit any energy in the thyroid. Enormous numbers of simulated particle histories would have to be constructed to obtain even a few thyroid scores. MC simulations, however, can still be used effectively by introducing biases in the simulation. For example, source photons could be forced to be emitted in directions toward the thyroid target and photon scattering could be biased so that scattered photons move preferentially toward the target. Such simulations no longer mimic nature and are called *nonanalog* MC simulations. Of course, if the introduction of biases in the simulation were all that was done in nonanalog simulations, the estimate PDF $f(E)$ would be totally wrong. The key for such biased simulations is to adjust the tally, or score, to undo the biases introduced into the physical models to obtain a correct score. The biased physics allows more photons to reach the thyroid target and the tally adjustments give correct tally values, thereby requiring far fewer particle histories to obtain statistically meaningful results compared to analog simulations. Some of the many tricks used in nonanalog MC are discussed in Chapter 5.

Finally, there are many MC simulations used to solve various mathematical problems that are not based on modeling natural processes. For example, in

Chapter 6, it is shown that Markov chain Monte Carlo can be used to solve integral equations and other interesting mathematical problems. Likewise, random walks of synthetic particles can be used to solve sets of linear algebraic equations, equations which lie at the heart of numerical solutions of heat conduction, fluid flow, diffusion problems, and many other important equations of science and engineering. We hint at such applications in Example 2.6.

Example 2.6 Solving Linear Algebraic Equations

To demonstrate how MC can be used to solve linear algebraic equations, consider the problem of heat conduction in a one-dimensional rod of length L. The temperature at the left end is $T(0)$ and at the right end it is $T(L)$. There are no heat sources in the rod. The temperature profile is given by the heat conduction equation

$$k\frac{d^2T(x)}{dx^2} = 0, \quad \text{whose solution is} \quad T(x) = T(0) + \frac{T(L) - T(0)}{L}x.$$

To solve such equations numerically, often some discretization scheme is used to find the temperatures at a finite number of distances $\{x_i\}$, i.e., $T_i \equiv T(x_i)$. For example, assume an equimesh of n nodes with $x_1 = 0$ and $x_n = L$. Integration of the heat conduction equation over the interval $(x_{i-1/2}, x_{i+1/2})$, where $x_{i+1/2} \equiv (x_i + x_{i+1})/2$, and use of finite differences yields

$$\int_{x_{i-1/2}}^{x_{i+1/2}} \frac{d^2T(x)}{dx^2} = \frac{dT}{dx}\bigg|_{x_{i+1/2}} - \frac{dT}{dx}\bigg|_{x_{i-1/2}} \simeq \frac{T_{i+1} - T_i}{\Delta x} - \frac{T_i - T_{i-1}}{\Delta x} = 0.$$

Thus, T_i, $i = 2, \ldots, (n-1)$, are the solutions of the following $(n-2)$ linear algebraic equations (with $T_1 = T(0)$ and $T_n = T(L)$ known):

$$T_{i-1} - 2T_i + T_{i+1} = 0, \quad i = 2, \ldots, (n-1),$$

or rearranging

$$T_i = T_{i-1}/2 + T_{i+1}/2, \quad i = 2, \ldots, (n-1). \tag{2.76}$$

The solution of these equations can be estimated by the following random walk "game":

1. Start a "particle" at x_2 with probability $P_2 = T(0)/(T(0)+T(L))$ or at x_{n-1} with probability $1 - P_2$.
2. At each step in the walk, move the particle to the left or right, each with a 50% probability.
3. Each time that a particle lands on node i increase a counter $c_i = c_i + 1$, where c_i records the cumulative number of visits to node i.
4. If the particle reaches x_1 or x_n, kill it and start a new particle at x_2 or x_{n-1}, as in step 1.
5. Perform steps 1–4 for $(T(0)+T(L))/2$ particles. This constitutes one game.

After playing M complete games,

$$T_i = \langle c_i \rangle \simeq \overline{c_i} = \frac{1}{M} \sum_{j=1}^{M} c_i^j,$$

where c_i^j is the cumulative number of particle visits to node i in game j. Equivalently, one could play the game for a large number N of source particles continuously accumulating c_i. The number of games represented by N is $N_g = 2N/(T(0) + T(L))$, so that

$$T_i = \langle c_i \rangle \simeq \frac{c_i}{N_g}.$$

Note that this equivalent approach does not require $(T(0) + T(L))/2$, the number of particles per game, to be an integer.

That $\langle c_i \rangle = T_i$ can be seen from the fact that the expected number of visits to node i equals one-half of the expected number of visits to nodes $i - 1$ and $i + 1$, i.e.,

$$\langle c_i \rangle = \langle c_{i-1} \rangle /2 + \langle c_{i+1} \rangle /2, \quad i = 2, ..., (n-1).$$

These are the same equations as Eq. (2.76), and hence $\langle c_i \rangle = T_i$. Example results are shown in Fig. 2.6.

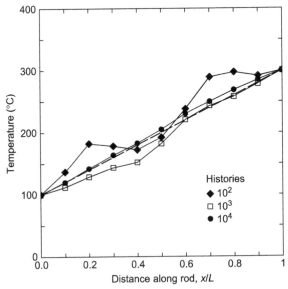

Figure 2.6 MC results for the one-dimensional heat conduction problem of Example 2.3. Here $T(0) = 100°C$ and $T(L) = 300°C$. The exact profile is shown by the heavy dashed line. Results based on three different numbers of particle histories are shown.

2.8 Summary

A PDF of a single continuous variable is a nonnegative function defined on
an interval and normalized so that its integral over that interval is unity. The
associated CDF of x is interpreted as the probability that a random sample has
a value less than or equal to x. These concepts extend naturally to discrete and
multidimensional random variables.

Generally, probability distributions have well-defined population means and
variances. These are explicit values that are characteristic of the distribution.
MC is a method that allows one to estimate the population mean and population
variance by the sample mean and sample variance. These concepts apply to
functions of a discrete variable, to functions of a continuous variable, and to
functions of many variables, whether discrete or continuous.

The MC estimates of the sample mean and sample variance almost surely
approach the population mean and population variance as the number of samples
(usually called "histories") get large. This important feature is a consequence of
the law of large numbers, which can be stated as follows. If

$$\langle z \rangle = \int_a^b z(x) f(x)\, dx \tag{2.77}$$

and

$$\bar{z} = \frac{1}{N} \sum_{i=1}^N z(x_i), \tag{2.78}$$

with x_i suitably sampled from $f(x)$, then, almost surely,

$$\lim_{N \to \infty} \bar{z} = \langle z \rangle. \tag{2.79}$$

The CLT then provides a prescription for estimating the uncertainty in the sample mean and sample variance.

It should be kept in mind that the term "standard deviation" is often used,
rather loosely, for any of the following:

- the *population standard deviation* $\sigma(z)$,
- the *sample standard deviation* $s(z) = [N(\overline{z^2} - \bar{z}^2)/(N-1)]^{1/2}$,
- the *standard deviation of the sample mean* $\sigma(\bar{z}) = \sigma(z)/\sqrt{N}$,
- the *estimation of the standard deviation of the sample mean* $s(\bar{z}) = s(z)/\sqrt{N} = [(\overline{z^2} - \bar{z}^2)/(N-1)]^{1/2}$.

The context must be employed to indicate which interpretation is meant.

Finally, MC can mean a wide variety of different calculational methods, but
all are based on the statistical power of the CLT, which guarantees expected
values can be approximated by averages and which allows confidence intervals
to be constructed for the averages. Although many different applications of MC
appear to be quite different, they all depend on this critical theorem.

Problems

2.1 What is the PDF for a plane angle ϕ, expressed in radians, distributed uniformly between 0 and 2π?

2.2 The discrete Poisson distribution has PDF

$$p(x) = \frac{\lambda^x}{x!} \exp(-\lambda), \tag{2.80}$$

where λ is a positive constant. (a) What is the CDF for this distribution? (b) What are the mean and variance of this distribution?

2.3 Radiation particles escaping into a vacuum from a plane surface are often observed to do so with an intensity proportional to the nth power of the cosine ω of the angle between the outward normal and the particle travel direction, i.e., proportional to ω^n, $0 \le \omega < 1$. For $n = 1$, this is known as *Lambert's law*. (a) What is the PDF $f(\omega)$ that describes this phenomenon, namely, that $f(\omega)d\omega$ is the probability that an escaping particle has a direction in $d\omega$ about ω? (b) What is the associated cumulative distribution? (c) What is the mean and variance of this distribution? (d) What fraction of escaping particles escape at angles greater than 45 degrees?

2.4 Each die of a pair of loaded dice has twice the probability of giving a "six" than any other number, all of which have equal probability. (a) Write a PDF that describes the probability of obtaining each possible outcome from a throw of this pair of dice. (b) What is the expected value of a throw? (c) What is the variance of the result of a throw? (d) Compare these results to those for a "fair" pair of dice.

2.5 A radioisotope decays exponentially with time according to

$$N(t) = N_0 e^{-\lambda t}, \tag{2.81}$$

where $N(t)$ is the number of radioactive atoms remaining at time t, N_0 is the number of radioactive atoms at time $t = 0$, and λ is called the decay constant. For a radioisotope for which $\lambda = 0.0693$ min^{-1}, what fraction of the atoms are expected to decay between the times $t = 15$ min and $t = 30$ min?

2.6 Verify Eq. (2.56).

2.7 Show that the covariance terms in Eq. (2.57) vanish if z_i are independent random variables.

2.8 Estimate the probability that $|\bar{z}_k - \langle z \rangle_k| \le \lambda\sigma/\sqrt{N}$ for $\lambda = 0.6745, 1, 2, 3$, and 4.

2.9 The weights of 20 football players are (in kg) 92, 94, 100, 86, 93, 91, 89, 86, 105, 96, 87, 90, 97, 87, 95, 93, 103, 88, 98, and 92. (a) Estimate the average weight of all the football players. (b) What is the 95% confidence interval for this estimate? (c) What is the probability that a player has a weight in excess of 120 kg? (d) What is the probability that a player has a negative weight? (e) What does this tell you about the CLT and real-world random variables?

2.10 Show that the joint PDF of Example 2.2 for isotropic scattering obeys the appropriate normalization condition.

2.11 What must be the value of the constant A in order to make $f(x)$ in the following expression a PDF:

$$f(x) = A \sin^2(x)?$$

2.12 What is the CDF of the PDF

$$f(x) = \frac{3}{4}\left(1 + x^2\right), \quad 0 \leq x \leq 1?$$

2.13 Consider the function

$$f(\theta) = A(1 - \cos\theta). \quad 0 \leq \theta \leq \pi.$$

(a) What is the value of A if $f(\theta)$ is to be a PDF? (b) What is the corresponding CDF?

2.14 Use first-order finite differences and an equimesh size Δx to approximate the boundary value problem

$$\frac{d^2 y(x)}{dx^2} - \kappa^2 y(x) = 0, \quad x \in [a, b], \quad a < b, \quad \kappa^2 > 0,$$

subject to boundary conditions $y(a) = y_a$ and $dy/dx = 0$ at $x = b$, by a set of linear algebraic equations.

2.15 In Example 1.4, a simulation of Buffon's needle problem was presented. In this simulation, needles were dropped so that the location x_i of one end of the needle was placed uniformly in the interval $(0, D)$. Discuss any potential difficulties that would arise if the closed interval $[0, D]$ were used.

2.16 Consider two perpendicular, one-way, one-lane roads that intersect each other. The intersection has no traffic control signage and cars of length L travel at constant speed v along the roads and through the intersection. Cars reach the intersection at random times at an average rate of N_1 and N_2 per hour along roads 1 and 2, respectively. Design a simulation, namely, the simplifications and models used, PDFs needed, and a list of the *rules of the game* that could be used to estimate the probability of an accident at the intersection as a function of N_1 and N_2.

References

Carter, L.L., Cashwell, E.D., 1975. Particle-Transport Simulation with the Monte Carlo Method. TID-26607. National Technical Information Service, U.S. Department of Commerce, Springfield, VA.

Dunn, P.F., 2005. Measurement and Data Analysis for Engineering and Science. McGraw Hill, New York, NY.

Fishman, G.S., 2003. Monte Carlo Concepts, Algorithms, and Applications. Springer-Verlag, New York, NY. Corrected printing, 2003.

Hayes, B., 2007. Fat tails. Sci. Am. 95, 200–204.

Kendall, M., Stuart, A., 1977. The Advanced Theory of Statistics, Volume 1 Distribution Theory, fourth ed. MacMillan Publishing Co., Inc., New York, NY.

Landau, D.P., Binder, K., 2002. A Guide to Monte Carlo Simulations in Statistical Physics. Cambridge University Press, Cambridge, U.K. Reprinted 2002.

Ross, S.M., 1970. Applied Probability Models with Optimization Applications. Holden-Day, San Francisco, CA.

Chapter 3

Pseudorandom Number Generators

Monte Carlo needs numbers quite random
but practicality has mostly banned 'em.
So in order to score
we need do no more
than generate them as pseudo*random!*

The many Monte Carlo (MC) applications explored in this book need random numbers (RNs) in vast profusion. The application user generally does not have to understand their origin and how they were generated but only that they are there when needed. Consequently, the details presented in this chapter about how random-like numbers are produced are often of little concern to those using MC methods. The reader may safely skip this chapter and can return to it when curious about how the RNs used in MC methods is actually generated.

This chapter is included to give one an appreciation of the many methods used to concoct RNs that are then used in the many MC applications that are described in subsequent chapters. It is not our intent to tell MC practitioners how to construct their own algorithms for generating RNs; rather the intent is to give the reader an appreciation of how technical and specialized the many proposed schemes to construct random-like numbers have become. Today the construction of algorithms for the so-called *random number generators* (RNGs) is a very active field and one in which the nonspecialist should be very hesitant to tread. Unless you really understand this highly specialized field and also are very confident, do not tweak/modify your RNG or, worse, write your own RNG.

3.1 Pseudorandom Numbers

At the core of all MC calculations are some mechanisms to produce a long sequence of random *numbers*[1] ρ_i that are uniformly distributed over the open interval $(0, 1)$. These random numbers are usually produced from random *integers* $x_i \in (0, L)$ as $\rho_i = x_i/L$. In digital computers, random integers are represented by random *bits*. Some references call these three quantities all "random num-

---

[1] Also sometimes called random *floats* for floating-point numbers.

Exploring Monte Carlo Methods. https://doi.org/10.1016/B978-0-12-819739-4.00011-1

55

bers," but here the distinction is used. By contrast and confusingly, a random *sequence* refers to a stream of random bits, integers, or numbers.

Digital computers, by design, are incapable of producing random results. A true random sequence could, in principle, be obtained by coupling to our computer some external device that would produce a random signal. For example, one could use the time interval between two successive clicks of a Geiger counter placed near a radioactive source or use the "white noise" in some electronic circuit to provide truly RNs.[2]

However, use of such a true RNG would be disastrous! First, the feasibility of having to couple some external device that generates a random signal to every computer used for MC calculations would be impractical. But more important would be the impossibility of writing and debugging an MC code if, on every run, a different sequence of RNs were used. What is needed is a sequence of numbers that has the properties of RNs but that is the same every time the program is run; this allows coding errors to be found and the same results to be produced when the same code is run on different computers.

One way to use true RNs that would have the same sequence every time a code is run would be to create somehow a table of random numbers and then read them into the computer as they are needed. In fact, Marsaglia (2016) has produced a CD-ROM that contains 4.8×10^9 truly RNs (generated by combining two pseudorandom number generators [PRNGs] and three devices that use random physical processes). However, this is not a good idea. First, in modern MC applications many more RNs than 10^9 often are needed, and second, and more important, the time to read a number from an external file is prohibitively long for modern MC calculations.

An alternative to producing the same sequence of RNs each time a program is run is to use a *pseudorandom* number generator (PRNG). Such a generator uses a deterministic algorithm that, given the previously generated numbers (often just the last few), the next number can be efficiently calculated. The numbers produced by a PRNG should be able to pass many tests for randomness, e.g., the ρ_i appear to be evenly distributed over the interval $(0, 1)$. The one randomness test pseudorandom numbers cannot pass is, by design, the unpredictability of a pseudorandom number, given the previous numbers.

Many PRNGs, some good but most poor, have been proposed and used over the years in a wide variety of MC applications. Designing better PRNGs is still a very active area of research. In this chapter, the basic properties of some important RNGs are discussed and several important RNGs are presented.[3]

The term *quasi-random* numbers is sometimes encountered. These numbers are designed not so much to be random as to be extremely uniform over the

---

[2] However, such physical RNGs still need corrections and filtering to remove nonideal characteristics, such as the finite resolving (or dead) time of a Geiger counter or the high frequency cutoff of a white noise source.

[3] The term "pseudo" is often dropped, and throughout this book the acronyms PRNG and RNG are used synonymously.

interval $(0, 1)$ so as to allow very accurate integration by the MC method. Although the generation of quasi-random numbers is not considered further here, the application of such numbers is considered in Section 5.11.2.

Although not reviewed in this chapter, there have been many highly deficient RNGs used in many studies. Some still continue to be used by researchers unaware of the biases being introduced into their calculations. Indeed, if all publications based on these questionable generators were to disappear, much valuable library shelf space would be recovered. When you embark on MC investigations, be well aware of the pedigree of the RNG you are using. Even more important, do not alter your RNG unless you are *very* confident your alterations are an improvement. The construction of RNGs is best left to the experts.

3.1.1 Origins of Random Number Generators

We immediately break our promise not to review bad RNGs. But for historical interest we must mention the first RNG, which was proposed by John von Neumann at a conference in 1949 (von Neumann, 1951). In his mid-square method, a random integer x_i with n digits is obtained by taking the middle portion of the square of the previous integer x_{i-1}. If the square has fewer than $2n$ digits, then leading zeros are added so the middle portion has n digits. For the method to work, n must be even for there to be a well-defined "middle portion." The maximum cycle length is 8^n, but often it is less because the integer chain becomes "stuck." For example, if the middle digits of a square are all zero, then subsequent integers are also zero. The method also can become stuck on a number other than zero such as for $n = 4$, 100, 2500, 3792, and 7600. Even when a long cycle is achieved there is a strong positive correlation in the $\rho_i = x_i/10^n$. Another problem is that the starting value x_0 (often called the "seed") can produce very short cycles such as the following: 0540, 2916, 5030, 3009, 0540, ...

Recently it has been found that combining a Weyl sequence with the original mid-square method prevents convergence to zero and the occurrence of short repeating sequences (Widynski, 2017). Specifically, the Weyl sequence is added to x_i^2. Widynski (2020) recently produced an extremely fast RNG called SQUARES which passes many statistical tests (see Section 3.13) for randomness.

3.1.2 Properties of Good RNGs

Beginning in the 1970s, a great many RNGs have been proposed, each claiming some superiority in some characteristic compared to that of other RNGs. Today the creation of RNGs, most variations of an earlier RNG, has become almost a "cottage industry." The theory and mathematics behind most of these new RNGs are often complex and opaque to most MC users so it is often difficult for a nonspecialist to know what RNG is appropriate for a given MC problem. In Section 3.13, the many statistical tests that a "good" RNG should be able to pass are briefly discussed; however, a good RNG should have the following properties.

Uniformity: Sorting the $N \gg M$ RNs ρ_i from the RNG into M contiguous equiwidth $(1/M)$ bins, covering the interval $(0, 1)$, should produce bin populations whose expected values are all close to the expected value N/M. This indicates the RNG produces RNs uniformly distributed over the interval $(0, 1)$.

Independence: The generated numbers should be independent of each other, i.e., there should be no serial correlation between the numbers and their successive numbers. This lack of serial correlation means that any two nonoverlapping substrings are uncorrelated.

Pass Tests Emphasizing Known Weaknesses: However, the number of possible tests for randomness are uncountably infinite. Thus, the finite number of statistical tests used to evaluate and find the "best" set of parameters for a particular generator should be those that emphasize known weaknesses of the generator. For example, tests that examine the effect of the lattice structure should be emphasized for congruential generators. Extensive methods for testing PRNGs are given by Knuth (1981) and Marsaglia (1985).

Coverage: This characteristic measures how well the RNG covers the output space $(0, 1)$ for any seed. Often an RNG has multiple cycles. Which cycle is used by the RNG is determined by the seed chosen. In this case, only certain numbers in the output space are produced by the RNG for a given seed. For this reason, full-cycle RNGs are desired because they give complete coverage.

Spectral Characteristics: A good RNG produces numbers each with the same frequency, i.e., there is no bias for one number over another. In other words, if substrings of k successive output values from the RNG are taken as coordinates of points in a k-dimensional unit hypercube, there should be no discernible pattern for $k = 2, 3, \ldots$.

Large Period: Every RNG has a period or cycle length, after which the RN sequence is repeated and, thus, is completely predictable. A good RNG should have a large period of at least 10^{19} for modern MC uses. For complex applications even longer periods are needed. Generators with very long periods up to 10^{170} are available. Such long-period generators are very useful for generating long independent subsequences so that many collaborators can independently work on subsimulations and later combine their results assured of statistical independence.

Reproducibility: An RNG must be reproducible, i.e., given the same initial *seed* or initial condition x_0, the same random sequence is produced. Such reproducibility is imperative for debugging and testing purposes.

Consistency: The above properties of the RNG should not depend on the seed used, except for special seeds known to produce degenerate sequences. For example, in many RNGs a seed $x_0 = 0$ results in a sequence of zeros.

Disjoint Subsequences: Subsequences produced by different seeds should have little or no correlation.

Portability: A good RNG should be portable, i.e., the algorithm for the RNG must work on every computer architecture so that, for the same seed, every computer produces the same output sequence. Generators written in machine language are generally not portable.

Programmability: The algorithm for the RNG should be programmable in a high-level computer language like FORTRAN or C. This requirement ensures the RNG does not depend on a particular architectural feature of one type of computer system.

Efficiency: This attribute refers to the speed with which an RN is generated. The faster, the better. It also refers to the amount of memory and computational overhead the RNG needs, both of which should be as small as possible. Good efficiency is needed to ensure most of the computational effort in an MC application is devoted to the application and not to the generation of the RNs. With modern computers, the generation time is almost insignificant, with the function call time usually exceeding the actual calculation time. If generation time is of concern, two methods can be used to reduce it. First, the generator can be coded in-line to avoid a function call. Second, each function call to the generator should return a large vector of RNs, thereby greatly reducing the number of function calls.

Cryptographically Secure: For cryptographic applications, the output stream can not be backcalculated to predict future output from the RNG. This property usually is lacking in algorithmic RNGs. Cryptographic RNGs are a special area for concocting RNGs and are not covered in this book, other than to mention many of the recent developments in creating RNGs for MC applications are based on cryptographic generators.

3.1.3 Classification of RNGs

The first widely used RNGs were based on linear recurrence modulo m where m is a large prime number. These generators, commonly called *linear congruential generators* (LCGs), had the properties of uniformity, independence, and relatively large periods. Lehmer (1951) introduced this type of RNG and, over the next 40 years, many variations of the LCG were proposed that had somewhat better properties. Many of these improved LCGs are summarized in Section 3.2.

In 1965, Tausworthe introduced another class of RNGs. These RNGs still used linear recurrences but in modulo 2. In essence, this type of generator treats the bits used by a digital computer to represent numbers as random entities. Tausworthe's approach quickly became popular because of the ease and efficiency with which it could be implemented by using a computer's binary arithmetic. The linear recurrence modulo-2 calculations are done mostly on a *linear feedback shift register* (LFSR). An LFSR can be efficiently implemented using the XOR and shift operations. This scheme and variations to it have been investigated and many RNGs using it have been proposed.

Another way to generate (pseudo)RNs was proposed by Stephen Wolfram (1986). He suggested that the chaotic-like behavior exhibited by certain one-dimensional cellular automata (CAs) could be used to generate strings of random bits from which random integers x_i and RNs ρ_i could be formed. Since Wolfram's seminal paper, many researchers have extended the method to two and three dimensions and produced many RNGs based on CA.

In the following three sections, LCG-based, LFSR-based, and CA-based RNGs are discussed.

3.2   Linear Congruential Generators

Perhaps the RNGs most widely used in the twentieth century were based on the *linear congruential generator* (LCG) proposed by Lehmer (1951), which has the general form[4]

$$x_{i+1} = (ax_i + c) \bmod m, \quad i \geq 0. \tag{3.1}$$

Here integer $a > 1$ is called the *multiplier*, integer c the *increment*, and integer m the *modulus* of the generator. Starting with some initial integer *seed* x_0, each integer in the sequence $x_1, x_2, \ldots$ must lie in the interval $0 \leq x_i < m$. Each random integer x_i is then scaled to the interval $[0, 1)$ by dividing by m, i.e.,

$$\rho_i = \frac{x_i}{m}, \tag{3.2}$$

to obtain RNs over the interval $[0, 1)$. In the most frequently programmed congruential generator, the increment c is set to zero and Eq. (3.1) becomes the so-called *multiplicative congruential generator* (MCG)

$$x_{i+1} = ax_i \bmod m, \quad i \geq 0. \tag{3.3}$$

Because of the modular arithmetic, the values of the integers x_i cannot exceed $m - 1$ and for a nonzero seed, x_i can never equal 0. Thus, the $\rho_i = x_i/m$ must lie in the open interval $(0, 1)$. For many MC simulations, it is important that the RNG never returns a ρ_i that is exactly 1 or 0.

Just how many distinct integer values $1 \leq x_i \leq m - 1$ are obtained depends very critically on the relation between a and m. This is the crux of a good generator: pick a and m such that a vast number of different values x_i are obtained and

---

[4] The notation mod m means that the value of $ax_i + c$ is divided by m and the *remainder* resulting from the division is then kept as x_{i+1}. Two integers a and b are said to be *congruent modulo* m if a mod $m = b$ mod m. Hence the name of this RNG. The actual generator Lehmer used was $x_{i+1} = 23x_i \bmod (10^8 + 1)$ which, while better than some successors, is not very good by today's standards.

that they and all pairs, triplets, quadruplets, and so on (i.e., all order k-tuples), "appear" to be distributed randomly.

If $k_i = [ax_{i-1}/m]$,[5] then Eq. (3.3) can be written in the alternative fashion:

$$x_1 = ax_0 - k_1 m$$

$$x_2 = a^2 x_0 - k_2 m - ak_1 m$$

$$x_3 = a^3 x_0 - k_3 m - ak_2 m - a^2 k_1 m$$

$$\vdots \qquad \vdots$$

$$x_j = a^j x_0 - m(k_j + ak_{j-1} + a^2 k_{j-2} + \ldots + a^{j-1} k_1).$$

Because $1 \leq x_i < m$, $1 \leq i \leq j$, the integer $(k_j + ak_{j-1} + \cdots + a^{j-1} k_1)$ must be the largest integer in $a^j x_0 / m$, i.e., it must equal $[a^j x_0 / m]$. Hence, an alternative representation of the generator of Eq. (3.3) is

$$x_j = a^j x_0 \bmod m, \quad j \geq 1. \tag{3.4}$$

The initial seed x_0 must not equal 0 or m because these generate an infinite sequence of zeros. Because there are at most $m - 1$ integers x_i in $0 < x_i < m$, the maximum *period* or *cycle length* of the multiplicative consequential generator is $m - 1$. However, for a given m and a, the period may be much less than $m - 1$. Clearly, for a practical RNG the modulus m must be very large and the multiplier a must be picked to produce a very long (if not full) period.

3.2.1   Structure of the Generated Random Numbers

All of the ρ_i produced by a full-period generator are uniformly distributed over the interval $(0, 1)$ because each possible integer $1, 2, \ldots, m - 1$ appears just once during a cycle. However, a useful RNG must produce *subsamples* which must also be distributed uniformly over $(0, 1)$. Moreover, there should be no correlation between successive numbers, i.e., each number should appear to be independent of the preceding number. Although such correlations can be minimized with proper choices for a and m, they are never completely eliminated. For a poor choice of a and m, the correlations become obvious.

For example, consider a full-period generator with $a = 2$ and a very large m. At some point in the cycle x_i will be returned as 1 (or some other small integer). The subsequent values of x_{i+1}, x_{i+2}, ... will then also be small integers. If $x_i = 1$, the subsequent sequence is 1, 2, 4, 8, 3, ... (see Example 3.1). Clearly this subsample is far from random.

---

[5] The notation $[x/y]$, where x and y are integers, means that only the integer portion of the quotient x/y is used, i.e., any fractional part is discarded.

Example 3.1 A Simple Multiplicative Congruential Generator

Consider the sequences produced by the simple generator $x_{i+1} = ax_i \bmod 13$ for $(1 \leq a < 13)$. All are started with $x_0 = 1$. The following results are obtained:

							x_i						
a	$i=0$	1	2	3	4	5	6	7	8	9	10	11	12
1	1	1											
2	1	2	4	8	3	6	12	11	9	5	10	7	1
3	1	3	9	1									
4	1	4	3	12	9	10	1						
5	1	5	12	8	1								
6	1	6	10	8	9	2	12	7	3	5	4	11	1
7	1	7	10	5	9	11	12	6	3	8	4	2	1
8	1	8	12	5	1								
9	1	9	3	1									
10	1	10	9	12	3	4	1						
11	1	11	4	5	3	7	12	2	9	8	10	6	1
12	1	12	1										

Note $a = 1$ produces only a sequence of ones, and hence a must always be > 1. Only four multipliers—namely, $a = 2, 6, 7, 11$—produce full periods. For other multipliers, cycles of lengths 6, 4, 3, and 2 are realized. Generally, the cycle length of the nonfull period sequences depends on the initial seed. Note also that, except for $a = 12$, the nontrivial sequences occur in pairs, one being the reverse of the other.

As observed by Marsaglia (1968) the sequence of values x_i produced by any LCG has an extremely rigid structure. If k RNs at a time (k-tuples) are used as coordinates of points in k-space, the points do not fill the space randomly, but rather they lie on a finite number—and possibly a small number—of parallel hyperplanes. There are at most $m^{1/k}$ such planes. Further, these points form a simple k-dimensional lattice.[6]

As a simple example, consider the generator

$$x_{i+1} = ax_i \bmod 29 \qquad \text{with} \quad x_0 = 1,$$

and plot the points $(x_0, x_1), (x_1, x_2), (x_2, x_3), \ldots$. In this two-dimensional plot, the hyperplanes become straight lines. For this generator, $a = 3, 8, 10, 11, 14, 15, 18, 19, 21, 26$, and 27 all produce full-period generators. However, the

---

[6] A k-dimensional lattice is defined by k independent vectors $\mathbf{v}_j$ such that any point on the hyperplanes (lattice) is given by $\sum_{j=1}^{k} n_j \mathbf{v}_j$, where n_j are integers.

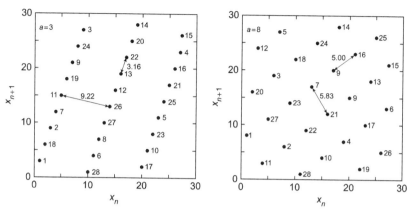

Figure 3.1 Two plots of successive numbers from $x_{n+1} = ax_n$ mod 29. The left-hand plot has a multiplier $a = 3$ and the right-hand $a = 8$. Both are full-period generators, but the right-hand generator fills the sample space much more uniformly than does the left-hand generator.

$m - 1 = 28$ points lie on different numbers of lines: two lines for $a = 14$, 15, and 27; three lines for $a = 3$, 10, 19, and 26; and six lines for $m = 8$, 11, 18, and 21. Two examples of the resulting 2-D plots are shown in Fig. 3.1.

The left-hand plot uses the generator $x_{n+1} = 3x_n$ mod 29 to produce the sequence 1 3 9 27 23 11 4 12 7 21 5 15 16 19 28 26 20 2 6 18 25 17 22 8 24 14 13 10 1.... The numbering of the points refers to the number pairs, i.e., 1 is the first pair (1, 3), 2 is the second pair (3, 9), and so on. This plot is somewhat alarming. All points in the lattice fall on just three lines with slope 3 and there are also 10 lines of slope $= -2/9$. The distance between adjacent points on these two sets of lines is very different, and the full "volume" (area) of the square is far from being uniformly covered by the points, a property we would expect if we use a good RNG.

The right plot in Fig. 3.1 uses the generator $x_{n+1} = 8x_n$ mod 29 to produce the sequence 1 8 6 19 7 27 13 17 20 15 4 3 24 18 28 21 23 10 22 2 16 12 9 14 25 26 5 11.... This is a much more reassuring picture. The points are more evenly distributed over the plot area and lie along six lines of slope 3/4 or along seven lines of slope $-5/3$. Moreover the distance between adjacent points of these two sets of lines are comparable (5.00 vs. 5.83), indicating a nearly equidistribution of points over the area of the square.

3.3 Tests for Linear Congruential Generators

Because of the lattice formed by LCGs, the effect of the lattice granularity is of great concern.[7] Not only should the k-dimensional hypercube be uniformly

---

[7] Today LCGs have mostly fallen out of favor among RN researchers, largely because of the RN granularity caused by the lattice hyperplanes inherent in this type of RNG. We believe this is short sighted because very adequate LCGs for practical use are available. More LCGs have one property

filled with lattice points, but the hyperplanes and points also should be close together to reduce the effect of the granularity. Here several of the statistical tests that have been used to test various congruent generators are reviewed. More details and the theory behind these tests can be found in Fishman (1996).

3.3.1 Spectral Test

The maximum period for a congruential generator with prime m is $t(m) = m - 1$, while for a generator with $m = 2^{\alpha}$, $t(m) = m/4$. For a given multiplier a and modulus m, the distance between adjacent hyperplanes $d_k(a, m)$ is (Cassels, 1959)

$$d_k(a, m) \geq d_k^{\min} \equiv \frac{1}{[t(m)]^{1/k}} \begin{cases} (3/4)^{1/4} & k = 2, \\ 2^{-1/6} & k = 3, \\ 2^{-1/4} & k = 4, \\ 2^{-3/10} & k = 5, \\ (3/64)^{1/12} & k = 6, \end{cases} \tag{3.5}$$

where $d_k^{\min}$ is the theoretical lower bound on the plane spacing. For a given modulus m, the "best" multiplier a would produce $d_k(a, m)$ that is closest to the minimum possible hyperplane separation. The statistic

$$S_{1,k} = \frac{d_k^{\min}}{d_k(a, m)} \tag{3.6}$$

is such that $0 < S_{1,k} \leq 1$ for $k \geq 2$ and those values of a that have $S_{1,k}$ as close to unity as possible for $2 \leq k \leq 6$ are sought.

3.3.2 Number of Hyperplanes

The number of hyperplanes $N_k(a, m)$ formed by the lattice in the k-dimensional hypercube is bounded by (Marsaglia, 1968)

$$N_k(a, m) \leq [k! t(m)]^{1/k}. \tag{3.7}$$

Then the statistic $S_{2,k} \equiv N_k(a, m)/[k! t(m)]^{1/k}$ has values between 0 and 1, and the best multiplier a from this test is the one that produces the largest values of $S_{2,k}$, i.e., the closest values to unity, for $2 \leq k \leq 6$.

---

that no other class of RNGs possesses, namely, the ability to easily skip ahead. This is an essential property if multiple users and machines are to be used to tackle a single difficult MC problem and where each machine must use a different substring of RNs in order to produce independent results.

3.3.3 Distance Between Points

The distance between the k-tuples in the k-dimensional hypercube should be as large as possible to ensure the best uniform coverage. Let $r_k(a, m)$ be the Euclidean distance between the nearest points. Then it can be shown that (Cassels, 1959)

$$r_k(a, m) \leq \frac{1}{d_k^{\min}(a, m)[t(m)]^{2/k}}, \tag{3.8}$$

where $d_k^{\min}$ is given by Eq. (3.5). The statistic $S_{3,k} \equiv r_k(a, m)d_k^{\min}(a, m)[t(m)]^{2/k}$ has values between 0 and 1, and the best multiplier a from this test is, again, the one that produces the largest values of $S_{3,k}$, i.e., the closest values to unity, for $2 \leq k \leq 6$.

3.3.4 Other Tests

Often a test for *discrepancy* is used to evaluate RNGs. This test quantifies how equidistributed the points are in the k-dimensional hypercube compared to that which would occur with truly RNs. It, thus, provides a benchmark that allows one to assess how closely the deterministic sequence $\{\rho_i\}$ looks random.

The lattice of k-tuples is defined by a set of k basis vectors $\mathbf{v_i}$. The closer the lengths $|\mathbf{v_1}|, \ldots, |\mathbf{v_k}|$ are to each other, the more equidistant are the k-tuples to each other in the k-dimensional hypercube. The *Beyer quotient* (Beyer et al., 1971) is

$$q_k(a, t(m)) \equiv \min_{1 \leq i \leq k} |\mathbf{v_i}| \bigg/ \max_{1 \leq i \leq k} |\mathbf{v_i}|, \quad k > 1. \tag{3.9}$$

Multipliers with a quotient close to unity are preferred.

Finally, various statistical tests of various hypotheses of samples of length n from the sequence $\{\rho_i\}$ should be made. These include hypotheses that (1) the sample is a sequence of independent identically distributed random variables, (2) the $\{\rho_i\}$, $1 \leq i \leq n$, are distributed uniformly over $(0, 1)$, (3) $\{\rho_{2i-1}, \rho_{2i}\}$, $1 \leq i \leq n/2$ are uniformly distributed over the unit square, (4) $\{\rho_{3i-2}, \rho_{3i-1}, \rho_{3i}\}$, $1 \leq i \leq (n - 2)/2$, are distributed uniformly over the unit cube, and (5) all of the previous hypotheses are true simultaneously. Tests for these hypotheses are provided by Fishman (1996).

3.4 Practical Multiplicative Congruential Generators

The congruential RNG is very fast, requiring very few operations to evaluate x_{i+1} and, consequently, it was widely used until about 20 years ago, when RNGs based on linear feedback shift registers began to be promoted by RNG researchers. With some care, the MCG generator can be written in high-level computer languages and made portable. To produce very long periods and to reduce the granularity between the RNs, the modulus m must be a very large integer, typically as large as the computer precision allows.

The generator of Eq. (3.3) can also be written in an alternate form by multiplying it by a^{-1}, the multiplicative inverse of a modulo m, defined as $a^{-1}a = 1 \bmod m$, to obtain

$$x_i = a^{-1}x_{i+1} \bmod m. \tag{3.10}$$

This means that for every sequence of RNs, there is another sequence that is the same but in reverse order provided $a \neq a^{-1} \bmod m$. This is what is observed in Example 3.1.

3.4.1   Generators with $m = 2^\alpha$

When digital computers were first introduced, computing efficiency was of major concern and often m was taken as some large power of 2, thereby enabling rapid calculation of the right-hand side of Eq. (3.3). Just as multiplying (dividing) a decimal number by 10^n is easily done by moving the decimal point n places to the right (left), so is multiplying (dividing) a base-2 binary number by 2^n simply a matter of moving the binary point n places to the right (left). Shifting the binary point of a number in a computer register is much faster than performing multiplication or division. However, such algorithms usually require machine language coding, and, thus, usually are not portable to different computer architectures. However, today the time needed to compute an RN is usually negligible. But $m = 2^\alpha$ is still of major interest, because with large values of α, very long periods can be obtained.

The maximum period of a generator with a modulus 2^α ($\alpha \geq 3$) is $m/4$ or $2^{\alpha-2}$. This maximum cycle length is realized for any multiplier if $a \bmod 8 = 3$ or 5 and the initial seed is an odd integer (Knuth, 1981). It is found that multipliers with $a = 5 \bmod 8$ produce numbers that are more uniformly distributed than those produced with $a = 3 \bmod 8$. That the seed must be odd is a nuance that a user must remember when using this type of generator.

What multipliers should be used for a given modulus? This is a topic which has received considerable attention with many early proposals having subsequently been shown to be somehow deficient. Fishman (1990) has exhaustively investigated congruential generators with $m = 2^{32}$ and with $m = 2^{48}$ by using the tests of Section 3.3 to find the "best" multipliers. A summary of his findings is given in Table 3.1. The first five multipliers for both $m = 2^{32}$ and $m = 2^{48}$ performed the best in all the statistical tests. For $m = 2^{32}$, the widely used multiplier 69,069 had significantly poorer test scores. Details and test scores are given by Fishman (1996).

An example of a multiplicative congruential generator with modulus $m = 2^{48} \simeq 2.81 \times 10^{14}$ is that employed by the widely used particle transport code MCNP (2003). This generator has a multiplier $a = 5^{19} = 19,073,486,328,125$

Table 3.1 Recommended multipliers for generators of the form $x_{i+1} = ax_i \bmod 2^\alpha$. The second multiplier in the pairs shown in the first column is $a^{-1} \bmod 2^{32}$ and hence produces the same sequence as a, but in reverse order. After (Fishman, 1996).

$m = 2^{32}$		$m = 2^{48}$	
$a = 5^n \bmod 2^{32}$	n	$a = 5^n \bmod 2^{48}$	n
1,099,087,573[a]	9,649,599	68,909,602,460,261[a]	528,329
4,028,795,517	93,795,525		
2,396,548,189[a]	126,371,437	33,952,834,046,453[a]	8,369,237
3,203,713,013	245,509,143		
2,824,527,309[a]	6,634,497	43,272,750,451,645[a]	99,279,091
1,732,073,221	96,810,627		
3,934,873,077[a]	181,002,903	127,107,890,972,165[a]	55,442,561
1,749,966,429	190,877,677		
392,314,069[a]	160,181,311	55,151,000,561,141[a]	27,179,349
2,304,580,733	211,699,269		
410,092,949[b]	–	44,485,709,377,909[c]	66,290,390,456,821
69,069[d]	–	19,073,486,328,125[e]	19

[a]Fishman (1990), [b]Borosh and Niederreiter (1983), [c]Durst as reported by Fishman (1996), [d]Marsaglia (1972), [e]Beyer as reported by Fishman (1996).

and has a period of $2^{46} \simeq 7.04 \times 10^{13}$. Beginning with MCNP5, an alternative, optional, 63-bit L'Ecuyer LCG is available that has a period of 9.2×10^{18}.

3.4.2 Prime Modulus Generators

Many congruential RNGs take m as a large prime number (or a number with a large prime number as a factor). With a prime number modulus, a full-period generator is obtained for a multiplier a that is a *primitive root* modulo m (Fuller, 1976).[8]

To find a primitive root for a given prime modulus, one can test successive values of $a = 2, 3, \ldots$ to find a relatively small primitive root a_1. However, this root is a poor choice for a prime modulus generator because a small value of x_i results in an initial ascending run of small integers, hardly what one would expect from RNs. To avoid such initial low-order correlations, a primitive root multiplier a should be chosen such that $a \lesssim \sqrt{m}$. To find larger primitive roots from a small primitive root a_1, it can be shown that (Fishman and Moore,

---

[8] If the smallest integer k that satisfies $a^k = 1 \bmod m$ is $k = m - 1$, then a is called a primitive root, modulo m.

1986)

$$a = a_1^n \bmod m \tag{3.11}$$

is also a primitive root if the integer n is relative prime to $m - 1$, i.e., if n and $m - 1$ have no factors in common (other than 1). An illustration of finding the primitive roots is given in Example 3.2.

A widely used prime modulus is the Mersenne prime $2^{31} - 1 = 2,147,483,647.$[9] For this modulus, the smallest primitive root is $k_1 = 7$. The factors of $2^{31} - 1$ are 2, 3, 7, 11, 31, 151, and 331. Thus $7^5 = 16,807$ is also a primitive root and produces a full-cycle length. First introduced by Lewis et al. (1969), the RNG

$$x_{i+1} = 7^5 x_i \bmod (2^{31} - 1) \tag{3.12}$$

has become well accepted. The modulus $m = 2^{31} - 1$ is an obvious choice for the 32-bit architecture of today's personal computers. This generator has passed many empirical tests for randomness, its theoretical properties have been well studied, and, more important, it has a vast history of successful application.

Example 3.2 Primitive Roots

Consider the simple generator $x_{i+1} = ax_i \bmod 13$. What values of a produce a full-cycle generator? For such a prime modulus generator all primitive roots produce full cycles. Thus, first find a small primitive root, i.e., find an a such that the smallest integer k that satisfies $a^k \bmod 13 = 1$ is $k = m - 1 = 12$. It is easily verified that $2^k \bmod 13 = 2, 4, 8, 3, 6, 12, 11, 9, 5, 10, 7, 1$ for $k = 1, 2, \ldots, 12$. Hence $a = 2$ is the smallest primitive root.

The factors of $m - 1 = 12$ are 3, 2, and 2. Then the values of $n < m - 1$ that have no factors (except 1) in common with 3 and 2 are $n = 5, 7,$ and 11. Hence the primitive roots for $m = 13$ are 2, $2^5 \bmod 13 = 6$, $2^7 \bmod 13 = 11$, and $2^{11} \bmod 13 = 7$. Thus, $a = 2, 6, 7,$ and 11 produce full-period generators, a result previously seen in Example 3.1.

3.5   A Minimal Standard Congruential Generator

Park and Miller (1988) have proposed Eq. (3.3) with $m = 2^{31} - 1$ and $a = 16,807$ as a "minimal standard" generator because it passes three important

---

[9] Mersenne primes M_p have the form $2^p - 1$, where p is a prime number. Many early mathematicians thought all numbers of the form $2^p - 1$ were prime but in 1536 Hudalricus Regius showed $2^{11} - 1 = 2047$ was not prime (23×89). Pietro Cataldi in 1603 showed $2^{17} - 1$ and $2^{19} - 1$ were prime but incorrectly stated $p = 23, 29, 31,$ and 37 produced prime values of M_p. Then Marin Mersenne (1588–1648), a French Minim friar who studied these numbers, wrote in the preface of his *Gogitata Physica-Mathematica* (1644) that $p = 2, 3, 5, 7, 13, 17, 19, 31, 67, 127,$ and 257 yielded prime numbers and conjectured that all other integers < 257 produced composite numbers. His conjecture was wrong but his name became attached to these numbers. Today 51 values of p are known to give Mersenne primes with the latest (Dec. 2018) being $p = 82,589,933$, which yields a Mersenne prime with an astounding 24,862,048 digits.

tests: (1) it is a full-period generator, (2) its sequence ... $x_1, x_2, \ldots, x_{m-1}, x_0, x_1,$ ... is "random," and (3) ax_i mod m can be efficiently implemented (with some clever coding discussed in Section 3.5.1) on a computer using 32-bit integer arithmetic. This is not to say that this particular generator is "ideal." In fact, it does have some flaws (see Section 3.5.2). But it has been widely used. This RNG is the one we have used for the many illustrative MC examples in this book.

Is the multiplier $a = 16,807$ the optimum multiplier for use with $m = 2^{31} - 1$? Probably not. There are about 534,600,000 primitive roots of $m = 2^{31} - 1$ which produce a full period. Hence, test 1 eliminates about 75% of the possible multipliers. Fishman and Moore (1986) examined the remaining 25% for passing test 2. To pass this test, the Euclidean distance between adjacent hyperplanes has to be no more than 25% of the theoretical minimum for dimensions $2 \leq k \leq 6$ (the *lattice test*). Only 410 multipliers met this criterion. The multiplier 16,807 is not among them (nor are other commonly used multipliers). However, *none* of these 410 optimal test-2 multipliers lend themselves to the clever programming trick of test 3 that permits coding the generator on machines using 32-bit integers. Park and Miller (1988) found that only 23,093 multipliers pass both tests 1 and 3. By replacing the original 25% criterion for test 2 by a 30% criterion, several good 32-bit multipliers were found, the two best being $a = 48,271$ and $a = 69,621$. Until further testing is done and confidence is gained with these two new multipliers, the original $a = 16,807$ will continue to be used widely.

3.5.1 Coding the Minimal Standard

Coding Park and Miller's minimal standard RNG,

$$x_{i+1} = ax_i \text{ mod } m \text{ with } a = 16,807 \text{ and } m = 2^{31} - 1 = 2,147,483,647,$$
$$(3.13)$$

for a computer based on 32-bit integers is not straightforward because, for $1 \leq x_i \leq m - 1$, the product ax_i often exceeds the maximum value expressible as a 32-bit integer. One could avoid this problem by using the extended precision option available in most computer languages. For example, in FORTRAN one could use

```
DOUBLE PRECISION FUNCTION randm(dseed)
DOUBLE PRECISION dseed
  dseed = DMOD(16807.d0*dseed,2147483647.d0)
  randm = dseed/2147483647.d0
RETURN
END
```

However, using double-precision floating-point arithmetic is considerably less efficient than integer arithmetic, especially because this function might be called billions of times in a typical MC calculation.

It is much better to use 32-bit integer arithmetic and a trick proposed by Schrage (1979) and improved in 1983 (Bratley et al., 1987) to integer overflow problem. In this method, the modulus m is approximately factored as

$$m = aq + r \quad \text{where} \quad q = [m/a] \quad \text{and} \quad r = m \bmod a. \tag{3.14}$$

With this factorization, the generator of Eq. (3.13) can be evaluated as (Press et al., 1996)

$$x_{i+1} = ax_i \bmod m = \begin{cases} a(x_i \bmod q) - rk & \text{if } \geq 0, \\ a(x_i \bmod q) - rk + m & \text{if above is } < 0, \end{cases} \tag{3.15}$$

where $k = [x_i/q]$. How does this reformulation avoid integer overflows? Recall that $1 \leq x_i < m - 1$. If $r < q$, then it can be shown (Park and Miller, 1988) that both $a(x_i \bmod q)$ and rk are in the range $0, \ldots, m - 1$. This factorization with $r < q$ is the requirement for a generator to pass Park and Miller's test 3. For the minimal standard generator ($m = 2^{31} - 1$ and $a = 16{,}807$), Schrage factorization uses $q = 127{,}773$ and $r = 2836$.

Here is a very efficient FORTRAN routine that implements this algorithm for the minimal standard generator. The value of x_i is input as iseed and the value of x_{i+1} is returned in iseed. The returned value of ran is $\rho_{i+1} = x_{i+1}/m$.

```
FUNCTION ran(iseed)
INTEGER iseed,ia,im,iq,it,k
REAL ran,am
PARAMETER (ia=16807,im=2147483647,am=1./im,
&            iq=127773,it=2836)
k = iseed/iq
iseed = ia*(iseed-k*iq)-it*k
IF(iseed.LT.0) iseed=iseed+im
ran = am*iseed
RETURN
END
```

It is easy to modify this program to use the two optimal multipliers found by Park and Miller (1988), namely, $a = 48{,}271$ (with $q = 44{,}488$ and $r = 3399$) and $a = 69{,}621$ (with $q = 30{,}845$ and $r = 23{,}902$).

One final note about programming the minimal standard generator in other computer languages. It is easy to make mistakes; and to check the proper operation of the code, you should start with an initial seed of iseed $= x_0 = 1$ and ensure that the generator returns $x_{10{,}000} = 1{,}043{,}618{,}065$. Implementation of this generator in other programming languages is considered in Appendix F. All

the examples in this book use this minimal standard LCG with the starting seed
iseed = 73,907.

3.5.2 Deficiencies of the Minimal Standard Generator

The minimal standard generator of Eq. (3.13) is not perfect. Indeed, a small
value of ρ_i, say less than 10^{-6}, is produced one time in 10^6. This RN is followed by another small number less than 0.017. For very improbable events,
this serial correlation of small ρ_i values could, in principle, cause a minor decrease in the accuracy of MC simulation results. Even more subtle low-order
serial correlations exist in this generator (Press et al., 1996). With any small
multiplier a, all congruential generators have low-order serial correlations.

To minimize small number correlations, the multiplier of a multiplicative
congruential generator should be approximately equal to the square root of the
modulus, which is more restrictive than the requirement that $r < q$ in order to
pass Park and Miller's test 3. However, for the application to the MC methods
discussed in this book, such deficiencies have negligible effect. See, for example, Problem 3.17 which directly addresses this issue.

Table 3.2 Recommended multipliers for prime modulus congruential random number generators. After (Fishman, 1996).

$m = 2^{31} - 1$	multipliers a with $r < q$			
multiplier a	modulus m	multiplier a	$q = \lfloor m/a \rfloor$	$r = m \bmod a$
742,938,285[a]	$2^{31} - 1$	48,271[b]	44,488	3399
950,706,376[a]	$2^{31} - 1$	69,621[b]	30,845	23,902
1,226,874,159[a]	$2^{31} - 1$	16,807[c]	127,773	2836
62,089,911[a]	$2^{31} - 1$	39,373[d]	54,542	1481
1,343,714,438[a]	$2^{31} - 85$	40,014[d]	53,668	12,211
630,360,016[e]	$2^{31} - 249$	40,692[d]	52,774	3791

[a]Fishman and Moore (1986), [b]Park and Miller (1988), [c]Lewis et al. (1969), [d]L'Ecuyer (1988),
[e]Payne et al. (1969).

3.5.3 Optimum Multipliers for Prime Modulus Generators

As Park and Miller (1988) suggest, the minimal standard's multiplier $a =$
16,807 is probably not the best multiplier for $m = 2^{31} - 1$. In a massive effort, Fishman and Moore (1986) exhaustively studied all possible primitive roots
for this modulus, including the values 48,271 and 69,621 suggested by Park
and Miller (1988) as being superior to the minimal standard multiplier. The
results of the "best" multipliers are summarized in Table 3.2. The multiplier
$a = 62,089,911$ was overall the best, with $a = 69,621$ the best for multipliers
with $r < q$. Although 16,807 is not unreasonable, it did significantly worse in
the tests for randomness than did all the other multipliers shown in this table.

3.5.4   Shuffling a Generator's Output

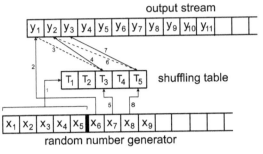

Figure 3.2 In the Bays and Durham shuffling scheme, a random number generator, such as the minimal standard, is used to populate any vacancies in the shuffling table, from which values are selected randomly and placed in the output stream for use in a Monte Carlo calculation. The numbered arrows refer to the steps in the algorithm described in the text.

The simplicity and ease of implementation of congruential RNGs have resulted in their widespread use despite known limitations such as low-order correlations and potentially imperfect equidistributedness. Generators with carefully chosen multipliers for a given modulus can produce k-tuples that are almost equidistributed. The following question arises: Is there some minor change that could be made in the generation process to make the k-tuples (or lattice) more equidistributed and random?

One simple idea is to shuffle the numbers produced by a generator so that all generated numbers appear in the output stream, but in an order different from the order in which they were generated. To accomplish this, one could use the output of a second (and perhaps simpler) generator to shuffle or permute the output sequence from the main generator. Such shuffling can increase the period because no longer does the same value follow a given value every time it occurs. Moreover, the shuffling breaks up the original lattice structure replacing it with a different one with a different number of hyperplanes. Another benefit of shuffling is that it removes low-order correlations.

Bays and Durham (1976) have shown that a single generator can be used to first fill a table (vector) with x_i values and then use subsequent values from the RNG to select and replace particular table entries. In this way, the output values from the RNG are shuffled and low-order correlations are removed. This shuffling can also increase the period of the generator.

The Bays and Durham shuffling algorithm is described below and illustrated in Fig. 3.2. The generator produces random values x_i and the output stream from the shuffling is denoted by y_i. The shuffling table (vector) has length k.

1. (Initialization) Fill the table vector T_j with x_j, $j = 1, \ldots, k$.
2. Place $y_1 = x_{k+1}$ as the first entry in the output stream and set $i = 1$.
3. Compute the table pointer $j = 1 + [ky_i/(m-1)]$.
4. Set $i = i + 1$ and place $T_j = y_i$ into the output stream.
5. Get the next number x_{k+i} and place it in T_j.
6. Loop back to step 3 to place another number into the output stream.

As an example of the effectiveness of this shuffling, consider the example shown in the left-hand plot of Fig. 3.1 where the full-cycle generator

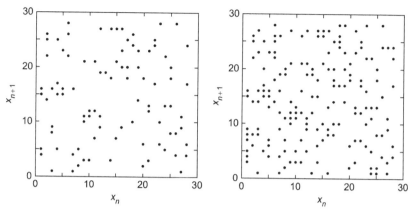

Figure 3.3 The effect of using the Bays and Durham shuffling scheme on the random number generator $x_{i+1} = 3x_i$ mod 29 (used to generate the pair distribution of the left-hand plot in Fig. 3.1) is shown in the left-hand plot above for the first 101 pairs and in the right-hand plot for the first 201 pairs. Besides a much longer-cycle length, a more uniform distribution is achieved.

$x_{i+1} = 3x_i$ mod 29 produces far from uniform distributions of 2-tuples over the square. With the Bays and Durham shuffling algorithm applied to this generator, the results shown in Fig. 3.3 are obtained. The cycle length is at least several thousand and the plot area is much more uniformly covered.

Shuffling programs for the minimal standard RNG are given by Press et al. (1996). With such shuffling, even more statistical tests for randomness are met.

The idea of shuffling the output of an RNG applies to all RNGs and not just LCGs. In fact, in a test comparison of 500 RNGs using the Diehard suite of statistical tests for RNGs (see Section 3.13), Klimasauskas (2002) reports that of the 20 highest scoring RNGs, 16 used a final Bays–Durham shuffle on the generator's output.

3.5.5   Refinement of the Minimal Standard

In response to criticisms of the minimal standard generator (Marsaglia, 1993; Sullivan, 1993), Park et al. (1993) reaffirm their original proposed generator, discussed above, but with two modifications. They suggest the use of the multiplier 48,271 ($r = 3399$ and $q = 44,488$) as being slightly better than their original choice of 16,807. This new choice is the "best" choice as found by Fishman and Moore (1986) (see Table 3.2). Such a change is easily made to the code in Section 3.4 and Appendix F.[10] They also recommend that the generator output be shuffled as described in the previous section by, for example, the Bays and Durham algorithm. Press et al. (1992a, 1992b, 1996) give code in FORTRAN77, FORTRAN90, C, and C++ that combines the minimal standard

---

[10] For an initial seed $x_0 = 1$, this revised multiplier yields $x_{10,000} = 399,268,537$.

and the Bays and Durham shuffler. One big advantage of such a pairing, besides eliminating the low-level correlation effects inherent in all multiplicative congruential generators, is a large increase in the cycle length of the generator. This is of major importance in today's use of MC methods on massively parallel computer architectures. This pairing also preserves the basic philosophy of the minimal standard: simplicity, good efficiency, and portability among different computer systems.

3.6   Skipping Ahead

Often it is useful to generate separate independent sequences of RNs from the same congruential generator. For example with a very long-period generator, independent substrings could be obtained so that different machines could work simultaneously on the same simulation and the independent results from each machine can, therefore, be combined.

It is not a good idea to simply use a different seed for each sequence because with bad choices the resulting sequences may overlap significantly. The best way to generate separate independent sequences is to skip ahead a specified distance in the sequence to start a new subsequence.

This skipping ahead is easily done for an LCG

$$x_{i+1} = ax_i \bmod m. \tag{3.16}$$

By recursive application of this relation, it is easily shown (see Problem 3.2) that

$$x_{i+k} = a^k x_i \bmod m. \tag{3.17}$$

With this relation, it is easy to start a new sequence some number of integers ahead of the start of the previous sequence.

Suppose the nth sequence started with x_n. Then the next sequence must begin with x_{n+s}, where s is the jump or *stride* between the two sequences. From Eq. (3.17), the $(n+1)$th sequence is begun with the integer

$$x_{n+s} = a^s x_n \bmod m = bx_n \bmod m, \quad \text{where} \quad b = a^s \bmod m. \tag{3.18}$$

The remainder of the new sequence is then obtained from Eq. (3.16).

In the past 30 years, LCGs have fallen out of favor by RNG specialists, primarily because of the inherent manner in which their pseudorandom numbers form hyperlattices. However, no other type of RNG is able to easily skip ahead and form nonoverlapping substrings of RNs. Consequently, computationally intensive codes that use multiple cores or CPUs and, thus, need long nonoverlapping substrings of independent RNs will continue to rely on LCGs.

3.7   Combining Generators

An obvious way to increase the period and randomness of an RNG is to use two or more generators and merge their outputs. For example, Collins (1987) suggested using many generators each producing an output stream of RNs. At each step, the output from one generator is used to select randomly one of the other generators whose number is selected as the RN for that step. Such a procedure would vastly increase the cycle length. However, it slightly decreases the efficiency of a generator because extra calculations are needed. Other methods for combining several LCGs have been proposed. These include addition using integer arithmetic (Wichmann and Hill, 1982), shuffling (Nance and Overstreet, 1978), and bitwise addition modulo 2 (Bratley et al., 1987). A note of caution. The period of combined RNGs is usually the least common multiple (lcm) of the periods of each generator. So if all the generators have the same period, then there is no augmentation of the period.

When one combines generators whose normalized output is in the open interval $(0, 1)$, it is possible because of the finite machine precision to obtain a 0 or a 1 (see Problem 3.12). Special care must be exercised when coding such generators to preclude the production of an endpoint. Finally, it should be mentioned that the beneficial increase in a combined generator's period applies not just to LCGs but to all RNGs. Indeed many of the best RNGs available today use such a strategy.

3.7.1 Bit Mixing

A very common way to improve randomness and increase cycle length is to combine the integers x_i and y_i produced by two different generators to generate a new integer $z_i = x_i \odot y_i$, where $\odot$ is some binary operator, typically addition or subtraction modulo 1, to produce a mixing of the bits in the base 2 representations of x_i and y_i. This is very efficient and easily implemented. Again this technique can be used with other RNGs and not just LCGs.

3.7.2 The Wichmann–Hill Generator

Wichmann and Hill (1982) proposed combining the random integers from the following three MCGs:

$$x_{i+1} = 171x_i \bmod 30{,}269, \quad y_{i+1} = 172y_i \bmod 30{,}307,$$
$$z_{i+1} = 170z_i \bmod 30{,}323,$$

to produce the ith RN in $(0, 1)$ as

$$\rho_i = \left(\frac{x_i}{30{,}269} + \frac{y_i}{30{,}307} + \frac{z_i}{30{,}323} \right) \bmod 1. \tag{3.19}$$

This generator requires three seeds (x_0, y_0, z_0) and, because the moduli are relatively prime, this generator cannot yield an exact zero (although finite machine precision requires careful programming to avoid 0). The period of this generator is about 10^{12}. This generator has been investigated for its long-range correlations and found to be comparable to other good generators (De Matteis and Pagnutti, 1993). Moreover, it is easily programmed for a 16-bit machine.

3.7.3 The L'Ecuyer Generator

L'Ecuyer (1988) has suggested a way to combine several different sequences with different periods so as to generate a sequence of which the period is the lcm of the constituent sequences. Such a combination greatly helps reduce the serial correlations present in each constituent generator. In particular, L'Ecuyer proposed combining the two RNGs (Gentle, 1998)

$$x_i = 40,001 x_{i-1} \bmod 2,147,483,563, \qquad y_i = 40,692 y_{i-1} \bmod 2,147,483,399$$

and taking $z_i = x_i - y_i$ as a random integer. Should z_i be negative, then adjust it as $z_i = z_i + 2,147,483,562$. The RN uniformly distributed in $(0, 1)$ is then given by

$$\rho_i = \frac{z_i}{21,474,883,589} = 4.656613 z_i \times 10^{-10}. \tag{3.20}$$

Because x_i and y_i are uniform distributions over almost the same interval, the difference is also uniform. Moreover, a value of $\rho_i = 1$ cannot be obtained because the normalizing constant is slightly greater than 2,147,483,536. The period of this generator is about 2.3×10^{18} and L'Ecuyer claims this generator produces a good sequence of RNs. It can also be programmed on a 32-bit machine (see James (1990) for a portable FORTRAN version). It has one minor limitation. It is not easy to generate separate independent sequences because it is not easy to skip ahead with this algorithm.

Press et al. (1996) offer the RAN2 program that is based upon this L'Ecuyer generator with a final Bays–Durham shuffle added. This generator produces a sequence with a period greater than 2.3×10^{18}. The authors claim this generator is "perfect."

3.8 Other Congruential Random Number Generators

Many variations to the congruential and related RNGs have been proposed. The basic idea is that by making a generator more complex, better randomness and longer periods can be achieved.[11] The properties of many of these generators are often not well established, and the problem of finding optimum generators (in some sense) is still an area of active research. In the following, a few of these generator variants are briefly outlined. For a more in-depth discussion and for summaries of even more generators, the reader is referred to Gentle (1998) and Fishman (1996).

---

[11] Although this is a widely held belief, it is just "folk lore." Complexity certainly makes the mathematical analysis more difficult, but empirical studies do not support this belief. There are a few very good RNGs that are quite simple and that require only a few lines of coding.

3.8.1 Multiple Recursive Generators

Instead of using only x_i to generate the next random integer x_{i+1}, one could use several earlier integers by using, for example,

$$x_i = (a_1 x_{i-1} + a_2 x_{i-2} + \ldots + a_n x_{i-n}) \bmod m. \qquad (3.21)$$

Some of these *multiple recursive* MCGs can have much longer cycles than a simple MCG. Such generators have been studied by L'Ecuyer et al. (1993), who made recommendations for various moduli (including $m = 10^{31} - 1$), gave a portable code for $n = 5$ that does not yield a 0 or 1, and showed how to skip ahead in the sequence.

3.8.2 Lagged Fibonacci Generators

Reducing the famous Fibonacci sequence $x_i = x_{i-1} + x_{i-2}$ modulo m has been proposed as the basis for an RNG. However, it has poor random properties. Much better results are obtained if two earlier results some distance apart are combined. Thus the *additive lagged Fibonacci congruential generator* is

$$x_i = (x_{i-j} + x_{i-k}) \bmod m. \qquad (3.22)$$

If the lags j and k and the modulus m are chosen properly, good random sequences can be obtained (Altman, 1989) that can have cycle lengths as large as $m^k - 1$ (Gentle, 1998). Knuth (1981) gives suggested values for the lags j and k that produce periods for $j > k$ of $(2^j - 1)(2^{m-1})$.

Marsaglia et al. (1990) have proposed a lagged Fibonacci generator with lags of 97 and 33 that has a period of $2^{144} \simeq 10^{43}$. Moreover, it is portable, giving bit-identical results on all 32-bit machines, and has passed extensive statistical tests despite limited theoretical understanding of this type of generator. James (1990) gives a very clever FORTRAN program RANMAR for this generator. A very useful feature of RANMAR is the ease with which independent subsequences can be obtained. The initialization of this generator requires a 32-bit integer. Each different integer produces an independent (nonoverlapping) sequence of average length $\simeq 10^{30}$, enough for very complex MC simulations.

The PRNG of Eq. (3.22) can be generalized to

$$x_i = (x_{i-j} \odot x_{i-k}) \bmod m, \qquad (3.23)$$

where $\odot$ is any *binary operator* such as addition, subtraction, multiplication, or bitwise exclusive-or (XOR) operator.[12] For a *multiplicative lagged Fibonacci generator*, the maximum cycle length is $(2^j - 1)2^{m-3}$ or 1/4 that of the maximum cycle length of an additive lagged Fibonacci generator. If the XOR binary

---

[12] b_m XOR b_n returns 0 if both binary bits b_n and b_m have the same value 1 or 0 and returns 1 if b_n and b_m are different. This binary operation on two binary numbers is very efficient in machine time and is used extensively in the rest of this chapter.

operator is used, the resulting generator is called a *two-tap generalized feedback shift register* and is a variant of the popular Mersenne Twister (MT) (see Section 3.10).

3.8.3 Add-with-Carry Generators

A variant of the lagged Fibonacci generator, introduced by Marsaglia and Zaman (1991), is the *add-with-carry* generator

$$x_i = (x_{i-j} + x_{i-k} + c_i) \bmod m, \tag{3.24}$$

where the "carry" $c_1 = 0$ and c_{i+1} is given by

$$c_{i+1} = \begin{cases} 0, & \text{if } x_{i-j} + x_{i-k} + c_i < m, \\ 1, & \text{otherwise.} \end{cases}$$

With proper choices for j, k, and m, very long-period generators can be obtained. Marsaglia and Zaman give values that produce periods of the order of 10^{43}. Moreover, this generator can be implemented in base 2 arithmetic. It has since been shown that the add-with-carry generator is equivalent to an LCG with a very large prime modulus (Tezuka and L'Ecuyer, 1992), which gives one some confidence in this generator and also shows how to program congruential generators with large prime moduli.

Other variations of this generator are the subtract-with-carry and the multiply-with-carry, also introduced by Marsaglia and Zaman (1991). Much work still needs to be done to determine the random properties and optimal parameter values for these generators.

James (1990) gives a FORTRAN program for 32-bit machines for the subtract-with-carry variant. He uses $j = 24$, $k = 10$, and $m = 2^{24}$ to produce a generator with a period of $\simeq 2^{570} \simeq 10^{171}$!

3.8.4 Inversive Congruential Generators

The multiplicative inverse x^- of an integer $x \bmod m$ is defined by $1 = x^- x \bmod m$. Eichenauer and Lehn (1986) proposed an RNG based on the multiplicative inverse, namely,

$$x_i = (a x_{i-1}^- + c) \bmod m. \tag{3.25}$$

This generator does not yield regular lattice hyperplanes as does an LCG, and it appears to have better uniformity and fewer serial correlation problems. However, calculation of multiplicative inverses is computationally expensive.

3.8.5 Nonlinear Congruential Generators

Knuth (1981) suggested the generator

$$x_i = (dx_{i-1}^2 + ax_{i-1} + c) \bmod m, \quad \text{with} \quad 1 \le x_i < m. \tag{3.26}$$

Other polynomials could be used as well. But the properties of such generators are not well established. Eichenauer et al. (1988) consider a more general variant, namely,

$$x_i = f(x_{i-1}) \bmod m, \tag{3.27}$$

with a wide variety of functions $f(x)$. Although such generators can produce very good uniformity properties, not enough is known yet for serious application of these generators. There needs to be much more study about nonlinear generators if they are to become useful.

3.9 RNGs Using Linear Feedback Shift Registers

The RNGs discussed so far are mostly LCG generators that manipulate base-10 numbers. Tausworthe (1965) suggested using an LCG-like generator that manipulated the bits (the 1s or 0s) used by computers to represent our base-10 numbers. The fundamental idea is to produce a very long string of 1s and 0s so that the string of these binary digits appears to be random but, given a short initial seed string, their placement is reproducible, i.e., their placement is pseudorandom. From this string of 1s and 0s, base-10 pseudorandom integers can then be obtained.

The two main approaches presently being pursued to produce RNGs at the bit-level are those that use primitive polynomials and LFSRs to manipulate the bits and those that use CA to generate random bits. In this section, the methods based on LFSRs are discussed, and in the next section the CA method is presented. Both approaches are capable of producing RNGs with long-cycle lengths and that can pass the most stringent of tests for randomness. But first the origin of the LFSR approach is described.

3.9.1 The Tausworthe Bit-Level RNG

In 1965, Tausworthe proposed the generation of a sequence of binary digits $b_1, b_2, \ldots, b_n$ from (Tausworthe, 1965)

$$\boxed{b_i = \left(\sum_{j=1}^{n} c_{n-j} b_{i-j} \right) \bmod 2, \quad i > n,} \tag{3.28}$$

where n is a positive integer and b_i and c_i are binary variables with values of 0 or 1. Note how Eq. (3.28) resembles Eq. (3.3) for the LCG. The evaluation of the arithmetic in Eq. (3.28) is done in modulo 2, so that addition/subtraction $\pm$ is done with the logical exclusive XOR or $\oplus$ operator and multiplication $\times$ uses the logical AND operator.[13] The evaluation of Eq. (3.28) is thus expressed as

$$b_i = c_{n-1}b_{i-1} \oplus c_{n-2}b_{i-2} \oplus c_{n-3}b_{i-3} \oplus \ldots \oplus c_0 b_{i-n}. \qquad (3.29)$$

Note that b_i depends on the n immediate prior values of b in the sequence. Such a *bit stream* or sequence is called an *autoregressive sequence* of order n.

Maximum Cycle Length

Just as an LCG has a period or cycle length before it begins to repeat, so does the above sequence. The maximum cycle length can be shown to be $2^n - 1$. For a 64-bit machine, one might pick $n = 63$ to give a maximum cycle length of 9.22×10^{18}. Under what conditions can this maximum cycle length be realized? To answer this question the concept of *characteristic polynomials* must be introduced.

Toward this end, define the *delay operator* $\mathcal{D}$ by $\mathcal{D}b_i = b_{i+1}$ so that $\mathcal{D}^m b_i = b_{i+m}$. Then Eq. (3.28) can be written as

$$\mathcal{D}^n b_{i-n} = c_{n-1}\mathcal{D}^{n-1}b_{i-n} + c_{n-2}\mathcal{D}^{n-2}b_{i-n} + \ldots + c_0 b_{i-n} = 0 \bmod 2,$$

or

$$[\mathcal{D}^n - c_{n-1}\mathcal{D}^{n-1} - c_{n-2}\mathcal{D}^{n-2} - \ldots - c_0]b_{i-n} = 0 \bmod 2 \qquad (3.30)$$

or

$$[\mathcal{D}^n + c_{n-1}\mathcal{D}^{n-1} + c_{n-2}\mathcal{D}^{n-2} + \ldots + c_0]b_{i-n} = 0 \bmod 2.$$

Then define the characteristic polynomial (with $c_0 = 1$ by convention) as

$$P(x) = x^n + c_{n-1}x^{n-1} + c_{n-2}x^{n-2} + \ldots + 1. \qquad (3.31)$$

It can be shown (Tausworthe, 1965) that (1) the period of the bit stream is the smallest positive integer n for which $x^n - 1$ is divisible by the characteristic polynomial and (2) the maximum possible period with a polynomial of order n is $2^n - 1$. The polynomials that give this maximum period are called *primitive polynomials*. Again this is reminiscent of an LCG generator needing a primitive root of the modulus to produce a full-period generator.

Primitive Polynomials

Before looking at the properties of modulo-2 primitive polynomials, a notation and modulo-2 arithmetic need to be introduced. One of the easiest shorthands to

---

[13] Here The AND operator is defined such that the binary variable $b = b_i \times b_j$ equals 1 if $b_i = b_j$ and equals 0 if $b_i \neq b_j$.

describe a polynomial like Eq. (3.31) is to specify the exponents of the nonzero monomial terms, i.e., terms for which $c_i = 1$. Thus,

$$x^8 + x^6 + x^5 + x^4 + 1$$

can be abbreviated as $(8, 6, 5, 4, 0)$.

An arbitrary nth-order polynomial such as that of Eq. (3.31) generally does not produce a full-period bit stream because the polynomial is not *primitive*. To be primitive, the polynomial must necessarily (a) be *irreducible*, i.e., it cannot be factored, and (b) the number of nonzero monomial terms x^k with $k \geq 1$ are even and their exponents are setwise *coprime*, i.e., there is no divisor other than 1 common to all their exponents. However, only a fraction of irreducible polynomials are primitive. For example, the polynomial $x^2 + 1 = (x + 1)(x + 1)$ mod 2 is reducible and, hence, is not primitive. Although $x^4 + x^3 + x^2 + x^1 + 1$ is irreducible, it is not primitive, whereas $x^4 + x + 1$ is both irreducible and primitive.[14]

Once a primitive polynomial of order n is found, a second primitive polynomial exists that is called its *dual* or *reciprocal* polynomial. For example, the eighth-order polynomial $P(x) = x^8 + x^6 + x^5 + x^4 + 1 = (8, 6, 5, 4, 0)$ is primitive. So its reciprocal polynomial, defined as $P^*(x) \equiv x^n P(1/x)$, is

$$P^*(x) = x^8(x^{-8} + x^{-6} + x^{-5} + x^4 + 1) = 1 + x^2 + x^3 + x^4 + x^8 = (8, 4, 3, 2, 0).$$

In his seminal paper (Tausworthe, 1965), Tausworthe considered only trinomial primitive polynomials although this initial restriction was later relaxed. A simple example of Tausworthe's RNG is given in Example 3.3.

Example 3.3 A Tausworthe Sequence and Associated RNG

Generate a full-cycle sequence of binary digits from an autoregressive sequence of order $n = 5$. Then with this sequence generate pseudorandom integers for the interval $(0, 32)$. Use $b_1 = b_2 = b_3 = b_4 = b_5 = 1$ as the initial *seed*.

For a fifth-degree polynomial, there are three primitive polynomials: (1) $x^5 + x + 1$, (2) $x^5 + x^4 + x^2 + x + 1$, and (3) $x^5 + x^4 + x^3 + x^2 + 1$. Choose the first one because it is the simplest. From the recursion relation of Eq. (3.30),

$$\mathcal{D}^5 b_i + \mathcal{D}^2 b_i + b_i \text{ mod } 2 \quad \text{or} \quad b_{i+5} + b_{i+2} + b_i = 0 \text{ mod } 2.$$

In terms of the XOR or $\oplus$

$$b_{i+5} \oplus b_{i+2} \oplus + b_i = 0 \quad \text{or} \quad b_{i-5} = b_{i+2} \oplus b_{i+5}, \quad i = 0, 1, 2, \dots.$$

---

[14] Comprehensive listings can be found on the internet. See, for example, https://www.partow.net/programming/polynomials/index.html.

Finally, substituting $i - 5$ for i gives

$$b_i = b_{i-3} \oplus b_{i-5}, \quad i = 6, 7, 8, \ldots . \tag{3.32}$$

Calculation of some of the b_is from Eq. (3.32) is shown below:

$b_6 = b_3 \oplus b_1 = 1 \oplus 1 = 0$	$b_{12} = b_9 \oplus b_7 = 1 \oplus 0 = 1$
$b_7 = b_4 \oplus b_2 = 1 \oplus 1 = 0$	$b_{13} = b_{10} \oplus b_8 = 1 \oplus 0 = 1$
$b_8 = b_5 \oplus b_3 = 1 \oplus 1 = 0$	$b_{14} = b_{11} \oplus b_9 = 0 \oplus 1 = 1$
$b_9 = b_6 \oplus b_4 = 0 \oplus 1 = 1$	$b_{15} = b_{12} \oplus b_{10} = 1 \oplus 1 = 0$
$b_{10} = b_7 \oplus b_5 = 0 \oplus 1 = 1$	$b_{16} = b_{13} \oplus b_{11} = 1 \oplus 0 = 1$
$b_{11} = b_8 \oplus b_6 = 0 \oplus 0 = 0$	$b_{17} = b_{14} \oplus b_{12} = 1 \oplus 1 = 0$
$\vdots$	$\vdots$

The resulting binary string is **1111** 1000 1101 1101 0100 0010 0101 100**1 1111** 0001 1011 1010 1000 .... Note the string repeats after $2^5 - 1 = 31$ entries.

But how does one use a binary string of 1s and 0s to generate pseudorandom numbers ρ that are uniformly distributed over the interval $(0, 1)$? One easy way is to break the binary string into a set of nonoverlapping substrings each consisting of L digits. Then interpret substring k as a base-2 expression of a positive base-10 integer x_k. Then take $\rho_k = x_k/2^L$.

To illustrate, let $L = 4$ so $2^L = 16$. Then use the binary string just obtained above. The conversion from base 2 to base 10 gives

$$(1111)_2 \ (1000)_2 \ (1101)_2 \ldots (1001)_2 \ (1111)_2 \ (0001)_2 \ (1011)_2 \ldots$$

$$\xrightarrow{\text{base 10}} 15 \ 8 \ 13 \ldots 9 \ 15 \ 1 \ 11 \ldots$$

The corresponding ρ_i are $15/16, \ 8/16, \ 13/16, \ldots, 9/16, \ 15/16, \ 1/16, \ 11/16, \ldots$. Note that, although the binary string repeats after 31 digits, ρ_i do not and can have cycle lengths much greater than that of the underlying binary string.

Tausworthe's Legacy

Although the Tausworthe RNG never achieved widespread use, it is important because it demonstrated one could produce very fast RNGs with long-cycle lengths. It is also interesting to note that in his landmark paper, Tausworthe never used the term "feedback" and used "register" but once. Yet what is known today as LFSR RNGs use autoregressive sequences similar to those introduced by Tausworthe. Because binary operations on bits in a computer register are typically very efficient, almost all modern RNGs operate, to some extent, at the bit level.

Moreover, RNGs based on LFSRs can be easily implemented on hardware and the resulting circuits can be very fast, cost-effective, and very efficient in terms of computer overhead. These RNG properties are basic requirements for applications with intensive RN needs, such as the testing of very large-scale integration (VLSI) circuits, pattern recognition, computer simulations, cryptography, signature analysis, digital broadcasting, and many more. For this reason, much of today's research on RNGs is based on variations of LFSR generators.

3.10 What Is a Linear Feedback Shift Register?

A *linear feedback shift register* (LFSR) takes an n-bit binary integer b_n, b_{n-1}, $b_{n-2}, \ldots, b_2, b_1$, often called the *state vector*, and endlessly transforms it by a specified *update rule*. At each transformation step (e.g., each clock cycle), the digits that are in the state vector are shifted one position toward the nth end of the vector, and a new b_1', formed from a linear combination of the old b_is, replaces the first component of the state vector. The old b_n passes out of the register and enters an output bit stream.

In mathematical terms, the update rule, as described above in words, is

$$
b_1' = \left(b_n + \sum_{j=1}^{n-1} c_j b_j \right) \bmod 2,
$$

$$
b_i' = b_{i-1}, \quad i = 2, 3, \ldots, n.
$$

(3.33)

Here b_i are c_i are binary variables with values of 0 or 1. Because the length of the LFSR is n, c_n must be 1; otherwise the state vector would have a length only up to the last nonzero coefficient in the c_js. The specification of which c_js are nonzero defines the update rule.

The properties of Eq. (3.33) and how the shift register is configured are determined by those of a characteristic polynomial over the integers modulo 2, namely

$$
P(x) = x^n + c_{n-1} x^{n1} + \cdots + c^2 + c_1 x + 1.
$$

(3.34)

For example, consider the polynomial

$$
x^8 + x^4 + x^3 + x^2 + 1, \quad \text{abbreviated as } (8, 4, 3, 2, 0).
$$

(3.35)

The LFSR defined by this polynomial is illustrated in the top half of Fig. 3.4. The operation of a simple 3-bit LFSR is illustrated in Example 3.4.

The manipulation of the bits in accordance to the characteristic polynomial uses modulo-2 arithmetic as indicated by the "linear" in the acronym LFSR. This means the logical XOR operation for $+$ and the logical AND operation for $\times$ are used. Modulo-2 arithmetic is illustrated in Example 3.5.

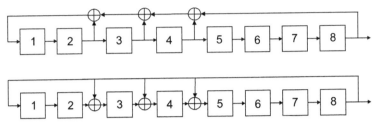

Figure 3.4 Two equivalent methods for generating pseudorandom bits from an 8-bit shift register based on the primitive polynomial $x^8 + x^4 + x^3 + x^2 + 1$. (top) The feedback used to create a new value of b_1 is taken from the *taps* at register cells 8, 4, 3, and 2 and combined modulo 2 (XOR or $\oplus$ operator) and the result is shifted in from the left. This design goes by several names: many-to-one, Fibonacci, and external feedback generators. It is well suited for hardware implementation but is inefficient for software emulation. (bottom) Here selected right shifting bits are XORed with the feedback bit entering the output stream. This design is called the modular, one-to-many, Galois, or internal XORs generator. It is easy to implement in software, but requires specialized hardware techniques.

Example 3.4 Operation of a 3 Flip-Flop LFSR

Consider two 3-bit shift registers, one (below right) configured for the characteristic primitive polynomial $x^3 + x^2 + 1 = (3, 2, 0)$ and the other for $x^3 + x + 1 = (3, 1, 0)$. These are dual polynomials. Start each register with the *seed* $b_1 b_2 b_3 = 100$.

The two registers are shown below with tables giving their states after each update which takes one clock cycle. The output stream from each register consists of the binary bits in column 4 of each table, i.e., the output sequence of b_3s. Note each register has a full period $7 = 2^3 - 1$ before repeating. Also note that if the initial seed were 000, an endless sequence of zeros would be output. This is true for any LFSR. See Fig. 3.5.

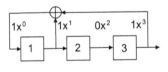
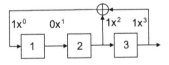

clock	b_1	b_2	b_3
–	1	0	0
1	1	1	0
2	1	1	1
3	0	1	1
4	1	0	1
5	0	1	0
6	0	0	1
7	1	0	0
8	1	1	0
9	1	1	1
⋮	⋮	⋮	⋮

clock	b_1	b_2	b_3
–	1	0	0
1	0	1	0
2	1	0	1
3	1	1	0
4	1	1	1
5	0	1	1
6	0	0	1
7	1	0	0
8	0	1	0
9	1	0	1
⋮	⋮	⋮	⋮

Figure 3.5 Two 3-bit XOR-based LFSRs with different tap selections.

The internal configuration of the shift register shown in the bottom half of Fig. 3.4, while technically not an LFSR (there are no feedback loops), can be shown (Press et al., 2007) to produce the same output stream of bits as that of the device in the top half of the figure, although in a different order. As the number of XOR gates increases, the external LFSR design becomes increasingly less efficient because the output of each XOR gate must be performed sequentially with each gate requiring one clock cycle. By contrast, with the internal design all gate operations are independent of each other and can be performed simultaneously, thereby allowing higher output speeds.

Finally, for computational efficiency or lower hardware cost, the fewer the number of XOR gates the better. Thus for a shift register with n bits, the nth-order characteristic primitive polynomial, of which there are generally many, should have the minimum number of nonzero monomials. As an example, for $n = 8, 16$, and 24, there are 16, 2048 and 276,480 primitive polynomials, respectively. One primitive polynomial of order $n = 2$ to 32 with the fewest nonzero terms is given in Table 3.3.

Example 3.5 Modulo-2 Polynomial Arithmetic

When dealing with LFSRs and the polynomials that determine how the feedback is configured, modulo-2 arithmetic is used because the coefficients of the polynomials are either 1 or 0. For such arithmetic, the key relation is

$$\text{modulo 2:} \quad x^n + x^n = x^n - x^n = 0.$$

Consequently, $-x^n = x^n$, i.e., subtraction is the same as addition modulo 2.

To demonstrate this modulo-2 arithmetic consider (a) $(x^4 + x^3 + x^2 + 1) \pm (x^3 + x^2 + x)$, (b) $(x^4 + x^3 + x^2 + 1) \times (x + 1)$, and (c) $(x^4 + x^3 + x^2 + 1)/(x + 1)$. The modulo-2 calculations are shown below.

Addition/Subtraction	Multiplication	Division
$x^4 + x^3 \quad\ +x^1\ +1$	$x^3 \quad\ +x^1 + 1$	$x^3\ +x\ +1$
$\underline{\quad\ x^3 + x^2 + x^1}$	$\underline{\times \qquad\qquad x+1}$	$x+1\ \overline{)\ x^4 + x^3 + x^2 + 1}$
$x^4 \qquad\ +x^2 \quad\ +1$	$x^3 \quad\ +x^1 + 1$	$\underline{x^4 + x^3}$
	$\underline{x^4 \qquad +x^2\ +x}$	$x^2 + 1$
	$x^4 + x^3 + x^2 \qquad +1$	$\underline{x^2 + x}$
		$x + 1$
		$\underline{x + 1}$
		0

3.10.1 Random Numbers From Random Bits

Once a long sequence of pseudorandom bits has been obtained, one still has to convert these bits into a base-10 random integer x_i from which a random vari-

Table 3.3 Primitive polynomials of degree 2 to 32 that require the fewest number of XOR components. Not all such polynomials are shown. Data are extracted from A. Partow (www.partow.net/programming/polynomials/index.html).

degree (n)	primitive polynomial
2, 3, 4, 6, 7, 15, 22	$x^n + x^1 + 1$
3, 5, 11, 20, 21, 29	$x^n + x^2 + 1$
4, 5, 7, 10, 17, 20, 25, 28, 31	$x^n + x^3 + 1$
9, 15	$x^n + x^4 + 1$
9, 17, 23	$x^n + x^5 + 1$
7, 17, 18	$x^n + x^6 + 1$
7, 17, 18	$x^n + x^7 + 1$
10, 15, 18	$x^n + x^8 + 1$
8, 19	$x^n + x^6 + x^5 + x^1 + 1$
13, 24	$x^n + x^4 + x^3 + x^1 + 1$
26, 27	$x^n + x^8 + x^7 + x^1 + 1$
30	$x^n + x^{23} + x^2 + x^1 + 1$
32	$x^n + x^{22} + x^2 + x^1 + 1$

ate ρ_i from $\mathcal{U}(0, 1)$ is derived. In most RNG studies of LFSRs, this conversion of bits to base-10 integers and thence to values of ρ_i is almost an afterthought.

The simplest approach is that used in Example 3.3, namely, break the output binary stream into substrings of L contiguous, nonoverlapping bits and treat the ith substring $(b_1 b_2 \ldots b_L)$ as a random base-2 integer x_i. Then in base 10

$$x_i = \sum_{\ell=1}^{L} b_\ell 2^{L-\ell} \quad \text{so that} \quad \rho_i = x_i / 2^L = \sum_{\ell=1}^{L} b_\ell 2^{-\ell}. \tag{3.36}$$

However, this simple approach generally does not produce very good results with the resulting x_i failing some standard statistical tests for randomness. A better result is usually obtained by using a register of size $n > L$ and using the lower significant L binary bits as x_i. However, even this approach usually produces poor results. It is especially a bad idea if L and $2^n - 1$ are not relatively prime, because in this case not all L-bit words are produced uniformly (Press et al., 2007).

Proposed methods to improve the randomness of the LFSR output include using a subset of the LFSR for an RN to increase the permutations of the binary numbers (Maxim Integrated, 2010). Shifting the LFSR more than once before obtaining an RN also improves the statistical properties of the output. And then one could shuffle the output stream in a manner similar or equal to the Bays–Durham shuffle of Section 3.5.4.

3.10.2 LFSR Extensions

Any successful method for an RNG invariably spawns extensions and variations that improve on the basic design. This is especially true of LFSR RNGs. In this section, a few notable extensions are mentioned, but the theoretical details are beyond the scope of this text. The interested reader who seeks these details should refer to the references in the RNG review by Bhattacharjee et al. (2018).

L'Ecuyer (1996) proposed combining J RNGs by XORing the output bits of J generators as described in Section 3.7. Equivalently, the RNs ρ_{ij} produced by generator j, $j = 1, \ldots, J$, can be combined as

$$\rho_i = (\rho_{i1} + \rho_{i2} + \cdots + \rho_{iJ}) \bmod 1.$$

This combining of generators not only improves the statistical quality of the ρ_i but increases the cycle period t_c of the compound generator to the lcm of the periods t_j of the J constituent RNGs, i.e.,

$$t_c = \text{lcm}(t_1, t_2, \ldots, t_J). \tag{3.37}$$

This period augmentation technique can be used with Tausworthe generators to increase the rather modest periods of the constituent generators to those needed for today's MC applications. For $J \geq 2$ LFSR RNGs, each based on a primitive polynomial of order n_j, the period of the resulting compound generator is $\text{lcm}(2^{n_1} - 1, 2^{n_2} - 1, \ldots, 2^{n_J-1})$. For example, consider two Tausworthe generators with $n_1 = 31$ and $n_2 = 32$ (both of which have maximum periods of the order of 10^9). Because $2^{31} - 1$ is prime and $2^{32} - 1 = 3 \cdot 5 \cdot 17 \cdot 257 \cdot 65537$, the period of the combined generator is the $\text{lcm}(2^{31} - 1 \times 2^{32} - 1) = (2,147,483,647 \times 65,537) \simeq 1.41 \times 10^{14}$. When using such a generator, it is best to use fewer bits than $\min(n_i)$ to obtain the random integers and numbers, such as 16 bits for the example just given (Maxim Integrated, 2010). Also note XORing two LFSRs of the same size does not increase the period of the RNG.

Lewis and Payne (1973) introduced a new type of LFSR-based RNG named the *generalized feedback shift register* (GFSR). This RNG produces multidimensional pseudorandom numbers, i.e., a vector of RNs. It was purported to have an arbitrarily long period, be very fast, and be machine-independent. However, this generator failed to reach the theoretical upper bound on the period (equal to the number of possible states) and has a very large memory requirement for 1980-era machines. The generator also depends on a well-chosen seed to obtain good randomness.

To ameliorate these problems with the GFSR, Matsumoto and Kurita (1992) introduced a variant called the *twisted* GFSR (TGFSR). This initial proposed generator evolved over the next few years into the now very popular RNG called the *Mersenne Twister* (MT), which is discussed in detail in Section 3.10.3. Incorporated into the output of the MT was a *tempering* step that improved the equidistribution property of the underlying PRNG. This tempering step uses elementary bitwise transformations such as XOR, AND, and shift operations.

In 2003, Marsaglia introduced the *xorshift* generator[15] and in 2006 Panneton et al. introduced another PRNG based on a TGFSR called *well-equidistributed long-period linear generator* (WELL). The idea behind these generators to obtain an RN is to first shift a block of bits a specified number of positions in one direction and then apply XOR on the original block with the shifted block.

All of the LFSR generators mentioned so far have been *linear*. But in the last 20 years, several attempts have been made to include some nonlinear operation into a linear RNG. However, we must refer the interested reader to the technical literature for these efforts and the mathematics behind these experimental advanced LFSR-based RNGs. And in recent years yet more RNGs based on LFSRs continue to be proposed.

3.10.3 Mersenne Twister

In 1998, Profs Matsumoto and Nishimura introduced a new PRNG that has become quite popular and is incorporated in many software packages. The most widely used implementation of this *Mersenne Twister* (MT) RNG is based on the Mersenne prime number $2^{19,937} - 1$ and can be implemented using a 32-bit word length. It is often called the MT19937-32 generator. A 32-bit seed is used to initialize the generator and produce the first *state*, which consists of 624 pseudorandom numbers, all but the last without a sign. The word *twister* is vaguely relevant to the algorithm itself, and possibly refers to the fact that the MT algorithm makes use of many binary operations like bit-shifting, which visually evoke a sense of torque or jumbling bits (0s and 1s) around. When RNs are wanted, a function, called the *temper*, is used to introduce logical operations to introduce additional randomness to the output, which consists of 624 32-bit RNs from each state. To obtain more RNs, the state is transformed to another state with a one-way function (called the *twister*). And then the temper function, which is reversible, is used to obtain another 624 RNs. This process continues creating one state after another with the twister and using the temper to generate 624 RNs from each state (see Fig. 3.6).

---

[15] This generator was quickly shown to give the same sequence of random bits as an ordinary LFSR generator but is faster and requires less computer memory (Brent, 2004).

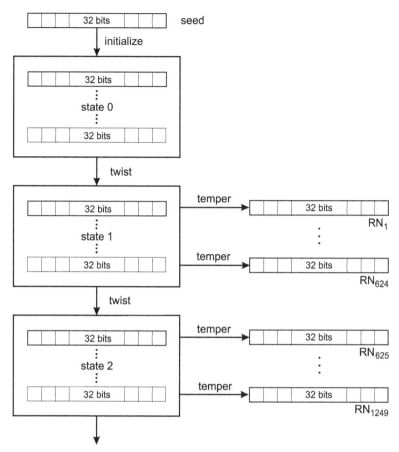

Figure 3.6 Schematic of MT19937 showing how the "twister" moves the generator from state to state and how the "temper" extracts 624 pseudorandom numbers from each state.

Theoretical analysis has shown the period of this PRNG is the Mersenne prime $2^{19,937} - 1 \simeq 10^{6002}$, which is very much longer than those of other RNGs.[16] This enormous period is one of the reasons for the popularity of the MT. It is also designed to give uniform coverage in a hypercube of dimension up to 623. It has also passed many tests for randomness, but not all. Indeed, S. Vigna (2019) in his paper identifies several deficiencies of the MT. Some of these deficiencies (such as the Marsaglia binary-rank test) plague other PRNGs and hence are not considered crucial. However, some are more important. Hamming-weight dependencies refer to the number of ones that appear in the binary output

---

[16] This ridiculously enormous number is much greater than a *googol* ($= 10^{100}$), which itself is much greater than the number of photons emitted by all stars since the Big Bang. Fortunately, it is smaller than a *googleplex* $= 10^{\text{google}} = 10^{10^{100}}$. So there is still room for improvement for those obsessed with PRNG cycle lengths.

of a PRNG. According to Vigna, the standard MT fails this test. The MT algorithm also fails the Marsaglia *birthday-spacing* test, which says that the spacings between random points in a sequence should be asymptotically exponentially distributed. Other disadvantages of the MT include the following: (1) it requires a relatively large buffer, (2) some initial seeds generate poor sequences for MC applications, and (3) seeds with many zeros, i.e., those not very random, produce RN sequences that also do not appear random. Vigna (2019) concludes, somewhat harshly, "knowledge accumulated in the last 20 years suggests that the Mersenne Twister has, in fact, severe defects, and should never be used as a general-purpose pseudorandom number generator."

The MT generator, however, does have three very desirable features. First, it is portable, meaning that it can be implemented on any machine of comparable word length and yield the same results for the same initial seed. Second, the MT generator is also efficient, meaning many pseudorandom numbers can be generated quickly and without consuming much memory. Third, it is efficient because it relies primarily on the binary representation of integers and bitwise operations to generate pseudorandom numbers. Much of MT's efficiency can be attributed to the use of these bitwise operations, which can be much faster than standard arithmetic operations (addition, subtraction, multiplication, and division) used by other PRNGs when implemented correctly. Bitwise operations are also used to cleverly bypass matrix operations that would be otherwise costly computationally. Bitwise operations are fast due to the elimination of *translation* time. Computers store data in strings of binary numbers; combinations of 0s and 1s. When a computer attempts integer addition, say $3 + 4$, it first internally translates the command into binary, evaluates in binary, and then translates that binary value back into an integer that is expressed in binary numbers. This translation time is negligible for a single, simple operation like $3 + 4$, but for many operations, like those needed to generate many pseudorandom numbers, translation time accumulates quickly. Reducing translation time by working in binary can greatly accelerate performance.

Although the MT algorithm has supplanted other PRNGs for many applications, the Park and Miller minimal standard PRNG, especially when combined with a Bays–Durham shuffle, is often a good PRNG to use as a point of comparison with newer PRNGs because it is consistent and easy to program on any platform. Codes in several languages for the Park and Miller minimal standard PRNG are presented in Appendix F and codes for the MT MT19937 are available on the web (Takano, 2021). A comparison of results obtained from these two codes is shown in Table 3.4.

3.11 RNGs Based on Cellular Automata

A CA is a discrete dynamical system consisting of a regular network of cells each of which is in one of a finite number of states such as "black" or "white" for a two-state system. The grid can be defined in any number of dimensions and

Table 3.4 Comparison of the sample mean and its standard deviation as a function of the numbers of samples drawn from a uniform distribution $\mathcal{U}(0, 1)$ using the minimal standard and the Mersenne Twister (MT19937-32) generators.

	Minimal standard		Mersenne Twister	
N	$\overline{x}$	$s(\overline{x})$	$\overline{x}$	$s(\overline{x})$
10^2	0.493708	2.83733×10^{-2}	0.504763	2.77304×10^{-2}
10^4	0.499749	2.88589×10^{-3}	0.498942	2.87597×10^{-3}
10^6	0.500122	2.88488×10^{-4}	0.499924	2.88742×10^{-4}
10^8	0.499960	2.88646×10^{-5}	0.500019	2.88664×10^{-5}

each cell has a *neighborhood* of cells with which the cell interacts as the system evolves in time. At each time step, the new state of each cell is determined by a fixed *rule*[17] (usually a formula) which depends on the current state of the cell and those in the cell's neighborhood. Typically, the updating rule is the same for each cell, independent of time, and is applied to the whole grid simultaneously. However, exceptions are known, such as the stochastic CA and asynchronous CA.

CAs were originally discovered in the 1940s by Stanislaw Ulam and John von Neumann while they worked together at Los Alamos National Laboratory. Although CA was investigated by a few academics during the 1950s and 1960s, it was not until the 1970s and Conway's introduction of the *Game of Life*, a two-dimensional CA, that interest in CA expanded rapidly beyond academia. In the 1980s, Stephen Wolfram (1986) undertook a systematic study of one-dimensional CAs and suggested that a CA could be used as an RNG with particular application to cryptography.

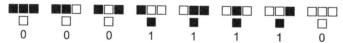

Figure 3.7 The graphical table of rule 30 for an ECA. The eight possible combinations of the three nearest cells in the previous generation are shown on the top line, and by choosing the state of the resulting next-generation cell (second row), the rule for a particular ECA is determined. At the bottom is the binary state for each choice, which, when interpreted as an 8-bit base-2 integer and converted to a base-10 integer, gives rule "30" for this case.

3.11.1 One-Dimensional Cellular Automata

To solidify the ideas behind CA, the simplest case of a one-dimensional CA is considered in this section. Wolfram (1986, 2002) calls this case elementary

---

[17] Also called the *next state function* or *local rule*.

cellular automata (ECAs). Each cell has two possible states (0 or 1) or, in graphic displays, "white" or "black," respectively. Each ECA has a rule in which the state s_i' of cell i depends on only the values s_{i-1}, s_i, and s_{i+1} of the three nearest cells in the previous generation, i.e.,

$$s_i' = f(s_{i-1}, s_i, s_{i+1}). \tag{3.38}$$

As a result, the evolution of an ECA can completely be described by a table specifying the state a given cell will have in the next generation based on the value of the cell to its left, the value of the cell itself, and the value of the cell to its right. Because there are $2 \times 2 \times 2 = 2^3 = 8$ possible binary states for the three cells neighboring a given cell, there are a total of $2^8 = 256$ ECAs, each of which can be indexed with an 8-bit binary number (Wolfram, 1986, 2002). For example, the table giving the evolution of the ECA by rule 30 ($30_{10} = 00011110_2$) is illustrated in Fig. 3.7. Equation (3.38) for rule 30 can be written as (Wolfram, 1986)

$$s_i' = s_{i-1} \text{ XOR } (s_i \text{ OR } s_{i+1}) \tag{3.39}$$

or, equivalently,

$$s_i' = (s_{i-1} + s_i + s_{i+1} + s_i s_{i+1}) \bmod 2. \tag{3.40}$$

The evolutionary pattern generated by rule 30 from a single black cell in generation 0 (row 0) is shown in Fig. 3.8. The ith row shows the system at the ith generation.

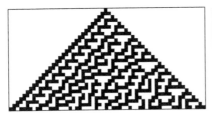 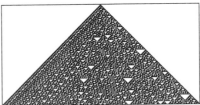

Figure 3.8 The ECA for rule 30 showing the first 30 rows or generations (left) and the first 141 rows (right).

Reflections and Complements

Although there are 256 possible rules for ECAs, they are not all fundamentally different. Several pairs are equivalent in that one is the same as the other when reflected through the vertical axis, the so-called *mirrored* rule. Such pairs exhibit the same behavior and hence are equivalent in a computational and behavioral sense. Rules that are the same as their mirrored rule, such as rule 30 and rule 86, are called *amphichiral*. Of the 256 ECAs, 64 are amphichiral.

Likewise several pairs are equivalent if the 1s and 0s are interchanged in one of the rules. The result of applying this transformation to a given rule is called the *complementary* rule. For example, if this transformation is applied to rule 110, one obtains rule 137. There are 16 rules which are the same as their complementary rules.

Finally, the previous two transformations can be applied successively to a rule to obtain the *mirrored complementary* rule. For example, the mirrored complementary rule of rule 110 is rule 193. There are 16 rules which are the same as their mirrored complementary rules.

So, of the 256 ECAs, there are 88 which are inequivalent under these transformations.

Characteristics of the ECAs

The primary classifications of CAs, as outlined by Wolfram (1986), are numbered one to four. They are as follows. Examples of the four classes are shown in Fig. 3.9.

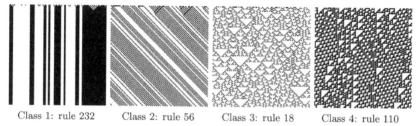

Class 1: rule 232   Class 2: rule 56   Class 3: rule 18   Class 4: rule 110

Figure 3.9 Example ECAs for the four CA classes. Here the initial state is a random string of 0s and 1s. Source: https://en.wikipedia.org/wiki/Elementary_cellular_automaton.

Class 1: Automata in this class converge rapidly to a uniform, fixed, homogeneous state. Any randomness in the initial pattern quickly disappears. Examples include rules 0, 8, 32, 136, 160, and 232.

Class 2: Automata in this class rapidly produce patterns that evolve into mostly stable or oscillating structures. However, even though some of the randomness in the initial pattern may die out, some usually remains. Examples include rules 4, 37, 56, 73, 108, 218, and 250.

Class 3: Automata in this class produce aperiodic patterns. Nearly all initial patterns evolve into seemingly endless pseudorandom or chaotic structures. Any stable patterns that appear are rapidly destroyed by the surrounding noise. Examples include rules 18, 22, 30, 45, 126, 146, 150, and 182. It is the pattern produced by rule 30 which Wolfram (1986) suggested for use as an RNG.

Class 4: Automata in this class produce (1) localized patterns which become extremely complex and may last for a long time and (2) areas of repetitive or stable states. Stable or oscillatory patterns, characteristic of class 2 ECAs, may, after a very long time, be the eventual outcome even when the initial pattern is relatively simple. Local alterations to the initial pattern may spread indefinitely. The classic example is the ECA produced by rule 110.

3.11.2 Random Number Generation From Cellular Automata

In Fig. 3.8, it is seen that rule 30 produces seemingly endless randomness despite the fact there is nothing random about how the pattern is created. The randomness undoubtedly has its origin in the simple nonlinearity between the neighboring states s_i and s_{i+1} in Eq. (3.40).[18] It is now known that very simple nonlinearities can lead to chaotic behavior and fractal geometries.

Stephen Wolfram (1986, 2002) proposed using the string of states on the centerline of the ECA produced by rule 30 as random bits (1s or 0s). Then extracting contiguous substrings of n bits each would give (in base 2) pseudorandom integers that represent base-10 integers x_i in the interval $[0, m - 1]$ where $m = 2^n$. Finally, the corresponding pseudorandom number is $\rho_i = x_i/m$, which lies in the interval $[0, 1)$. Wolfram studied rule 30 extensively, and found that the RNs produced from the centerline bits (or even many off centerline bits) could pass many of the standard tests for randomness. A simple example of generating RNs in this fashion is given in Example 3.6.

The dynamic response of chaotic systems generally depends sensitively on the initial condition. For ECAs, the initial condition is the distribution of 0s and 1s used in the first row (the "zeroth generation") from which the full ECA evolves. For example, ECAs for a random initial distribution of 1s and 0s are shown in Fig. 3.9. When the bits of an ECA are used to generate RNs, the initial distribution is the *seed* for the RNG.

Example 3.6 Generating Random Numbers From ECA

Generate random 8-bit integers and numbers from an ECA based on rule 30 using the values of the centerline states.

Solution:

With a computer program based on Eq. (3.40), the following central portion of the ECA (with infinite boundaries) is obtained.

row		row		row		row		row		row	
0	00000100000	34	01000100101	68	01110001010	102	10100100110	136	01101110110	170	11001100100
1	00001110000	35	11101111101	69	11001011010	103	00111111101	137	01001000100	171	00110011110
2	00011001000	36	00001000001	70	00111010010	104	11100000001	138	01111101111	172	11100010000
3	00110111100	37	10011100011	71	11100011110	105	10010000011	139	11000001000	173	10010111001
4	01100100010	38	01110010110	72	00010110000	106	11111000110	140	10100011100	174	11110100111
5	11011110111	39	11001110101	73	00110101001	107	10000101100	141	00110110010	175	10000111100
6	10010000100	40	00111000101	74	01100101111	108	11001101011	142	01100101110	176	11001100011
7	01111001111	41	11100101101	75	01011101000	109	10111001010	143	11011101000	177	00111010110
8	01000111000	42	10011101001	76	11010001101	110	00100111011	144	00010001101	178	01100010101
9	11101100100	43	01110001111	77	00011011001	111	01111100010	145	10111011001	179	01010110101
10	00001011100	44	01001011000	78	10110010111	112	01000010111	146	00100010111	180	11010100101
11	10011010000	45	11111010101	79	00101110100	113	01100110100	147	01110110100	181	00010111101
12	01110011001	46	00000010101	80	11101000110	114	01011100111	148	01000100110	182	00110100001

---

[18] This nonlinearity arises, in turn, from the inherent nonlinearity in the Boolean OR operator in Eq. (3.39).

13	11001110111	47	00000110101	81	00001101100	115	11010011100	149	01101111101	183	11100110011
14	10111000100	48	00001100101	82	00011001011	116	00011110010	150	11001000001	184	00011101110
15	10100101111	49	00011011101	83	10110111010	117	10110001110	151	10111100011	185	10110001000
16	00111101000	50	00110010001	84	10100100011	118	10101011000	152	10100010110	186	00101011101
17	11100001100	51	01101111011	85	10111110110	119	00101010101	153	10110110101	187	01101010001
18	10010011011	52	01001000010	86	00100000101	120	11010101010	154	00100100101	188	11001011011
19	01111110010	53	11111100111	87	11110001101	121	10001010101	155	11111111101	189	00111010010
20	11000001111	54	10000011100	88	10001011001	122	11011010101	156	00000000001	190	01100011111
21	00100011000	55	11000110010	89	01011010111	123	10010010101	157	00000000011	191	01010110000
22	01110110100	56	10101101111	90	11010010100	124	11111110101	158	00000000110	192	01010101000
23	01000100111	57	00101001000	91	00011110110	125	10000000101	159	00000001101	193	11010101100
24	01101111100	58	01101111100	92	00110000100	126	01000001101	160	00000011001	194	10010101010
25	11001000010	59	11001000011	93	01101001111	127	01100011001	161	00000110111	195	11110101010
26	10111100110	60	00111100110	94	11001111000	128	11010110111	162	10001100100	196	10000101010
27	10100011100	61	01100011101	95	00111000100	129	10010100100	163	11011011111	197	01001101010
28	10110110011	62	01010110001	96	01100101110	130	11110111110	164	10010010000	198	11111001010
29	00100101110	63	11010101011	97	11011101000	131	00000100000	165	11111111000	199	00000111010
30	01111101001	64	00010101010	98	00010001101	132	00001110001	166	00000000101	200	00001100010
31	11000001111	65	00110101011	99	10111011001	133	10011001011	167	00000001101		
32	00100011000	66	11100110010	100	10100010111	134	01110111010	168	00000011001		
33	01110110101	67	10011101010	101	10110110100	135	01000100010	169	10000110111		

Subdivide the center column bit string into 8-bit substrings starting at row 1 to obtain

10111001	10001011	00100111	01011100	11101010
11000011	00101011	01010111	11100001	11100010
10111000	00100101	10001110	00110110	11010000
00010001	11110111	01001110	00111010	11100000
11001000	11001111	00111111	00000011	11111011

Next convert these 25 base-2 integers to the desired base-10 random integers x_i:

185	139	39	92	234
195	43	87	225	226
184	37	142	54	208
17	247	78	58	224
200	207	63	3	251

Finally divide the random integers by $2^8 = 256$ to get the corresponding RNs ρ_i:

0.72266	0.54297	0.15234	0.35938	0.91406
0.76172	0.16797	0.33984	0.87891	0.88281
0.71875	0.14453	0.55469	0.21094	0.81250
0.06641	0.96484	0.30469	0.22656	0.87500
0.78125	0.80859	0.24609	0.01172	0.98047

Although rule 30 produces randomness for many initial distributions, there are also a large number of initial distributions that produce repeating patterns in the ECA. The trivial example is an initial distribution, or seed, consisting of all zeros. Such a seed (as with most other types of RNGs) produces only zeros. A less trivial example, found by Matthew Cook, who was Stephen Wolfram's research assistant, is any input distribution consisting of infinite repetitions of the pattern "00001000111000," with repetitions optionally being separated by six 1s. Other input distributions have since been found that produce repeating patterns in the ECA. So, as with many RNGs, the initial seed must be chosen carefully.

3.11.3 Calculating Elementary Cellular Automata

A CA is defined on an infinite grid of cells on which each cell evolves according to the same local rule. In 1-D each generation is an infinite row of cells. In 2-D each generation is an infinite plane of cells, while in 3-D each generation is an infinite universe of cells. Computers do not deal well with an infinity of cell states. So, for example, in simulations of 1-D, ECA boundaries are placed on the lateral extent of the number N of cells used. But in so doing the cells on the boundary can no longer follow the chosen rule for the automaton. Boundary conditions are needed. The two most common boundary conditions are (1) to fix the state of boundary cells at a constant value (typically the null or "0" state) and (2) to use a *periodic* boundary condition such that for the two boundary cells ($i = 1$ and $i = N$) Eq. (3.38) is

$$s_1' = f(s_N, s_1, s_2) \quad \text{and} \quad s_N' = f(s_{N-1}, s_N, s_1). \tag{3.41}$$

These two boundary conditions are illustrated in Fig. 3.10. Other boundary conditions have been proposed such as (3) keep the states of boundary cells fixed at a "0" or (4) declare boundary cells to have only two nearest neighbors and adjust the propagation rule for these cells. The use of a finite number of cells in each generation and the selected boundary condition affects the ECA pattern somewhat compared to that obtained with an infinite grid.

However, the use of N cells for each generation to simulate an ECA means that the ECA pattern must necessarily repeat after some number of generations. Because each cell has two possible states, the number of different row compositions is 2^N. Hence the maximum cycle length of the bits in any column of the ECA is at best $2^N - 1$ and, for many ECAs, often very much less. Nevertheless, there are ECAs that have this maximum period (Bhattacharjee et al., 2018).

Although limiting the number of cells N in each generation eliminates one infinity, the complete ECA diagram has an infinite number of generations or rows, thereby making the planar grid infinitely long. But because the state of any cell depends on only the states of the three nearest cells in the previous generation, one could use a finite number of rows, M, so that, once the states in row M have all been calculated, the subsequent states of their progeny can be placed in row 1 since and state information that was in row 1 is no longer needed. After calculation of the state for a new generation, the bit information in the column(s) of the ECA that are to be used to form RNs can be extracted to form bit strings from which random integers can be obtained (see Example 3.6). In the simulation of an ECA, the replacement of an infinite planar grid with small finite grid makes the simulation efficient and one with greatly reduced memory requirements.

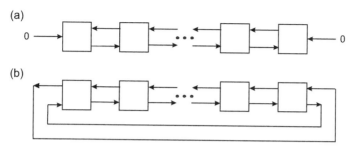

Figure 3.10 Boundary conditions for a finite number of cells. (a) Boundary cells are coupled to a "0" state imaginary cell adjacent to and outside the boundary. (b) For a periodic boundary condition, the left boundary is coupled to the right boundary cell and the right boundary cell is coupled to the left boundary cell. After (Bhattacharjee et al., 2018).

3.11.4 Some CA-Based RNGs

Not surprisingly, Stephen Wolfram, who did the pioneering studies on ECAs, proposed the first ECA-based (rule 30) RNG (Wolfram, 1986). About the same time, Hortensius et al. (1989) introduced an RNG that combined rules 30 and 45 ECAs to produce 32-bit random integers. They also produced another RNG similar to the first but, by combining different ECAs, a generator with maximum cycle length was obtained. In both RNGs the null boundary condition was used. In their combined generators, the idea of taking CA values from nonadjacent cells (*site spacing*) and from different generations (*time spacing*) was implemented. To optimize the rules by which the CAs evolved, Wang et al. (2008) used a particle swarm algorithm (somewhat akin to a genetic algorithm) and produced PRN sequences of high quality.

Other 1-D RNGs were introduced by Das and Sikdar (2010), who offered a 45-cell, null boundary, 2-state, 3-neighborhood, nonlinear generator, and more recently by Bhattacharjee et al. (2017), who proffered a 51-cell, 3-state, 3-neighborhood, periodic boundary RNG. By now 1-D CA generation of RNs has been extensively studied (see Chaudhuri et al. (1997) for a good review).

Higher-dimensionality CAs have also been used in RNGs. Chowdhury et al. (1994) proposed the use of two-dimensional CAs. The randomness of the resulting numbers directly depended, as expected, on the definition of what constituted neighboring cells and the rules used to form the CA. The 2-D CA with special boundary conditions and optimized rules produced "better" RNs compared to those from a 1-D ECA. However, finding the best rules and boundary conditions is not straightforward. Tomassini et al. (2000) used a genetic algorithm to optimize a 5-neighbor, 2-D CA with periodic boundaries to obtain higher-quality RNs than those from Chowdhury's generator which relied on manual tweaks for its optimization. To avoid some of the difficulties encountered with Chowdhury's 2-D RNG, Koikara (2017) proposed a 3-D CA-based RNG which used incremental boundary conditions and combined several rules.

This RNG is claimed to have passed all the Diehard, ENT, and NIST test suites of statistical tests (see Section 3.13).

3.12 "Recent" Random Number Generators

Since about 1995, there has been a Cambrian explosion of different methods and implementations of RNGs. Concomitant with this flurry of activity has been the creation of an alphabet soup of acronyms and names (often whimsical) for the new RNG methods and tests to evaluate them. The overall trend seems to be one that avoids the simplicity of LCGs and their kin; rather the trend favors bit-level manipulations that capitalize on efficient features of modern computer architectures. Calls for "in-line coding," to squeeze the processing time, once again can be heard. Although improvements in the "randomness" of the generated pseudorandom numbers have surely been achieved, as demonstrated with a huge battery of tests designed to ferret out any weaknesses, the effect of the improvements on practical MC problems, such as those discussed in the following chapters, is largely unknown. It is the authors' opinion that the addition of an extra angel or two to those already dancing on the head of the pin that generates RNs has little, if any, effect on MC simulations that involve a large number of RNs per scoring history.

The implementation of these improved RNG, while still avoiding machine-language coding, often uses little known features of high-level languages such as FORTRAN and C++. For example, one highly recommended modern RNG written in C++ is (Press et al., 2007)

```
struct Ranq1 { ran.h
Ullong v;
Ranq1(Ullong j) : v(4101842887655102017LL) {
v ^= j;
v = int64();
}
inline Ullong int64() {
v ^= v >> 21; v ^= v << 35; v ^= v >> 4;
return v * 2685821657736338717LL;
}
inline Doub doub() { return 5.42101086242752217E-20 * int64(); }
inline Uint int32() { return (Uint)int64(); }
};
```

Most users of MC techniques are not computer scientists and most would be hard pressed to understand this bit of coding, let alone translate it into their workaday computer language, even after multiple readings of the accompanying explanation. But at least this somewhat opaque example shows that a good RNG need not be complex and require hundreds of lines of code.

3.12.1 Some RNG Developments in the Last 30 Years

Space limitations preclude a detailed description of the many proposed methods and RNGs developed since the early 1990s. Rather a listing by year of some notable achievements is presented below. This incomplete listing is intended to give the reader a sense of the renewed interest in RNGs and of some of the many acronyms encountered. Reference for these various codes can be found in (Bhattacharjee et al., 2018) and on the web at en.wikipedia.org/wiki/List_of_random_number_generators.

1991 G. Marsaglia and A. Zaman introduced the add-with-carry and subtract-with-borrow RNGs. Both are generalizations of the lagged Fibonacci generator. The subtract-with-borrow generator is the basis for the **RAN-LUX** generator, widely used, for example in particle physics.

G. K. Savvidy and N. G. Ter-Arutyunyan-Savidy created **MIXMAX**, a member of the class of matrix LCGs. It is based on results of ergodic theory and classical mechanics.

1992 R. A. J. Matthews proposed a method based on number theory. Never widely used in practical applications.

1993 G. Marsaglia created **KISS**, which uses a prototypical method for combining RNGs.

1994 G. Marsaglia and Koç use the multiply-with-carry (MWC) method to generate pseudorandom numbers.

1997 R. Couture and P. L'Ecuyer introduced the complementary-multiply-with-carry.

1998 M. Matsumoto and T. Nishimura created the MT, which is discussed in Section 3.10.3. This LFSR-based RNG adds a twist and temper to the algorithm to produce a very long-period RNG. It is widely used today.

1999 P. L'Ecuyer offered two LFSR style RNGs based on the combining of 4 or 5 Tausworthe RNGs: **LFSR113** with a period of $\approx 2^{113}$ and **LFSR258** with a period of $\approx 2^{258}$.

2000 D. E. Knuth proposed an LCG of the form of Eq. (3.1) with

$$a = 63{,}641{,}362{,}238{,}467{,}930{,}005, \ m = 2^{64},$$
$$c = 1{,}442{,}695{,}040{,}888{,}963{,}407.$$

This RNG has a period of $2^{64} \simeq 1.84 \times 10^{19}$ and produces normalized RNs ρ_i.

P. L'Ecuyer and R. Tonzin offered the mixed congruential generator **MRG31k3p** which combined two mixed congruential generators of order 3 to produce an RNG with a period of $2^{185} \simeq 4.90 \times 10^{55}$.

2003 G. Marsaglia created **Xorshift**, which is a very fast variant of LFSR RNGs. Marsaglia also proposed **Xorwow** as an improved generator, in which the output of a xorshift generator is added to a Weyl sequence.

2006 F. Panneton, P. L'Ecuyer, and M. Matsumoto introduced the "well-equidistributed long-period linear" (**WELL**) generator to ameliorate some of the deficiencies of the MT RNG.

2007 Bob Jenkins published a small, fast, noncryptographic RNG **JSF** with an average cycle length up to 2^{126}.

W. Press, S. Teukolsky, W. Vetterling, and B. Flannery proposed **RAN.h**, which they claim is "a suspenders-and-belt, full-body-armor, never-any-doubt" RNG. It is a replacement for their previous recommended Park and Miller LCG to which a Bays–Durham shuffle was added.

2008 M. Saito and M. Matsumoto introduced the code **SMFT**, whose acronym stands for "single instruction multiple data (SIMD)-oriented Fast Mersenne Twister." It uses all features of MT together with multistage pipelines and SIMD. It has the same enormous period as the MT RNG.

2010 S. Das and B. K. Sikdar used a nonlinear 2-state CA to generate RNs for testing of multicore chips.

2011 J. Salmon, M. Moraes, R. Dror, and D. Shaw introduced the "Advanced Randomization System" (**ARS**). It is a simplification of the *AES block cipher* and is very fast with very long cycle lengths of at least 2^{128}.

J. Salmon, M. Moraes, R. Dror, and D. Shaw also introduced **Threefry**. This is a simplified version of Threefish block cipher, suitable for use on GPUs.

J. Salmon, M. Moraes, R. Dror, and D. Shaw created **Philox**, which is a simplification and modification of **Threefish** with the addition of an **S-box**.

2014 G. Steele, D. Les, and C. Flood discussed **SplitMix**, an ENG based on the final mixing function of Murmurhash3, which is part of the Java Development Kit.

M. E. O'Neal described **Permuted Congruential**, which is a modification of an LCG to improve the randomness of the output.

2016 H. Cookman introduced the "Random Cycle Bit Generator" (**RCB**). This RNG is a bit pattern generator design to overcome some of the deficiencies of the **MT** and the short periods/bit length restrictions of shift/modulo generators.

2017 S. Widynski revisited J. von Neumann's original middle-square method and produces the "Middle Square Weyl Sequence PRNG." This generator may be the fastest one that passes all statistical tests.

S. Vigna explored the capabilities of four xorshift RNGs developed by Marsaglia, who used several shift operations moving different numbers of bits in opposite directions.

K. Bhattacharjee, D. Paul, and S. Das demonstrated a 3-bit 3-neighborhood CA with periodic boundaries can be used to generate both 32- and 64-bit binary RNs.

2018 D. Blackman and S. Vigna modified Marsaglia's Xorshift generator to produce **Xoroshiro128+**, which is one of the fastest on modern 64-bit CPUs.

S. Harase and T. Kimoto introduced 64-bit **MELG**, which implements 64-bit maximally equidistributed F_2-linear generators with a Mersenne prime period.

2020 B. Widynski proposed **Squares RNG**, which is a *counter-based* version of "Middle Square Weyl Sequence PRNG." It is claimed to be one of the fastest counter-based generators.

3.13 Assessment Tests for RNGs

Once any RNG is proposed for widespread use, it should undergo a suite of different statistical tests to tease out any inherent weaknesses in the generator. Recall the tests that were used to optimize an LCG, as discussed earlier in this chapter, to minimize, for example, the spacing between hyperplanes. There are several test suites available that have been widely used to quantify how random the output of a given RNG is and to rank the many different RNGs in terms of ability to mimic randomness. The most widely used test suite is called jokingly the Diehard/Dieharder tests and are discussed below.

3.13.1 Basic Statistical Tests

Before subjecting an RNG to Diehard's "tough" tests, it first must pass some very basic tests: A sample of 3×10^9 is used to calculate the (a) average $\overline{\rho_i}$, (b) skewness $\tilde{\mu}_s$, and (c) serial correlation R_k. The average has an expected value of 0.5 and the last two have an expectation of zero. If a generator fails any one of these basic tests, it should not be used.[19] A test of 500 RNGs gave (a) $\overline{\rho_i} = 0.5 \pm 0.004$, (b) $\tilde{\mu}_s = 0 \pm 0.001$, and (c) $R_k < 10^{-7}$ (Klimasauskas, 2002).

3.13.2 The Diehard Test Suite

Although there are several collections of different specialized statistical tests that can be used to assess the randomness of the output of an RNG, the whimsically named Diehard battery of tests, introduced by Marsaglia in 1995, is the one most widely used. The various tests are listed below. A more complete description of each test can be found at https://en.wikipedia.org/wiki/Diehard_tests. The entire Diehard set is available on a CD-ROM (Marsaglia, 2016).

---

[19] Skewness is a measure of the asymmetry of the probability distribution about its mean and is defined as $\tilde{\mu}_3 = \langle ((\rho - \mu)/\sigma)^3 \rangle$. For a uniform distribution, $\tilde{\mu}_3 = 0$. Serial correlation, also called autocovariance at lag k, is $R_k = \text{covar}(\rho_i, \rho_{i+k})$, which, for large i, is normally distributed with variance $1/[144(i-k)]$.

Birthday spacings: Choose random points on a large interval. The spacings between the points should be asymptotically exponentially distributed. The name is based on the birthday paradox.

Overlapping permutations: Analyze sequences of five consecutive RNs. The 120 possible orderings should occur with statistically equal probability.

Ranks of matrices: Select some number of bits from some number of RNs to form a matrix over $(0, 1)$, and then determine the rank of the matrix. Count the ranks.

Monkey tests: Treat sequences of some number of bits as *words*. Count the overlapping words in a stream. The number of words that do not appear should follow a known distribution. The name is based on the infinite monkey theorem.

Count the 1s: Count the 1 bits in each of either successive or chosen bytes. Convert the counts to "letters," and count the occurrences of five-letter "words."

Parking lot test: Randomly place unit circles in a 100×100 square. A circle is successfully parked if it does not overlap an existing successfully parked one. After 12,000 tries, the number of successfully parked circles should follow a certain normal distribution.

Minimum distance test: Randomly place 8000 points in a $10,000 \times 10,000$ square, and then find the minimum distance between the pairs. The square of this distance should be exponentially distributed with a certain mean.

Random spheres test: Randomly choose 4000 points in a cube of edge 1000. Center a sphere on each point, whose radius is the minimum distance to another point. The smallest sphere's volume should be exponentially distributed with a certain mean.

Squeeze test: Multiply 2^{31} by random numbers on $(0, 1)$ until 1 is reached. Repeat this 100,000 times. The number of RNs needed to reach 1 should follow a certain distribution.

Overlapping sums test: Generate a long sequence of RNs on $(0, 1)$. Add sequences of 100 consecutive RNs. The sums should be normally distributed with a characteristic mean and variance.

Runs test: Generate a long sequence of RNs on $(0, 1)$. Count ascending and descending runs. The counts should follow a certain distribution.

Craps test: Play 200,000 games of craps, counting the wins and the number of throws per game. Each count should follow a certain distribution.

3.13.3 The Dieharder Test Suite

In this test suite, several additional tests have been added to the original Diehard collection, a more friendly user interface has been added, and the entire suite has been recoded in order to (1) correct errors, (2) add capability to the various tests such as changing the test parameters (p-values, input options, etc.), and (3) avoid licensing problems. This bigger and better test suite is called

Dieharder (Brown et al., 2020) and is freely available from Brown's website https://webhome.phy.duke.edu/~rgb/General/dieharder.php. In general, this improved version should be used when assessing RNGs.

3.13.4 Other Test Libraries for RNGs

The TestU01 battery of tests developed by Pierre L'Ecuyer and Richard Simard (2007) consists of many tests for randomness and includes almost all the tests in Diehard. The suite was designed to allow some of the limitations of Diehard such as the ability to change test parameters, e.g., *p*-values, and to include new tests. A subset, called the *rabbit* tests, consists of 26 tests designed specifically for testing binary sequences of random bits from a generator. It has been observed (Bhattacharjee et al., 2017) that some Diehard tests are more difficult to pass. For example, an RNG with a specified seed, even if it passes all of the rabbit tests, can fail to pass the overlapping permutations Diehard test.

Another set of randomness tests is provided by the U.S. National Institute of Standards and Technology (NIST). The 15 tests in the NIST test suite (Rukhin et al., 2001) are specifically intended to test RNGs for cryptographic applications. This test suite has three main purposes: (1) investigate the distribution of 1s and 0s, (2) analyze the harmonics of a bit stream using spectral methods, and (3) detect patterns using methods from probability and information theory. Cryptographic RNGs must also be extremely fast because of the vast number needed in real-time to encrypt communications, such as pay-for-view TV shows. Also the output RN stream must be such that the underlying algorithm is difficult to determine or decode.[20]

3.14 Summary

Although MC methods depend critically on algorithms to generate efficiently a sequence of pseudorandom numbers that exhibit many of the qualities of true RNs, a vast number of studies have relied on RNGs that are now known to be highly deficient. Many computer languages and code packages have built-in RNGs and, often, they are far from optimal. Indeed, it would be very interesting to see how many studies based on these inferior generators would have produced significantly different results if they were based on modern proven RNGs. It also seems to be a firmly held belief by most RNG specialists that only the "best" RNG should be used regardless of its ultimate application.

However, it is the authors' opinion that not many practical MC calculations, such as those addressed in this book, would produce significantly different results if a modern gold-plated RNG had been used instead of a reasonably "good but not great" RNG. Many MC applications typically use hundreds to millions (or even more) RNs per history to produce a single data point (or score). So, for

---

[20] By contrast, a few sequential pseudorandom numbers from an LCG allow the multiplier and modulus of the generator to be easily deduced.

example, an occasional sequential correlation among a few adjacent pseudorandom numbers becomes lost in the complexity of a single history or simulation. Further, because the MC answer to many problems is usually needed only to three or four significant figures, low-level bit correlations are generally of little practical concern. Finally, and probably most important, the various approximations made to equations and simulations (e.g., finite difference approximations of differentials, numerical quadrature for integrals, viral lifetimes under various environmental conditions, and so forth) usually introduce more uncertainty and bias into the MC calculations than those caused by some small deficiency in the RNG.

Because of the high rate at which new RNGs are being introduced, it is difficult for nonspecialists to make decisions about what generator to use or, for most of the new generators, even how to code them. The MC user's lament "Give me something I can understand, implement and port... it needn't be state-of-the-art, just make sure it's reasonably good and efficient" motivated Park and Miller (1988) to offer the minimal standard generator in response. Although it has been eclipsed by modern RNGs in randomness tests, we continue to use it in this second edition of our book because it is easy to code and use and gives good results for the practical MC applications addressed in this book. Further, the MC results of the many examples scattered throughout this book agree well with analytic answers. However, the MC methods presented in this text are totally independent of the RNG and computer used.

Last Words on Random Numbers

All numbers produced by a digital computer have a certain granularity caused by the finite number of binary digits used to represent a number. Thus, the infinite number of irrational numbers in the interval (0, 1) and an infinite number of the smaller infinite set of rational numbers in the same interval, such as $1/3$, can be only approximated by a finite number of binary digits. Further, the number of tests for randomness is also infinite, e.g., all n-tuplets of the integers $n = 1, \ldots, \infty$ must be equidistributed in the infinite stream of pseudorandom integers. Likewise an infinity of pseudorandom numbers ρ_i must be equidistributed in a unit hypercube of dimension $n = 1, 2, \ldots, \infty$. This conundrum of infinities and the impossibility of calculating true RNs are said to have led von Neumann to quip at a 1949 conference "Anyone who considers arithmetical methods of producing random digits is, of course, in a state of sin."

Extreme purists would recognize the futility of generating true, but deterministic, RNs on a computer and would thus eschew MC methods for obtaining solutions to their models and equations (which, of course, are themselves mostly approximations of reality). Pragmatists would, however, realize one can get close enough to the unachievable ideal of randomness and that MC methods can be a very practical way to obtain solutions, albeit approximate, to their models and equations. It is for these realists that this book is intended.

Problems

3.1 Repeat the analysis for Example 3.1 using different seeds.

3.2 All variables in this problem are integers. Two integers a and b are *congruent modulo n* if

$$a \equiv b(\bmod m).$$

The parentheses mean that $(\bmod n)$ applies to the entire equation, not just to the right side. This notation is not to be confused with the notation $a = b \bmod m$ (without parentheses), in which a is the remainder of b when divided by m. A key consequence of the congruence $a \equiv b(\bmod m)$ is that $(b - a)$ is divisible by m. For example, $-7 \equiv 8(\bmod 5)$ since $(-7 - 8) = -15$ is divisible by 5. However $8 \bmod 5 = 3$.
For the case $a \equiv b(\bmod m)$, show that

$$ia = ib(\bmod m) \qquad \text{and} \qquad a^j = b^j(\bmod m).$$

3.3 Show that the congruential generator $x_{i+1} = ax_i \bmod m$ implies the ability of the generator to skip ahead, namely, $x_{i+k} = a^k x_i \bmod m$. Illustrate with a simple example.

3.4 In Fig. 3.1, overlapping pairs of numbers $(x_1, x_2), (x_2, x_3), \ldots$ were used to define the points on the plane. What would the plots look like if adjacent pairs $(x_1, x_2), (x_3, x_4), \ldots$ were used?

3.5 What is the minimum distance between hyperplanes in the hypercube of dimensions $k = 2, \ldots, 6$ for a congruential generator with modulus (a) $m = 2^{31} - 1$ and (b) $m = 2^{48}$? Comment of the granularity of the lattice points.

3.6 What is the maximum number of hyperplanes in the hypercube of dimensions $k = 2, \ldots, 6$ for a congruential generator with modulus (a) $m = 2^{31} - 1$ and (b) $m = 2^{48}$? What do these results tell you about the equidistributedness of the lattice points?

3.7 Using C or FORTRAN, write a multiplicative congruential generator based on the Mersenne prime $m = 2^5 - 1 = 31$. (a) What multipliers produce full periods? (b) Make plots of overlapping pairs of numbers similar to those in Fig. 3.1 and determine which multiplier(s) give the most uniform covering of the plane.

3.8 Determine the primitive roots for the generator $x_{i+1} = ax_i \bmod 17$ that produce full-cycle generators.

3.9 Cycle lengths of RNGs are often reported as 2^m, where m is an integer. Convert such a number to a multiple of a power of 10, i.e., show that

$$2^m = 10^x = a10^n,$$

where $x = m \log_{10} 2$, $n = [m \log_{10} 2]$, and $a = 10^{m \log_{10} 2 - n}$.

3.10 For many years, the program RANDU was widely used. This generator is defined by

$$x_{i+1} = 65{,}539 x_i \bmod 2^{31} \qquad \text{and} \qquad \rho_{i+1} = \frac{x_{i+1}}{2^{31}-1}.$$

Although this generator has good random properties for the two-dimensional lattice, it has a very poor three-dimensional lattice structure because all triplets (x_i, x_{i+1}, x_{i+2}) lie on only 15 planes. Clearly, if such a triplet were used to define a starting location for a source particle in a transport calculation, the source would be poorly modeled. Write a FORTRAN or C program to implement this generator and generate a sequence of length 30,002. For all triplets in this sequence, $(\rho_i, \rho_{i+1}, \rho_{i+2})$, in which $0.5 \leq \rho_{i+1} \leq 0.51$, plot ρ_{i+2} versus ρ_i. What does this plot tell you about this generator? After (Gentle, 1998).

3.11 Consider the Tausworthe RNG discussed in Section 3.9.1 and illustrated in Example 3.3. Another primitive polynomial for the case $q = 5$ is $f(x) = x^5 + x^4 + x^2 + x + 1$. Repeat the analysis of Example 3.3 based on this primitive polynomial. (a) Take the initial seed as $b_1 = b_2 = b_3 = b_4 = b_5 = 1$. (b) Repeat the analysis with a different initial seed of your choosing.

3.12 Perform the analysis shown in Example 3.3 for the Tausworthe RNG for the case $q = 7$. For a seventh-degree polynomial one of the simplest primitive polynomials (of which there are nine) is $x^7 + x^3 + 1$. Take the initial seed to be $b_1 = b_2 = b_3 = b_4 = b_5 = b_6 = b_7 = 1$.

3.13 Consider the combined generator of Eq. (3.19). Because of the relative primeness of the moduli, an exact 0 cannot be obtained. But could a code using integer arithmetic produce a 0? First consider a computer with a very small binary integer, say 2 bits; then extend to larger integer precisions. For a 32-bit computer, what is the probability a 0 is obtained? After (Gentle, 1998).

3.14 In Section 3.11.1, it was mentioned that because the number of different "0" and "1" patterns that a two-state, 1-D CA with N cells per generation was 2^N, the maximum cycle length for the CA is $2^N - 1$. (a) Why is there the factor of -1? (b) What are the maximum cycle lengths for 2-D and 3-D two-state CAs? (c) What are the maximum cycle lengths for three-state CAs in one, two, and three dimensions? Summarize your findings in a plot of maximum cycle length versus N.

3.15 In Fig. 3.6, it is seen that for large $N = 10^{2m}$ the sample standard deviation $s(\overline{x})$ appears to be converging asymptotically to $0.2886\ldots \times 10^{-m}$. Explain why this is so.

3.16 Write a code that generates $N = 10^4$ samples from:

(a) the minimal standard generator with $a = 16{,}807$, $q = 127{,}773$, and $r = 2836$ (see Section 3.5.1 and Appendix F),

(b) the minimal standard generator with $a = 48,271$, $q = 44,488$, and $r = 3399$,

(c) the minimal standard generator with the parameters of part (a) but shuffled using the Bays–Durham approach. In this case, use a shuffling vector/table of length $k = 32$.

In all three cases start with a seed $x_0 = 1$, calculate the average value of the samples as $\overline{x} = 1/N \sum_{i=1}^{N} x_i$, and give the value of the final pseudorandom number, $x_{10,000}$.

3.17 Write a program to generate the evolution of a one-dimensional CA. In particular, generate figures similar to those of Fig. 3.8 for a rule 45 automaton.

$$s_i' = s_{i-1} \text{ XOR } (s_i \text{ OR } (\text{ NOT } s_{i+1}))$$

or, equivalently,

$$s_i' = (1 + s_{i-1} + s_{i+1} + s_i s_{i+1}) \bmod 2.$$

3.18 Repeat the analysis of Example 3.6 but for an ECA based on rule 45. See the previous problem for how to implement this rule.

3.19 Repeat the analysis of Example 3.6, but this time use a finite grid containing only 40 cells on each side of the center cells. At the grid's lateral boundaries use a periodic boundary condition.

3.20 The so-called "low-order correlation" is often a criticism made of Park and Miller's minimum standard generator. Repeat the analysis of Example 1.3 but start with the seed=iseed $x_0 = 1$ so that the first few RNs are $\rho_1 = 4.6566 \times 10^{-10}$, $\rho_2 = 7.8264 \times 10^{-5}$, $\rho_3 = 0.13154$, and $\rho_4 = 0.75561$. Does this low-order correlation have a significant effect? Explain your reasoning.

References

Altman, N.S., 1989. Bit-wise behavior of random number generators. SIAM J. Sci. Stat. Comput. 9, 839–847.

Bays, C., Durham, S.D., 1976. Improving a poor random number generator. ACM Trans. Math. Softw. 2, 59–64.

Beyer, W.A., Roof, R.B., Williamson, D., 1971. The lattice structure of multiplicative congruential pseudorandom vectors. Math. Comput. 25, 345–363.

Bhattacharjee, K., Maitya, K., Das, S., 2018. A search for good pseudorandom number generators: survey and empirical studies. arXiv:1811.04035v1 [cs.CR]. 3 Nov 2018.

Bhattacharjee, K., Paul, D., Das, S., 2017. Pseudorandom number generation using a 3-state cellular automaton. Int. J. Mod. Phys. C 28 (06), 1750078.

Borosh, I., Niederreiter, H., 1983. Optimal multipliers for pseudorandom number generation by the linear congruential method. BIT 23, 65–74.

Bratley, P., Fox, B.L., Schrage, E.L., 1987. A Guide to Simulation. Springer-Verlag, New York.

Brent, R.P., 2004. Note on Marsaglia's Xorshift random number generators. J. Math. Softw. 11 (5), 1–4.

Brown, R.G., Eddelbuettel, D., Bauer, D., 2020. Dieharder: a random number test suite. ver. 3.31.1. https://webhome.phy.duke.edu/~rgb/General/dieharder.php. (Accessed 12 December 2020).

Cassels, J.W.S., 1959. An Introduction to the Geometry of Numbers. Springer-Verlag, Berlin.

Chaudhuri, P.P., Chowdhury, D.R., Nandi, S., Chattopadhyay, S., 1997. Additive Cellular Automata: Theory and Applications, vol. 1. IEEE CS Press, Los Alamitos, CA.

Chowdhury, D.R., Sangupta, I., Chaudhuri, P.P., 1994. A class of two-dimensional cellular automata and their applications in random pattern testing. J. Electron. Test. 5 (1), 67–82.

Collins, B.J., 1987. Compound random number generators. J. Am. Stat. Assoc. 82, 525–527.

Das, S., Sikdar, B.K., 2010. A scalable test structure for multicore chip. IEEE Trans. Comput.-Aided Des. Integr. Circuits Syst. 29 (1), 127–137.

De Matteis, A., Pagnutti, S., 1993. Long-range correlation analysis of the Wichmann-Hill random number generator. Stat. Comput. 3, 67–70.

Eichenauer, J., Grothe, H., Lehn, J., 1988. Marsaglia's lattice test and nonlinear congruential pseudo random number generators. Metrika 35, 241–250.

Eichenauer, H., Lehn, J., 1986. A nonlinear congruential pseudo random number generator. Stat. Hefte 27, 315–326.

Fishman, G.S., 1990. Multiplicative congruential random number generators with modulus 2^{β}: an exhaustive analysis for $\beta = 32$ and a partial analysis for $\beta = 48$. Math. Comput. 54, 331–334.

Fishman, G.S., 1996. Monte Carlo: Concepts, Algorithms, and Applications. Springer-Verlag, New York.

Fishman, G.S., Moore, L.R., 1986. An exhaustive analysis of multiplicative congruential random number generators with modulus $2^{31} - 1$. SIAM J. Sci. Stat. Comput. 7, 24–25.

Fuller, A.T., 1976. The period of pseudorandom numbers generated by Lehmer's congruential method. Comput. J. 19, 173–177.

Gentle, J.E., 1998. Random Number Generation and Monte Carlo Methods. Springer-Verlag, New York.

Hortensius, P.D., McLeod, R.D., Pries, W., Miller, D.M., Card, H.C., 1989. Cellular automata-based pseudorandom number generators for built-in self-test. IEEE Trans. Comput.-Aided Des. Integr. Circuits Syst. 8 (8), 842–859.

James, F., 1990. A review of pseudorandom number generators. Comput. Phys. Commun. 60, 329–344.

Klimasauskas, C., 2002. Not knowing your random number generator could be costly: random generators—why are they important. PC AI Mag. 16 (3). Phoenix, AZ.

Knuth, D.E., 1981. Seminumerical algorithms, vol. 2. In: The Art of Computer Programming, 2nd ed. Addison-Wesley, Reading, MA.

Koikara, R., 2017. SAC:G: 3-D cellular automata based PRNG. ACM 978-1-4503-3738-0. Available at https://src.acm.org/binaries/content/assets/src/2016/rosemarykoikara.pdf. (Accessed 5 September 2020).

L'Ecuyer, P., 1988. Efficient and portable combined random number generators. Commun. ACM 31, 742–749.

L'Ecuyer, P., Blouin, F., Couture, R., 1993. A search for good multiplicative recursive random number generators. Math. Comput. 65, 203–223.

L'Ecuyer, P., 1996. Maximally equidistributed combined Tausworthe generators. ACM Trans. Model. Comput. Simul. 3, 87–98.

L'Ecuyer, P., Simard, R., 2007. TestU01: a C library for empirical testing of random number generators. ACM Trans. Math. Softw. 33 (4), 22:1–22:40. https://doi.org/10.1145/1268776.1268777.

Lehmer, D.H., 1951. Mathematical methods in large-scale computing units. Annu. Comput. Lab. Harvard Univ. 26, 141–146.

Lewis, T.G., Payne, W.H., 1973. Generalized feedback shift register pseudorandom number algorithm. J. ACM 20 (3), 456–468.

Lewis, P.A.W., Goodman, A.S., Miller, J.M., 1969. A pseudo-random number generator for the system/360. IBM Syst. J. 8, 136–146.

Marsaglia, G., 1968. Random numbers fall mainly in the plane. Proc. Natl. Acad. Sci. 61, 25–28.

Marsaglia, G., 1972. The structure of linear congruential sequences. In: Zaremba, S.K. (Ed.), Applications of Number Theory to Numerical Analysis. Academic Press, New York.

Marsaglia, G., 1985. A current view of random number generators. In: Billard, L. (Ed.), Computer Science and Statistics: The Interface. Elsevier, Amsterdam.

Marsaglia, G., 1993. Remarks on choosing and implementing random number generators. Commun. ACM 36 (7), 105–108.

Marsaglia, G., 2003. Xorshift RNGs. J. Stat. Softw. 8 (14), 1–6.

Marsaglia, G., 2016. The Marsaglia Random Number CD-ROM, Including the DIEHARD Battery of Tests of Randomness. Dept. Statistics, Florida State University, Tallahassee, Florida. Archived from the original on 2016-01-25.

Marsaglia, G., Zaman, A., Tsang, W.-W., 1990. Stat. Probab. Lett. 9, 35.

Marsaglia, G., Zaman, A., 1991. A new class of random number generators. Ann. Appl. Probab. 1, 462–480.

Matsumoto, M., Kurita, Y., 1992. Twisted GFSR generators. ACM Trans. Model. Comput. Simul. 2 (3), 179–194.

Matsumoto, M., Nishamura, T., 1998. Mersenne twister: a 623-dimensional equidistributed uniform pseudorandom number generator. ACM Trans. Model. Comput. Simul. 8 (1), 3–30.

Maxim Integrated, 2010. Pseudo Random Number Generation Using Linear Feedback Shift Registers. Application Note 4400.

MCNP, 2003. A General Monte Carlo N-Particle Transport Code, Version 5, vol. I: Overview and Theory. LA-UR-03-1987. Los Alamos National Lab.

Nance, R.E., Overstreet Jr, C., 1978. Some experimental observations on the behavior of composite random number generators. Oper. Res. 26 (5), 915–935.

Panneton, F., L'Ecuye, P., Matsomoto, M., 2006. Improved long-period generators based on linear recurrence modulo 2. ACM Trans. Math. Softw. 32 (1), 1–16.

Park, S.K., Miller, K.W., 1988. Random number generators: good ones are hard to find. Commun. ACM 31, 1192–1201.

Park, S.K., Miller, K.W., Sockmeyer, P.K., 1993. Response. Commun. ACM 36 (7), 108–110.

Press, W.H., Teukolsky, S.A., Vetterling, W.T., Flannery, B.P., 1992a. Numerical Recipes in C, 2 ed. Cambridge Univ. Press, Cambridge.

Press, W.H., Teukolsky, S.A., Vetterling, W.T., Flannery, B.P., 1992b. Numerical Recipes in FORTRAN, 2 ed. Cambridge Univ. Press, Cambridge.

Press, W.H., Teukolsky, S.A., Vetterling, W.T., Flannery, B.P., 1996. Numerical Recipes in FORTRAN90, 2 ed. Cambridge Univ. Press, Cambridge.

Press, W.H., Teukolsky, S.A., Vetterling, W.T., Flannery, B.P., 2007. Numerical Recipes in C++, 3 ed. Cambridge Univ. Press, Cambridge.

Payne, W.H., Rabung, J.R., Bogyo, T.P., 1969. Coding the Lehmer pseudorandom number generator. Commun. ACM 12, 85–86.

Rukhin, A., Soto, J., Nechvatal, J., Smid, M., Barker, E., 2001. A Statistical Test Suite for Random and Pseudorandom Number Generators for Cryptographic Applications. Tech. rep., DTIC Document.

Schrage, L., 1979. A more portable FORTRAN random number generator. ACM Trans. Math. Softw. 5, 132–138.

Sullivan, S.J., 1993. Another test for randomness. Commun. ACM 36 (7), 108.

Takano, H., 2021. MT in various languages. Available at http://www.math.sci.hiroshima-u.ac.jp/m-mat/MT/VERSIONS/eversions.html. (Accessed 14 February 2021).

Tausworthe, R.C., 1965. Random numbers generated by linear recurrence modulo two. Math. Comput. 19 (90), 201–209.

Tezuka, Shu, L'Ecuyer, P., 1992. Analysis of add-with-carry and subtract-with-borrow generators. In: Proc. 1992 Winter Simulation Conference. Assoc. Comp. Mach., New York, pp. 443–447.

Tomassini, M., Sipper, M., Perrenoud, M., 2000. On the generation of high-quality random numbers by two-dimensional cellular automata. IEEE Trans. Comput. 49 (10), 1146–1151.

Vigna, S., 2019. It is high time we let go of the Mersenne Twister. On-line as arXiv:1910.06437v2.

von Neumann, J., 1951. Various techniques used in connection with random digits. J. Res. Natl. Bur. Stand. Appl. Math. Ser. 12, 36–38.

Wang, Q., Yu, S., Ding, W., Leng, M., 2008. Generating high-quality random numbers by cellular automata with PSO. IEEE Trans. Fourth Int. Conf. Nat. Comput., 430–433. https://doi.org/10.1109/ICNC.2008.560.

Wichmann, B.A., Hill, I.D., 1982. An efficient and portable pseudorandom number generator. Appl. Stat. 31, 188–190. Corrections, Appl. Stat. 33, 123 (1984).

Widynski, B., 2017. Middle square Weyl sequence RNG. arXiv:1704.00358v5.

Widynski, B., 2020. Squares: a fast counter-based RNG. arXiv:2004.06278v2.

Wolfram, S., 1986. Random sequence generation by cellular automata. Adv. Appl. Math. 7, 123–169.

Wolfram, S., 2002. A New Kind of Science. Wolfram Media, Inc, Champaign, IL.

Chapter 4

Sampling, Scoring, and Precision

Everyone thought Pat was a bore;
with them Pat had little rapport.
But studying at nights
Monte Carlo's delights,
Pat finally learned how to score.

The Monte Carlo (MC) method relies inherently on the concepts of *sampling* and *scoring*. Sampling is the process of picking a particular value of a random variable from its governing distribution. For example, suppose a random variable is governed by the exponential distribution. Then, sampling that variable generates numerical values of the variable such that a large number of samples approximately reproduce the exponential distribution. Scoring is the process whereby the expected value of some random variable, such as $z(x)$, is estimated. In many cases, the score of an MC history is the calculation of $z(x_i)$ that results from the ith history. In the simplest case, the score is zero or one (failure or success). The overall score, $\bar{z}$, is the cumulative average of N history scores. Some of these concepts and methods for implementing them are explored in this chapter. Another formal method of sampling that is sometimes used when the probability density function (PDF) is highly multidimensional or otherwise difficult to sample by straightforward means is based on Markov chains. This method has evolved into its own procedure, called Markov chain Monte Carlo (MCMC), which is treated separately in Chapter 6. However, MCMC is in essence just another way of sampling in order to estimate expectations from sample averages.

4.1 Sampling

A variety of techniques can be used to generate values of a random variable that has a specific distribution. Perhaps the most frequently used method to sample from a discrete distribution that has two equally probable values is to flip a coin. Tossing a die can be used to generate values of a random variable that has six equally probable outcomes. Many lotteries select a winning number by randomly picking numbered tiles or balls from a cage in which the contents are mixed. Picking numbered pieces of paper from a hat can also be used to generate random numbers. Throwing darts at the stock pages of the *Wall Street Journal* can be used to randomly select stocks for your portfolio. Dropping matchsticks onto a hardwood floor can be used as a physical simulation of Buffon's needle problem.

Many random physical processes also can be used to generate random numbers with various distributions. "White noise" in electronic circuits can be used to generate random values from a uniform distribution. The time interval between successive decays in a radioactive source can be shown to have an exponential distribution. The hiss from the cosmic microwave background radiation pervading our universe likewise can be used to generate sequences of random numbers.

However, computer-generated random numbers are almost always used in MC calculations, in which thousands, millions, or even billions of random numbers are routinely needed. Such computer-generated numbers are not truly random, but proper algorithms can generate extremely long sequences of numbers that pass many statistical tests for randomness. These so-called *pseudorandom* number generators were discussed in Chapter 3.

All the sampling techniques discussed in this chapter for generating values of a random variable x with a specific distribution are based on having a pseudorandom number generator that can produce long sequences of pseudorandom numbers ρ_i that are uniformly distributed on the interval $(0, 1)$.[1] How this is done computationally was discussed in Chapter 3. The symbol ρ_i is used in this book to refer to a pseudorandom number sampled from the uniform distribution on the unit interval (also called the unit rectangular distribution).

4.1.1   Inverse CDF Method for Continuous Variables

There is a one-to-one correspondence between any two cumulative distribution functions (CDFs), because CDFs are monotone increasing functions between zero and unity. This result indicates that the probability that x, whose PDF is $f(x)$ and CDF is $F(x)$, has a value less than x_i is equal to the probability that y, whose PDF is $g(y)$ and CDF is $G(y)$, has a value less than y_i, i.e.,

$$F(x_i) = G(y_i). \tag{4.1}$$

Now, the unit rectangular PDF $\mathcal{U}(0, 1)$ (see Appendix A) has the CDF $U(\rho) = \int_0^\rho d\rho' = \rho$. Replace $G(y_i)$ with $U(\rho_i) = \rho_i$ in Eq. (4.1) to obtain

$$F(x_i) = \rho_i. \tag{4.2}$$

The inverse CDF method begins by selecting a number ρ_i uniformly between 0 and 1. Then the desired value of x_i, distributed as $f(x)$, is obtained by solving

$$x_i = F^{-1}(\rho_i), \tag{4.3}$$

---

[1] The notation $\mathcal{U}(a, b)$ is used to represent a uniform distribution over the interval (a, b). That ρ_i are random numbers uniformly distributed on the interval $(0, 1)$ that can be written more compactly as $\rho_i \leftarrow \mathcal{U}(0, 1)$.

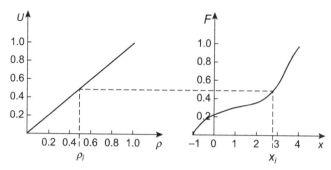

Figure 4.1 A pseudorandom number ρ_i selected from the unit rectangular distribution corresponds to one and only one value x_i of another random variable whose CDF is given by F.

where F^{-1} is the function obtained by solving Eq. (4.2) for x_i. Although this is a very efficient method for generating samples of random variables (or variates) with a given distribution $f(x)$, it is limited to those distributions for which Eq. (4.2) can be solved for x_i, i.e., to distributions whose inverse F^{-1} is known. Inverse CDF straightforward sampling is illustrated schematically in Fig. 4.1.

Example 4.1 Inverse CDF Method for a Uniform Distribution

Consider how to obtain a sample from a uniform distribution defined on the interval $[a, b]$. From Eq. (4.2) one can write

$$\rho_i = F(x_i) = \frac{1}{b-a} \int_a^{x_i} du = \frac{x_i - a}{b - a}.$$

Solving for x_i then yields

$$x_i = F^{-1}(\rho_i) = a + \rho_i(b - a).$$

Note that if $\rho_i = 0, x_i = a$ and if $\rho_i = 1, x_i = b$.

Example 4.2 Inverse CDF for an Exponential Distribution

Consider sampling from an exponential distribution $f(x) = \alpha e^{-\alpha x}$ with $x \in [0, \infty)$ and $\alpha > 0$. The CDF for this distribution with parameter α can be written

$$F(x) = \int_0^x \alpha e^{-\alpha x'} dx' = 1 - e^{-\alpha x}.$$

The random variable x_i is then found by solving the equation

$$\rho_i = 1 - e^{-\alpha x_i},$$

to obtain

$$x_i = -\frac{1}{\alpha}\ln(1 - \rho_i).$$

Because the random quantity $1 - \rho_i$ is also distributed according to $\mathcal{U}(0, 1)$, a more computationally efficient result is

$$x_i = -\frac{1}{\alpha}\ln(\rho_i),$$

thereby saving one subtraction operation each time a sample is generated.

Example 4.3 Inverse CDF Method for a Displaced Exponential Distribution

Consider an exponential distribution that is displaced to the right of $x = 0$ by an amount a. To use the inverse CDF method to sample from this distribution, the form of the PDF must first be determined. Assume

$$f(x) = A\alpha e^{-\alpha x}, \ x \geq a,$$

where A is a normalization constant. To obtain A, apply the normalization condition $\int_a^\infty f(x)dx = 1$. This leads to

$$A\int_a^\infty \alpha e^{-\alpha x}dx = A\alpha \left[\frac{e^{-\alpha x}}{-\alpha}\right]_a^\infty = Ae^{-\alpha a} = 1,$$

from which it is determined that $A = e^{\alpha a}$. To sample from this PDF, select a pseudorandom number ρ_i and set it equal to the CDF, i.e.,

$$\rho_i = F(x_i) = A\alpha \int_a^{x_i} e^{-\alpha x'}dx' = A[e^{-\alpha a} - e^{-\alpha x_i}].$$

Substituting for A and solving for x_i then yields

$$x_i = -\frac{1}{\alpha}\ln[e^{-\alpha a}(1 - \rho_i)].$$

Again, the quantity $(1 - \rho_i)$ can be replaced with ρ_i, because both are similarly distributed, so that

$$x_i = -\frac{1}{\alpha}\ln[\rho_i e^{-\alpha a}].$$

Note that if $\rho_i = 1$, $x_i = \frac{\alpha a}{\alpha} = a$ and if $\rho_i = 0$, $x_i \to \infty$, as desired.

The validity of the inverse CDF method, or any other sampling procedure, can be established by showing that the probability of obtaining a value of x within the interval between some fixed value x and $x + dx$ is $f(x)\,dx$. For the inverse CDF method, one notes that the value of x_i corresponds to the random number ρ_i, and the probability of x_i being between x and $x + dx$ is the same as

$$\text{Prob}\{x < F^{-1}(\rho_i) < x + dx\} = \text{Prob}\{F(x) < \rho_i < F(x + dx)\}$$
$$= F(x + dx) - F(x)$$
$$= \int_x^{x+dx} f(x')\,dx' = f(x)\,dx. \qquad (4.4)$$

The inverse CDF method for single continuous random variables is almost always preferred when the inverse function, F^{-1}, is easily found. Methods for sampling from discrete random variables using the inverse CDF method are discussed in the next subsection. Sampling from joint distributions is discussed in Section 4.1.9.

4.1.2 Inverse CDF Method for Discrete Variables

The inverse CDF method also can be used for discrete variables as illustrated in Fig. 4.2. Suppose that the random variable x has only I discrete values x_i with probabilities f_i. In other words, the PDF is

$$f(x) = \sum_{i=1}^{I} f_i \delta(x - x_i), \qquad (4.5)$$

so that the CDF is given by Eq. (2.38):

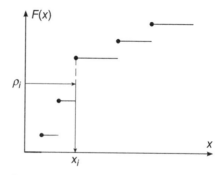

Figure 4.2 Inverse CDF method for a discrete distribution.

$$F(x) = \sum_{j=1}^{i} f_j. \qquad (4.6)$$

To generate values of x that have this distribution, first select a pseudorandom number ρ uniformly from $\mathcal{U}(0, 1)$, and then find the number x_i such that

$$x_i = \begin{cases} x_1 & \text{if } 0 < \rho \le f_1, \\ x_i & \text{if } \sum_{j=1}^{i-1} f_j < \rho \le \sum_{j=1}^{i} f_j. \end{cases} \qquad (4.7)$$

Equivalently, this can be written as

$$F(x_{i-1}) < x_i \le F(x_i) \qquad (4.8)$$

with $F(0) = 0$. To find x_i, the following simple methodical search algorithm can be used.

1. Pick ρ from $\mathcal{U}(0, 1)$.
2. Set $i = 1$.
3. If $\rho \leq F(x_i)$, return x_i.
4. Otherwise, set $i = i + 1$ and go to step 3.

For many discrete distributions, such as the Poisson, binomial, and geometric distributions (see Appendix A for details), the mass points x_i, i.e., the values of x for which $f(x) \neq 0$, are the integers $0, 1, 2, \ldots$. Moreover, for these distributions, the ratio of successive PDF values $R(x) = f(x + 1)/f(x)$ is often a simple function of x which can be used to construct values of $F(x)$ while searching for the x that satisfies if $\rho \leq F(0)$, then $x = 0$; otherwise if

$$F(x_i - 1) < \rho \leq F(x_i), \tag{4.9}$$

then $x = x_i$. The above search algorithm is thus modified as follows.

1. Pick ρ from $\mathcal{U}(0, 1)$.
2. Set $x = 0$, $\Delta F = f(0)$, and $F = \Delta F$.
3. If $\rho \leq F(x)$, return x.
4. Otherwise, set $x = x + 1$, $\Delta F = R(x)\Delta F$, $F = F + \Delta F$, and go to step 3.

4.1.3 Sampling by Table Lookup

Using the inverse CDF method for a discrete distribution is essentially a table lookup procedure; the search can be performed sequentially, as in the above algorithms, or by some tree traversal. Faster sampling times can be achieved if nodes with high probability are tested before nodes with low probability. For example, in unimodal discrete distributions with a large number of nodes, it is better to start at the node with the greatest probability. Various other methods have been proposed for speeding a table search (see Marsaglia (1963); Norman and Cannon (1972); Chen and Asau (1974), and Edwards et al. (1991)).

Sometimes the CDF of a continuous random variable is presented as a table of cumulative values because an analytic expression for the CDF is unknown over all or part of the range. For example, an analyst may be able to assign probabilities in certain regions of the distribution but be unable to supply an underlying functional form. Or the continuous distribution may be produced by a previous MC calculation and is known only in a "binned" or histogram form. Sometimes a very complex and computationally expensive function must be evaluated billions of times, and it is more efficient to tabulate the function before embarking on an MC calculation and then use a fast table lookup method during the MC calculation rather than compute the function over and over again. To sample from such tabulated distributions, some approximate interpolating function is fit to histogram data (a piecewise linear function is easiest, although quadratic and cubic functions have been used) and the inverse CDF method is then used to obtain random variates.

Discrete Alias Sampling

A. J. Walker (1974) described an alternative way to sample x_i from a discrete distribution with a finite number I of values, i.e., $i = 1, 2, \ldots, I$. His method is called the *discrete alias method*. In this method, the distribution of the I outcomes, which generally have different probabilities f_i, is recast into a distribution of outcomes g_i that have equal probabilities, all equal to $1/I$. To do this, one adds to any outcome i, for which $f_i < 1/I$, a probability increment δ_k from an outcome k that has excess probability, i.e., an outcome for which $f_k > 1/I$, to make a new PDF for which each outcome has probability $g_i = 1/I$. Those outcomes x_i with $f_i > 1/I$ are called *donors* because they donate some probability and those with $f_i < 1/I$ are called *recipients* because they receive probability from donors. Of course, if $f_i = 1/I$ for an outcome, it is neither a donor nor a recipient. The outcome f_i that has a probability less than $1/I$ is called the *nonalias probability component* and the probability component δ_k donated by a donor is called the *alias probability component*. If donation of δ_k from outcome k reduces that outcome's probability below $1/I$, then that outcome becomes a recipient.

After the process of conversion from the original PDF f to the uniform PDF g is complete, one must account for the various probability contributions to each g_i using

$$q_k = \frac{\delta_k}{1/I} = I\delta_k, \tag{4.10}$$

the probability of an alias component from state k. The probability of the nonalias component is called $p_i, i = 1, 2, \ldots, I$. If there is no alias component in g_i, then

$$p_i = 1; \tag{4.11}$$

if there is one alias component coming from state k,

$$p_i = 1 - q_k. \tag{4.12}$$

If there are k alias components, then the probability of the nonalias component is

$$p_i = 1 - \sum_{j=1}^{k} q_j = 1. \tag{4.13}$$

It can be shown that the event x_i is properly chosen during history s by using the following algorithm:

1. Pick a pseudorandom number ρ_{1s} on the unit interval.
2. Pick an outcome i based on $g_{i-1} < \rho_{1s} \leq g_i$ with $g_0 = 0$.
3. Pick a second pseudorandom number ρ_{2s}.
4. If $\rho_{2s} < p_i$, set $x_s = x_i$, the nonalias component.

5. Otherwise, choose the appropriate alias component, δ_k, and let $x_s = x_k$, the outcome from which the alias was donated. This last step is straightforward if there is only one alias component; if there are more than one, then find the m for which $p_i + \sum_{j=1}^{m-1} < \rho_{2s} \leq p_i + \sum_{j=1}^{m}$ and set $k = m$ and $x_s = x_k$.

This procedure is generally more efficient than other table lookup methods and has the added advantage that it lends itself to vectorization.

It is always possible to construct a uniform PDF g that consists only of states i that have either no alias components or only one (Fishman, 2003). In these cases, every $g_i = 1/I$ is composed of either only a nonalias component for which $p_i = 1$ and $q_i = 0$ or of two components for which $0 < q_i < 1$ and $p_i = 1 - q_i$. Even though this is true, the "alias table" that consists of the p_i and q_i and k_i is not unique. Smith and Jacobson (2005) give algorithms that can be used to generate alias tables with only one p_i and one q_i for each state i but it is not guaranteed to be optimal. Other algorithms may exist and some may lead to more efficient alias sampling. Because alias sampling generally is more efficient than table lookup, the specific method used to construct the alias table is of concern in only a few complicated cases. In order to determine which outcome is chosen if $\rho_{2s} > p_i$, it is necessary to keep track of which state k contributed the alias probability q_i.

Thus, in discrete alias sampling, a probability table is created in which the original PDF f is converted into an equal-probability PDF g and an alias table is made that lists the states i and their corresponding value of k_i, p_i, and q_i. This time-consuming conversion is done prior to any MC calculation. Then, during the MC calculation, samples are obtained by using a random number to select g_i from the equiprobable states and then using a second random number to select between the nonalias state with probability p_i or the alias state with probability q_i and then using k_i to identify from which state the alias contribution was donated. An example illustrates implementation of this method.

Example 4.4 Discrete Alias Sampling of a Binomial

Consider the binomial distribution (see Appendix A)

$$f(x) = \frac{n!}{x!(n-x)!} p^x (1-p)^{n-x}, \quad x = 0, 1, 2, \ldots, n,$$

where n is the number of trials, p is the constant probability of success on each trial, and x is the number of successes. In order to implement alias sampling, it is convenient to recast this in the form

$$f(x) = \frac{(I-1)!}{(x-1)!(I-x)!} p^{x-1} (1-p)^{I-x}, \quad x = 1, 2, \ldots, I,$$

where $I = n + 1$. For the specific case $n = 4$ and $p = 0.4$, the following table can be generated:

i	x_i	f_i	F_i
1	1	0.1296	0.1296
2	2	0.3456	0.4752
3	3	0.3456	0.8208
4	4	0.1556	0.9744
5	5	0.0256	1.0000

It is possible to convert this PDF into a PDF, g, with equal probabilities $1/5 = 0.20$ by a procedure such as the following.

1. Take $\delta_2 = 0.0774$ from state 2 and add it to state 1, making state g_1 have a value of 0.20.
2. Take $\delta_2 = 0.0464$ from state 2 and add it to state 4, making state $g_4 = 0.20$.
3. Then take $\delta_3 = 0.1744$ from state 3 and add it to state 5, making $g_5 = 0.20$.
4. Finally, take $\delta_4 = 0.0288$ from state 2 and add it to state 3, making $g_2 = g_3 = 0.20$.

This procedure is summarized in the following combined probability-alias table.

x_i	f_i	Step 1	Step 2	Step 3	Step 4 (g_i)	p_i	k	q_k
1	0.1296	0.2000	0.2000	0.2000	0.2000	0.648	2	0.352
2	0.3456	0.2752	0.2288	0.2288	0.2000	1.000		0.000
3	0.3456	0.3456	0.3456	0.1712	0.2000	0.856	2	0.154
4	0.1556	0.1556	0.2000	0.2000	0.2000	0.768	2	0.232
5	0.0256	0.0256	0.0256	0.2000	0.2000	0.128	3	0.872

Note that all g_i values involve only zero or one alias component. Employing the sampling algorithm previously described, the following $\overline{f}_i$ and $s(\overline{f}_i)$ values were obtained for $N = 10^6$ histories:

i	f_i	$\overline{f}_i$	$s(\overline{f}_i)$
1	0.1296	0.1294	0.0011
2	0.3456	0.3445	0.0015
3	0.3456	0.3456	0.0015
4	0.1556	0.1545	0.0011
5	0.0256	0.0260	0.0005

In this example, all sample means were within one standard deviation of the true mean but this is not guaranteed in all cases. The time saving using alias sampling in this simple example is minimal but it may be significant when sampling binomials or other discrete PDFs with large I.

Alias sampling of finite discrete distributions is popular when sampling efficiency is important. See Walker (1977), Kronmal and Peterson (1979), or Smith and Jacobson (2005) for further discussion of this technique.

Other Monte Carlo Applications of Table Lookups

Sometimes, a continuous function, $f(x)$, $x \in (a, b)$, is tabulated at discrete points $x_0 = a, x_1, x_2, \ldots, x_J = b$ with the tabulated values denoted by $f_j \equiv f(x_j)$. Often a very computationally expensive function must be evaluated very many times in an MC calculation, and it is better to tabulate the function and use the much faster table lookup method. It is assumed here that a sufficiently fine grid $\{x_j\}$ is used so that all significant features of $f(x)$ are captured. To sample from such tabulated functions, some approximate interpolating function is fit to the tabulated data. A piecewise linear function is easiest, although quadratic and cubic functions can be used. For linear interpolation

$$f(x) \simeq f_{j-1} + m(x - x_{j-1}) \quad \text{with slope} \quad m = \frac{f_j - f_{j-1}}{x_j - x_{j-1}}, \quad j = 1, 2, \ldots J.$$

If $f(x)$ is a tabulated continuous PDF from which samples x_i are to be generated, first form the (approximate) CDF $F(x)$ by using some interpolation function and then apply the inverse CDF method to generate the desired samples. For example, if linear interpolation is chosen, the CDF for $x_{j-1} < x \le x_j$ the CDF assumes a piecewise quadratic shape, i.e.,

$$F(x) = F_{j-1} + \int_{x_{j-1}}^{x} f(x)\, dx = F_{j-1} + \frac{m}{2}(x^2 - x_{j-1}^2) + (x - x_{j-1})[f_{j-1} - m x_{j-1}],$$

in which $F(x_0) \equiv F(a) = 0$. Upon setting $x = x_j$ and simplification, the following recursion relation is obtained:

$$F_j = F_{j-1} + f_j x_j - f_{j-1} x_{j-1}.$$

However, numerical errors introduced by the finite tabulation grid and the assumed linearity of f, means $F(x)$ is usually not quite a properly normalized CDF because $F(x_J) \equiv F(b) \ne 1$. To generate samples with the inverse CDF method, it is essential that the CDF equals unity at end point x_J. To obtain a properly normalized CDF $G(x)$ define $G(x) = \xi F(x)$, where the *normalization correction factor* $\xi = 1/F(x_J)$.

Edwards et al. (1991) proposed a more efficient way to sample from a piecewise linear PDF. First a random number ρ_1 is used to select from which interval $(x_{j-1}, x_j]$ the sample x_i is to be generated, namely, $j = \lceil J\rho_1 \rceil$.[2] Then use a second random number ρ_2 to set

$$x' = (1 - \rho_2)x_{j-1} + \rho_2 x_j \quad \text{and} \quad x'' = \rho_2 x_{j-1} + (1 - \rho_2)x_j.$$

---

2 The ceiling function $\lceil y \rceil$ is the smallest integer greater than or equal to y.

Finally, with a third pseudorandom number ρ_3, choose the sample as

$$x_i = \begin{cases} x', & \text{if } \rho_3(f_{j-1} + f_j) \leq (1 - \rho_3)f_{j-1} + \rho_3 f_j, \\ x'', & \text{otherwise.} \end{cases}$$

This is quite efficient and yields samples that are distributed in a piecewise linear fashion.

4.1.4 Rejection Method

The CDF method presents difficulties if the PDF $f(x)$ cannot be integrated analytically to obtain $F(x)$ or, even if $F(x)$ can be obtained easily, $F^{-1}(\rho)$ is difficult to obtain analytically. An alternative approach in such cases, first proposed by von Neumann (1951), is the *rejection method*. Consider Fig. 4.3, which illustrates such a PDF $f(x)$, defined analytically within the range $[a, b]$ and known to be less than or equal to M. Two steps are required to select x_i from the $f(x)$ distribution:

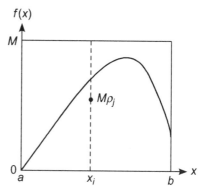

Figure 4.3 Basic rejection method.

1. Select a random value of $x \in [a, b]$ as $x_i = a + \rho_i(b - a)$.
2. Select another random number ρ_j. If $\rho_j M \leq f(x_i)$, then accept x_i. Otherwise, return to step 1.

Example 4.5 Sampling From a Discrete Distribution

Consider the flipping of a coin. Let x be the random variable that describes whether a head ($x_1 = 1$) or tail ($x_2 = 0$) is obtained on a flip. The head outcome has probability $f_1 = 0.5$ (assuming a "fair" coin) and the tail outcome has probability $f_2 = 0.5$. To generate each sample x_i, a pseudorandom number ρ_i, uniformly distributed over $(0, 1)$, is generated. If $\rho_i < 0.5$, then set $x_i = 0$; otherwise set $x_i = 1$. This procedure is an example of the generation of a random sample from a Bernoulli distribution (see Appendix A.2.1).

This process of selecting values of x is simple. The efficiency, or probability of not rejecting x_i, is given by the ratio of the areas $\int_a^b dx\, f(x)$ and $M(b - a)$, namely, $1/[M(b - a)]$. This efficiency may be very low if M is chosen inordinately large or if $f(x)$ is highly peaked. Improvements in efficiency may be achieved as follows.

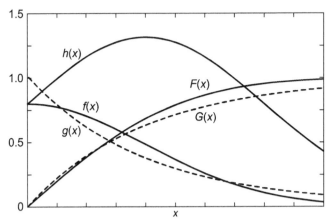

Figure 4.4 Graphs of $f(x)$ (solid line), the more tractable PDF $g(x)$ (dashed line), the ratio $h(x) = f(x)/g(x)$, and the CDFs $F(x)$ and $G(x)$. The value of h_{max} is any finite bound of $h(x)$, but in practice should not greatly exceed the maximum value of $h(x)$.

Consider Fig. 4.4, which illustrates a PDF $f(x)$ along with a similar PDF $g(x)$ defined within the same limits and suitable for the CDF method, that is, to have an easily obtainable CDF $G(x)$, which can in turn be easily inverted to give $G^{-1}(\rho)$. Note that

$$F(x) = \int_a^x dx' \, h(x')g(x'), \quad \text{where} \quad h(x) \equiv \frac{f(x)}{g(x)}. \tag{4.14}$$

If $g(x)$ is a close approximation to $f(x)$, then $h(x)$ does not deviate markedly from unity. At any rate, $g(x)$ must be chosen so that $h(x)$ does not exceed some maximum bounding value, h_{max}, in the interval a to b. It is more efficient, but not mandatory, to make h_{max} the actual maximum value of h between the limits a and b. The rejection technique proceeds as follows:

1. Select a pseudorandom number ρ_1.
2. Obtain a tentative value of x from the equation $x_t = G^{-1}(\rho_1)$.
3. Select another pseudorandom number ρ_2.
4. If $\rho_2 > h(x_t)/h_{max}$, reject the value x_t and return to step 1.[3]
5. If $\rho_2 < h(x_t)/h_{max}$, accept x_t as a valid sample of x.

To verify that this procedure does produce values of x_t that are distributed according to $f(x_t) = h(x_t)g(x_t)$, first consider the probability that ρ_2 is less

---

[3] The probability that ρ_2 exactly equals $h(x_t)/h_{max}$ is ignored here and throughout this section. This probability is negligible, in principle; but in practical calculations using pseudorandom numbers, the situation may arise on very rare occasions. In such case, it is arbitrary whether to group the "equal" case with the "less than" or "greater than" case.

than $h(x)/h_{max}$, namely,

$$P\{\rho_2 < h(x_t)/h_{max}\} = \int_0^{h(x_t)/h_{max}} d\rho_2 = h(x_t)/h_{max}. \qquad (4.15)$$

Let $u(x_t, \rho_2)$ be the joint probability distribution of the two random variables x_t and ρ_2. The conditional probability of x_t given that $\rho_2 < h(x_t)/h_{max}$ is, from Bayes' theorem of Eq. (2.49) and from Eq. (4.15),

$$P\{x_t | \rho_2 < h(x_t)/h_{max}\} = \frac{u(x_t, \rho_2) P\{\rho_2 < h(x_t)/h_{max}\}}{\int_a^b u(x_t, \rho_2) P\{\rho_2 < h(x_t)/h_{max}\} dx_t}$$

$$= \frac{u(x_t, \rho_2)[h(x_t)/h_{max}]}{\int_a^b u(x_t, \rho_2)[h(x_t)/h_{max}] dx_t}. \qquad (4.16)$$

But x_t and ρ_2 are independent random variables, so $u(x_t, \rho_2) = g(x_t) U(\rho_2) = g(x_t)$ because the density function of the unit distribution is simply unity. Substitution of this result into Eq. (4.16) gives the desired result, namely,

$$P\{x_t | \rho_2 < h(x_t)/h_{max}\} = \frac{g(x_t)[h(x_t)/h_{max}]}{\int_a^b g(x_t)[h(x_t)/h_{max}] dx_t}$$

$$= \frac{g(x_t)h(x_t)}{\int_a^b f(x_t) dx_t} = g(x_t)h(x_t). \qquad (4.17)$$

Example 4.6 Sampling Uniformly Over a Disk

Consider the problem of sampling (x_i, y_j) points uniformly over the disk of radius R shown to the left. One method to generate such coordinate pairs is to uniformly sample from the bounding square containing the disk and to use the rejection technique to obtain samples inside the disk. Specifically, the algorithm is:

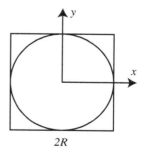

1. Select ρ_i and ρ_j from $U(0, 1)$.
2. Set $x_i = R(1 - 2\rho_i)$ and $y_j = R(1 - 2\rho_j)$.
3. Accept this pair if $x_i^2 + y_j^2 < R^2$.

Clearly the efficiency of this approach is the disk-area/square-area $= \pi/4$. To obtain a better efficiency, the inverse CDF method could be used. The probability a coordinate pair is in dr at a distance r from the center is $f(r) = 2\pi r\, dr/\pi R^2 = 2r\, dr/R^2$. The associated CDF is $F(r) = r^2/R^2$. Then the following procedure is used to generate (x_i, y_j) pairs:

1. Set $\rho_i = F(r_i) = r_i^2/R^2$.
2. Solve for $r_i = R\sqrt{\rho_i}$.
3. Sample a polar angle from $U(0, 2\pi)$, i.e., choose $\phi_j = 2\pi\rho_j$.
4. Set $x_i = r_i \cos\phi_j$ and $y_j = \sin\phi_j$.

Although this second approach has 100% efficiency, the calculation of sines and cosines is very computationally expensive compared to the arithmetic involved with the rejection method approach, so the rejection method, although less efficient, may be preferred.

4.1.5 Composition Method

This method is also known as the alternative-PDF or *probability mixing* method. It sometimes happens that the PDF for a process is cumbersome to use but that it can be split into separate functions, each of which is easier to sample. Suppose, for example, that

$$f(x) \equiv A_1 f_1(x) + A_2 f_2(x), \qquad a \le x \le b, \tag{4.18}$$

where the functions $f(x)$, $f_1(x)$, and $f_2(x)$ are all normalized to unit area between a and b, i.e., they are all proper PDFs. Because $f(x)$ is a PDF, the sum of the constants A_1 and A_2 must be unity.

A value of x may be sampled from $f(x)$ by the following procedure:

1. Select a random number ρ_i.
2. If $\rho_i < A_1$, use $f_1(x)$ as the PDF for selecting a value of x.
3. If $\rho_i > A_1$, use $f_2(x)$ as the PDF for selecting a value of x.

The extension to an arbitrary number of separate PDFs is straightforward. The constants A_i are just the probabilities that $f_i(x)$ is used to generate a random sample. This extension is considered in the next section.

The great advantage of this method is that sampling from a very complex distribution can be achieved by sampling from much simpler distributions using the inverse CDF or acceptance-rejection methods. Another advantage of this method is that it is often possible to give high sampling probabilities A_i to the PDFs that are inexpensively sampled and lower probabilities to those functions that are expensive to sample (see Example 3.5).

Example 4.7 Use of the Composition Method

Generate random variates from the PDF

$$f(x) = \frac{7}{8}[1 + (x-1)^6], \qquad 0 \le x \le 1.$$

This PDF can be decomposed into

$$f(x) = \frac{7}{8} f_1(x) + \frac{1}{8} f_2(x),$$

where

$$f_1(x) = 1, \ [\mathcal{U}(0,1)], \qquad f_2(x) = 7(x-1)^6, \qquad 0 \le x \le 1.$$

Thus, with ρ_1 and ρ_2 from $\mathcal{U}(0,1)$, the sampling scheme is

$$x = \begin{cases} \rho_2, & \text{if } \rho_1 \le 7/8, \\[2mm] 1 + \sqrt[7]{\rho_2 - 1}, & \text{if } \rho_1 > 7/8. \end{cases}$$

Note that the difficult sampling scheme from $f_2(x)$ is used only one out of eight times, on average.

4.1.6 Rectangle-Wedge-Tail Decomposition Method

One of the fastest methods for generating random variables from a continuous PDF is based on the rectangle-wedge-tail method (MacLauren et al., 1964). This method, which is a special application of the composition method, decomposes the area under the PDF into a contiguous series of k rectangles, k wedges, and one tail region, as shown in the figure to the right.[4] By using many narrow rectangles most of the area under the PDF is contained in the rectangles, so that by using the composition method of the previous section, there is a high probability

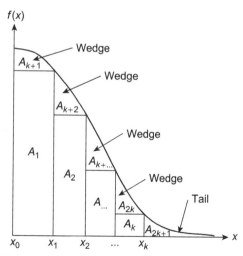

Figure 4.5 Decomposition of a PDF into a series of rectangles, wedges, and a tail.

of sampling from a rectangular PDF, a very easy and fast procedure. Also, the PDFs for the wedges become almost linear, and, as shown in the next section,

---

[4] Here it is assumed the PDF is on an infinite interval (x_0, ∞). This method is easily extended to the interval $(-\infty, \infty)$ by including a second tail region.

there is a fast method for sampling from a nearly linear PDF. The tail region takes the longest time to sample, so the decomposition tries to minimize the probability of taking a sample from the tail, i.e., area $A_{2k+1} << 1$. See Fig. 4.5. The PDF $f(x)$ is, thus, decomposed in a generalization of Eq. (4.18) as

$$f(x) = \sum_{i=1}^{k} A_i f_i(x) + \sum_{i=1}^{k} A_{k+i} f_{k+i}(x) + A_{2k+1} f_{2k+1}(x). \qquad (4.19)$$

The probabilities of sampling from the various regions are given by

$$A_i = f(x_i)(x_i - x_{i-1}), \qquad A_{k+i} = \int_{x_{i-1}}^{x_i} f(x)\,dx - A_i, \qquad i = 1, 2, \ldots, k,$$

and
$$A_{2k+1} = \int_{x_k}^{\infty} f(x)\,dx.$$

Because the sum of all the A_i is the area under the PDF, $\sum_{i=1}^{2k+1} A_i = 1$, as required for the composition method. The PDFs for each region are

$$f_i(x) = 1/(x_i - x_{i-1}), \qquad\qquad x_{i-1} \le x < x_i, \quad i = 1, 2, \ldots, k,$$

$$f_{k+i}(x) = [f(x) - f(x_i)]/A_{k+i}, \qquad x_{i-1} \le x < x_i, \quad i = 1, 2, \ldots, k,$$

$$f_{2k+1}(x) = f(x)/A_{2k+1}, \qquad\qquad x \ge x_k.$$

4.1.7 Sampling From a Nearly Linear PDF

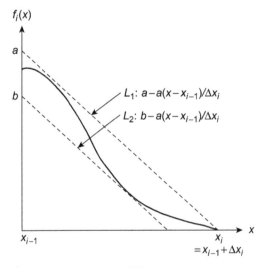

Figure 4.6 A nearly linear PDF.

By decomposing a PDF into many rectangles and wedges and a tail, the sampling in a wedge region is often from a PDF $f_i(x)$, $x \in [x_{i-1}, x_i]$, that varies almost linearly from some value at one end of the interval to zero at the other end of the interval, as illustrated in Fig. 4.6. The interval width is denoted by $\Delta x_i \equiv x_i - x_{i-1}$. Also shown on this figure are two straight parallel lines that "squeeze" the PDF. Rather than use the L_1 as a limiting function for the rejection method (see Section 4.1.4), the two parallel lines are used to limit the sampling volume and give a much higher chance of success. Knuth, as reported by Gentle (1998), gives the following simple algorithm to sample from a nearly linear PDF such as that shown in Fig. 4.6.

1. Generate ρ_1 and ρ_2 from $\mathcal{U}(0, 1)$; set $u = \min(\rho_1, \rho_2)$, $v = \max(\rho_1, \rho_2)$, and $x = x_{i-1} + u\Delta x_i$.
2. If $v \leq b/a$, then return x as an acceptable sample.
3. Else, if $v \leq u + f_i(x)/a$, then return x as an acceptable sample.
4. Otherwise no acceptable value is found, and return to step 1 to try again.

4.1.8 Composition-Rejection Method

A generalization of the rejection method is the composition-rejection method of sampling. Suppose that the PDF $f(x)$ can be written

$$f(x) = \sum_{j=1}^{n} A_j f_j(x) g_j(x), \tag{4.20}$$

in which all A_j are positive, all $g_j(x)$ are PDFs with analytical invertible CDFs, and all $f_j(x)$ have the property $0 \leq f_j(x) \leq 1$. A value of the random variate x may be selected as follows:

1. Select a random number and a corresponding value of i, where the probability of selecting i is proportional to A_i, i.e., with probability $A_i / \sum_{j=1}^{n} A_j$.
2. Select a test variate x_t from the PDF $g_i(x)$ using the inverse cumulative distribution method, i.e., $x_t = G_i^{-1}(\rho_i)$.
3. If $\rho_{i+1} \leq f_i(x_t)$ accept $x = x_t$; otherwise reject x_t and return to step 1.

The efficiency of the composition-rejection method is $1/\sum_i A_i$.

4.1.9 Ratio of Uniforms Method

Suppose $h(z)$ is some integrable nonnegative function on the interval $(-\infty, \infty)$ and one wants to sample from its PDF

$$f(z) = h(z) \bigg/ \int_{-\infty}^{\infty} h(z)\, dz. \tag{4.21}$$

A rather remarkable result by Kinderman and Monahan (1977) often allows fast generation of samples from this distribution.

Consider the region V in the xy-plane defined by the values of x and y that satisfy

$$V: \ 0 \leq x^2 \leq h(y/x). \tag{4.22}$$

If random points x and y are uniformly sampled over V, then it can be shown (Fishman, 2003) that the ratio $z \equiv y/x$ is a random variable whose PDF is $f(z)$.

The region V may be complex; but if $h(x)$ and $x^2 h(x)$ are bounded in V, the region V can be embedded in a simple rectangle region R and the rejection technique can be used to sample from V. Such a rectangular region R is the region containing those x- and y-values that satisfy

$$R: \quad 0 \le x \le b \quad \text{and} \quad c \le y \le d, \tag{4.23}$$

where

$$b = \max \sqrt{h(x)}, \quad c = \min x\sqrt{h(x)}, \quad \text{and} \quad d = \max x\sqrt{h(x)}. \tag{4.24}$$

The following algorithm implements this ratio-of-uniforms sampling scheme.

1. Generate ρ_1 and ρ_2 from $\mathcal{U}(0, 1)$.
2. Set $x = b\rho_1$, $y = c + (d - c)\rho_2$, and $z = y/x$.
3. If $x^2 \le h(x)$, accept z as a sample; otherwise go to step 1 and try again.

This very simple algorithm is easy to program and generally very fast, provided $h(x)$ is easy to evaluate.

4.1.10 Sampling From a Joint Distribution

To sample from a joint PDF (see Section 2.3), one usually samples first from the marginal PDF for a single variable and then works through the remaining variables by sampling from the appropriate conditional PDFs. For example, the joint PDF $f(x, y, z)$ can be decomposed as

$$f(x, y, z) = f(x|y, z) f(y|z) f_z(z). \tag{4.25}$$

A sample of z first is picked from $f_z(z)$, then a sample of y is selected from the conditional PDF $f(y|z)$, and finally a sample of x is obtained from the conditional PDF $f(x|y, z)$ given the other two values. In each case, standard sampling techniques are used. This sampling technique is illustrated in Example 4.8. An alternative sampling technique for joint PDFs is MCMC, which is covered in Chapter 6.

4.1.11 Sampling From Specific Distributions

Over the last half-century, there have been literally many hundreds of papers on how to compute random values from specific distributions, each claiming some new wrinkle or advantage over previous methods. Many of these papers present little new, a consequence of the large diversity of disciplines that use MC techniques, so that users in one area are unaware of techniques used in another. Likewise, there is little intellectual investment needed to make some subtle modification in one step of an existing algorithm and claim some improvement. Almost all algorithms depend on using random numbers uniformly distributed on $(0, 1)$ and some combination of the methods discussed earlier in this section. Both the algorithm efficiency and speed, which often depend on the computing platform, are important considerations. In some cases, the most efficient algorithm in one range of the distribution is not the most efficient in another range.

Example 4.8 Sampling From a Joint Distribution

Consider the joint PDF

$$f(\theta, \phi) = \frac{1}{K} \sin \theta, \qquad \frac{\pi}{4} \leq \theta \leq \frac{\pi}{2}, \quad 0 \leq \phi \leq \theta,$$

where $K = 1 + \sqrt{2}(\pi - 4)/8 \simeq 0.8483$ to ensure proper normalization. PDFs of this general type arise naturally when sampling from a solid angle subtended by a surface at a point. A procedure to sample from this PDF is developed. Begin by constructing the marginal PDFs

$$f_\theta(\theta) = \int_0^\theta f(\theta, \phi) \, d\phi = \frac{\theta \sin(\theta)}{K},$$

$$f_\phi(\phi) = \frac{1}{K} \int_{\pi/4}^{\pi/2} \sin(\theta) \, d\theta = \frac{-\cos(\pi/2) + \cos(\pi/4)}{K} = \frac{1}{\sqrt{2}K} \simeq 0.8335.$$

The conditional PDFs are

$$f(\theta|\phi) = \frac{f(\theta, \phi)}{f_\phi(\phi)} = \sqrt{2} \sin \theta \qquad \text{and} \qquad f(\phi|\theta) = \frac{f(\theta, \phi)}{f_\theta(\theta)} = \frac{1}{\theta}.$$

The joint PDF can then be sampled in two ways; both are considered here, although it is seen that one is preferable.

One way is to sample θ from $f_\theta(\theta)$ and then ϕ from $f(\phi|\theta)$. A random value θ_i is found from

$$\rho_i = \frac{1}{K} \int_{\pi/4}^{\theta_i} \theta \sin \theta \, d\theta = \frac{\sin \theta_i - \theta_i \cos \theta_i - \sqrt{2}(4 - \pi)/8}{K}.$$

This is a transcendental equation that first must be solved for θ_i (e.g., by trial and error) before sampling for ϕ. Clearly this approach is computationally expensive.

Alternatively, sample ϕ from $f_\phi(\phi)$ and then θ from $f(\theta|\phi)$. A value of ϕ_i is found from

$$\rho_{1i} = F_\phi(\phi_i) = \frac{1}{\sqrt{2}K} \int_0^{\phi_i} d\phi = \frac{\phi_i}{\sqrt{2}K} \simeq 0.8335 \phi_i.$$

This result can easily be inverted to give $\phi_i = \rho_{1i}/0.8335$. Then a value θ_i is picked from the conditional PDF $f(\theta|\phi_i)$ as follows:

$$\rho_{2i} = \int_{\pi/4}^{\theta_i} f(\theta|\phi) \, d\theta = \sqrt{2} \int_{\pi/4}^{\theta_i} \sin \theta \, d\theta = 1 - \sqrt{2} \cos \theta_i,$$

whose solution, after replacing $(1 - \rho_{2i})$ by ρ_{2i}, is found to be $\theta_i = \cos^{-1}(\rho_{2i}/\sqrt{2})$.

Finding the "best" algorithm for a given distribution is not easy. Simple algorithms can be programmed easily, but often the best algorithms require many lines of code and have multiple variate subranges, break points, empirical constants, and different sets of mixing functions. In Appendix A, some simple methods are provided for calculating random variates from commonly encountered distributions. The reader is referred to McGrath et al. (1975), Devroye (1986), Gentle (1998), and Fishman (2003) for more thorough treatments of optimum algorithms for sampling from particular distributions.

4.2　Scoring

In MC calculations, the results, or *scores*, of a large number N of histories are used to obtain an estimate of some random variable $z(x)$, which may or may not be known explicitly a priori. By generating values x_i of the random variable x from its governing PDF $f(x)$, the estimate of $\langle z \rangle$ is found from

$$\langle z \rangle \equiv \int_a^b z(x) f(x) dx \simeq \bar{z} \equiv \frac{1}{N} \sum_{i=1}^N z(x_i). \tag{4.26}$$

Before any histories are simulated, various *accumulators* are set to zero. Then, after each history, the accumulators are incremented by the value produced by each history for the quantity being accumulated. For example, to find $\bar{z}$, an accumulator $\mathcal{S}_1$ is used to sum the values of $z(x_i)$ produced by each history, i.e., after each history the value of $z(x_i)$ is added to $\mathcal{S}_1$. After N histories, the estimate of $\langle z \rangle$ is given by $\bar{z} = \mathcal{S}_1/N$.

Besides computing the sample mean $\bar{z}$ from the accumulated value of $\mathcal{S}_1$, other statistics of the histories are also accumulated throughout the MC calculation. For example, an accumulator $\mathcal{S}_2$ sums the $z^2(x_i)$ for each history so that $\overline{z^2} \equiv (1/N) \sum_{i=1}^N z^2(x_i) = \mathcal{S}_2/N$. With this result and Eq. (2.60), an estimate of the standard deviation of the sample mean is obtained, namely,

$$s(\bar{z}) \simeq \frac{s(z)}{\sqrt{N}} = \frac{1}{\sqrt{N}} \sqrt{\frac{N}{N-1}(\overline{z^2} - \bar{z}^2)} = \frac{1}{\sqrt{N-1}} \sqrt{\overline{z^2} - \bar{z}^2} \tag{4.27}$$

$$= \frac{1}{N\sqrt{N-1}} \sqrt{N\mathcal{S}_2 - \mathcal{S}_1^2}. \tag{4.28}$$

Because $s(\bar{z})$ is proportional to $1/\sqrt{N-1} \simeq 1/\sqrt{N}$ for large N, four times the number of histories must be used to halve the standard deviation on the estimate of $\bar{z}$. This is an inherent weakness of the MC method for, in order to obtain an accurate value of $\bar{z}$, one must reduce its standard deviation to be acceptably small. Another approach to reduce $s(\bar{z}) = s(z)/\sqrt{N}$ is to make

$s(z)$ smaller, thereby reducing the spread of the tally results. Such reduction of $s(z)$ can be accomplished by using various *variance reduction techniques* that modify the MC simulation. Some of these techniques are discussed in Chapter 5.

Example 4.9 Estimation of a Definite Integral

Consider the definite integral

$$Q = \int_0^1 x^m (\ln(1/x))^m dx = \frac{\Gamma(n+1)}{(m+1)^{n+1}}, \quad m > -1, \quad n > -1$$

for $m = 1$ and $n = 2$. Because $\Gamma(3) = 2! = 2$, the value of Q is 0.25. Give MC estimates for $\overline{Q}$ and $s(\overline{Q})$.

A simple way to proceed is to pick uniform samples from the unit interval so that $x_i = \rho_i$ and score each history with $z_i = x_i(\ln(1/x_i))^2$. Use accumulators $S_1 = \sum_{i=1}^N z_i$ and $S_2 = \sum_{i=1}^N z_i^2$. Then,

$$\overline{Q} = S_1/N \text{ and } s(\overline{Q}) = [S_2/N - \overline{Q}^2/(N-1)]^{1/2}.$$

The results shown below were obtained.

N	$\overline{Q}$	$s(\overline{Q})$
100	0.2503	0.0189
10,000	0.2498	0.0019
1,000,000	0.2500	0.0002

Note that as the number of histories increase by a factor of 100, the standard deviation is reduced by a factor of approximately 10, as expected.

4.2.1   Use of Weights in Scoring

In many problems most histories contribute little, if any, to $\overline{z}$. For example, in the Buffon's needle drop problem, if the separation distance D between lines is much greater than the needle length L, a needle whose either end is a distance more than L from the nearest line has no chance to fall on the line. To continue the history by calculating the coordinates of the other end and then to determine if this end is on the other side of a line (as was done in Example 1.3) is a waste of computer time. Similarly, if the integral $I = \int_a^b g(x)dx$ has an integrand with highly peaked but narrow features, then uniform sampling over the interval (a, b) produces many integrand evaluations that contribute little to the value of I.

The great power of MC is that many methods have been developed over the past half-century to greatly enhance the number of histories that produce a high score and to minimize histories that contribute little to $\bar{z}$. These methods, therefore, *bias* the MC calculation to produce histories that yield a large score or contribution to $\bar{z}$. Of course, the resulting value of $\bar{z}$ would be totally different from a value of $\bar{z}$ that would have been obtained without bias, and hence would be a very bad estimate of $\langle z \rangle$. To correct the score or *tally* of such biased histories to give an unbiased value of $\bar{z}$, each history is assigned a *weight* W, which may change throughout the history to compensate for the biases introduced at various steps of the history.

For example, if, in estimating $\langle z \rangle = \int_a^b z(x) f(x) dx$ by MC, some subintervals of $[a, b]$, in which $z(x)$ is highly peaked, were sampled 10 times as frequently as would be done if x_i were truly sampled from $f(x)$, then the weight of such biased values of x_i should be $W_i = 1/10$. The weights for x_i outside this biased subinterval are $W_i = 1$. In this way, the score for the ith history is $\zeta_i = W_i z(x_i)$ and an unbiased tally is

$$\bar{z} = \frac{1}{N} \sum_{i=1}^{N} W_i z(x_i) = \frac{1}{N} \sum_{i=1}^{N} \zeta_i. \tag{4.29}$$

Similarly, in Buffon's needle problem with widely spaced lines, one could drop the end of needles only within a distance L of a line, but then reduce the weight of such needles to account for the fact that needle ends should be uniformly dropped over $0 < x < D$. This biased approach to Buffon's problem is illustrated in the example below.

Example 4.10 A Biased Buffon's Needle Simulation

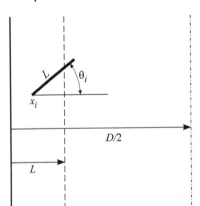

In this example, a biased Buffon's needle simulation is considered in which the grid separation distance D is much greater than the length L of the needle. In an analog simulation, the needle end is dropped uniformly over the interval $(0, D)$. However, needle ends falling in $(L, D - L)$ cannot overlap a grid line, resulting in many wasted simulated needle drops. Problem symmetry can be used to simplify the simulation (see adjacent). Clearly, the probability P_{cut} that a needle whose one end lands in $(0, D/2)$ lands on a grid line is the same as that for a needle whose end lands in $(D/2, D)$. Moreover, needle ends dropped in $(L, D/2)$ cannot cut the grid line at $x = 0$. Thus, in this biased simulation, needle ends are dropped uniformly over the subinterval $(0, L)$. However, the probability an end would actually land in $x_i \in (0, L)$ in an analog simulation is $L/(D/2)$. So the weight of these dropped needles is set to $W_i = L/(D/2)$ to remove the bias in this sampling. In the next segment of a needle's analog history, an angle θ_i of the needle with respect to the x-axis is uniformly sampled over $(0, 2\pi)$. But, only a needle whose θ_i angle is in the range $(\pi/2, 3\pi/2)$ can possibly intersect the grid line at $x = 0$. To eliminate needle drops that do not intersect the grid line, the selection of θ_i is biased to occur only in $(\pi/2, 3\pi/2)$, thereby sampling this range twice as often as would be done in an analog simulation. Accordingly, to keep the score unbiased, the weight of the needle W_i is halved. In the next segment of a needle's history, the x-coordinate of the needle's other end is calculated as $x_i^o = x_i + L\cos\theta_i$, and if $x_i^o < 0$ the simulated needle drop has crossed a grid line. Its contribution to the tally is recorded as W_i (here $z_i = 1$ if the grid line is intersected and 0 if not). In Fig. 4.7, results for the analog and biased (nonanalog) simulations are shown. Clearly, biased simulations, in which the targeted outcome is frequent, require much less computational effort.

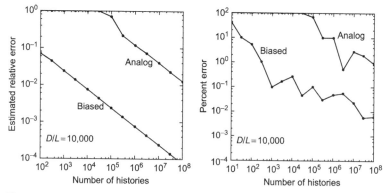

Figure 4.7 The left-hand plot shows the estimated relative error $R = s(\bar{z})/\bar{z}$ for the biased and analog simulations. For $D/L = 10,000$ used in this example, Eq. (1.1) gives $P_{cut} = 2L/(\pi D) \simeq 6.36620 \times 10^{-5}$. The right-hand figure shows the actual percent error for the two Monte Carlo estimates.

4.2.2   Scoring for "Successes-Over-Trials" Simulation

The simplest scoring is for problems in which each history results in a "success," $z(x_i) = 1$, or a "failure," $z(x_i) = 0$. In such an MC problem, the accumulator S_1 is the number of successes and the number of histories N is the number of trials. Thus, an estimate of the probability of success is S_1/N, i.e., the number of successes divided by the number of trials and, hence, the adjective *successes-over-trials*. Buffon's needle problem, used in many examples in this book, is such a success-over-trials problem. Another example of a success-over-trials problem is a simulation of the popular game of craps (see Example 4.11).

Example 4.11 Craps as a Success-Over-Trials Game

Consider the roll of a pair of dice in the game of craps. What is the probability of obtaining a winning "seven"? If $\mathbf{x} \equiv (J, K)$, $J, K = 1, \ldots, 6$, are the numbers of dots on each die, then a winning seven is obtained from the six combinations $(1, 6)$, $(2, 5)$, $(3, 4)$, $(4, 3)$, $(5, 2)$, $(6, 1)$. Of the possible $6 \times 6 = 36$ possible outcomes, the probability of a seven is $P_7 = 6/36 \simeq 0.1666667$.

In an MC analysis, the outcomes J and K are sampled from the discrete PDF $f_i = 1/6$, $i = 1, \ldots, 6$. From Section 4.1.2, the sampled value of J is set to

$$J = n \quad \text{if} \quad (n-1)(1/6) \le \rho_i < n(1/6), \quad n = 1, \ldots, 6.$$

A similar procedure is used to select a value of K. Then, the *score* of a simulated roll (*history*) is the success of the roll, i.e.,

$$z(\mathbf{x}_i) = \begin{cases} 1 & \text{if } J + K = 7, \\ 0 & \text{otherwise.} \end{cases}$$

After N simulated dice rolls, the composite score of successes $S_1 = \sum_{i=1}^{N} z(\mathbf{x}_i)$ is formed. From Eq. (4.28), the estimate of the probability of rolling a seven is $\overline{P}_7 \simeq S_1/N$ and the estimate of the standard deviation of this probability is, in the limit of large N,

$$\sigma(\overline{z}) \simeq s(\overline{P}_7) \equiv \frac{1}{\sqrt{N}} \sqrt{\frac{S_2}{N} - \overline{P}_7^2}.$$

From such an MC analysis, one can loosely say that the probability of rolling a seven, on any individual roll of the dice, is $\overline{P}_7 \pm s(\overline{P}_7)$, whereas, of course, what is really meant is that the interval $[\overline{P}_7 - s(\overline{P}_7), \overline{P}_7 + s(\overline{P}_7)]$ has a 68% chance of containing the true probability of rolling a seven on a throw of the dice. Results of such a simulation are shown below.

N	$\overline{P}_7$	$s(\overline{P}_7)$
10	0.2	0.1
100	0.21	0.04
1000	0.156	0.011
10,000	0.1626	0.0037
100,000	0.1652	0.0012

4.2.3 Scoring for Multidimensional Integrals

An interesting question arises in relation to multiple integrals. Consider a multidimensional integral of the form

$$\langle z \rangle = \int_V z(\mathbf{x}) f(\mathbf{x}) \, d\mathbf{x}, \tag{4.30}$$

where f is a joint PDF of, say, n stochastic variables $\mathbf{x} = \{x_j, j = 1, 2, \ldots, n\}$. In principle, it is possible to construct an MC estimate of the form

$$\bar{z}_a = \frac{1}{N_1} \sum_{i_1=1}^{N_1} \frac{1}{N_2} \sum_{i_2=1}^{N_2} \cdots \frac{1}{N_n} \sum_{i_n} z(x_{1,i}, x_{2,i}, \ldots, x_{n,i}), \qquad (4.31)$$

where $x_{j,i}$ is the ith sample of the jth component of $\mathbf{x}$ and these are sampled from the appropriate marginal and conditional PDFs, as discussed in Section 4.1.11. This prescription samples $\prod_{i=1}^{n} N_i$ histories, but only N_1 values of x_1 and for each of these N_2 values of x_2, and so forth. A simpler and more efficient MC estimator takes the form

$$\bar{z}_b = \frac{1}{N} \sum_{i=1}^{N} z(\mathbf{x}_i), \qquad (4.32)$$

which involves N histories each using n samples to form each $\mathbf{x}_i$. This latter approach is more efficient because each component of $\mathbf{x}$ is sampled in each history, providing better coverage of the sample space.

4.2.4 Statistical Tests to Assess Results

To assess how good an estimate the mean $\bar{z}$ is of the expected value $\langle z \rangle$ several statistical measures should also be calculated to provide confidence in the result. For example, the standard deviation of the mean $s(\bar{z})$ is one such statistic. Ideally, a good result for $\bar{z}$ should have a small $s(\bar{z})$. However, how small is small? To provide confidence in the precision of an MC result, several other statistics generated from the histories should be computed and examined carefully. The analysis of such statistical measures of an MC calculation is very important to ensure that a precise estimate of the expected value has been obtained. Sometimes, through bad luck (i.e., an incomplete sampling of all possible histories), a very imprecise result is obtained. The statistical analysis of an MC calculation is a very critical aspect of the analysis. Several important MC statistics are discussed below.

Relative Error and Figure of Merit
The *relative error* is defined as $R \equiv s(\bar{z})/\bar{z}$ and is a useful number because it gives a measure of the statistical precision of $\bar{z}$ as a fractional result with respect to the estimated mean. From Eqs. (4.26) and (4.27)

$$R = \frac{1}{\sqrt{N-1}} \frac{\sqrt{\bar{z^2} - \bar{z}^2}}{\bar{z}} = \frac{1}{\sqrt{N-1}} \sqrt{\frac{N}{N-1} \frac{S_2}{S_1^2} - \frac{1}{N-1}}. \qquad (4.33)$$

For large N, $\sqrt{N-1} \simeq \sqrt{N}$ and the above simplifies to

$$R = \sqrt{\frac{S_2}{S_1^2} - \frac{1}{N}}. \tag{4.34}$$

Some important observations. First, if all $z_i \equiv z(x_i)$ are equal, then $R = 0$, i.e., low-variance solutions minimize the spread of z_i. Second, if all $z_i = 0$, R is defined as 0. If there is only one nonzero z_i, then $R \to 1$ as $N \to \infty$. Finally, if z_i all have the same sign, then $\overline{z^2} > \overline{z}^2$ so that $s(\overline{z}) < \overline{z}$ and $R < 1$. Thus, R, for z_i of the same sign, is always between zero and unity. For cases with z_i of both signs, this is not true. The smaller R, the better is the precision of $\overline{z}$. Typically, a good result should have $R \leq 0.05$ (MCNP, 2003).

Another useful statistic for an MC calculation is the *figure of merit* (FOM), defined as FOM $\equiv 1/(R^2T)$, where T is the run-time needed to produce N histories. Because T is proportional to N and R^2 is proportional to $1/N$ (see Eq. (4.33)), the FOM should be approximately constant throughout an MC run, save for statistical fluctuations early in the run. This statistic has several important uses. First, if the FOM is not roughly constant (except at the beginning of a run), the estimated confidence limits for $\overline{z}$ may not overlap the expected value $\langle z \rangle$. Second, short runs with difference methods for estimating $\overline{z}$, for example, with different variance reduction techniques, allow a quick determination of which is the best method, namely, the one that produces, for the same computing time, the largest FOM (smallest R). Finally, the computing time needed to achieve a specified relative error can be estimated as $T \simeq 1/(R^2\text{FOM})$.

Variance of the Variance

Higher moments $z^n(x_i)$ are also often accumulated in order to find other statistical measures of precision of MC results. One such statistic is the variance of the variance (VOV) (Estes and Cashwell, 1978; Pederson, 1991). The relative VOV is like the relative error R of the mean, except that it is a measure of the variance of R. Specifically, it is defined as (Pederson, 1991)

$$\text{VOV} = \frac{s^4(\overline{z})}{s^2(s^2(\overline{z}))} = \frac{\sum_{i=1}^{N}(z_i - \overline{z})^4}{\left[\sum_{i=1}^{N}(z_i - \overline{z})^2\right]^2} - \frac{1}{N}. \tag{4.35}$$

To compute this statistic, four accumulators are needed: $\mathcal{S}_m = \sum_{i=1}^{N} z_i^m$, $m = 1, 2, 3, 4$. Then expansion of Eq. (4.35) in terms of the sums of powers of z_i gives

$$\text{VOV} = \frac{S_4 - 4S_1S_3/N + 6S_2S_1^2/N^2 - 3S_1^4/N^3}{\left[S_2 - S_1^2/N\right]^2} - \frac{1}{N}$$

$$= \frac{S_4 - 4S_1S_3/N + 8S_2S_1^2/N^2 - 4S_1^4/N^3 - S_2^2/N}{\left[S_2 - S_1^2/N\right]^2}. \tag{4.36}$$

From many statistical tests (MCNP, 2003), it has been found that the VOV should be less than 0.1 to increase the chance that reliable confidence intervals are obtained. Moreover, multiplication of the numerator and denominator of Eq. (4.36) by $1/N$ shows that the VOV should decrease monotonically as $1/N$.

Estimating the Underlying PDF

In addition to calculating various statistics from the histories, the scores of each history are usually binned according to their values in order to estimate the (usually unknown) underlying PDF $f(z)$. From this estimate of $f(z)$, an important statistical test can be made about the reliability of the estimated PDF to quantify how well the N histories have sampled very rare but very important high scoring events. The central limit theorem requires the first two moments of $f(z)$ to exist, so that if proper sampling of high scoring events has occurred, the largest values of z_i should reach an upper bound (if one exists) or should decrease faster than $1/z^3$ so that $\langle z^2 \rangle = \int_{-\infty}^{\infty} z^2 f(z) dz$ exists. Thus, if an MC analysis has properly sampled high-z histories, the high-z portion of the empirical distribution $f(z)$ should have a well-defined upper bound or decrease at least as rapidly as $1/z^3$. For many difficult simulations, this requirement on $f(z)$ is the hardest to satisfy.

4.3   Accuracy and Precision

In common parlance, the terms *precision* and *accuracy* are often used rather loosely. However, there is an important distinction between the two. In MC, the *precision* of the calculation is limited by the uncertainty in $\bar{z}$ caused by the statistical fluctuations in z_i and is measured by the standard deviation $s(\bar{z})$. As more histories are run, $s(\bar{z}) \simeq s/\sqrt{N}$ becomes smaller and a more precise estimate is obtained for $\bar{z}$. By contrast, the *accuracy* of $\langle z \rangle \simeq \bar{z}$ in MC is how far $\bar{z}$ is from the physical quantity being estimated. The difference between the true physical value $\langle z \rangle$ and the estimate $\bar{z}$, in the limit as $N \to \infty$, is the *systematic error*, an error that is rarely known. It is quite possible to obtain a very precise MC result (by running many histories) that is not very accurate, perhaps because approximations or errors were made in the models of the physical phenomenon being studied.

The same concepts apply to measurement theory and analysis of physical systems. If a given physical quantity is measured several times, by multiple people, or even by the same person employing the same instrumentation, a range of values are usually obtained. For instance, a class of third graders each measuring the height of a fence with rulers are liable to get similar but different numbers. The precision of their measurements is measured by how far the individual measurements differ from the average. The accuracy of their measurement is a measure of how far their average is from the true height of the fence and is determined by how good their rulers are and how good their measurement technique is. Similarly, a good marksman is capable of both high accuracy (the average of the bullet pattern is close to the bulls eye) and high precision (the spread in the pattern is small). However, if someone misaligns the gun's sights without the marksman's knowledge (and the marksman does not check his result after each firing), the result may still be precise but is less accurate.

Because MC is, in essence, an experiment with numbers used to simulate a real experiment, it is subject to the imprecision of actual experiments. Just as scientists trying to understand nature have used crude instruments and made grossly inaccurate measurements, which are usually improved as methods evolve, MC using poor models can lead to inaccurate results. In these cases, it is customary to interpret MC results by their precision, as measured by the sample standard deviation, and say these results are accurate to within the assumptions made in the simulation model. These concepts are explored in more detail below.

4.3.1 Factors Affecting Accuracy

The accuracy of an MC analysis can be compromised by two main factors: (1) code errors and (2) user errors. The computer code may have large uncertainties or errors in the physical constants used in the physical models. Likewise, the physical models used in the analysis may be too simplistic or even wrong. A poor random number generator may also introduce errors. Finally, there may be coding errors (bugs) that, for instance, can occasionally produce an incorrect high score for an unusual history. These are factors that can only be remedied by benchmarking results against measurements, theory, or other independent calculations to find and identify any source of disagreement.

User Errors

Accuracy can be affected by user errors. For instance, a user may misuse a previously developed MC code. An incomplete understanding of the code's features and options allows users, particularly beginners, to make mistakes, resulting in flawed output. Simple errors in an input file can be made. A mistyped density or an incorrect unit (e.g., cm for m) can produce results that, while apparently reasonable, are inaccurate. To avoid this source of inaccuracies, careful review of the input and output is required. For complex problems it is often useful to develop the complex model by first using simpler models upon which to build the complex model. At each stage in the model development, the changes in the output should be plausible from the changes made to the model.

Random Number Generators

Many early MC calculations used pseudorandom number generators that, in hindsight, were far from ideal. The resulting inaccuracies introduced into the results by such generators are largely unknown. However, with modern random number generators, the compromise of accuracy due to the inadequacies of the generators has largely been eliminated. But one potential danger still exists. The

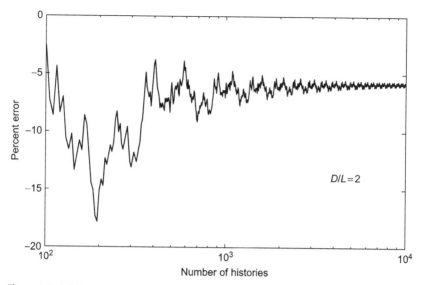

Figure 4.8 A false convergence in a simulation on Buffon's needle problem. Between each simulated needle drop using the method described in Example 1.4, a large number of random numbers were discarded so that the cycle length of the random number generator was exceeded.

Achilles heel of all pseudorandom number generators is that they have a finite cycle length, i.e., the same sequence of random numbers eventually repeats. If a Monte Carlo calculation uses many more random numbers than the cycle length of the generator, then inaccuracies are introduced by using the same sequence of random numbers multiple times.

The repeated use of the same subsequence of random numbers can lead to *false convergence*. In Fig. 4.8, results of the Buffon's needle simulation used in Example 1.4 are shown for the case $D = 2L$. However, in this simulation a great many random numbers were discarded between needle drops so that after about 500 simulated needle drops, the cycle length of the random number generator was exceeded. Even though the estimate of P_{cut} appears to converge, it converges to a value about 5% below the true value of $P_{cut} \simeq 0.318310$. This false convergence is easily identified by the same repeated pattern (with decreasing amplitude) produced by each cycle of the random number generator. Fortunately it is easy to avoid such false convergences by ensuring that the cycle length of your random number generator is never exceeded. A good rule of thumb is never use more than 10–20% of the random numbers in a cycle.

4.3.2 Factors Affecting Precision

The factors affecting the precision of an MC result mostly arise from user-made choices for the calculation. Different ways of scoring a tally can affect the precision of the tally. Similarly, the use of variance reduction techniques (see

Chapter 5) can profoundly affect the precision of a tally, as does use of problem symmetry to simplify the MC model. Finally, the precision of a tally is proportional to $1/\sqrt{N}$. Quadrupling the number of histories increases the precision by a factor of approximately two. However, relying on N to increase precision is a brute-force approach. It is often preferred to try to reduce $s(\overline{z})$, i.e., the spread among the scores, by methods discussed in the next chapter.

Round-Off and Overflow

In principle, any arbitrary precision can be obtained in a score if a sufficient number of histories are run. However, it turns out that round-off or truncation errors associated with representing real numbers on digital computers, among other things, limit the precision. Round-off and truncation are similar in that a rational number is limited in some fashion to a finite number of digits. We will use round-off as a term that applies to either method. In a perfect world, all calculations would have perfect precision. However, digital computers use finite collections of bits to represent numbers. The number 2/3 must be represented in a digital computer by a finite number of digits (either rounded to 0.666...7 or truncated to 0.666...6), thus introducing a small error. No matter how carefully an MC simulation is conducted, it can never reconstruct the value 2/3 exactly.

Overflow is even worse. In all digital computers, no matter what finite precision is used, eventually there is a point at which an accumulator S can get so large that $S + \epsilon < S$ no matter how small is ϵ. Once an accumulator has reached the maximum value it can represent, adding any score to the accumulator gives a smaller number. Eventually, after a large number of histories, S can become so large that adding a small score no longer increases the value of S. As examples, consider the two summations $\mathcal{S}_1 = \sum_{i=1}^{N}(1/N)$ and $\mathcal{S}_2 = (1/N)\sum_{i=1}^{N} 2\rho_i$, where ρ_i is uniformly sampled on $(0, 1)$. Clearly $\mathcal{S}_1 = 1$, no matter the value of N and, because $\langle \rho_i \rangle = 1/2$, $\mathcal{S}_2$ should converge to unity as $N \to \infty$. Results for these two summations are shown in Table 4.1 for single precision values of $\mathcal{S}_1$ and $\mathcal{S}_2$ on a 32-bit machine. In these examples, the effects of round-off errors become increasingly severe as N exceeds 10^7.

Because every history score is limited by the computer's ability to represent real numbers, round-off errors accumulate. In many MC calculations, in which summation accumulators are used, the estimate $\overline{z}$ may appear to be converging to $\langle z \rangle$ as N increases. But often, as N becomes very large, $\overline{z}$ begins to decrease or $s(\overline{z})$ stops decreasing as expected. This is a sure sign that round-off and/or overflow errors have become important and that using more histories will not produce increasingly precise results. If such large numbers of histories are to be used, extended precision arithmetic must be used for the summation accumulators, and even then there is a limit to the attainable precision. The use of double precision is always preferable to using single precision, unless it inordinately increases the computation time. Whether or not double precision increases, the compute time depends on the architecture of the computer.

Table 4.1 Results of using single precision on a 32-bit machine to estimate $\mathcal{S}_1$ and $\mathcal{S}_2$, both of whose expected values are unity. After about 10^7 histories, the round-off error becomes evident.

N	$\mathcal{S}_1 = \sum_{i=1}^{N}(1/N)$	$\mathcal{S}_2 = (1/N)\sum_{i=1}^{N} 2\rho_i$
10	1.0000000	1.1705571
30	1.0000000	0.9968364
1×10^2	1.0000000	0.9874166
3×10^2	1.0000000	1.0246525
1×10^3	1.0000000	0.9853277
3×10^3	1.0000000	1.0023125
1×10^4	1.0000000	0.9994981
3×10^4	1.0000000	1.0018269
1×10^5	1.0000000	0.9994974
3×10^5	1.0000000	1.0000912
1×10^6	1.0000000	1.0002490
3×10^6	1.0000000	0.9999786
1×10^7	1.0000000	1.0001853
3×10^7	0.5592405	0.9998386
1×10^8	0.1677722	0.3355443
3×10^8	0.0559241	0.1118481
1×10^9	0.0167772	0.0335544

Another way to mitigate round-off errors in situations in which tally scores can vary widely in magnitude is to accumulate the largest scores in separate accumulation bins and to accumulate the sum of only small scores in a single variable. Then $\bar{z}$ is computed by adding all the scores, from smallest to largest, so that small scores are rounded correctly when the largest scores are added. The binning of the largest tallies has an additional benefit. From these binned tallies, a histogram is obtained of the high-end tail of the underlying PDF $f(z)$. Several statistical tests can then be performed to ensure that the high-end tail has been sufficiently sampled.

4.3.3 The Use of Batches

When overflow is a concern, MC calculations can be performed in batches. Instead of running N independent histories, the histories are run in K batches, each batch using a different substring of $N_k = N/K$, $k = 1, 2, \ldots, K$, histories, where the value of N_k avoids round-off or overflow errors within batches. The batch estimates

$$\bar{z}_k = \frac{1}{N_k} \sum_{i=1}^{N_k} z_i, \; k = 1, 2, \dots, K, \tag{4.37}$$

are then averaged to obtain the desired result, i.e.,

$$\bar{z} = \frac{1}{K} \sum_{k=1}^{K} \bar{z}_k. \tag{4.38}$$

The proper way to estimate $s(\bar{z})$, the standard deviation of the estimate of $\bar{z}$, is to estimate $s(\bar{z}_k)$ along with $\bar{z}_k$ during each batch. Then, by propagation of errors,

$$s_1(\bar{z}) = \left[\frac{1}{K} \sum_{k}^{K} s^2(\bar{z}_k) \right]^{1/2}. \tag{4.39}$$

For $K > 1$, one could treat each $\bar{z}_k$ as an independent sample and estimate $\sigma(\bar{z})$ directly from

$$s_2(\bar{z}) \simeq \frac{1}{\sqrt{K}} \left[\frac{1}{K-1} \sum_{k=1}^{K_k} (\bar{z}_k - \bar{z})^2 \right]^{1/2}. \tag{4.40}$$

For large N, these two estimators give similar results, as is shown in the example below. The use of batches is not common for simple integrals. However, for some integrals in which the integrand changes rapidly over the integration range, the smaller contributions to the accumulator sum terms may be swamped by the larger terms, due to round-off or truncation. Alternatively, the number of histories can grow so large that one or more of the accumulators overflow. The use of batches can improve the accuracy and precision of the result. The next example illustrates the use of batches in estimating an integral.

Example 4.12 The Use of Batches

Consider the definite integral

$$Q = \int_0^{\pi/2} \frac{dx}{[a^2 \sin^2 x + b^2 \cos^2 x]^2} = \frac{\pi(a^2 + b^2)}{4a^3 b^3}, \quad a, b > 0.$$

For $a = 1$ and $b = 2$, $Q = 0.490\,873\,852 \dots$. An MC estimate of Q using K batches can be written as

$$\bar{Q} = \frac{1}{K} \sum_{k=1}^{K} \frac{1}{N_k} \sum_{i=1}^{N_k} \frac{1}{[\sin^2(x_i) + 4\cos^2(x_i)]^2},$$

where $x_i = \rho_i \frac{\pi}{2}$. The following results, for both single- and double-precision arithmetic, were obtained using this procedure.

K	N_K	$\overline{Q}$	$s_1(\overline{Q})$	$s_2(\overline{Q})$
		Single precision		
1	10^6	0.490477	4.72×10^{-4}	–
10	10^5	0.491299	4.72×10^{-4}	1.96×10^{-4}
100	10^4	0.490430	4.72×10^{-3}	4.75×10^{-4}
1	10^8	0.263548	5.9×10^{-5}	–
100	10^6	0.490561	4.7×10^{-5}	4.4×10^{-5}
1000	10^5	0.490869	4.7×10^{-5}	4.7×10^{-5}
		Double precision		
1	10^6	0.490775	4.72×10^{-4}	–
10	10^5	0.491308	4.72×10^{-4}	1.97×10^{-4}
100	10^4	0.490430	4.72×10^{-4}	4.75×10^{-4}
1	10^8	0.490764	4.7×10^{-5}	–
100	10^6	0.490878	4.7×10^{-5}	4.4×10^{-5}
1000	10^5	0.490878	4.7×10^{-5}	4.7×10^{-5}

It is apparent that, with single-precision arithmetic, overflow occurred between $N = 10^6$ and $N = 10^8$ histories, for both $\overline{Q}$ and $s_1(\overline{Q})$. The use of 1000 batches of 10^5 histories gave four-significant-figure agreement for both single- and double-precision arithmetic. It is interesting to note that the value of $s_2(\overline{Q})$ actually increases slightly as more batches are used for both single- and double-precision cases. This occurs because there are fewer histories per batch and the discrepancies among $\overline{z}_k$ and $\overline{z}$ actually increase slightly.

4.4 Summary

MC simulations are performed by repeatedly sampling from probability distributions and scoring the results to form average quantities. The sampling process generally begins by generating pseudorandom numbers uniformly on the unit interval (i.e., between 0 and 1). A sample ρ_i from the unit interval is then transformed to a random sample over the range of interest and distributed according to the relevant probability distribution. When possible, this is usually achieved by what is called straightforward sampling, which involves inverting the appropriate CDF, namely,

$$x_i = G^{-1}(\rho_i). \tag{4.41}$$

When this inversion is difficult or impossible, other methods, such as rejection sampling or the composition method, can be used.

For a joint probability distribution, the PDF is decomposed into a product of marginal and conditional PDFs, e.g.,

$$f(x, y, z) = f_1(x|y, z) f_2(y|z) f_3(z). \tag{4.42}$$

First the marginal PDF is sampled, in this case $f_3(z)$, and then the other variables are sampled from their conditional distributions, using the previously sampled values, for example, y from $f_2(y|z)$ given the sample for z and then sample x from $f_3(x|y, z)$ given the samples for y and z.

The scoring process is simply bookkeeping. To account for nonanalog or biased sampling, each sample is assigned a weight W_i that corrects for any bias. The quantity

$$\zeta_i = W_i z(x_i) \tag{4.43}$$

is scored and accumulators for ζ_i and ζ_i^2 are incremented. After all histories have been completed, the averages

$$\bar{z} = \frac{1}{N} \sum_{i=1}^{N} \zeta_i \quad \text{and} \quad \overline{z^2} = \frac{1}{N} \sum_{i=1}^{N} \zeta_i^2 \tag{4.44}$$

are calculated. The quantity $\bar{z}$ is the sample mean, the estimate of $\langle z \rangle$. The precision of this result is given by the standard deviation of the sample mean

$$s(\bar{z}) = \frac{s(z)}{\sqrt{N}} = \frac{1}{\sqrt{N-1}} \sqrt{\overline{z^2} - \bar{z}^2}. \tag{4.45}$$

Other statistics, such as the VOV, skewness, kurtosis, etc., in addition to $s(\bar{z})$, are often used to give further guidance in assessing the reliability of $\bar{z}$ as an estimate of $\langle z \rangle$.

Of course, most actual MC simulations involve more complex integrands and PDFs of several random variables. But the general procedure just outlined is followed, with appropriate modifications to account for all of the variables.

Whereas the strong law and the central limit theorem indicate that, in principle, a result can be obtained to any desired precision, in fact the precision is limited by the number of bits used to represent numbers in the computer. For this reason, it is generally advisable to employ extended-precision arithmetic in MC calculations on digital computers whenever feasible. Although MC cannot produce results to arbitrary precision, the good news is that it can often produce results with several significant figures of precision with modest effort.

Problems

4.1 Consider a unit square and insert a circle whose center is coincident with the center of the square and whose radius is 0.5 units. Let A_c be the area of the circle and let A_s be the area of the square. Their ratio is given by

$$\frac{A_c}{A_s} = \frac{0.5^2\pi}{1} \text{ or } \pi = 4\frac{A_c}{A_s}.$$

Thus, one can randomly sample points within the square and let the number N_s of successes be the number of points that are within the circle and the number N of trials be the number of points in the square. Use this approach to estimate the integral above and estimate π for $N = 10^2, 10^4$, and 10^6 histories. Give estimates of the standard deviations for these estimates.

4.2 Show that for the discrete distribution with M mass points (i.e., number of f_i values), all with equal probability, the inverse CDF method to generating random variates x reduces to $x = \lceil \rho M \rceil$.[5]

4.3 In principle, one could solve $\rho_i = F(x_i)$ numerically for x_i for distributions $F(x)$ whose inverse cannot be obtained analytically. Why is this not a practical method for generating random variates?

4.4 You wish to obtain random samples x_i from a distribution whose PDF is given by $g(x) = Cx^2$ defined on the interval $x \in (2, 3)$. (a) What is the value of the constant C needed to make $g(x)$ a proper PDF on the specified interval? (b) Divide the interval $(2, 3)$ into 20 equiwidth, contiguous subintervals and sort the samples obtained with the inverse CDF method into the appropriate subinterval or bin. Show graphically that for a large number of samples, the PDF is recovered as a histogram.

4.5 It can be shown that

$$\sqrt{\pi} = \int_0^\infty \frac{e^{-x}}{\sqrt{x}} dx.$$

Sample from the PDF $f(x) = e^{-x}$ and score with $\zeta_i = \frac{1}{\sqrt{x}}$ to estimate $\overline{\pi}$ and $s(\overline{\pi})$ for $N = 10^2, 10^4$, and 10^6 histories.

4.6 Consider a spherical shell with inner and outer radii a and b and the problem of how to find the radius r of points uniformly distributed within the shell. Show that the PDF is

$$f(r) = \frac{3r^2}{b^3 - a^3}, \qquad a \le r \le b,$$

and the CDF

$$F(r) = \frac{r^3 - a^3}{b^3 - a^3},$$

and that the random radii are given by $r_i = [a^3 + \rho_i(b^3 - a^3)]^{1/3}$.

---

[5] The ceiling function $\lceil y \rceil$ gives the smallest integer $\ge y$.

4.7 Devise a computer program to select radii for points uniformly distributed within a sphere of unit radius using the following two methods. (a) Show that the probability the radius is in dr about r is given by $p(r)\, dr = 3r^2\, dr$ and then use the simple rejection technique, with $M = 3$. (b) Embed the sphere in a cube with sides of length 2 and sample points uniformly in the cube and accept the point if it is inside the sphere. Which method is more efficient and why is it more efficient?

4.8 It is desired to generate the sine and cosine of a random angle ψ in the interval $(0, 2\pi)$ using a method attributed to von Neumann (1951). Show that this can be accomplished as follows: Select two random numbers, ρ_i and ρ_j, and set $\xi_i = 2\rho_i - 1$ and $\xi_j = 2\rho_j - 1$. Then, in those cases for which $\xi_i^2 + \xi_j^2 < 1$,

$$\cos \psi = \frac{\xi_i^2 - \xi_j^2}{\xi_i^2 + \xi_j^2} \quad \text{and} \quad \sin \psi = \frac{2\xi_i \xi_j}{\xi_i^2 + \xi_j^2}.$$

Show that the efficiency of this rejection method is $\pi/4$.

4.9 Consider the integral given by

$$Q = \int_0^{\pi/2} \frac{dx}{[a^2 \sin^2 x + b^2 \cos^2 x]^2} = \frac{\pi(a^2 + b^2)}{4a^3 b^3}, \quad a > 0, b > 0.$$

For $a = 1$ and $b = 2$, $Q = 0.490\,873\,852\ldots$. Obtain MC estimates of Q and its standard deviation for $N = 100$, $N = 10^4$, and $N = 10^6$ histories.

4.10 Verify by calculation the following rejection technique for sampling values of x distributed according to the Gaussian PDF (Kahn, 1954):

$$f(x) = \sqrt{2/\pi} \exp[-x^2/2], \quad x \geq 0.$$

(1) Select random numbers ρ_1 and ρ_2 and let $x = -\ln \rho_1$ and $y = -\ln \rho_2$. (2) Accept x if $0.5(x - 1)^2 \leq y$; otherwise, return to step 1. (3) Select random numbers ρ_3 and assign to x the negative sign if $\rho_3 \leq 0.5$. Verify that the efficiency is $\sqrt{\pi/2} \exp[-1/2] = 0.760\,173\,451\ldots$.

4.11 Reconsider the integral of Example 4.9, i.e.,

$$Q = \int_0^1 x^m (\ln(1/x))^m dx = \frac{\Gamma(n + 1)}{(m + 1)^{n+1}}, \quad m, n > -1,$$

but this time let $m = 3$ and $n = 1$. The known value of Q is $1/16 = 0.0625$. (a) Estimate $\overline{Q}$ and $s(\overline{Q})$ for $N = 10^6$ histories. (b) Compare estimates of $\overline{Q}$ and $s(\overline{Q})$ for 10^9 histories using one batch, 100 batches, and 10,000 batches.

4.12 Verify, by calculation, the following technique (Carter and Cashwell, 1975) for sampling energies of prompt fission neutrons given by the PDF

$$f(E) = \sqrt{4E/\pi T^3} \exp[-E/T],$$

where E is the neutron energy and T is a characteristic energy (both in units of MeV). Select three random numbers ρ_1 through ρ_3 and then set

$$E = T\left[-\ln\rho_1 - (\ln\rho_2)\cos^2(\pi\rho_3/2)\right].$$

4.13 Apply the rectangle-wedge-tail method to the problem of sampling from a unit exponential distribution $f(x) = e^{-x}$, $x \geq 0$. Use an equimesh such that $x_i - x_{i-1} = \delta$, so that $x_i = i\delta$, $i = 1, 2, \ldots, k$. Show that

$$A_i = \delta e^{-i\delta}, \qquad\qquad i = 1, 2, \ldots, k,$$
$$A_{k+i} = e^{-i\delta}[e^{\delta} - 1 - \delta], \quad i = 1, 2, \ldots, k,$$
$$A_{2k+1} = e^{-k\delta},$$

and that corresponding PDFs for each region are

$$f_i(x) = 1/\delta, \qquad\qquad x_{i-1} \leq x < x_i, \quad i = 1, 2, \ldots, k,$$
$$f_{k+i}(x) = \frac{e^{-x+i\delta} - 1}{[e^{\delta} - 1 - \delta]}, \qquad x_{i-1} \leq x < x_i, \quad i = 1, 2, \ldots, k,$$
$$f_{2k+1}(x) = e^{-x+k\delta}, \qquad\qquad x \geq x_k.$$

Write a program to implement sampling using this rectangle-wedge-tail method, and compare the speed of this sampling scheme to that based on the inverse CDF method (see Example 3.2).

4.14 Verify by calculation the following rejection technique for sampling energies of prompt fission neutrons according to the Cranberg formula, with PDF

$$f(E) = 0.453\exp[-E/0.965]\sinh\sqrt{2.29E},$$

where E is the neutron energy in MeV. (1) Select random numbers ρ_1 and ρ_2 and let $x = -\ln\rho_1$ and $y = -\ln\rho_2$. (2) If $-4.57257x + [y - 1.06918(x + 1)]^2 \leq 0$, set $E = 1.99657x$ MeV; otherwise, return to step 1. According to Kalos et al. (1968), the efficiency of this method is 0.74.

4.15 Suppose that the angular distribution of a neutron-scattering cross section is represented by the truncated Legendre polynomial series

$$f(\omega) = \frac{1}{2} + \frac{1}{2}\sum_{n=1}^{N}(2n + 1)f_n P_n(\omega).$$

Discuss the advantages and disadvantages of sampling ω using **(a)** the composition method and **(b)** the ratio of uniforms method.

4.16 Write a general code to estimate definite integrals of the form

$$Q = \int_0^{\pi} \frac{dx}{a + b\cos(x)} = \frac{\pi}{\sqrt{a^2 - b^2}}, \qquad a > 0, b \geq 0.$$

Consider the case $a = 2$ and $b = 1$, for which $Q = 1.81380$. (a) Estimate $\overline{Q}$ and $s(\overline{Q})$ for one batch of $N = 10^8$ histories. (b) Estimate $\overline{Q}$ and $s(\overline{Q})$ for 100 batches of $N = 10^6$ histories. (c) Estimate $\overline{Q}$ and $s(\overline{Q})$ for 1000 batches of $N = 10^5$ histories.

4.17 It can be shown that

$$\int_0^\infty \frac{e^{-x}}{\sqrt{x}} dx = \sqrt{\pi}.$$

The integral can be estimated by MC by sampling values of x from the PDF $f(x) = e^{-x}, x > 0$. An MC estimate of π thus can be expressed as

$$\overline{\pi} = \left[\frac{1}{N} \sum_{i=1}^N \frac{1}{\sqrt{x_i}} \right]^2.$$

Use this estimator to estimate $\overline{\pi}$ and $s(\overline{\pi})$ for $N = 10^2$, $N = 10^4$, and $N = 10^6$ histories.

4.18 Consider the discrete PDF given by $f_1 = 0.05$ for $x_1 = 1$, $f_2 = 0.10$ for $x_2 = 2$, $f_3 = 0.50$ for $x_3 = 3$, $f_4 = 0.30$ for $x_4 = 4$, and $f_5 = 0.05$ for $x_5 = 5$. Convert this to a PDF of the form $g_i = 0.20$, $i = 1, 2, \ldots, 5$, and give your alias table. Apply discrete alias sampling to generate $N = 10^6$ samples. Use these to recover the original PDF and give the standard deviations in your estimates of f_i, $i = 1, 2, \ldots, 5$.

4.19 Use the methods described in Appendix A to estimate the integral

$$Q = \int_{-\infty}^\infty \int_{-\infty}^\infty \frac{1}{xy} f(x, y) dy dx,$$

where f is the bivariate normal distribution given by

$$f(x, y) = \frac{1}{2\pi \sigma_1 \sigma_2 \sqrt{1 - \rho^2}} \exp\left[-\frac{\xi}{2(1 - \rho^2)} \right],$$

with

$$\xi = \frac{(x - \mu_1)^2}{\sigma_1^2} - \frac{2\rho(x - \mu_1)(y - \mu_2)}{\sigma_1 \sigma_2} + \frac{(y - \mu_2)^2}{\sigma_2^2},$$

$\mu_1 = 0$, $\sigma_1 = 2$, $\mu_2 = 3$, $\sigma_2 = 2$, and $\rho = 4$. To solve this problem, it is acceptable to ignore samples of $x_i < -10$ or $x_i > 10$ and samples of $y_i < -7$ or $y_i > 13$. Also estimate the standard deviation of your estimate of the integral.

References

Carter, L.L., Cashwell, E.D., 1975. Particle-Transport Simulation with the Monte Carlo Method. U.S. Energy Research and Development Administration. Available from National Technical Information Service, Springfield, VA.

Chen, H.C., Asau, Y., 1974. On generating random variates from an empirical distribution. AIIE Trans. 6, 163–166.

Devroye, L., 1986. Non-Uniform Random Variate Generation. Springer-Verlag, New York.

Edwards, A.L., Rathkopf, J.A., Smidt, R.K., 1991. Extending the alias Monte Carlo sampling method to general distributions. In: Am. Nucl. Soc. Intern. Topical Meeting. April 28 to May 1, 1991, Pittsburgh, PA. Available from https://www.osti.gov/servlets/purl/6023539, 1991.

Estes, G., Cashwell, E., 1978. MCNP1B Variance Error Estimator. TD-6-27-78. Los Alamos Nat. Lab., NM.

Fishman, G.S., 2003. Monte Carlo Concepts, Algorithms, and Applications. Corrected Printing. Springer, New York.

Gentle, J.E., 1998. Random Number Generation and Monte Carlo Methods. Springer, New York.

Kahn, H., 1954. Applications of Monte Carlo. AECU-3259. U.S. Atomic Energy Commission. Available from Technical Information Service Extension, Oak Ridge, TN.

Kalos, M.H., Nakache, F.R., Celnik, J., 1968. Monte Carlo methods in reactor computations. In: Greenspan, H., Kelber, C.N., Okrent, D. (Eds.), Computing Methods in Reactor Physics. Gordon and Breach, New York.

Kinderman, A.J., Monahan, J.F., 1977. Computer generation of random variables using the ratio of uniform deviates. ACM Trans. Math. Softw. 3, 257–260.

Kronmal, R.A., Peterson, A.V., 1979. On the alias method for generating random variables from a discrete distribution. Am. Stat. 33, 214–218.

Marsaglia, G., 1963. Generating discrete random variable in a computer. Commun. ACM 6, 37–38.

MacLauren, M.D., Marsaglia, G., Bray, T.A., 1964. A fast procedure for exponential random variables. Commun. ACM 7, 298–300.

McGrath, E.J., Irving, D.C., Basin, S.L., Burton, R.W., Jaquette, S.J., Kelter, W.R., Smith, C.A., 1975. Techniques for Efficient Monte Carlo Simulation, vol. II: Random Number Generation for Selected Probability Distributions. ORNL-RSIC-38. Radiation Shielding Information Center, Oak Ridge Nat. Lab., TN.

MCNP, 2003. A General Monte Carlo N-Particle Transport Code, Version 5, vol. I: Overview and Theory. LA-UR-03-1987. Los Alamos National Lab.

Norman, J.E., Cannon, L.E., 1972. A computer program for the generation of random variables from any discrete distribution. Comput. Simul. 1, 331–348.

Pederson, S.P., 1991. Mean Estimation in Highly Skewed Samples. Report LA-12114-MS. Los Alamos Nat. Lab., NM.

Smith, J.C., Jacobson, S.H., 2005. An analysis of the alias method for discrete random-variate generation. J. Comput. 17 (3), 321–327.

von Neumann, J., 1951. Various techniques used in connection with random digits. Bur. Stand. Math. Ser. 12.

Walker, A.J., 1974. New fast method for generating discrete random numbers with arbitrary frequency distributions. Electron. Lett. 10, 127–128.

Walker, A.J., 1977. An efficient method for generating discrete random variables with general distributions. ACM Trans. Math. Softw. 3 (3), 253–256.

Chapter 5

Variance Reduction Techniques

There was a young student who'd fret,
if ever an answer she'd get.
The events were so rare,
few histories got there;
but she hadn't tried Russian roulette.

The law of large numbers and the central limit theorem provide a powerful quadrature prescription for calculating expected values, namely, one samples from the governing probability distributions and form appropriate averages to estimate the expected values. In the last chapter, methods were discussed for performing sampling from various distributions and for scoring the results.

So what's left? Is that all there is to Monte Carlo (MC) calculations? The answer is a resounding NO. One of the features of Monte Carlo is that it is so flexible and adaptable that it can be used to solve a given problem in many ways. Sometimes even, Monte Carlo can be applied to very complex problems for which one cannot formulate the underlying distribution whose expected value is sought. However, the real power of Monte Carlo is that the sampling procedure can be intentionally biased toward the region where the integrand is large or to produce simulated histories that have a better chance of creating a rare event, such as Buffon's needle falling on widely spaced lines. Of course, with such biasing, the scoring then must be corrected by assigning weights to each history in order to produce a corrected, *unbiased* estimate of the expected value. In such *nonanalog* MC analyses, the sample variance $s^2(\bar{z})$ of the estimated expectation value $\bar{z}$ can be reduced compared to that obtained by an unbiased or *analog* analysis. In Example 4.10, biasing a simulation of Buffon's problem with widely spaced lines to force all dropped needles to have one end within a needle's length of a grid line was seen to reduce the relative error by two orders of magnitude over that of a purely analog simulation.

An MC calculation seeks an estimator $\bar{z}$ to some expected value $\langle z \rangle$. The goal of variance reduction methods is to produce a more precise estimate of some expected value than could be obtained in a purely analog calculation using the same computational effort. For "successes-over-trials" problems in which a success is very unlikely (e.g., flipping 10 heads in a row), variance reduction tries to bias the calculation so that more successes are obtained. For problems in which a continuum of outcomes is possible, variance reduction strives to reduce the spread among the history results and bring them closer to the mean.

Exploring Monte Carlo Methods. https://doi.org/10.1016/B978-0-12-819739-4.00013-5

In this chapter, several methods will be considered to reduce the variance of the estimator $\bar{z}$ for

$$\langle z \rangle \equiv \int_V z(\mathbf{x}) f(\mathbf{x}) \, d\mathbf{x} \simeq \bar{z} = \frac{1}{N} \sum_{i=1}^{N} z(\mathbf{x}_i), \qquad (5.1)$$

where the $\mathbf{x}_i$ are sampled from the joint probability density function (PDF) $f(\mathbf{x})$. The standard deviation of the sample mean is given by Eq. (2.61), namely,

$$\sigma(\bar{z}) = \frac{1}{\sqrt{N-1}} \sqrt{\overline{z^2} - \bar{z}^2}. \qquad (5.2)$$

Because both $\overline{z^2}$ and $\bar{z}^2$ must always be positive, $\sigma\bar{z}$ can be reduced by reducing their difference. Thus, the various variance reduction techniques discussed in this chapter are directed, ultimately, at minimizing the quantity $\overline{z^2} - \bar{z}^2$. Note that, in principle, it is possible to attain zero variance, if $\overline{z^2} = \bar{z}^2$, which occurs if every history yields the sample mean. This is not, unfortunately, very likely. However, it is possible to reduce substantially the variance among histories by introducing various biases into a Monte Carlo calculation.

Over the years, many clever variance reduction techniques have been developed for performing biased MC calculations. The introduction of variance reduction methods into MC calculations can make otherwise difficult MC problems more easily solvable. However, use of these variance reduction techniques requires skill and experience. Nonanalog Monte Carlo, despite having a rigorous statistical basis, is in many ways an "art" form and cannot be used blindly.

Before looking at specific techniques for reducing the variance of an estimator, several basic approaches are summarized. These range from very basic to very esoteric.

Transform the Problem: In some calculations, problem symmetry or other features of the problem can be used to create an equivalent problem that has the same expected value but that can be treated by MC much more efficiently. Often such a transformation produces far better variance reduction than any other method. However, it takes considerable understanding of the fundamental processes governing the problem to find such a transformation. A good example for a radiation shielding problem is given in the next section. Unfortunately, transformations that increase the calculational efficiency are not readily evident for many problems.

Truncation Methods: This is the simplest of all variance reduction techniques. In this approach, the underlying physical models from which random samples are taken are simplified so that each history or sample takes, on average, less computer time. Thus, for a given computational time more histories or samples can be taken, and because $s(\overline{z})$ varies as $1/\sqrt{N}$,[1] the variance of the estimator is reduced. For example, some detail in the geometry far from the scoring region generally has little effect on $\langle z \rangle$ and hence need not be modeled. Equivalently, a detailed model of the low-probability tail of the PDF sampling distribution is not needed if $z(\mathbf{x})$ is small in the tail region.

Model simplification is a "brute-force" approach because quadrupling the number of histories only halves the variance. Moreover, physical insight into a particular problem is required to introduce simplifications that reduce the calculational complexity for parts of the sampling space that have little effect on the expected value. Also, the only way to verify that some simplification has negligible effect on the estimate of $\langle z \rangle$ is to perform the calculation with and without the simplification.

Modified Sampling Methods: By far, most variance reduction techniques modify the underlying sampling PDFs to bias the sampling. Such biasing is used to increase the likelihood that a history produces a nonzero score, i.e., a nonzero $z(\mathbf{x}_i)$. Biased sampling can still produce an unbiased score provided each history is assigned a "weight" that is adjusted to compensate for any biases introduced at the various steps in the history. By recording a history's weight in the tally, an unbiased value of $\overline{z}$ is achieved.

History Control Methods: A very powerful variance reduction technique is to use *Russian roulette* and *splitting* to alter, on the fly, the number of histories that can potentially produce a nonzero tally. As a history moves through phase space, it is possibly killed if it enters a subregion from which it is unlikely to produce a nonzero score, or it is possibly split into multiple histories if a history leaving this subregion is likely to yield a nonzero score. Similarly, in biased sampling, Russian roulette and splitting can be used to eliminate histories that have too small a weight or split if the history obtains too large a weight. This technique is often used in combinations with other variance reduction methods.

Modify the Tally: In some problems, the function $z(\mathbf{x})$ can be modified to allow more nonzero values of $z(\mathbf{x}_i)$ without significantly changing $\overline{z}$. For example, several different $z(\mathbf{x})$ may have nearly the same expected value, so the average of all such $z(\mathbf{x})$ should be used to increase the likelihood that a history produces a nonzero score. Problem symmetry often suggests ways to modify the tally so that more histories give nonzero values.

Incorporate Partially Deterministic Methods: These sophisticated techniques circumvent normal random sampling at certain stages of a history by substituting deterministic or analytical results. For example, if a history reaches a certain subregion of the problem phase space, an analytical

---

[1] From Eq. (4.27), it is seen that $s(\overline{z})$ varies as $1/\sqrt{N-1}$; but from here on it is assumed that N is large and the distinction between N and $N-1$ can be ignored.

estimate can be made of the eventual value of the tally. These techniques include next-event estimators or they control the random number sequence.

Several of the fundamental methods for variance reduction are reviewed and several examples are given below.

5.1 Use of Transformations

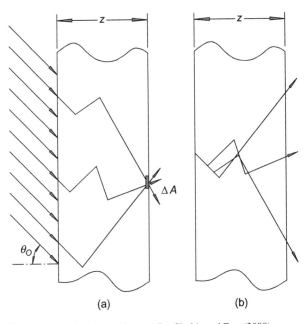

(a)   (b)

Figure 5.1 Equivalent problems. After Shultis and Faw (2000).

Although not a traditional variance reduction technique, use of problem symmetries or transformations can often be very effective in reducing the variance of some estimator. Consider, for example, the problem shown in Fig. 5.1(a) to the left. Here monodirectional radiation particles are incident uniformly over the left surface of a radiation shield. The MC simulation problem is to estimate the number of particles transmitted through a small unit area ΔA on the right surface. If this problem was simulated by sampling incident particles uniformly over the whole left surface, hardly any particles would ever reach the scoring area ΔA.

A much more efficient, but equivalent, simulation can be realized if it is recognized that the probability a particle eventually reaches the right surface is independent of where on the left surface the particle is incident. Thus, a far superior simulation is to start all incident particles at the same point on the left surface and then use the entire right surface as the scoring surface, as shown in Fig. 5.1(b). The ratio of the number of transmitted particles to the number of incident particles in Fig. 5.1(b) is the same as the ratio of the number of transmitted particles per unit surface area to the number of incident particles per unit surface area in Fig. 5.1(a).

Similarly, when using MC to evaluate an integral $\int_a^b f(x)dx$ whose integrand $f(x)$ varies widely over the range of integration, the use of a change of variables can often produce an equivalent integral $\int_c^d g(s)ds$ with an integrand that has much less variation. This technique is closely related to importance sampling, discussed below. Example 5.1 provides an illustration of the effectiveness of using an integral transformation.

Example 5.1 Transforming an Integral

Consider the evaluation of the integral

$$I = \int_0^1 f(x)e^{\alpha x}dx, \quad \alpha > 0,$$

where $f(x)$ is some slowly varying function. As α increases, the integrand becomes more and more peaked near the upper limit, and sampling uniformly over $[0, 1]$ produces many values of the integrand that contribute little to the value of $\bar{I}$. To avoid this, use the change of variables $s = e^{\alpha x}/\alpha$. Then $ds = e^{\alpha x}dx$ and $x = (1/\alpha)\ln \alpha s$. The transformed integral is thus

$$I = \int_{1/\alpha}^{e^{\alpha}/\alpha} f\left(\frac{1}{\alpha}\ln \alpha s\right)ds.$$

For the case $f(x) = x^2$ and $\alpha = 20$, both integrands, normalized by dividing by their maximum value at the upper limit, are shown in Fig. 5.2. The transformed integrand is seen to be much more uniform over the integration range and, thus, is more amenable to efficient evaluation by Monte Carlo. For this example, the first

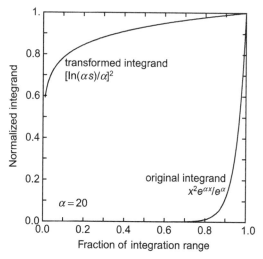

Figure 5.2 The normalized original and transformed integrands. The integrands are divided by their maximum value.

form of the integral can readily be evaluated analytically to give $I = e^{\alpha}[(1/\alpha) - (2/\alpha^2) + (2/\alpha^3)] - (2/\alpha^3) \simeq 2.195372 \times 10^7$ for $\alpha = 20$. The percentage error as a function of the number of histories is shown in Fig. 5.3 for both the analog case (sampling uniformly over the interval $[0, 1]$) and the transformed case. It is rare in MC calculations to know the exact or expected value, but in this example it is known. Unlike relative errors, which decrease monotonically, actual errors exhibit a stochastic decrease with the number of histories.

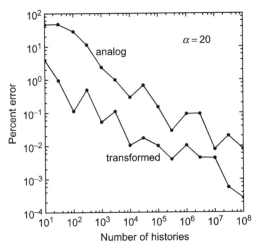

Figure 5.3 Percentage absolute error as a function of the number of Monte Carlo histories used.

5.2 Importance Sampling

Suppose the expected value of a multidimensional function $z(\mathbf{x})$ is sought, i.e.,

$$\langle z \rangle = \int_V z(\mathbf{x}) f(\mathbf{x}) d\mathbf{x} \simeq \overline{z} = \frac{1}{N} \sum_{i=1}^{N} z(\mathbf{x}_i), \qquad (5.3)$$

where the $\mathbf{x}_i$ are sampled from the PDF $f(\mathbf{x})$. Often, however, the function $z(\mathbf{x})$ is extremely small over most of V and almost all of the value of $\langle z \rangle$ comes from a small subregion of V where the PDF samples $\mathbf{x}$ infrequently. Many samples contribute little and, thus, represent a waste of computer time. It seems reasonable that better results could be obtained, with the same number of histories, if the sampling were preferentially made in regions where the integrand $|z(\mathbf{x}) f(\mathbf{x})|$ is large in magnitude. This idea was first proposed by Marshall (1956). To effect such a biased sampling, the integral can be transformed by introducing an arbitrary PDF $f^*(\mathbf{x})$ as

$$\langle z \rangle = \int_V \frac{z(\mathbf{x}) f(\mathbf{x})}{f^*(\mathbf{x})} f^*(\mathbf{x}) d\mathbf{x} \equiv \int_V z^*(\mathbf{x}) f^*(\mathbf{x}) d\mathbf{x} = \langle z^* \rangle, \qquad (5.4)$$

where $z^*(\mathbf{x}) = z(\mathbf{x})W(\mathbf{x})$ and the *weight* or *importance* function is

$$W(\mathbf{x}) \equiv f(\mathbf{x})/f^*(\mathbf{x}). \tag{5.5}$$

Thus, z and z^* have the same expectation, i.e.,

$$\langle z \rangle = \langle z^* \rangle \simeq \overline{z^*} = \frac{1}{N}\sum_{i=1}^{N} z(\mathbf{x}_i)W(\mathbf{x}_i), \tag{5.6}$$

where the $\mathbf{x}_i$ are sampled from $f^*(\mathbf{x})$.

Now consider the variance of the two expected values $\langle z \rangle$ and $\langle z^* \rangle$, namely,

$$\sigma^2(z) = \langle z^2 \rangle - \langle z \rangle^2 \quad \text{and} \quad \sigma^2(z^*) = \langle z^{*2} \rangle - \langle z^* \rangle^2 = \langle z^{*2} \rangle - \langle z \rangle^2, \tag{5.7}$$

where

$$\langle z^{*2} \rangle = \int_V z^{*2}(\mathbf{x})f^*(\mathbf{x})d\mathbf{x} = \int_V z^2(\mathbf{x})W^2(\mathbf{x})f^*(\mathbf{x})d\mathbf{x}$$

$$= \int_V z^2(\mathbf{x})W(\mathbf{x})f(\mathbf{x})d\mathbf{x} \neq \int_V z^2(\mathbf{x})f(\mathbf{x})d\mathbf{x} = \langle z^2 \rangle. \tag{5.8}$$

Hence, the two variances are, in general, different. If the variance for importance sampling, $\sigma^2(z^*)$, is smaller than that for straightforward sampling, $\sigma^2(z)$, then importance sampling gives a better estimate of the sample mean for the same number of histories.

When does this improvement occur? If $f^*(\mathbf{x})$ is chosen such that the weight function $W(\mathbf{x}) < 1$ over regions in which $z(\mathbf{x})$ makes a large contribution to the expected value, then from Eq. (5.8)

$$\langle z^{*2} \rangle = \int_V z^2(\mathbf{x})W(\mathbf{x})f(\mathbf{x})d\mathbf{x} < \int_V z^2(\mathbf{x})f(\mathbf{x})d\mathbf{x}. \tag{5.9}$$

For this choice of $W(\mathbf{x})$, it follows that $\sigma^2(z^*) < \sigma^2(z)$. In fact, if $f^*(\mathbf{x})$ is chosen as

$$f^*(\mathbf{x}) = f(\mathbf{x})z(\mathbf{x})/\langle z \rangle, \tag{5.10}$$

then $W(\mathbf{x}) = f(\mathbf{x})/f^*(\mathbf{x}) = \langle z \rangle/z(\mathbf{x})$, and Eq. (5.9) reduces to

$$\langle z^{*2} \rangle = \int_V z^2(\mathbf{x})W(\mathbf{x})f(\mathbf{x})d\mathbf{x} = \langle z \rangle \int_V z(\mathbf{x})f(\mathbf{x})d\mathbf{x} = \langle z \rangle^2, \tag{5.11}$$

which, from Eq. (5.7), yields $\sigma(z^*) = 0$!

This result indicates that, if the expected value is known, a Monte Carlo procedure can be developed that has zero variance. Of course $\langle z \rangle$ is not known or there would be no need to perform a Monte Carlo calculation. But if a PDF f^* can be formed that behaves more like the product $z(x)f(x)$ than does $f(x)$, $\langle z \rangle$ can be estimated with a smaller sample variance.

Thus, it is seen that importance sampling can reduce the sample variance. In practice, this can be achieved by using knowledge of the underlying processes to choose an appropriate f^* that favors histories that lead to successes (or to results close to the mean). Although sampling from f^* introduces a bias, this is accounted for by the weight factor $W = f/f^*$ in Eq. (5.6), so that the estimate $\bar{z}$ is an unbiased estimator for $\langle z \rangle$.

Example 5.2 Buffon's Needle Problem with Importance Sampling

As an application of importance sampling with a discontinuous PDF, consider Buffon's needle problem in which the distance D between the grid lines is much larger the length L of the needles. The probability a randomly dropped needle lands on a grid line is given by Eq. (1.6), namely,

$$P_{cut} = \langle z \rangle = \int_0^D z(x) f(x)\, dx = \int_0^D \left(\frac{s_1(x) + s_2(x)}{2\pi L} \right) \left(\frac{1}{D} \right) dx, \qquad (5.12)$$

where the functions $s_1(x)$ and $s_2(x)$ are given by Eqs. (1.4) and (1.5). This result is of the form of Eq. (1.8) and thus could be evaluated by the MC quadrature method described in Section 1.3 in which random values of x_i that are uniformly distributed on the interval $[0, D]$ are used to calculate

$$P_{cut} = \langle z \rangle \simeq \bar{z} = \frac{1}{N} \sum_{i=1}^{N} z(x_i),$$

as was done in Example 1.2.

However, when $D \gg L$, this is very inefficient because most x_i will fall in the interval $[L, D - L]$, for which $z(x_i) = 0$, and, thus, make no contribution to the score $\bar{z}$. Rather than sample from the uniform PDF $f(x) = 1/D$, $x \in [0, D]$, it is better to sample from the discontinuous PDF

$$f^*(x) = \begin{cases} \dfrac{1}{2L}, & 0 < x < L \text{ or } D - L < x < D, \\ 0, & \text{otherwise.} \end{cases}$$

Then P_{cut} is given by

$$P_{cut} = \int_0^D \frac{z(x) f(x)}{f^*(x)} f^*(x)\, dx = \int_0^D z(x) W(x) f^*(x)\, dx, \simeq \frac{1}{N} \sum_{i=1}^{N} z(x_i) W(x_i),$$

where the weight function is

$$W(x) = \frac{f(x)}{f^*(x)} = \begin{cases} \dfrac{2L}{D}, & 0 < x < L \quad \text{or} \quad D - L < x < D, \\ 0, & \text{otherwise} \end{cases}$$

and the x_i are sampled from $f^*(x)$. Results are shown in Fig. 5.4 with and without importance sampling for the case $D/L = 10,000$. For this case, Eq. (1.1) gives the exact value of $P_{cut}^{exact} = 2/(10^4\pi) = 6.36620\ldots \times 10^{-5}$.

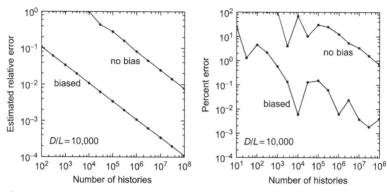

Figure 5.4 Results with and without importance sampling for the Buffon's needle integral of Eq. (5.12). The left-hand figure shows the relative error $R = s(\bar{z})/\bar{z}$, and the right-hand figure gives the percentage error $100|\bar{z} - P_{cut}^{exact}|/P_{cut}^{exact}$.

Example 5.3 Importance Sampling for an Important Subinterval

Consider an example of the use of an important subinterval. Let x be distributed according to the PDF $f(x)$, which is defined on the interval $[a, b]$. Suppose that it is desired to preferentially select from the "important" subinterval $[c, d]$ of $[a, b]$. This can be accomplished in the following way: Form the function

$$f^*(x) = \begin{cases} \alpha f(x), & x \in [c, d], \\ \beta f(x), & x \in [a, b], x \notin [c, d]. \end{cases}$$

In order to favor the preferred subinterval, choose a value for β from the interval $[0, 1]$. To make f^* a PDF, choose α to satisfy the condition $\int_a^b f^*(x)\,dx = 1$. Thus,

$$\alpha = \frac{1 - \beta[1 + F(c) - f(d)]}{F(d) - F(c)}.$$

Select x from $f^*(x)$ and form the weight factor

$$W(x) = \begin{cases} \dfrac{f(x)}{f^*(x)} = \dfrac{1}{\alpha}, & x \in [c, d], \\[2mm] \dfrac{1}{\beta}, & x \in [a, b] \quad x \notin [c, d]. \end{cases}$$

This procedure can be quite useful, although it involves a discontinuous PDF.

5.2.1   Application to Monte Carlo Integration

The general approach of using importance sampling to reduce the standard deviation of the estimated mean value can be applied to the Monte Carlo evaluation of a multidimensional integral

$$I \equiv \int_V g(\mathbf{x})\, d\mathbf{x}. \tag{5.13}$$

Example 5.1, which shows how to transform an integral to produce a more uniform integrand, is an example of importance sampling. In this example, the term $f(x)dx = e^{\alpha x}dx$, which caused the integrand to vary rapidly, was identified as an exact differential ds that was integrable analytically and whose integral could be inverted. Often it is difficult to find some function that reproduces the large variation in an integrand that satisfies these two properties. However, importance sampling can still be used.

Consider the problem of evaluating, by *straightforward* Monte Carlo, the general integral of Eq. (5.13) by uniformly sampling over V, namely,

$$\langle g \rangle \pm \sigma(g) \equiv \int_V g(\mathbf{x})d\mathbf{x} \pm \sqrt{\int_V [g - \langle g \rangle]^2 d\mathbf{x}} \simeq V\overline{g} \pm V\sqrt{\frac{\overline{g^2} - \overline{g}^2}{N}}, \tag{5.14}$$

where the $\pm$ term gives the 1-sigma standard deviation. Rather than sample $\mathbf{x}_i$ uniformly in V, one could use a biased sampling from a PDF $f^*(\mathbf{x})$. Thus, the integral becomes

$$I = \int_V g(\mathbf{x})d\mathbf{x} = \int_V \left(\frac{g}{f^*}\right) f^* d\mathbf{x} \simeq \overline{\left(\frac{g}{f^*}\right)} \pm \sqrt{\frac{\overline{(g^2/f^{*2})} - \overline{(g/f^*)}^2}{N}}. \tag{5.15}$$

But what choice to make for f^*? Intuitively, a good choice would be one that makes g/f^* relatively constant over V. More rigorously, the numerator in the square-root term in Eq. (5.15) should be made as small as possible. In particular, one seeks to minimize

$$\sigma^2(g/f^*) \equiv \left\langle \frac{g^2}{f^{*2}} \right\rangle - \left\langle \frac{g}{f^*} \right\rangle^2 = \int_V \frac{g^2}{f^{*2}} f^* d\mathbf{x} - \left[\int_V \frac{g}{f^*} f^* d\mathbf{x}\right]^2$$

$$= \int_V \frac{g^2}{f^*} d\mathbf{x} - \left[\int_V g\, d\mathbf{x}\right]^2. \tag{5.16}$$

Hence to minimize $\sigma^2(g/f^*)$, f^* must be chosen to make the integral $\int_V (g^2/f^*)d\mathbf{x}$ as small as possible. This minimization is achieved when

$$f^*(\mathbf{x}) = |g(\mathbf{x})| \Big/ \int_V |g(\mathbf{x})|\, d\mathbf{x}. \tag{5.17}$$

To prove this result, apply Schwartz' inequality[2] to the integral

$$\left(\int_V |g(\mathbf{x})| \, d\mathbf{x}\right)^2 = \left(\int_V \frac{|g(\mathbf{x})|}{\sqrt{f^*(\mathbf{x})}} \sqrt{f^*(\mathbf{x})} \, d\mathbf{x}\right)^2$$

$$\leq \left(\int_V \frac{g^2(\mathbf{x})}{f^*(\mathbf{x})} \, d\mathbf{x}\right)\left(\int_V f^*(\mathbf{x}) \, d\mathbf{x}\right) = \int_V \frac{g^2(\mathbf{x})}{f^*(\mathbf{x})} \, d\mathbf{x}, \quad (5.18)$$

where the fact that f^* is a PDF has been used to set $\int_V f^* \, d\mathbf{x} = 1$. The integral on the right-hand side is precisely the integral that must be minimized. Further, when f^* is chosen according to Eq. (5.17), the right-hand integral equals that on the left-hand side, i.e., the equality in Eq. (5.18) holds. In other words, the integral $\int_V (g^2/f^*) d\mathbf{x}$ is minimized and the minimum of $\sigma^2(g/f^*)$ is given by

$$\sigma(g/f^*)_{\min} = \left(\int_V |g(\mathbf{x})| \, d\mathbf{x}\right)^2 - \left(\int_V g(\mathbf{x}) \, d\mathbf{x}\right)^2. \quad (5.19)$$

In particular, if g does not change sign in V, $\sigma(g/f^*) = 0$!

This estimator of I with a greatly reduced variance (or even a zero variance if g is of one sign) is achieved if the importance is chosen as $f^* = |g|/\int_V |g| \, d\mathbf{x}$. But to evaluate $\int_V |g| \, d\mathbf{x}$ is tantamount to evaluating I, and there would be no point in using MC to find I. But all is not lost. The variance can be substantially reduced by choosing an importance function f^* that has a shape similar to that of g. With such a choice, importance sampling tries to smooth the values of g/f^* in regions where g varies rapidly. How effective importance sampling is depends on how effective g/f^* is in reducing variations in the integrand. Also when choosing such an approximate f^*, careful consideration must be given to the problem of how to sample values of $\mathbf{x}$ from it. This sampling can be difficult if g is not a well-behaved function, e.g., if g is a function with many rapid fluctuations.

5.3 Systematic Sampling

Estimation of $\langle z \rangle$ often can be obtained more accurately by subdividing the integration volume V into M contiguous mutually exclusive subregions V_m such that

$$V = \sum_{m=1}^{M} V_m. \quad (5.20)$$

---

[2] For those who have forgotten, Schwartz' inequality for two functions u and w is

$$\left(\int_V uw \, d\mathbf{x}\right)^2 \leq \left(\int_V u^2 \, d\mathbf{x}\right)\left(\int_V w^2 \, d\mathbf{x}\right).$$

Let p_m be an assigned probability of sampling from region m, where, of course, $\sum_{m=1}^{M} p_m = 1$. In systematic sampling, p_m is the probability that $\mathbf{x}_i$ lies in V_m, namely,

$$p_m = \int_{V_m} f(\mathbf{x})\,d\mathbf{x} \Big/ \int_V f(\mathbf{x})\,d\mathbf{x} = \int_{V_m} f(\mathbf{x})\,d\mathbf{x}, \tag{5.21}$$

because $f(\mathbf{x})$ is a PDF over V. Now define

$$f_m(\mathbf{x}) = \begin{cases} f(\mathbf{x})/p_m, & \mathbf{x} \in V_m, \\ 0, & \text{otherwise.} \end{cases} \tag{5.22}$$

Note that

$$\int_{V_m} f_m(\mathbf{x})\,d\mathbf{x} = \frac{1}{p_m} \int_{V_m} f(\mathbf{x})\,d\mathbf{x} = \frac{p_m}{p_m} = 1, \tag{5.23}$$

as must be the case for a PDF.

The population mean is the same regardless of the way in which sampling is performed to obtain the sample mean. Nevertheless, the population mean can be written in terms of p_m, as follows:

$$\langle z \rangle = \int_V z(\mathbf{x}) f(\mathbf{x})\,d\mathbf{x} = \sum_{m=1}^{M} p_m \int_{V_m} z(\mathbf{x}) f_m(\mathbf{x})\,d\mathbf{x}, \tag{5.24}$$

or, more concisely,

$$\langle z \rangle = \sum_{m=1}^{M} p_m \langle z \rangle_m, \tag{5.25}$$

where $\langle z \rangle_m$ is defined by

$$\langle z \rangle_m = \int_{V_m} z(\mathbf{x}) f_m(\mathbf{x})\,d\mathbf{x}. \tag{5.26}$$

Similarly, the population variance can be expressed as

$$\sigma^2(z) = \int_V [z(\mathbf{x}) - \langle z \rangle]^2 f(\mathbf{x})\,d\mathbf{x} = \int_V [z^2(\mathbf{x}) - 2z(\mathbf{x})\langle z \rangle + \langle z \rangle^2]\,d\mathbf{x}. \tag{5.27}$$

By Eq. (5.22), $f(\mathbf{x}) = p_m f_m(\mathbf{x})$, $\mathbf{x} \in V_m$, and hence

$$\sigma^2(z) = \sum_{m=1}^{M} p_m \int_{V_m} [z^2(\mathbf{x}) - 2z(\mathbf{x})\langle z \rangle + \langle z \rangle^2] f_m(\mathbf{x})\,d\mathbf{x}$$

$$= \sum_{m=1}^{M} p_m [\langle z^2 \rangle_m - 2\langle z \rangle\langle z \rangle_m + \langle z \rangle^2], \tag{5.28}$$

where

$$\langle z^2 \rangle_m = \int_{V_m} z^2(\mathbf{x}) f_m(\mathbf{x}) \, d\mathbf{x}. \tag{5.29}$$

Systematic sampling assumes p_m are known and defines

$$N_m = p_m N, \tag{5.30}$$

where N is the total number of histories. The systematic sampling estimator of $\langle z \rangle$ can be expressed, using Eq. (5.30), as

$$\bar{z}_{sys} = \sum_{m=1}^{M} \frac{p_m}{N_m} \sum_{i_m=1}^{N_m} z(\mathbf{x}_{i_m}), \tag{5.31}$$

where the $\mathbf{x}_{i_m}$ are sampled from V_m according to the PDF $f_m(\mathbf{x})$. This can be written more concisely in the form

$$\bar{z}_{sys} = \sum_{m=1}^{M} p_m \bar{z}_m, \tag{5.32}$$

where

$$\bar{z}_m = \frac{1}{N_m} \sum_{i_m=1}^{N_m} z(\mathbf{x}_{i_m}). \tag{5.33}$$

It is clear that Eq. (5.32) is an unbiased estimator of $\langle z \rangle$ because, by the law of large numbers, each $\bar{z}_m$ approaches each $\langle z_m \rangle$.

The variance of this estimator is $\langle (\bar{z}_{sys} - \langle z \rangle)^2 \rangle$. Substitution of Eq. (5.25) and Eq. (5.32) into this expression yields

$$\sigma^2(\bar{z}_{sys}) = \left\langle \left\{ \sum_{m=1}^{M} \frac{p_m}{N_m} \sum_{i_m=1}^{N_m} \left[z(\mathbf{x}_{i_m}) - \langle z \rangle_m \right] \right\}^2 \right\rangle$$

$$= \sum_{m=1}^{M} \frac{p_m^2}{N_m^2} \sum_{i_m=1}^{N_m} \langle [z(\mathbf{x}_{i_m}) - \langle z \rangle_m]^2 \rangle$$

$$+ \text{ cross-product terms that average to zero}$$

$$= \sum_{m=1}^{M} \frac{p_m^2}{N_m^2} \sum_{i_m=1}^{N_m} \sigma_m^2 = \sum_{m=1}^{M} \frac{p_m^2}{N_m} \sigma_m^2. \tag{5.34}$$

But in systematic sampling $p_m = N_m/N$ and, thus, the variance of the estimator is

$$\sigma^2(\bar{z}_{sys}) = \frac{1}{N} \sum_{m=1}^{M} p_m \sigma_m^2(z), \tag{5.35}$$

where

$$\sigma_m^2(z) \equiv \int_{V_m} [z(\mathbf{x}) - \langle z \rangle_m]^2 f_m(\mathbf{x}) \, d\mathbf{x}$$

$$\simeq \frac{1}{N_m} \sum_{i_m=1}^{N_m} z^2(\mathbf{x}_{i_m}) - \left[\frac{1}{N_m} \sum_{i_m=1}^{N_m} z(\mathbf{x}_{i_m}) \right]^2. \tag{5.36}$$

5.3.1 Comparison to Straightforward Sampling

In straightforward sampling, N values of $\mathbf{x}_i$ are sampled from the PDF $f(\mathbf{x})$ over V and the expected value of z is estimated as (Kahn, 1954)

$$\bar{z}_{sf} = \frac{1}{N} \sum_{i=1}^{N} z(\mathbf{x}_i) \simeq \langle z \rangle = \int_V z(\mathbf{x}) f(\mathbf{x}) \, d\mathbf{x}. \tag{5.37}$$

The probability a sample point $\mathbf{x}_i$ is in V_m is simply p_m. The variance of this estimator is calculated as follows:

$$\sigma^2(\bar{z}_{sf}) = \frac{1}{N} \sigma^2(z) = \frac{1}{N} \langle [z(\mathbf{x}) - \langle z \rangle]^2 \rangle = \frac{1}{N} \int_V [z(\mathbf{x}) - \langle z \rangle]^2 f(\mathbf{x}) \, d\mathbf{x}$$

$$= \frac{1}{N} \sum_{m=1}^{M} p_m \int_{V_m} [z(\mathbf{x}) - \langle z \rangle]^2 f_m(\mathbf{x}) \, d\mathbf{x}$$

$$= \frac{1}{N} \sum_{m=1}^{M} p_m \int_{V_m} [(z(\mathbf{x}) - \langle z \rangle_m) + (\langle z \rangle_m - \langle z \rangle)]^2 f_m(\mathbf{x}) \, d\mathbf{x}$$

$$= \frac{1}{N} \sum_{m=1}^{M} p_m \left\{ \int_{V_m} [z(\mathbf{x}) - \langle z \rangle_m]^2 f_m(\mathbf{x}) \, d\mathbf{x} \right.$$

$$+ 2 \int_{V_m} [z(\mathbf{x}) - \langle z \rangle_m][\langle z \rangle_m - \langle z \rangle] f_m(\mathbf{x}) \, d\mathbf{x}$$

$$\left. + \int_{V_m} [\langle z \rangle_m - \langle z \rangle]^2 f_m(\mathbf{x}) \, d\mathbf{x} \right\}.$$

The first integral is just $\sigma_m^2(z)$, the second integral integrates to zero, and the integrand of the third integral is the constant $[\langle z \rangle_m - \langle z \rangle]^2$. Thus

$$\sigma^2(\bar{z}_{sf}) = \frac{1}{N} \sum_{m=1}^{M} p_m \sigma_m^2(z) + \frac{1}{N} \sum_{m=1}^{M} p_m [\langle z \rangle_m - \langle z \rangle]^2. \qquad (5.38)$$

Comparison of Eq. (5.35) to Eq. (5.38) shows that the estimator for systematic sampling almost always has a smaller variance than does that for straightforward sampling. Only in the unlikely case that the means of z in all subregions are the same does systematic sampling produce no advantage. The amount of reduction is seen to equal the variance in the average values of z in the subregions, and for cases in which there is a large variation in z among the different regions, systematic sampling is effective. For cases in which p_m are known, systematic sampling should always be done. But often it is difficult to obtain good estimates of p_m.

5.3.2 Systematic Sampling to Evaluate an Integral

The following multidimensional integral can be estimated as

$$I = \int_V g(\mathbf{x}) \, d\mathbf{x} = V \langle g \rangle \simeq V \bar{g}. \qquad (5.39)$$

To apply systematic sampling to estimate $\langle g \rangle$, the results of Section 5.3 can be used directly by making the following changes: $z(\mathbf{x}) \to g(\mathbf{x})$, $f(\mathbf{x}) \to 1/V$, $f_m(\mathbf{x}) \to 1/V_m$, and $p_m \to V_m/V$. Thus, the systematic estimator of $\langle g \rangle$ becomes

$$\bar{g}_{sys} = \sum_{m=1}^{M} p_m \bar{g}_m = \sum_{m=1}^{M} \frac{p_m}{N_m} \sum_{i_m=1}^{N_m} g(\mathbf{x}_{i_m}), \qquad (5.40)$$

where the $\mathbf{x}_{i_m}$ are uniformly sampled from V_m. The variance of this estimator is

$$\sigma^2(\bar{g}_{sys}) = \frac{1}{N} \sum_{m=1}^{M} p_m \sigma_m^2(g), \qquad (5.41)$$

where

$$\sigma_m^2(g) \equiv \frac{1}{V_m} \int_{V_m} [g(\mathbf{x}) - \langle g \rangle_m]^2 \, d\mathbf{x} \simeq \frac{1}{N_m} \sum_{i_m=1}^{N_m} g^2(\mathbf{x}_{i_m}) - \left[\frac{1}{N_m} \sum_{i_m=1}^{N_m} g(\mathbf{x}_{i_m}) \right]^2. \qquad (5.42)$$

5.3.3   Systematic Sampling as Importance Sampling

Like many variance reduction schemes, systematic sampling can be viewed as a special case of importance sampling, because it involves sampling from a slightly modified PDF, namely,

$$f^*(\mathbf{x}) = \frac{f(\mathbf{x})}{p_m}, \quad \mathbf{x} \in V_m, \quad m = 1, \ldots, M. \tag{5.43}$$

The corresponding importance function is thus

$$W(\mathbf{x}) = \frac{f(\mathbf{x})}{f^*(\mathbf{x})} = p_m, \quad \mathbf{x} \in V_m, \quad m = 1, \ldots, M. \tag{5.44}$$

5.4   Stratified Sampling

Stratified sampling, also sometimes called *quota* sampling, is akin to systematic sampling in that a predetermined number of samples are taken from each of M subregions, but the method of selection N_m is quite different. As with systematic sampling, one seeks

$$\langle z \rangle = \int_V z(\mathbf{x}) f(\mathbf{x}) \, d\mathbf{x} = \sum_{m=1}^{M} \int_{V_m} z(\mathbf{x}) f(\mathbf{x}) \, d\mathbf{x}. \tag{5.45}$$

Proceeding exactly as in systematic sampling (see Section 5.3), the stratified sampling estimator of $\langle z \rangle$ is

$$\bar{z}_{strat} = \sum_{m=1}^{M} p_m \bar{z}_m = \sum_{m=1}^{M} \frac{p_m}{N_m} \sum_{i_m=1}^{N_m} z(\mathbf{x}_{i_m}). \tag{5.46}$$

From Eq. (5.34), the variance of this estimator is

$$\sigma^2(\bar{z}_{strat}) = \sum_{m=1}^{M} \frac{p_m^2}{N_m} \sigma_m^2(z), \tag{5.47}$$

with $\sigma_m^2(z)$ given by Eq. (5.36).

So far this is the same as systematic sampling. But in stratified sampling, the N_m are not predetermined but are chosen so as to minimize the variance of the estimator. To minimize $\sigma^2(\bar{z}_{strat})$ with the restraint that $\sum_{m=1}^{M} N_m = N$, one seeks the unrestricted minimum of

$$\sum_{m=1}^{M} \frac{p_m^2 \sigma_m^2(z)}{N_m} + \lambda \sum_{m=1}^{M} N_m, \tag{5.48}$$

where λ is a Lagrange multiplier. The variation of this quantity with N_j is then set to zero so that

$$0 = \frac{\delta}{\delta N_j}\left\{\sum_{m=1}^{M}\left[\frac{p_m^2\sigma_m^2}{N_m}+\lambda N_m\right]\right\} = \sum_{m=1}^{M}\frac{\delta}{\delta N_j}\left[\frac{p_m^2\sigma_m^2}{N_m}+\lambda N_m\right] = -\frac{p_j^2\sigma_j^2}{N_j^2}+\lambda.$$

(5.49)

From this quantity, the optimum number of samples from each region is found to be $N_j = p_j\sigma_j/\sqrt{\lambda}$. Rearranging and summing, this result gives

$$\sqrt{\lambda}\sum_{j=1}^{M}N_j = \sqrt{\lambda}N = \sum_{j=1}^{M}p_j\sigma_j \equiv \langle\sigma\rangle,$$

(5.50)

from which, the optimum number of samples that minimize $\sigma^2(\overline{z}_{strat})$ is

$$N_j = \frac{p_j\sigma_j(z)}{\langle\sigma\rangle}N = \frac{p_j\sigma_j(z)}{\sum_{j=1}^{M}p_j\sigma_j(z)}N.$$

(5.51)

Thus, in stratified sampling, the number of points sampled in each subregion is proportional to the variance of z in the region. Regions in which z varies rapidly are sampled more often than regions in which z is relatively constant. Substitution of this result (with j replaced by m) into Eq. (5.47) gives the minimum variance of $\overline{z}_{strat}$, namely,

$$\sigma^2(\overline{z}_{strat}) = \frac{1}{N}\left(\sum_{m=1}^{M}p_m\sigma_m\right)^2 = \frac{1}{N}\langle\sigma_m\rangle^2.$$

(5.52)

Unfortunately, usually the parameters $\sigma_j(z)$ and $\langle\sigma_j\rangle$ are not known a priori. It is *not* a good idea to use some of the same histories used to determine $\overline{z}_{strat}$ to also determine σ_j because these results are correlated and very strange results can result. One approach to estimate the optimum values of N_j is to estimate σ_j from a small preliminary run to obtain rough estimates, with consideration given to the trade-off between the desired precision and the cost of sampling.

A more serious limitation for stratified sampling is the subdivision of V for high-dimensional integrals, say $d \geq 5$. Division of each axis into k segments causes the number of subvolumes, M^d, to explode with increasing k and greatly confounds accurate estimates of the needed variances for each subregion.

5.4.1 Comparison to Straightforward Sampling

Stratified sampling almost always produce a more efficient estimate of $\langle z\rangle$ than does straightforward sampling, i.e., a smaller variance in the estimator. Subtraction of Eq. (5.52) from Eq. (5.38) yields

$$
\sigma^2(\overline{z}_{sf}) - \sigma^2(\overline{z}_{strat}) = \frac{1}{N} \sum_{m=1}^{M} p_m \sigma_m^2(z) + \frac{1}{N} \sum_{m=1}^{M} p_m [\langle z \rangle_m - \langle z \rangle]^2 - \frac{1}{N} \langle \sigma_m \rangle^2
$$

$$
= \frac{1}{N} \left[\langle \sigma_m^2 \rangle - \langle \sigma_m \rangle^2 \right] + \frac{1}{N} \sum_{m=1}^{M} p_m [\langle z \rangle_m - \langle z \rangle]^2
$$

$$
= \frac{1}{N} \left[\sum_{m=1}^{M} p_j (\sigma_m - \langle \sigma_m \rangle)^2 + \sum_{m=1}^{M} p_m (\langle z \rangle_m - \langle z \rangle)^2 \right].
$$

$$
(5.53)
$$

This last result is always nonnegative, so stratified sampling almost always is more efficient. Only in the case that both all the variances and all the means of z in each subregion are equal is no advantage realized, a very unlikely possibility.

5.4.2 Importance Sampling Versus Stratified Sampling

The differences between importance sampling and stratified sampling are quite distinct. First, importance sampling usually uses a continuous importance function to flatten the integrand, while stratified sampling always breaks the integration volume into subvolumes. Second, in importance sampling, points are concentrated in regions where the magnitude of the integrand $|z(\mathbf{x}) f(\mathbf{x})|$ is large, while in stratified sampling, points are concentrated in regions where the variance is large. These two techniques, thus, seem at odds with each other.

Importance sampling requires an approximate PDF f^* that has a shape similar to $z(\mathbf{x}) f(\mathbf{x})$ and that can also be efficiently sampled. With an imperfect f^*, the error decreases only as $N^{-1/2}$, and in regions in which f^* cannot match rapid variations in $z(\mathbf{x}) f(\mathbf{x})$, the sampled function $z(\mathbf{x}) f(\mathbf{x}) / f^*(\mathbf{x})$ has a large variance. The key to importance sampling is to smooth the sampled function, and only if this is achieved is importance sampling effective.

Stratified sampling, by contrast, requires no knowledge about the values of $z(\mathbf{x}) f(\mathbf{x})$ but, rather, the variances for each subregion are required. It works well if good estimates of the variances in all subregions are available and if subregions can be constructed that reduce the variance of $z(\mathbf{x}) f(\mathbf{x})$ in each subregion compared to the variance of $z(\mathbf{x}) f(\mathbf{x})$ over the whole volume V. This requires different knowledge about $z(\mathbf{x}) f(\mathbf{x})$ than does importance sampling.

In many cases, $z(\mathbf{x}) f(\mathbf{x})$ is small everywhere in V except in some small subregion. In such a subregion, the standard deviation of $z(\mathbf{x}) f(\mathbf{x})$ is often comparable to the mean value, so that both methods produce comparable results. As pointed out by Press et al. (1996), these two methods can also be used together by using one method on a coarse grid and the other in each grid cell.

Press et al. (1996) also describe (1) how importance sampling can be combined with stratified sampling in an adaptive algorithm for multidimensional integrals and (2) how stratified sampling can be used recursively to refine the subdivision of V into subregions. Although these uses of variance reduction are beyond the scope of this book, they do reveal the high level of sophistication that is presently available for MC analyses.

5.5 Correlated Sampling

An extremely powerful method for reducing the variance in a certain class of MC problems is that of *correlated sampling*. In many problems, the difference between two almost equal situations is sought, for example, the effect of a small change in some underlying model, geometry, or other system parameter. Rather than perform two independent MC analyses for each problem and then subtract one result from the other to obtain an estimate of the difference, it is often far better to perform both calculations simultaneously and use the same random number sequence for both. By using the same random numbers, the two results will be highly correlated, which, as is shown, results in a much smaller variance in the estimate of the difference.

Suppose $\langle z_1 \rangle$ and $\langle z_2 \rangle$ are given by

$$\langle z_1 \rangle = \int_{V_1} z_1(\mathbf{x}) f_1(\mathbf{x}) \, d\mathbf{x} \quad \text{and} \quad \langle z_2 \rangle = \int_{V_2} z_2(\mathbf{y}) f_2(\mathbf{y}) \, d\mathbf{y}. \tag{5.54}$$

What is sought is $\langle \Delta z \rangle = \langle z_1 \rangle - \langle z_2 \rangle$. First, generate $\{\mathbf{x}_i\}$, $i = 1, \ldots, N$, and $\{\mathbf{y}_i\}$, $i = 1, \ldots, N$. Second, estimate $\langle \Delta z \rangle$ as

$$\langle \Delta z \rangle \simeq \frac{1}{N} \sum_{i=1}^{N} z_1(\mathbf{x}_i) - \frac{1}{N} \sum_{i=1}^{N} z_2(\mathbf{y}_i) = \frac{1}{N} \sum_{i=1}^{N} \Delta z_i. \tag{5.55}$$

Now the variance of this estimator of the difference is

$$\sigma^2(\langle \Delta z \rangle) = \sigma_1^2 + \sigma_2^2 - 2\mathrm{cov}(\overline{z_1}, \overline{z_2}), \tag{5.56}$$

where

$$\overline{z_1} = \frac{1}{N} \sum_{i=1}^{N} z_1(\mathbf{x}_i), \qquad \overline{z_2} = \frac{1}{N} \sum_{i=1}^{N} z_2(\mathbf{y}_i), \qquad \sigma_i^2 = \langle (z_i - \langle z_i \rangle)^2 \rangle, \tag{5.57}$$

and

$$\mathrm{cov}(\overline{z_1}, \overline{z_2}) = \langle (\overline{z_1} - \langle z_1 \rangle)(\overline{z_2} - \langle z_2 \rangle) \rangle. \tag{5.58}$$

If the two estimators $\overline{z_1}$ and $\overline{z_2}$ were calculated independently, then $\mathrm{cov}(\overline{z_1}, \overline{z_2}) = 0$. But if the random variables $\mathbf{x}$ and $\mathbf{y}$ are positively correlated and if $f_1(\mathbf{x})$ and $f_2(\mathbf{y})$ are very similar in shape, then $\overline{z_1}$ and $\overline{z_2}$ are strongly correlated so that $\mathrm{cov}(\overline{z_1}, \overline{z_2}) > 0$ and $\sigma^2(\langle \Delta z \rangle) < \sigma_1^2 + \sigma_2^2$.

From this analysis, it is apparent that to reduce the variance of the difference between $\langle z_1 \rangle$ and $\langle z_2 \rangle$, the correlation between these two quantities must be positive and as large as possible. There is no general prescription to ensure this; however, a good way is to control the use of random numbers. In parts of the MC simulation that are the same or very similar, the same sequence of random numbers $\{\mathbf{x}_i\}$ and $\{\mathbf{y}_i\}$ should be used. A simple example of using correlated sampling is given in Example 5.4.

5.5.1 Correlated Sampling with One Known Expected Value

Sometimes, when trying to estimate the expected value

$$\langle z \rangle \equiv \int_V z(\mathbf{x}) f(\mathbf{x}) \, d\mathbf{x}, \tag{5.59}$$

the exact expectation of a very closely related problem, often some idealization of the problem of interest, is known, i.e., one knows a priori

$$\langle z_o \rangle = \int_V z_o(\mathbf{x}) f(\mathbf{x}) \, d\mathbf{x}. \tag{5.60}$$

Here $z_o(\mathbf{x})$ is some close approximation to $z(\mathbf{x})$, so that $\langle z \rangle$ is most likely close to $\langle z_o \rangle$. Thus, rather than trying to estimate $\langle z \rangle$ by calculating $\overline{z}$ directly, it is usually better to estimate the difference $\Delta = \langle z \rangle - \langle z_o \rangle$ and then add this estimate to the known $\langle z_o \rangle$.

More generally, another estimator of $\langle z \rangle$, for any α, is

$$\overline{z(\alpha)} \equiv \overline{z} - \alpha(\overline{z_o} - \langle z_o \rangle). \tag{5.61}$$

This is an unbiased estimate because, in the limit as $N \to \infty$, $\langle \overline{z(\alpha)} \rangle = \langle z \rangle - \alpha(\langle z_o \rangle - \langle z_o \rangle) = \langle z \rangle$ for any α. However, the variance of this estimator does depend on α and is given by

$$\sigma^2(\overline{z(\alpha)}) = \frac{1}{N}\left[\sigma^2(z) - 2\alpha \, \mathrm{cov}(z, z_o) + \alpha^2 \sigma^2(z_o)\right]. \tag{5.62}$$

Here the variances and covariance are

$$\sigma^2(z) \equiv \int_V (z - \langle z \rangle)^2 f(\mathbf{x}) \, d\mathbf{x} = \langle z^2 \rangle - \langle z \rangle^2 \simeq \overline{z^2} - \overline{z}^2, \tag{5.63}$$

$$\sigma^2(z_o) \equiv \int_V (z_o - \langle z_o \rangle)^2 f(\mathbf{x}) \, d\mathbf{x} = \langle z_o^2 \rangle - \langle z_o \rangle^2 \simeq \overline{z_o^2} - \overline{z_o}^2, \qquad (5.64)$$

$$\mathrm{cov}(z, z_o) \equiv \int_V (z - \langle z \rangle)(z_o - \langle z_o \rangle) f(\mathbf{x}) \, d\mathbf{x}$$

$$= \langle z \, z_o \rangle - \langle z \rangle \langle z_o \rangle \simeq \overline{z \, z_o} - \overline{z} \, \overline{z_o}. \qquad (5.65)$$

From Eq. (5.62), it is apparent that if $2\alpha \mathrm{cov}(z, z_o) > \alpha^2 \sigma^2(z_o)$, the variance is reduced, i.e., $\sigma^2(\overline{z(\alpha)}) < \sigma^2(z)$.

From Schwartz' inequality, it follows that

$$\int_V [z(\mathbf{x}) - \langle z \rangle]^2 f(\mathbf{x}) \, d\mathbf{x} \int_V [z_o(\mathbf{x}) - \langle z_o \rangle]^2 f(\mathbf{x}) \, d\mathbf{x} \geq$$

$$\left[\int_V [z(\mathbf{x}) - \langle z \rangle][z_o(\mathbf{x}) - \langle z_o \rangle] f(\mathbf{x}) \, d\mathbf{x} \right]^2, \qquad (5.66)$$

or

$$\sigma^2(z)\sigma^2(z_o) \geq \mathrm{cov}^2(z, z_o). \qquad (5.67)$$

From this result, it follows that the *correlation coefficient*, defined as

$$\rho(z, z_o) \equiv \frac{\mathrm{cov}(z, z_o)}{\sigma(z)\sigma(z_o)}, \qquad (5.68)$$

has the property $-1 \leq \rho(z, z_o) \leq 1$. In terms of this correlation coefficient,

$$\sigma^2(\overline{z(\alpha)}) = \frac{1}{N} \left[\sigma^2(z) - 2\alpha\sigma(z)\sigma(z_o)\rho(z, z_o) + \alpha^2\sigma^2(z_o) \right]. \qquad (5.69)$$

The optimum value of α that minimizes $\sigma^2(\overline{z(\alpha)})$ is found by requiring

$$\frac{d\sigma^2(\overline{z(\alpha)})}{d\alpha} = 0. \qquad (5.70)$$

From this requirement it is easily found that

$$\alpha_{opt} = \sigma(z)\rho(z, z_o)/\sigma(z_o) = \frac{\overline{z \, z_o} - \overline{z} \, \overline{z_o}}{\overline{z_o^2} - \overline{z_o}^2} \qquad (5.71)$$

With the optimum value the minimum variance is

$$\sigma^2(\overline{z(\alpha_{opt})}) = \frac{1}{N}\sigma^2(z)[1 - \rho^2(z, z_o)]. \qquad (5.72)$$

Hence, the more z is correlated with z_o, the smaller is the variance of the estimator of Eq. (5.61). Using this approach requires estimating $\overline{z}$, $\overline{z_o}$, $\overline{z z_o}$, and $\overline{z_o^2}$ with the same sampled values (i.e., using correlated sampling), but does lead to a reduced value of $\Delta = \langle z \rangle - \langle z_0 \rangle$. An example of correlated sampling with one known expected value is shown in Example 5.4.

Example 5.4 A Modified Buffon's Needle Problem

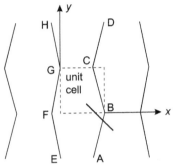

To illustrate correlated sampling, consider the Buffon's needle problem with the parallel grid lines replaced by zigzag grid lines shown to the left. A unit cell of dimension $D \times D$ is shown by the dashed lines and each straight line segment has one end displaced in the x-direction by an amount δ with respect to the other end. Because the grid line length is greater with the zigzag grid, it is to be expected that P_{cut}^{zigzag} should increase slightly over $P_{cut}^{\parallel}$ of the standard case of parallel grid lines, with the same spacing D and needle length L. For the parallel line grid, $P_{cut} = 2L/(\pi D)$ (see Example 1.1).

Unlike the standard Buffon's needle problem, no simple analytical solution is available, and to estimate $\Delta P_{cut} = P_{cut}^{zigzag} - P_{cut}^{\parallel}$ the difference between two simulations, one for each type of grid lines, is used. The simulation for the parallel grid lines is discussed in Example 1.3. For the zigzag grid lines, the simulation proceeds as follows. First coordinates (x_i, y_i) of a needle end are selected uniformly distributed in the unit cell. The angle of the needle with respect to the positive x-axis is then chosen uniformly in $(0, 2\pi)$. Then the coordinates of and distance to the intersection of a line through the needle and the lines through grid segments AB, BC, CD, EF, FG, and GH are computed. If any of the intersection distances are less than L and the intersection point is within the vertical coordinates of these grid segments, the needle intersects the grid. For small line displacements δ, a simulated needle that intersects the zigzag grid also has a good chance of intersecting the parallel-line grid, so that the results of the two simulations have a large positive correlations.

Examples of results of the difference between simulations using the same needle drops for both grids (correlated samplings) and independent drops for each grid are shown in Fig. 5.5.

5.5.2 Antithetic Variates

Consider the simple problem of estimating the integral

$$I = \int_0^1 g(x)\, dx \simeq \frac{1}{N} \sum_{i=1}^{N} g(x_i) = I_1, \tag{5.73}$$

where x_i are uniformly distributed in $(0, 1)$. Hammersley and Morton (1956) proposed that if two unbiased estimators I_1 and I_2 for some quantity I could

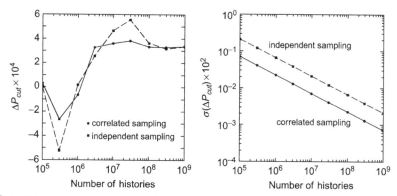

Figure 5.5 Results for correlated and independent Buffon's needle simulations to estimate the change ΔP_{cut} when the straight grid lines are replaced by zigzag grid lines. Results are for $D = 2L = 20\delta$. (Left) The convergence of ΔP_{cut} to 0.000334. (Right) The estimate of the standard deviation of the estimated ΔP_{cut}. As expected, correlated sampling has a smaller standard deviation.

be found that have a strong negative correlation, a better estimator of I may be $[I_1 + I_2]/2$. The variance of this estimator is

$$\sigma^2\left(\frac{I_1 + I_2}{2}\right) = \frac{1}{4}\sigma^2(I_1) + \frac{1}{4}\sigma^2(I_2) + \frac{1}{2}\text{cov}(I_1, I_2). \qquad (5.74)$$

If $\text{cov}(I_1, I_2)$ is strongly negative, then the average of two negatively correlated estimators may have a much reduced variance.

This integral can also be estimated as

$$I = \int_0^1 g(1 - x)\, dx, \simeq \frac{1}{N}\sum_{j=1}^{N} g(1 - x_j) = I_2, \qquad (5.75)$$

where x_j are also uniform in $[0, 1]$. Then the compound estimator

$$I_{12} \equiv \frac{I_1 + I_2}{2} \simeq \frac{1}{2N}\sum_{i=1}^{N}[g(x_i) + g(1 - x_i)] \qquad (5.76)$$

is an unbiased estimator because both I_1 and I_2 are unbiased estimators. But, because the same x_i are used for both estimators, there is a negative correlation between the two estimators, and the variance of the compound estimator is usually less than that for I_1. In fact, it can be shown for any function $g(x)$, $\sigma^2(I_{12}) \leq \sigma^2(I_1)$ (Kahn, 1954). This is not too surprising because the estimator I_{12} uses twice as many values of $g(x_i)$ than does the I_1 estimator for the same N.

However, because I_{12} requires twice the computational expense, this estimator is favored only if

$$\sigma^2(I_{12}) \leq \frac{\sigma^2(I_1)}{2}. \tag{5.77}$$

The inequality is indeed the case if $g(x)$ is a continuous monotonic function with a continuous first derivative, as is now demonstrated. The variance of I_{12} is

$$\sigma^2(I_{12}) = \frac{1}{N} \left\{ \frac{1}{4}\sigma^2(I_1) + \frac{1}{4}\sigma^2(I_2) + \frac{1}{2}\text{covar}(I_1, I_2) \right\}$$

$$= \frac{1}{N} \left\{ \frac{1}{4} \left[\int_0^1 g^2(x)\,dx - \langle I_1 \rangle^2 \right] + \frac{1}{4} \left[\int_0^1 g^2(1-x)\,dx - \langle I_2 \rangle^2 \right] \right.$$

$$\left. + \frac{1}{2} \left[\int_0^1 g(x)g(1-x)\,dx - \langle I_{12} \rangle^2 \right] \right\}$$

$$= \frac{1}{N} \left\{ \frac{1}{2} \int_0^1 g(x)g(1-x)\,dx + \left[\int_0^1 g^2(x)\,dx - I^2 \right] \right\}, \tag{5.78}$$

where the last line results from the fact that $\langle I_1 \rangle = \langle I_2 \rangle = \langle I_{12} \rangle = I$. The second term of Eq. (5.78) in square brackets is $\sigma^2(I_1)$. Hence, we have the following relation, which holds for any $g(x)$:

$$2\sigma^2(I_{12}) - \sigma^2(I_1) = \int_0^1 g(x)g(1-x)\,dx - I^2. \tag{5.79}$$

To prove that $\sigma^2(I_{12}) \leq \sigma^2(I_1)/2$, it is sufficient to show that the right-hand side of Eq. (5.79) is ≤ 0. Toward this end, consider the function

$$h(x) = \int_0^x g(1-x')\,dx' - xI, \qquad 0 \leq x \leq 1, \tag{5.80}$$

whose derivative is

$$h'(x) = g(1-x) - I. \tag{5.81}$$

Now, if $g(x)$ is a monotonic decreasing function of x, then $g(1-x)$ and $h'(x)$ are monotonic increasing functions of x. Because I is the average value of the monotonic decreasing function $g(x)$ in the interval $[0, 1]$, it follows that $h'(0) < 0$ and $h'(1) > 0$, so that $h(x) \leq 0$ for $x \in [0, 1]$. Then, because in the interval $[0, 1]$, $h(x) \leq 0$ and $g'(x) \leq 0$,

$$\int_0^1 h(x)g'(x)\,dx \geq 0. \tag{5.82}$$

Integration by parts and use of the boundary values $h(0) = h(1) = 0$ yields the relation

$$\int_0^1 h'(x)g(x)\,dx \le 0. \tag{5.83}$$

Substitution of Eq. (5.81) into this result yields

$$\int_0^1 g(1-x)g(x)\,dx - \int_0^1 Ig(x)\,dx = \int_0^1 g(1-x)g(x)\,dx - I^2 \le 0. \tag{5.84}$$

A similar analysis produces the same result if $g(x)$ is a monotonic increasing function of x. Thus, the right-hand side of Eq. (5.79) is indeed ≤ 0, and the relation of Eq. (5.77) is proved.

Generalization

A more general problem is to evaluate

$$I = \int_{-\infty}^{\infty} g(x)f(x)\,dx, \tag{5.85}$$

where $f(x)$ is a PDF with a cumulative distribution $F(x)$. Then by analogy to Eq. (5.76), this integral can be estimated as

$$I \simeq \frac{1}{2N}\sum_{i=1}^{N}[g(x_i) + g(y_i)], \tag{5.86}$$

where $x_i = F^{-1}(\rho_i)$ and $y_i = F^{-1}(1 - \rho_i)$. The pair x_i and y_i are negatively correlated because of the use of the same ρ_i to generate both. Thus, this estimator may have a smaller variance than that of the sample-mean estimator. The use of antithetic variates is demonstrated in Example 5.5.

Example 5.5 Buffon's Needle Problem with Antithetic Sampling

To illustrate antithetic sampling, consider the Buffon's needle problem using a square grid of vertical and horizontal lines as shown to the left. Each cell is $D \times D$ and needles of length L are randomly dropped on this grid. After dropping N needles, the number N_h cutting a horizontal grid line and the number N_v cutting a vertical grid line are recorded. Then, two independent estimators of P_{cut} are

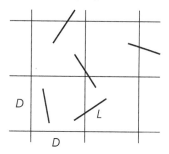

$$\overline{P_1} = \frac{N_h}{N} \quad \text{and} \quad \overline{P_2} = \frac{N_v}{N}.$$

These two estimators, especially for needle lengths L comparable to the grid spacing D, are negatively correlated. For example, needles landing close to the horizontal have a good chance of cutting a vertical grid line but not a horizontal one. Similarly vertically oriented needles often intersect horizontal lines but not the vertical lines. The compound estimator,

$$\overline{P}_{12} \equiv \frac{1}{2}[\overline{P_1} + \overline{P_2}] = \frac{1}{2}\left[\frac{N_v + N_h}{N}\right],$$

thus has a smaller variance than either of the separate estimators. An example of how the relative error varies with the number of needle drops for the $\overline{P}_{12}$ estimator and for the $\overline{P}_1$ or $\overline{P}_2$ estimator is shown in Fig. 5.6.

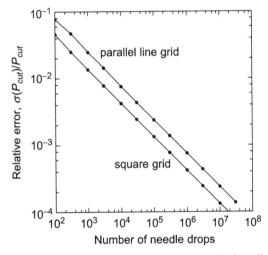

Figure 5.6 Results for a square grid with $D = L$ and for a grid of parallel lines. For the square grid, the $\overline{P}_{12}$ estimator based on the antithetic variates N_v and N_h is used, and is seen to have a relative error of almost one-half of that of the $\overline{P}_1$ or $\overline{P}_2$ estimator. For this example, $\mathrm{var}(\overline{P}_1) \simeq 3.2\,\mathrm{var}(\overline{P}_{12})$.

5.6 Partition of Integration Volume

Sometimes the integration volume V can be broken into two or more parts, and the integration over some of the subregions can be performed analytically. It is almost always better to use this partial exact information. For example, suppose $V = V_1 \cup V_2$, so that

$$I = \int_V z(\mathbf{x}) f(\mathbf{x})\,d\mathbf{x} = \int_{V_1} z(\mathbf{x}) f(\mathbf{x})\,d\mathbf{x} + \int_{V_2} z(\mathbf{x}) f(\mathbf{x})\,d\mathbf{x}, \tag{5.87}$$

where $I_1 = \int_{V_1} z(\mathbf{x}) f(\mathbf{x})\,d\mathbf{x}$ can be calculated analytically. Now define a truncated PDF for subregion V_2 as

$$h(x) = \begin{cases} \dfrac{f(\mathbf{x})}{1 - P}, & \text{if } \mathbf{x} \in V_2, \\ 0, & \text{otherwise,} \end{cases} \tag{5.88}$$

where $P = \int_{V_1} f(\mathbf{x}) \, d\mathbf{x}$. Equation (5.87) can, thus, be written as

$$I = I_1 + (1 - P) \int_{V_2} z(\mathbf{x}) \frac{f(\mathbf{x})}{1 - P} \, d\mathbf{x}$$

$$= I_1 + (1 - P) \int_{V_2} z(\mathbf{x}) h(\mathbf{x}) \, d\mathbf{x} = I_1 + (1 - P)\langle z \rangle_2. \tag{5.89}$$

An unbiased estimator of I is thus

$$I_{part} = I_1 + \frac{(1 - P)}{N} \sum_{i=1}^{N} z(\mathbf{x}_i), \tag{5.90}$$

where $\mathbf{x}_i$ are sampled from $h(\mathbf{x})$. It can be shown that the variance of this estimator is less than that of Eq. (5.3) (Rubinstein, 1981).

5.7 Reduction of Dimensionality

Suppose that the n-dimensional integral

$$I = \int_V z(\mathbf{x}) f(\mathbf{x}) \, d\mathbf{x} \tag{5.91}$$

can be decomposed as

$$I = \int_{V_1} \int_{V_2} z(\mathbf{u}, \mathbf{v}) f(\mathbf{u}, \mathbf{v}) \, d\mathbf{v} \, d\mathbf{u}, \tag{5.92}$$

where $\mathbf{u} = (x_1, x_2, \ldots, x_m)$ and $\mathbf{v} = (x_{m+1}, \ldots, x_n)$, and that the integration over $\mathbf{v}$ can be performed analytically so that the marginal PDF

$$f_{\mathbf{u}}(\mathbf{u}) = \int_{V_2} f(\mathbf{u}, \mathbf{v}) \, d\mathbf{v} \tag{5.93}$$

and the conditional expectation

$$\langle z(\mathbf{v}|\mathbf{u}) \rangle = \int_{V_2} z(\mathbf{u}, \mathbf{v}) f(\mathbf{v}|\mathbf{u}) \, d\mathbf{v} \tag{5.94}$$

can be calculated analytically. The conditional probability function is found from Eq. (2.51) as

$$f(\mathbf{v}|\mathbf{u}) = \frac{f(\mathbf{u}, \mathbf{v})}{f_{\mathbf{u}}(\mathbf{u})}. \tag{5.95}$$

Finally, I can be estimated as follows. From Eq. (5.92) and Eq. (5.95)

$$\begin{aligned}
I &= \int_{V_1} \int_{V_2} z(\mathbf{u}, \mathbf{v}) f(\mathbf{v}|\mathbf{u}) f(\mathbf{u}) \, d\mathbf{v} \, d\mathbf{u} \\
&= \int_{V_1} f_{\mathbf{u}}(\mathbf{u}) \left[\int_{V_2} z(\mathbf{u}, \mathbf{v}) f(\mathbf{v}|\mathbf{u}) \, d\mathbf{v} \right] d\mathbf{u} \\
&\equiv \int_{V_1} f_{\mathbf{u}}(\mathbf{u}) \left\langle \int_{V_2} z(\mathbf{v}|\mathbf{u}) \right\rangle d\mathbf{u} \\
&\simeq \frac{1}{N} \sum_{i=1}^{N} \langle z(\mathbf{v}|\mathbf{u}_i) \rangle, \tag{5.96}
\end{aligned}$$

where $\mathbf{u}_i$ are sampled from the marginal PDF $f_{\mathbf{u}}(\mathbf{u})$.

This method of reducing the dimensionality of the problem is sometimes referred to as the *expected value* and was introduced by Buslenko (Schreider, 1966). The estimator of Eq. (5.96) always has a smaller variance than the usual sample-mean estimator of Eq. (5.3) (Rubinstein, 1981).

5.8   Russian Roulette and Splitting

In many variance reduction schemes, the integration region is subdivided into M contiguous subregions V_m, i.e.,

$$\langle z \rangle = \int_V z(\mathbf{x}) f(\mathbf{x}) \, d\mathbf{x} = \sum_{m=1}^{M} \int_{V_m} z(\mathbf{x}) f(\mathbf{x}) \, d\mathbf{x}, \tag{5.97}$$

and the sampling from each region is modified over that predicted from $f(\mathbf{x})$. For example, in both the importance and stratified sampling techniques, regions with a large influence on $\bar{z}$, or its variance, are sampled more frequently than unimportant regions, each subregion having an assigned probability p_m that a random value selected from the PDF $f(\mathbf{x})$ will be taken from that subregion.

However, it is often difficult to do this. Often p_m and not known a priori, or p_m results from many levels of sampling so that it is difficult to make the number of samplings taken from V_m to be proportional to p_m.

To address such cases, J. von Neumann and S. Ulam developed a technique called *Russian roulette and splitting*, named after a supposedly popular game of chance played in the Russian army. In this technique, each subregion is classified as *computationally attractive* or *computationally unattractive*. In computationally attractive subregions either the variance of $z(\mathbf{x})$ is large or the cost of picking $\mathbf{x}_i$ and evaluating $z(\mathbf{x}_i)$ is relatively small. Such regions should be sampled often. The exact opposite is true in computationally unattractive subregions, i.e., the variance of $z(\mathbf{x})$ is small or the cost of computing $z(\mathbf{x}_i)$ is large, and to save computer time unattractive regions should be avoided. As an extreme example, if $z(\mathbf{x})$ were constant in subregion m, a single sample would give $\langle z \rangle_m$, and hence use of multiple samplings from this subregion would be a waste of time.

To implement Russian roulette and splitting, a particular subregion V_m is picked. If the selected subregion is unattractive, then Russian roulette is used. A value of $\mathbf{x}_i$ for this region is made only some of the time, with a predetermined probability q_m, and a weight of $z(\mathbf{x})/q_m$ is recorded; otherwise, with a probability $1 - q_m$, no sample is taken. The more unattractive the subregion, the smaller is q_m. As an example, if $q_m = 0.75$ no sample is made 25% of the time and 75% of the time a sample is taken.

By contrast, if an attractive subregion is selected, splitting is used. To implement this, each attractive region is assigned a value $r_m > 1$. The more attractive the region, the larger is r_m. Denote by $k_m = [r_m]$ the largest integer in r_m, and set $q_m = r_m - k_m < 1$. Then, with probability q_m, $k_m + 1$ independent samples z_i are taken from V_m, and, with probability $1 - q_m$, k_m samples are used. For example, if $r_m = 2.25$, 75% of the time the sample is split 3 for 1 and 25% of the time the split is 2 for 1. To remove the sampling bias, the weight for each sample is set to $z(\mathbf{x}_i)/r_m$.

Russian roulette combined with splitting is one of the oldest variance reduction schemes and is widely used in MC analyses, particularly in MC simulations. When used properly, considerable computer time can be saved because, in unattractive regions, fewer samples are taken (in order to save computer time) and, in attractive regions, more samples are taken (in order to produce a better estimate of $\langle z \rangle$). However, oversplitting can waste computer time because, although splitting decreases the variance per selected subregion, it increases computing time. By contrast, Russian roulette increases the variance per initial sample but decreases computing time.

5.8.1 Application to Monte Carlo Simulation

Russian roulette and splitting are widely used in MC simulations of problems for which $z(\mathbf{x})$ is not known explicitly and whose calculation involves a complex geometry and many physical models.

Suppose, for example, one is trying to estimate the expected number of annual traffic accidents at a particular intersection in a town. Many models are needed in such a simulation to specify the probabilities of when and where car trips are started, weather conditions, speeds and accelerations, the many possible turns drivers may take, etc. For each car trip history, the time the car reaches the intersection in question is recorded. An analog simulation would require millions of histories to even find a few near coincidental arrivals at the intersection. Most histories would never reach the intersection in question and produce a zero score. For such a simulation, Russian roulette and splitting would be very effective. Histories in which a car leaves the town on a highway have a low probability of returning and reaching the intersection. Such histories can be terminated by Russian roulette. Then histories in which cars turn onto streets leading to the intersection can be split to produce more arrivals at the intersection. Likewise, trips that begin between 1 and 4 in the morning can be terminated by Russian roulette because of the low traffic density, while trips that begin at rush hour should be split. Clearly, the decision of when to kill or split a history depends on the insight of the analyst.

In such simulations, a history is given some initial weight. The history is tracked as it moves through the various subregions in the problem space. Under certain adverse conditions, one may choose to terminate histories with probability p. Those histories not terminated (with probability $1 - p$) continue with modified weight

$$W' = \frac{W}{1 - p}, \tag{5.98}$$

where W is the incoming weight and W' is the outgoing weight. Under other favorable conditions, one may choose to split histories, replacing one having weight W with m, each having weight

$$W' = \frac{W}{m}. \tag{5.99}$$

Although splitting and Russian roulette can be extremely effective in problems in which very few histories ever produce a nonzero score, it is often difficult to subdivide the problem phase space into contiguous subregions and then to assign effective importance values to each region. For simple problems of low dimensionality, regions near scoring regions are given larger importances than regions far from the scoring regions. Also as a general rule, large changes in importance between adjacent regions should be avoided.

5.9 Combinations of Different Variance Reduction Methods

Until now, each variance reduction technique has been examined as if it were the only biased technique being used for a particular problem. By showing that the expected value of the tally in the biased game is the same as that for the unbiased (analog) game, the biased game is shown to be a "fair game." Then from examination of the second moments of the biased and unbiased game, conditions can be developed so that the biased game produces a tally with a smaller variance than that of the unbiased game.

In many complex MC analyses, particularly those based on simulations of nature, it is observed empirically that combining different variance reduction techniques can produce tallies with smaller variances than can be obtained using each single variance reduction technique alone. Yet, until relatively recently, no rigorous proof was available to guarantee that an arbitrary combination of variance reduction techniques, each of which alone produces a fair game, also yields a fair game. In other words, is any combination of fair games also a fair game?

Proofs of fairness were available for specialized cases. Lux and Koblinger (1991) showed that the combination of splitting and biased kernels gave unbiased scores. Booth (1985) showed that splitting can be combined with *any* variance reduction techniques provided the techniques do not depend on the particle weight. Booth and Pederson (1992) extended this result to show that any arbitrary combination of variance reduction techniques produces an unbiased tally provided the techniques are again weight-independent. This proviso means that if no weight-dependent games are played, the particle random walks are independent of weight and particles with different weights have the same random walks for the same random number sequence and produce the same score for a constant weight factor.

However, some of the most powerful variance reduction techniques do depend heavily on the particle weight (e.g., weight windows and energy cutoffs discussed in Chapter 10 for radiation transport). Booth and Pederson (1992) finally provided a theoretical basis to show that all possible combinations of variance reduction techniques (each of which produces an unbiased score) also gives an unbiased tally. Thus, any combination of fair games is itself a fair game.

5.10 Biased Estimators

Consider the estimation of

$$I = \int_V z(\mathbf{x}) f(\mathbf{x}) \, d\mathbf{x} = \int_V g(\mathbf{x}) \, d\mathbf{x}, \qquad (5.100)$$

where $f(\mathbf{x})$ is a PDF. To decrease the variance of the estimate of I, importance sampling is often used to recast the problem as

$$I = \int_V \frac{g(\mathbf{x})}{f^*(\mathbf{x})} f^*(\mathbf{x}) \, d\mathbf{x}, \qquad (5.101)$$

where $f^*(\mathbf{x})$ is a PDF that is approximately proportional to $g(\mathbf{x})$. An unbiased estimator of I is the straightforward sample-mean estimator

$$I_{sf} = \frac{1}{N} \sum_{i=1}^{N} \frac{g(\mathbf{x}_i)}{f^*(\mathbf{x}_i)}. \qquad (5.102)$$

Sometimes the PDF $f^*(\mathbf{x})$ from which the random variables are to be generated is very complicated and the sampling is difficult. For example, in the above importance sampling, if $g(\mathbf{x})$ is a complicated function, then $f^*(\mathbf{x})$ can also become complicated and sampling from it can be difficult. Powell and Swann

(1966) suggest a method called *weighted uniform sampling* that samples from a uniform distribution so as to avoid such difficulties.

Another estimator of I, albeit a biased one, is

$$\overline{I}_{bias} = \frac{\sum_{i=1}^{N} g(\mathbf{u}_i)}{\sum_{i=1}^{N} f^*(\mathbf{u}_i)}, \tag{5.103}$$

where the $\mathbf{u}_i$ are sampled uniformly in V. It is easy to show that $\overline{I}_{bias}$ is a biased estimator, namely,

$$\langle \overline{I}_{bias} \rangle = \left\langle \frac{\sum_{i=1}^{N} g(\mathbf{u}_i)}{\sum_{i=1}^{N} f^*(\mathbf{u}_i)} \right\rangle \neq \frac{V \langle \sum_{i=1}^{N} g(\mathbf{u}_i) \rangle}{V \langle \sum_{i=1}^{N} f^*(\mathbf{u}_i) \rangle}. \tag{5.104}$$

Clearly, $\lim_{N\to\infty} V \sum_{i=1}^{N} g(\mathbf{u}_i) = V \langle g(\mathbf{u}) \rangle = I$ and $\lim_{N\to\infty} V \sum_{i=1}^{N} f^*(\mathbf{u}_i) = 1$. Thus, $\overline{I}_{bias}$ is a *consistent* estimator because

$$\lim_{N\to\infty} \overline{I}_{bias} = \lim_{N\to\infty} \left[\frac{V \sum_{i=1}^{N} g(\mathbf{u}_i)}{V \sum_{i=1}^{N} f^*(\mathbf{u}_i)} \right] = \left[\frac{\lim_{N\to\infty} V \sum_{i=1}^{N} g(\mathbf{u}_i)}{\lim_{N\to\infty} V \sum_{i=1}^{N} f^*(\mathbf{u}_i)} \right] = I. \tag{5.105}$$

In this method, then, only uniform sampling from V is required, a great improvement if $f^*(\mathbf{x})$ is complex and difficult to sample. Powell and Swann (1966) show that, for large N, this method is $\sqrt{N}$ times more efficient than the sample-mean method.

5.11 Improved Monte Carlo Integration Schemes

In the usual sample-mean Monte Carlo integration of

$$I = \int_a^b z(x) f(x) \, dx \equiv \int_a^b g(x) \, dx \simeq \overline{I} \equiv \frac{b-a}{N} \sum_{i=1}^{N} g(x_i), \tag{5.106}$$

in which points x_i are sampled uniformly from (a, b), the error in the estimator $\overline{I}$ decreases as $1/\sqrt{N}$. However, this $1/\sqrt{N}$ behavior for MC quadrature is not inevitable. Quadrature schemes can be found that have much faster convergence. For example, if an equispaced Cartesian grid was superimposed on the interval (a, b) and each grid point was sampled exactly once, in any order, the relative error decreases at least as fast as $1/N$. But a grid scheme is not very attractive because one has to decide in advance how fine to make the grid, and all of the grid must be used. Unlike Monte Carlo, one cannot stop a grid-based calculation when a "good enough" result is obtained. All predetermined N points must be evaluated. The advantage of Monte Carlo methods is that N can be increased "on the fly" until a prescribed precision is reached. Two MC approaches for quadrature are presented below that achieve a faster convergence than that based on the uncorrelated random ordinates of the sample-mean method.

5.11.1   Weighted Monte Carlo Integration

A simple MC quadrature scheme that has faster convergence than the sample-mean approach is the so-called *weighted Monte Carlo* scheme (Yakowitz et al., 1978). In this approach, N values of $g(x_i)$ are calculated, where x_i are uniformly distributed over the interval (a, b). Then $\{x_i\}$ are ordered in increasing order as $x_{(1)}, \ldots, x_{(N)}$ and the trapezoid quadrature approximation is used to approximate the integral between adjacent pairs of $x_{(i)}$ to give the estimate

$$\bar{I} = \frac{1}{2} \sum_{i=0}^{N} [g(x_{(i)}) + g(x_{(i+1)})][x_{(i+1)} - x_{(i)}]$$

$$= \frac{1}{2} \left\{ \sum_{i=1}^{N} g(x_{(i)})[x_{(i+1)} - x_{(i-1)}] + g(x_{(0)})[x_{(1)} - x_{(0)}] \right.$$

$$\left. + g(x_{(N+1)})[x_{(N+1)} - x_{(N)}] \right\}, \quad (5.107)$$

Example 5.6 Buffon's Problem: Weighted Monte Carlo Quadrature

From Eq. (1.6), the probability a dropped needle cuts a grid line is

$$P_{cut} = \int_0^D \frac{s_1(x) + s_2(x)}{2\pi L D} dx \equiv \int_0^D g(x)\, dx.$$

The functions $s_1(x)$ and $s_2(x)$ are given by Eq. (1.4) and Eq. (1.5), respectively. In the usual sample-mean Monte Carlo approach, P_{cut} is estimated as

$$P_{cut} \simeq \frac{D}{N} \sum_{i=1}^{N} g(x_i),$$

where the random $\{x_i\}$ are uniformly distributed in the interval $(0, D)$.

With a uniform grid of x_i values over $(0, D)$, with $\Delta x = D/N$, P_{cut} can be estimated from the mid-interval values of $g(x)$ as

$$P_{cut} \simeq \Delta x \sum_{i=1}^{N} g\left(\left[i - \frac{1}{2}\right]\Delta x\right) = \frac{D}{N} \sum_{i=1}^{N} g\left(\left[i - \frac{1}{2}\right]\Delta x\right).$$

Finally, P_{cut} can be obtained using the weighted MC scheme of Eq. (5.107). A comparison of these three methods is shown in Fig. 5.7, where it is seen that both the Cartesian-grid and weighted MC schemes have an error that decreases as $1/N$, much faster than the $1/\sqrt{N}$ of the usual sample-mean MC method.

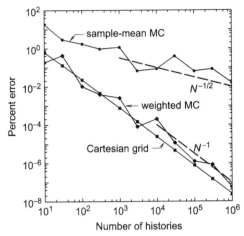

Figure 5.7 Comparison of three different quadrature schemes for the Buffon's needle problem with $L = D/2$.

where $x_{(0)} = a$ and $x_{(N+1)} = b$. In this scheme, the relative error decreases at least as fast as $1/N$. Moreover, if a more accurate result is needed, additional x_i and $g(x_i)$ values are calculated, included with the previously calculated values, the entire set is then resorted, and then Eq. (5.107) is used to calculate a refined value of $\overline{I}$. In this way, previous values of $f(x_i)$ are not discarded when increasing N. By contrast, if N is changed in deterministic numerical quadrature schemes, all the $\{g(x_i)\}$ must be recomputed, a terrible penalty if the evaluation of $f(x)$ is computationally expensive.

Although the weighted MC method has at least $1/N$ convergence, it has the extra computational overhead of sorting the x_i ordinates whenever another refined estimate of the integral is desired. However, very efficient sorting routines are available (Press et al., 1996) and, generally, most of the computational effort is associated with computing $g(x)$. The major drawback with the weighted MC integration scheme is that it is difficult to apply to multidimensional cases.

5.11.2 Monte Carlo Integration with Quasi-Random Numbers

It is possible to improve greatly the convergence of the sample-mean approach by using a different type of sequence for selecting x_i. This special type of sequence is based not on pseudorandom numbers that simulate true randomness, but on numbers that strive for uniformity of coverage over the integration volume. Generators that produce such numbers are intentionally designed to be *sub*random so that generated values, although appearing random, try to avoid previous values, thereby producing a more even-distributedness of the generated values. In other words, numbers from a quasi-random generator try to fill in the gaps left by the previous numbers.

By using various tricks to represent numbers in different prime bases, for example, it is possible to produce long sequences of quasi-random numbers that more uniformly fill the integration volume than do true random numbers (Halton, 1960). Several other ways to generate quasi-random number sequences have been proposed, notably that of Sobol' (1967). An excellent review of such quasi-random number generators is given by Bratley and Fox (1988), and practical implementations are provided by Press et al. (1996).

Can quadrature with these quasi-random number generators achieve a true $1/N$ convergence? The answer is almost yes. For integration of a smooth function in k dimensions, a Sobol' sequence has a convergence rate proportional to $(\ln N)^k/N$, which is almost as fast as $1/N$.

5.12 Summary

There are various ways to modify the sampling and scoring procedures of Monte Carlo that do not change the sample mean (in the limit) but do change the sample variance. If these variance reduction procedures are carefully implemented, the sample variance of a Monte Carlo estimate can be reduced, often significantly. However, the variance reduction schemes cannot be applied blindly or automatically. Often, several variance reduction approaches can be used simultaneously to good effect. But sometimes combining different schemes or misusing a single scheme can produce totally wrong results. The effective implementation of variance reduction for a particular problem is part of the "art" of using Monte Carlo.

Problems

5.1 Estimate the value of the integral

$$I = \int_0^1 e^{7x} dx$$

by straightforward sampling, i.e., with x_i uniformly sampled over the interval $[0, 1]$. The function $3^{7x}/2$ has a shape similar to that of the above integrand. Convert this function to a PDF $f^*(x)$ and use it with importance sampling to estimate I. Estimate the standard deviation of both estimates. Finally, plot the accuracy of your estimates as a function of the number of samples used.

5.2 Use stratified sampling to estimate the value of the integral $I = \int_V g(\mathbf{x}) d\mathbf{x}$, where

$$f(\mathbf{x}) = f(x, y) = \begin{cases} x^2 y^2 \sin^2 x \sin^2 y, & x > 0, y > 0, \\ x^2 y^2/25, & x \leq 0 \text{ or } y \leq 0. \end{cases}$$

Here V is the interior of the circle $x^2 + y^2 = 25$. Subdivide V into the four quadrants of the circle and estimate the number of samples to be taken from each quadrant to produce an estimate of I with the smallest variance. Compare your results with that obtained by using the straight-forward sample-mean result for a single region.

5.3 If ρ_i is a random number uniformly sampled from $[0, 1]$, show that the random numbers $x_i = a\rho_i + b$ and $x_i' = a(1 - \rho_i) + b$ have a correlation coefficient equal to -1.

5.4 Consider the problem of estimating the small difference between the two integrals

$$\Delta = \int_0^b \frac{\sin \sqrt{x}}{\sqrt{x}} dx - \int_0^b \left[1 - \frac{x}{3!} + \frac{x^2}{5!} - \frac{x^3}{7!} \right] dx.$$

The right-hand integral can be analytically integrated to give

$$I_o = b - \frac{b^2}{12} + \frac{b^3}{360} - \frac{b^4}{20,160}.$$

For $b = 5$, use the correlated sampling estimators of Eq. (5.55) and Eq. (5.61) to evaluate Δ. Also estimate the standard deviations of your estimates. Calculate the optimum value of α and use the same sampled values of x for both integrals.

5.5 Evaluate the integral

$$I = \int_1^3 x^4 dx$$

by the usual sample-mean method (see Eq. (5.106)) and the antithetic sampling estimator of Eq. (5.76). Plot the percent error and the estimated standard deviation of the estimators as a function of the number of samples used.

5.6 Prove Eq. (5.84) is still valid if $g(x)$ is a monotonic increasing function of x over the interval $[0, 1]$

5.7 Consider evaluation of the integral $\int_V z(\mathbf{x}) f(\mathbf{x}) d\mathbf{x}$ using the volume partitioning method of Section 5.6 for the function

$$z(\mathbf{x}) = z(x, y) = \begin{cases} x^2 y^2 \sin x \sin y, & x > 0, y > 0, \\ 1 & x \le 0, \text{ or } y \le 0. \end{cases}$$

Here V is the interior of the circle $x^2 + y^2 = 25$ and the PDF $f(\mathbf{x}) = 1/(25\pi)$, i.e., points $\mathbf{x} = (x, y)$ are uniformly distributed in V. Compared your estimated precision with that of the straightforward sample-mean method using a single integration volume for different number of samples N.

5.8 Consider the integral

$$I = \int_0^\infty x^2 \cos x \, e^{-x/2} \, dx.$$

(a) Use the straightforward sample-mean method with exponential sampling to estimate this integral. (b) Use exponential antithetic samples to estimate this integral. (c) With both methods obtain 200 estimates using 1000 samples for each estimate, and compare the variance of the two estimators.

5.9 Consider the integrand of Problem 5.1. Use the biased estimator of Section 5.10 to estimate the value of the integral and compare the accuracy with the number of samples used.

5.10 Use the weighted MC integration method as well as the straightforward sample-mean method to evaluate

$$I = \int_{-1}^1 \cos^2 x \, dx$$

and plot the accuracy of the results as a function of the number of samples used.

References

Bratley, P., Fox, B.L., 1988. Algorithm 659: implementing Sobol's quasirandom sequence generator. ACM Trans. Math. Softw. 14, 88–100.

Booth, T.E., 1985. Monte Carlo variance comparison for expected value versus sample splitting. Nucl. Sci. Eng. 89, 305–309.

Booth, T.E., Pederson, S.P., 1992. Unbiased combinations of nonanalog Monte Carlo techniques and fair games. Nucl. Sci. Eng 110, 254–261.

Halton, J.H., 1960. On the efficiency of certain quasi-random sequences of points in evaluating multi-dimensional integrals. Numer. Math. 2, 88–90.

Hammersley, J.M., Morton, K.W., 1956. A new Monte Carlo technique: antithetic variates. Proc. Camb. Philol. Soc. 52, 449–474.

Kahn, H., 1954. Applications of Monte Carlo. AECU-3259. U.S. Atomic Energy Commission. Available through Technical Information Service Extension, Oak Ridge, TN.

Lux, I., Koblinger, L., 1991. Monte Carlo Particle Transport Methods: Neutron and Photon Calculations. CRC Press, Boca Raton, FL.

Marshall, A.W., 1956. The use of multi-stage sampling schemes in Monte Carlo computations. In: Meyer, M.A. (Ed.), Symposium on Monte Carlo Methods. Wiley, New York, pp. 123–140.

Powell, M.I.D., Swann, I., 1966. Weighted uniform sampling—a Monte Carlo technique for reducing variance. J. Inst. Math. Appl. 2, 228–238.

Press, W.H., Teukolsky, S.A., Vetterling, W.T., Flannery, B.P., 1996. Numerical Recipes, 2 ed. Cambridge Univ. Press, Cambridge.

Rubinstein, R.V., 1981. Simulation and the Monte Carlo Method. Wiley, New York.

Schreider, Y.A. (Ed.), 1966. The Monte Carlo Method (the Method of Statistical Trials). Pergamon, Elmsford, New York.

Shultis, J.K., Faw, R.E., 2000. Radiation Shielding. American Nuclear Society, La Grange Park, IL.

Sobol', I.M., 1967. Distribution of points in a cube and approximate evaluation of integrals. USSR Comput. Math. Math. Phys. 7 (4), 86–112.

Yakowitz, S., Krimmel, J.E., Szidarovsky, F., 1978. Weighted Monte Carlo integration. Soc. Ind. Appl. Math. J. Numer. Anal. 15 (6), 1289–1300.

Chapter 6

Markov Chain Monte Carlo

A rose by any other name
smells as sweet and looks the same,
and chains that are Markov
can be the start of
just another MC game.

At this stage in our exploration of Monte Carlo methods, it is instructive to review what we have discovered and to see if there is a more general framework that can alleviate some inherent problems with the approaches taken so far. The sampling methods discussed in Chapters 4 and 5 are quite effective for many cases but not for all. The goal of a typical Monte Carlo calculation is to estimate the expected value of a real-valued function $h(\mathbf{x})$, i.e., to evaluate

$$\mu(h) \equiv \langle h \rangle = \int_V h(\mathbf{x}) f(\mathbf{x}) \, d\mathbf{x}, \qquad (6.1)$$

where $\mathbf{x}$ is a vector of random variables and f is the probability density function (PDF) defined on a "volume" or domain V that describes how $\mathbf{x}$ is distributed in V. This general formulation applies whether the random variable $\mathbf{x}$ is one- or multidimensional, whether it is continuous or discrete, and whether or not the integral is explicit or implicit.

6.1 Review of the Ordinary Monte Carlo Method

Here, we assume that a source of pseudorandom numbers is available. The goal of Monte Carlo calculations is to use these pseudorandom numbers to solve one or both of the following problems.

Problem 1: Generate independent samples $\mathbf{x}_1, \mathbf{x}_2, \mathbf{x}_3, \dots, \mathbf{x}_N$ from a known (either explicitly or from simulation) PDF $f(\mathbf{x})$. Such random variables $\mathbf{x}_n$ are said to be *independent identically distributed* (i.i.d.) random variables.

Problem 2: Estimate the expectation of $h(\mathbf{x})$ (i.e., approximate Eq. (6.1)) and its standard deviation.

The first problem is usually the more difficult one (as discussed in this chapter) because, if Problem 1 is solved, then the solution to the second problem can be

Exploring Monte Carlo Methods. https://doi.org/10.1016/B978-0-12-819739-4.00014-7

estimated as

$$\mu(h) \simeq \overline{\mu}(h) \equiv \frac{1}{N} \sum_{n=1}^{N} W_n h(\mathbf{x}_n), \qquad (6.2)$$

where the weights W_n are either unity or account for any bias introduced into the sampling. The law of large numbers (LLN) guarantees that $\overline{\mu}(h)$ is a consistent unbiased estimator of $\mu(h)$ as $N \to \infty$.[1] Further, the central limit theorem (CLT) says $\overline{\mu}(h)$ is approximately normally distributed for large N, provided $\langle h^2 \rangle$ is finite. In particular, $\sigma^2(h)$ can be estimated as

$$\sigma^2(h) \simeq s^2(h) \equiv \frac{1}{N-1} \sum_{n=1}^{N} [h(\mathbf{x}_n) - \overline{\mu}(h)]^2, \qquad (6.3)$$

from which confidence intervals for $\mu(h)$ can be constructed, for large N, from the approximating normal distribution. More important, from Eq. (2.60), the variance of $\overline{\mu}(h)$ is $\sigma^2(\overline{\mu}) = \sigma^2(h)/N$. In other words, $\overline{\mu}(h)$ is asymptotically and normally distributed as $\mathcal{N}(\mu(h), \sigma^2(h)/N)$. This result is extremely important because it says *the accuracy of the Monte Carlo estimate of $\mu(h) \simeq \overline{\mu}(h)$ depends on only the variance of h and not on the dimensionality of the space sampled.* So regardless of the dimensionality of V, from which random samples are drawn, a modest number of samples may give an estimate of $\mu(h)$ that is sufficiently accurate.

Monte Carlo methods that solve Problem 2 with i.i.d. random variables, as just described, are called *ordinary* Monte Carlo (OMC), or i.i.d. MC, or other variants.

Why Is Problem 1 Difficult?

In principle, one can estimate $\langle h \rangle$ by any of a number of Monte Carlo sampling methods described in Chapters 4 and 5. Indeed, this flexibility is one of the main virtues of Monte Carlo. For instance, one can use the marginal and conditional formulation implied by Eq. (4.25) for a three-dimensional PDF and pick samples of each variable sequentially by the inverse CDF method or by rejection sampling. Alternatively, one can pick samples from some other distribution and then weight the samples using importance sampling, as discussed in Section 5.2. These standard sampling techniques, however, may not work well in many cases for either or both of two reasons.

---

[1] Recall that a *consistent* estimator is one that approaches the true value as the sample size increases. The expected value of an *unbiased* estimator is the true value.

First, one may know the shape of $f(\mathbf{x})$ but not the normalizing constant C needed to make $f(\mathbf{x})$ a PDF. For instance, one might hypothesize that the variables are distributed according to complex functions that are difficult to reduce to normalized PDFs, i.e., the shapes of the distributions may be presumed, but the normalization constants required to make the shapes into PDFs may be difficult to determine. A one-dimensional example is given below and plotted in Fig. 6.1.

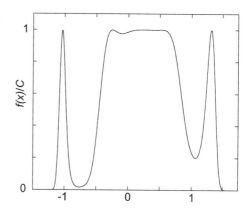

Figure 6.1 Plot of the PDF given below, divided by its normalization constant.

$$f(x) = C \exp[-5(x - 0.5)^4 \sin^2(4x)].$$

Even if the normalizing constant C was known approximately, say from the use of numerical quadrature, sampling from $f(x)$ is still computationally expensive.

Second, even if the normalization constant C was known, generating samples from $f(\mathbf{x})$ can be a daunting task, especially in high-dimensional spaces. Proper samples should come mostly from regions in $\mathbf{x}$-space in which $f(\mathbf{x})$ is large and less frequently from regions in which $f(\mathbf{x})$ is small or zero. Also, to reduce $\sigma^2(\overline{h})$, one should sample preferentially from regions in which large values of $f(\mathbf{x})$ vary rapidly. But how can one identify important regions in V without evaluating $f(\mathbf{x})$ *everywhere*?

6.2 Markov Chains to the Rescue

Metropolis et al. (1953) recognized that Markov chains could be applied to difficult sampling problems, and Hastings (1970) constructed what now is referred to as the *Metropolis–Hasting* (M-H) algorithm for performing such samplings. Interest in what has come to be known as *Markov chain Monte Carlo* (MCMC) has exploded since then. The concept of Markov chains was introduced by Markov[2] (1906) and these chains were given the name *Markov chains* by Bernstein (1926). MCMC is simply Monte Carlo in which the samples are obtained from a Markov chain. But what, you may ask, is a Markov chain? First, we will consider the case in which there is a single random variable x, which can assume a finite number M of *states*.

---

[2] Andrey Andreyevich Markov (1856–1922) was a Russian mathematician best known for his work on stochastic processes. A primary subject of his research later became known as Markov chains and Markov processes.

6.2.1   A Discrete Random Variable

A Markov chain is a sequence of random samples x_i, $i = 1, 2, \ldots, N$, for which the value x_{i+1} depends only on the previous value x_i and not on any of the values x_k, where $k < i$. The terminology often is expressed in a temporal analogy, in which the "next state" (random sample) x_{i+1} depends on the "current state" x_i but not on any earlier states. A chain of states (sequence of random samples) for which this dependence is true is called a Markov chain.

In the stochastic process the discrete random variable x, which specifies the state of the system, moves from the current state x_i to the new state x_j with a certain *transition probability* T_{ij}, which depends only on the current state x_i. Because this probability is a conditional probability, T_{ij} is sometimes written as $T_{i|j}$. If there are M possible states, then an $M \times M$ transition matrix

$$\mathbf{T} = \{T_{ij}\}, \quad i, j = 1, 2, \ldots, M, \tag{6.4}$$

gives the various probabilities of transitioning among all possible states. Because after each transition one of the m states must be selected, this matrix is normalized, such that $\sum_{j=1}^{M} T_{ij} = 1$, for any i. Note that T_{ii} is the probability of remaining in state x_i and also that certain transition probabilities may be zero, i.e., some transitions may be prohibited. Each transition from one state to another depends only on the current state and not on how that state happened to evolve. Thus, a sequence of states sampled from $\mathbf{T}$ form a Markov chain.

A square matrix with nonnegative elements and each row of which sums to 1 is called a *Markov* matrix or a left *stochastic* matrix. In a right stochastic matrix, the columns sum to 1. Such matrices have several interesting properties. A Markov matrix always has a maximal eigenvalue 1 and its associated eigenvector has real positive components. All other eigenvalues have an absolute value smaller than or equal to 1. These results are a consequence of the Perron–Frobenius theorem (Bapat and Raghavan, 1997): *a real square matrix whose elements are positive has a unique maximal eigenvalue whose eigenvector has positive components.*

The transition matrix is inextricably related to the discrete PDF for the random variable x. Let $p_{ij}^{(n)}$ be the probability of transitioning from state x_i to state x_j after n steps and, for simplicity, let $p_{ij}^{(1)} = p_{ij}$, where p_{ij} is the probability of transitioning from state x_i to state x_j in one step. Now, define the row vector

$$\boldsymbol{\pi}^{(n)} = \{\pi_1^{(n)}, \pi_2^{(n)}, \ldots, \pi_M^{(n)}\}, \tag{6.5}$$

where $\pi_i^{(n)}$ is the probability that x has the value x_i on the nth step. Such a vector with nonnegative components that sum to 1 is called a *stochastic* vector. The probability that the chain assumes state x_i at step $n+1$ is given by

$$\pi_i^{(n+1)} = \sum_{k=1}^{M} \pi_k^{(n)} T_{ki}. \tag{6.6}$$

The matrix form of this equation is[3]

$$\boldsymbol{\pi}^{(n+1)} = \boldsymbol{\pi}^{(n)} \mathbf{T}, \tag{6.7}$$

and can be used to infer that

$$\boldsymbol{\pi}^{(n)} = \boldsymbol{\pi}^{(n-1)} \mathbf{T} = (\boldsymbol{\pi}^{(n-2)} \mathbf{T}) \mathbf{T} = \boldsymbol{\pi}^{(n-2)} \mathbf{T}^2. \tag{6.8}$$

Thus, the probability of transitioning from state x_i to state x_j in one step is just $p_{ij} = T_{ij}$ and the probability of going from state x_i to state x_j in two steps is given by

$$p_{ij}^{(2)} = \sum_{k=1}^{M} T_{ik} T_{kj}, \tag{6.9}$$

which is just the product of the ith row of $\mathbf{T}$ with the jth column of $\mathbf{T}$, or the ijth element of $\mathbf{T}^2$.

This result easily can be generalized to show that the probability of transitioning from state x_i to state x_j after n steps, i.e., $p_{ij}^{(n)}$, is simply the ijth element of $\mathbf{T}^n$. This is equivalent to the result, derivable from Eq. (6.7), that

$$\boldsymbol{\pi}^{(n)} = \boldsymbol{\pi}^{(0)} \mathbf{T}^n. \tag{6.10}$$

Eventually, as n is increased, $\boldsymbol{\pi}^{(n)}$ approaches or converges to the *stationary* distribution $\boldsymbol{\pi}$ or the discrete PDF on which the transition matrix is based, as does each row of $\mathbf{T}^n$, i.e., $\pi_i \equiv f_i = \lim_{n\to\infty} \pi_i^{(n)}$. Upon convergence, Eq. (6.7) becomes

$$\boldsymbol{\pi} = \boldsymbol{\pi} \mathbf{T}, \tag{6.11}$$

from which the stationary PDF is seen to be the (left) eigenvector associated with the eigenvalue $\lambda = 1$ of $\mathbf{T}$. This result verifies the discussion earlier in this section about Markov matrices.

---

[3] Why many MCMC practitioners persist in using left matrix multiplication is unknown to the authors. The right matrix multiplication convention, usually used in linear algebra, when applied to Eq. (6.7) gives

$$\boldsymbol{\pi}^{(n+1)} = \mathbf{T}^T \boldsymbol{\pi}^{(n)},$$

where $\boldsymbol{\pi}$ is now a column vector. But we continue with the left multiplication practice.

The initial state x_i is specified by the row vector $\pi^{(0)}$, all elements of which are zero except the ith element, whose value is unity. For instance, if the initial state is x_i for $i = 3$ out of $m = 5$ states, the initial row vector would be $\pi^{(0)} = (0, 0, 1, 0, 0)$. Multiplication of $\mathbf{T}^n$ by the initial state row vector produces the ith row of $\mathbf{T}^n$, which gives the probabilities of transitioning from state x_i to each of the M states. Example 6.1 demonstrates some of these concepts for a particular situation.

Example 6.1 The Weather in Transylvania

Consider that the daily weather in Transylvania is categorized into $m = 3$ possibilities, namely, "sunny" ($x_1 = s$), "cloudy" ($x_2 = c$), and "rainy" ($x_3 = r$). Assume that the transition matrix is known and is given by

$$\mathbf{T} = \begin{pmatrix} T_{ss} & T_{sc} & T_{sr} \\ T_{cs} & T_{cc} & T_{cr} \\ T_{rs} & T_{rc} & T_{pr} \end{pmatrix} = \begin{pmatrix} 0.100 & 0.500 & 0.400 \\ 0.500 & 0.100 & 0.400 \\ 0.600 & 0.300 & 0.100 \end{pmatrix},$$

where the first subscript on T_{ij} indicates the initial state and the second subscript indicates the final state.

Today is a cloudy day in Transylvania, i.e., $\pi^{(0)} = (0, 1, 0)$. Out of the next three days, on which day is it least likely to rain? This is easily determined by comparing $p_{23}^{(n)}$ for $n = 1, 2, 3$. From $\mathbf{T}$, the probability of rain tomorrow is 40.0%; from

$$\mathbf{T}^2 = \begin{pmatrix} 0.500 & 0.220 & 0.280 \\ 0.340 & 0.380 & 0.280 \\ 0.270 & 0.360 & 0.370 \end{pmatrix},$$

the probability of rain on the second day is 28.0%; and from

$$\mathbf{T}^3 = \begin{pmatrix} 0.3280 & 0.3560 & 0.3160 \\ 0.3920 & 0.2920 & 0.3160 \\ 0.4290 & 0.2820 & 0.2890 \end{pmatrix},$$

the probability of rain on the third day is 31.6%. (Thus, you might plan your golf outing for the second day.) Continue in this fashion until on the 15th day from now one obtains

$$\mathbf{T}^{15} = \begin{pmatrix} 0.37912 & 0.31319 & 0.30769 \\ 0.37912 & 0.31319 & 0.30769 \\ 0.37912 & 0.31319 & 0.30769 \end{pmatrix},$$

each row of which gives the stationary PDF $\pi_i(\mathbf{x})$, $i = 1, 2, 3$, to five decimal places.

Multiplication of $\mathbf{T}^2$ by the initial state row-vector $\pi^{(0)} = (0, 1, 0)$, i.e., $\pi^{(0)}\mathbf{T}^2$ produces the second row of $\mathbf{T}^2$. Thus the weather probabilities on "golf-day" (two days from now) are sunny (34.0%), cloudy (29.2%), or rainy (28.0%). Finally, notice that 15 days from now the probabilities of the weather being sunny, cloudy, and rainy are, to within five decimal places, independent of today's weather. The Markov chain has "forgotten" its initial state.

6.2.2 A Continuous Random Variable

Not all random variables are discrete and Markov chains can be formed for continuous random variables as well. When x is continuous, the transition probability is a function $T(x \to u)$ such that the quantity $T(x \to u)du$ gives the probability of transitioning from state x to a state that is within du about u in one step. If $x, u \in [a, b]$, then, because some state must be selected in each transition, this function is normalized such that $\int_a^b T(x \to u)\, du = 1$. The probability of going from state x to state u in two steps is given by

$$T^{(2)}(x \to u) = \int_a^b T(x \to v_1)T(v_1 \to u)dv_1 \tag{6.12}$$

and the probability of transitioning from state x to state u after n steps is given by the integral over the $n - 1$ intervening steps

$$T^{(n)}(x \to u) = \int_a^b T(x \to v_1)T(v_1 \to v_2) \int_a^b T(v_2 \to v_3) \ldots$$
$$\int_a^b T(v_{n-2} \to v_{n-1}) \int_a^b T(v_{n-1} \to u)dv_1\, dv_2 \ldots dv_{n-2}\, dv_{n-1}. \tag{6.13}$$

As in the discrete case, the transition function $T(x \to u)$ inherently contains information about $\pi(x)$, the PDF of x. In particular, it can be shown (Tierney, 1994) that the stationary or invariant PDF is

$$\pi(u) = \lim_{n \to \infty} T^{(n)}(x \to u). \tag{6.14}$$

In general, a state may be specified by a vector $\mathbf{x}$ of n random variables, each component of which may be either discrete or continuous. The possible values of $\mathbf{x}$ define the states of the system and, again, transition probabilities of going from one state to another can be defined and used to create Markov chains. Each discrete variable has a transition matrix and each continuous variable has a transition function. Methods to sample from multidimensional random variables are discussed later in Section 6.2.8.

Ergodic Markov Chains

Every Markov chain is based either on a single distribution or on a cycle of distributions in the sense that the chain samples converge to a single PDF or to multiple PDFs. An *ergodic* Markov chain is one for which the probability of transitioning from one state to another is independent of when the transition takes place (MacKay, 2007). In other words, a Markov chain for which every time a specific state x is encountered, the probability of transitioning to any state u (for discrete variables) or to any state within du about u (for continuous variables) is the same is an ergodic Markov chain. In this case, the underlying distribution does not change and the samples defining every ergodic Markov chain converge to a single distribution, often called the *stationary* or *invariant* distribution. Every ergodic Markov chain, thus, is based on a single underlying stationary PDF. Samples produced from an ergodic Markov chain whose stationary distribution is $\pi(x)$ are in fact random samples from $\pi(x)$.

Construction of an ergodic Markov chain with $\pi(x)$ as the stationary distribution is the key task of MCMC. But given $\pi(x)$, how does one create a Markov chain whose stationary distribution is $\pi(x)$? The answer to this question is surprisingly simple. One uses the M-H algorithm, discussed in Section 6.2.4, to generate samples from the specified $\pi(x)$.

An important feature of MCMC is that it provides a relatively simple way to sample from complex distributions. In fact, MCMC is typically used to sample multidimensional distributions and can be applied when only the shape of the distribution is known. This ability is a very powerful characteristic of MCMC, because the normalizing factors in even simple joint and conditional PDFs sometimes may be difficult to determine. MCMC allows one to sample from *unnormalized* distributions. Nevertheless, MCMC is, in essence, another way to obtain Monte Carlo samples and, as such, could have been included in Chapter 4. Indeed, MacKay (2007) treats MCMC in just this way. However, MCMC has such unique features and such a devoted user base that it deserves to be discussed as its own MC method.

6.2.3 MCMC Versus OMC

MCMC is very similar to OMC but with a major generalization. The MCMC samples $x_1, x_2, \ldots, x_N$ need be neither independent nor even have the same distribution. Rather, as suggested by the name, they are generated as a Markov chain. Ekvall and Jones (2019) point out that i.i.d. random variables also form a Markov chain and, thus, OMC can be considered a special case of MCMC. In particular, if x_i are i.i.d., the Markov chain is *stationary* and *reversible*, a topic discussed later in this chapter.

Under certain restrictions on the Markov chain, the theory of MCMC is very similar to that for OMC, whose theory was summarized in Section 6.1. The power of MCMC is that the properties of $\overline{\mu}(h)$, resulting from the LLN and the CLT for a Markov chain of i.i.d. random variables, continue to hold for much more general chains. Because such general chains can be constructed in many situations, for which i.i.d. sampling is difficult or infeasible, MCMC is often more useful than OMC.

First consider a single random variable, i.e., $\mathbf{x} \to x$. The estimate of the expected value of $\mu(h)$ is still given by Eq. (6.2). The complication in MCMC is in estimating the variance of the estimated mean. If the initial distribution $\pi(x)$ and transition kernel (or matrix) T are such that the distribution for x_2 is the same as that of x_1, then the initial distribution is said to be invariant for the kernel. More generally, a distribution $\pi(x)$ is invariant for the given kernel T if $x_i \sim \pi(x)$ implies[4] $x_{i+1} \sim \pi(x)$ for $i \geq 1$. From this definition it follows that, if the initial distribution is invariant for T, then every x_i, $i \geq 1$, has the same distribution (Ekvall and Jones, 2019). Such a Markov chain is said to be *stationary* and they are indeed stationary in the usual sense for a stochastic process.

Now suppose $\pi(x)$ is the equilibrium (invariant) distribution for the kernel T and a Markov chain is constructed with $\pi(x)$ as the initial distribution. Then from the above discussion $x_1, \ldots, x_N$ are possibly dependent but identically distributed random variables distributed as $\pi(x)$. As a consequence $\overline{\mu}(h)$ is an unbiased estimator of $\mu(h)$. Moreover, it can be shown $\overline{\mu}(h)$ is consistent and asymptotically normal with variance $\kappa^2(h)/N$, where (Jones, 2004)

$$\kappa^2(h) = \sigma^2(h) + 2 \sum_{k=1}^{\infty} \text{cov}[h(x_i), h(x_{i+k})]. \tag{6.15}$$

Note if the x_i are i.i.d. samples, the autocovariances vanish and $\kappa^2(h)$ reduces to $\sigma^2(h)$, the same result as obtained for OMC.

However, generating stationary Markov chains is never done because it requires sampling x_1 from the invariant distribution $\pi(x)$. The principal reason for using MCMC is to avoid such sampling! Fortunately, it turns out that if $\overline{\mu}(h)$ is a consistent estimator and satisfies a CLT when the initial distribution is $\pi(x)$, then any other initial distribution also produces a consistent estimator (Meyn and Tweedie, 1993). Thus, if the chain is long enough (i.e., N is sufficiently large), the starting value x_1 is unimportant and can be randomly chosen or set to some fixed value. This observation means that the asymptotic variance is the same for all initial distributions. With the sampling methods discussed in the next sections, it is fairly straightforward to construct Markov chains with a desired invariant or stationary distribution and from which the estimator $\overline{\mu}(h)$ for $\mu(h)$ is both consistent and asymptotically normally distributed as $\mathcal{N}(\mu(h), \kappa^2(h)/N)$.[5]

---

[4] The notation $x_i \sim \pi(x)$ means x_i is a sample of a random variable x that is distributed in accordance to $\pi(x)$.

[5] However, not all chains with correct invariant distributions give a consistent and asymptotically normal estimate of $\overline{\mu}(h)$. Conditions to avoid such an undesired outcome are technically complex and subject to case-by-case analysis (Jones, 2004).

The Multivariate Case

Suppose one seeks to estimate $\mu = \langle \mathbf{h}(\mathbf{x}) \rangle$, where $\mathbf{x}$, μ, and $\mathbf{h}$ are vectors with components x_k, μ_k, and h_k, respectively. The Monte Carlo estimator is still given by Eq. (6.2), which now is a vector equation because $\mathbf{h}(\mathbf{x})$ is a vector. Then the multivariate Markov chain CLT says (Geyer, 2011)

$$\lim_{N \to \infty} \overline{\mu}(\mathbf{h}) = \mathcal{N}(\mu, \Sigma/N), \qquad (6.16)$$

where

$$\Sigma = \sigma^2(\mathbf{h}) + 2 \sum_{\ell=1}^{\infty} \mathrm{cov}[\mathbf{h}(\mathbf{x}_i), \mathbf{h}(\mathbf{x}_{i+\ell})]. \qquad (6.17)$$

Although the right sides of Eqs. (6.15) and (6.17) are the same, their interpretation is different. In Eq. (6.17), $\sigma^2(\mathbf{h})$ is the square matrix whose nmth component is $\mathrm{cov}[h_n(\mathbf{x}_i), h_m(\mathbf{x}_i)]$. Likewise, $\mathrm{cov}[\mathbf{h}(\mathbf{x}_i), \mathbf{h}(\mathbf{x}_{i+k})]$ denotes the matrix with components $\mathrm{cov}[h_n(\mathbf{x}_i), h_m(\mathbf{x}_{i+k})]$.

The necessary conditions for the univariate CLT to guarantee the asymptotic convergence of the estimator $\overline{\mu}(h)$ to a normal distribution are essentially the same as for the multivariate case (Geyer, 2011). Because the multivariate CLT follows almost directly from the univariate CLT, the former is seldom discussed. Nevertheless, the multivariate CLT can be applied to multivariate Markov chain analyses and should be used when needed.

Calculating the Uncertainty in $\overline{\mu}$

The variance $\sigma^2(h)$ is determined solely by the invariant distribution, whereas $\kappa^2(h)$ also depends on the joint distribution of all the variables in the chain. Because $\kappa^2(h)$ directly determines the uncertainty in the estimator $\overline{\mu}(h)$, an algorithm for constructing the MCMC chain should be used that produces small (or negative) autocovariances. Two stationary chains with the same invariant distribution but different autocovariances produce the same $\sigma^2(h)$ but possibly significantly different $\kappa^2(h)$. Finally, it should be mentioned that Eq. (6.15) is only used to estimate $\kappa^2(h)$ for simple idealized "toy" problems. In real analyses, factors such as noise, missing data, etc., preclude its use and many alternative methods have been developed (Flegal and Jones, 2010; Geyer, 1992a; Robert and Casella, 2013).

Nonoverlapping Batch Means

To illustrate a practical alternative to Eq. (6.15) for estimating $\kappa^2(h)$, consider the method of nonoverlapping batch means. A batch is a subsequence of b consecutive iterates of the Markov chain $x_1, x_2, \ldots x_N$. If the chain is assumed to be stationary, then all batches of the same length have the same joint distribution. Further, assume the batch length b divides N evenly so the chain consists of exactly $K = N/b$ nonoverlapping batches. The estimator of the mean for batch

k is simply

$$\overline{\mu}_{b,k} = \frac{1}{b} \sum_{i=b(k-1)+1}^{bk} h(x_i), \quad k = 1, 2, \ldots, K. \tag{6.18}$$

If $\overline{\mu}_N$ is the estimator of the mean using all N chain iterates, then

$$\frac{\kappa^2(h)}{b} \simeq \frac{1}{K} \sum_{k=1}^{K} (\overline{\mu}_{b,k} - \overline{\mu}_N)^2. \tag{6.19}$$

6.2.4 The Metropolis–Hastings Algorithm

Metropolis et al. (1953) sought to simulate a liquid in equilibrium with its gas phase.[6] This is a highly multidimensional problem whose dimensionality is proportional to the number of molecules being considered. Previously, statistical sampling had been used to estimate the *minimum* potential energy of a collection of particles. In principle, one could investigate the thermodynamic equilibrium by starting with a specific spatial distribution of molecules, calculating the energy of this state, changing particle positions (either individually or all together) in accordance with the dynamics of the system, calculating the potential energy of this new state, and retaining whichever state has the lower energy. The minimum-energy state, or equilibrium state, is identified after repeating this process until no new state with a lower energy is found.

However, the Metropolis group, at the urging of renowned physicist Edward Teller who was one of the authors of the 1953 paper, had a brilliant insight (a rare "eureka" moment) that tamed this formidable problem. The investigators realized that they did not have to simulate the exact dynamics of the system and let the simulation run until equilibrium was reached. Rather, they needed only to simulate some Markov chain that had the same equilibrium or stationary distribution.

Suppose the *average* energy state (or some other quantity such as average pressure) of a collection of particles is sought. The procedure used was the first MCMC calculation conducted on a digital computer and is outlined as follows. Metropolis and colleagues started with a collection of $N = 224$ particles at specified positions in a two-dimensional cell. The potential energy of the system can be easily calculated for this, and any, known state. They assumed a completely ordered initial state, in which each row had 14 particles at equal spacing d_0, and the even rows were displaced to the right by $d_0/2$. However, each of the N particles can move in random directions (within the plane), and thus the *expected* potential energy E requires evaluation of an integral over $2N$ dimensions (because each of the N particles can move in either of two directions). This high-dimensional integral is difficult to evaluate by means other than Monte

---

[6] Metropolis also co-authored the first paper (Metropolis and Ulam, 1949) that formally introduced the term "Monte Carlo" for statistical sampling.

Carlo. This difficulty increases catastrophically as N gets larger and is some-times called the *curse of dimensionality*.

The Metropolis group proposed a Monte Carlo procedure in which a new position of each of the particles was sampled and then the potential energy of the new state was calculated. An acceptance-rejection criterion then was applied. If the energy E' of the new system was lower than the original energy, the particles were moved to the new sampled positions. If the energy of the new system was higher than the original, then the particles were moved only with a certain probability; otherwise, the move was rejected and a new history was run. In their case, the acceptance probability was given by $e^{-E'/(kT)}/e^{-E/(kT)} = e^{-\Delta E/(kt)}$, where $\Delta E = E' - E$ is the change in potential energy between the two states, T is the temperature of the system, and k is the Boltzmann constant.

It is reasonable to suppose that the next state would depend only on the pre-vious state, i.e., the next state will be independent of how the last state happened to evolve. This, then, is a Markov process, although Metropolis and colleagues did not use the term "Markov" in their paper. In their simulation, the Metropolis group ran "something less than sixteen cycles" (Metropolis et al., 1953), before beginning their estimation of the sample mean, in order to "forget" the original ordered state. It is interesting that they obtained useful results running thereafter "about forty-eight to sixty-four cycles" (Metropolis et al., 1953). Years later, Popp and Wilson (1996) more formally considered the issue of the number of samples to discard in such equilibrium simulations.

The Algorithm

Almost two decades after Metropolis and colleagues introduced their novel sim-ulation approach, Hastings (1970) expanded their work and formalized MCMC into the procedure now known as the *Metropolis–Hastings algorithm* (M-H algorithm). The procedure is summarized as follows for a single continuous random variable. Extension to discrete variables is straightforward and methods to deal with multidimensional variables are considered later in this chapter.

Let $f(x)$, defined on some interval (or domain) $D = [a, b]$, be the stationary PDF from which random samples are desired.[7] The M-H algorithm for sampling from a normalized or unnormalized nonnegative $f(x)$ proceeds as follows.

1. Choose *any* PDF $q(u|x) > 0$ defined on the domain D. The PDF so cho-sen is called the *proposal* PDF, the *proposal* distribution, or the *proposal* function. The proposal distribution can be conditioned on the current state x but need not be.[8] Selection of the proposal distribution is quite

---

[7] Because the M-H algorithm is widely used to generate samples from any PDF or unnormalized, nonnegative function as well as from a stationary distribution $\pi(x)$ to produce an MCMC chain, the more general function notation $f(x)$ is used here.

[8] In the original Metropolis algorithm, the conditional proposal distribution had to be symmetric, i.e., $q(u|x) = q(x|u)$. It was Hastings (1970) who later generalized the algorithm to allow any proposal distribution, thereby making the M-H algorithm a keystone of the MCMC method.

arbitrary, although it should be simple enough so that generating samples from it is computationally easy.

2. Specify a starting value within D. This value should not be one that is highly unlikely for reasons discussed below. Treat this value as the starting state x_1.

3. Pick a sample u_i from q using any appropriate procedure. For instance, if the CDF of h can be inverted, use the inverse CDF method.

4. Evaluate the quantity R, generally called the *Metropolis ratio* but also sometimes the *Hastings ratio*,

$$R(x_i, u_i) = \frac{f(u_i)}{f(x_i)} \frac{q(x_i | u_i)}{q(u_i | x_i)}. \tag{6.20}$$

5. Set

$$\alpha(x_i, u_i) = \min[1, R(x_i, u_i)], \tag{6.21}$$

and pick a value ρ from the uniform PDF on the unit interval, $\mathcal{U}(0, 1)$.

6. Accept u_i as the new sample with probability α or reject it with probability $1 - \alpha$, i.e., set

$$x_{i+1} = \begin{cases} u_i, & \text{if } \rho \leq \alpha, \\ x_i, & \text{if } \rho > \alpha, \end{cases} \tag{6.22}$$

where ρ is a random number drawn from $\mathcal{U}(0, 1)$.

7. Replace i by $i + 1$ and repeat steps 3 to 7 until all the desired samples have been generated.

If $R \geq 1$, the proposed state is always accepted, whereas if $R < 1$, the proposed state is accepted with probability $\alpha = R$. In effect, the transition probability from state x to state u in MCMC is given by

$$T(x \rightarrow u) = q(u|x)\alpha(x, u). \tag{6.23}$$

Equation (6.22) is the M-H update and says that, with probability α, the new state is the proposed state u_i. Alternatively, with probability $1 - \alpha$, the proposed state is rejected and remains at the same state x_i.

Note that the Metropolis ratio R is undefined if $f(x_i) = 0$, a value produced, for example, by a computer underflow calculation if u_i is deep into a very low probability tail of $f(u)$ such as those produced by normal and exponential distributions. Hence, one must ensure the initial state x_1 is one for which $f(x_1) > 0$. In this way, the denominator of the Hastings ratio is nonzero. However, there is no problem if $f(u_i) = 0$. In this case, $R(x_i, u_i) = 0$ and the proposed state u_i is accepted with probability zero, i.e., u_i is always rejected. As a result, the M-H update can never move to an impossible state x_i, for which $f(x_i) = 0$.

The M-H algorithm automatically rejects impossible proposed states. No special programming is needed to avoid these states!

There are several other interesting features about the M-H algorithm that should be noted:

- The acceptance-rejection criterion depends only on the current state x_i, the critical requirement for the samples to form a Markov chain.
- The algorithm contains elements of importance sampling because samples are chosen from a distribution q other than f and, in essence, weights of the form f/q are formed, although they are not used in the same way as in traditional importance sampling.
- All moves that produce a larger value of f are accepted, whereas all moves that give a lower value of f are potentially rejected. Thus, the accept-reject mechanism biases the random walks in favor of larger values of f.
- The algorithm also exhibits qualities of rejection sampling because samples are accepted or rejected based on a condition involving a random number ρ selected from the uniform distribution on the unit interval. In the rejection sampling discussed in Chapter 4, the rejection criterion can be applied many times before a new sample x_{i+1} is obtained. In the M-H algorithm, however, a new sample is obtained on the first application of the rejection criterion (as long as the sample is inside the correct interval); sometimes the new sample is updated and sometimes the new sample is the same as the previous sample. In fact, many samples, one after the other, may be exactly the same!
- MCMC provides a relatively simple way to sample a stationary PDF $f(x) = \pi(x)$ that is difficult to sample by other means because, with the M-H algorithm, one samples from a proposal distribution $q(u|x)$ of one's choosing in order to generate samples from the stationary PDF $\pi(x)$. If the stationary PDF is easy to sample directly, there is no need to employ the M-H algorithm, which requires generating at least two pseudorandom numbers to obtain each new random sample.

Samples drawn according to the M-H algorithm form a Markov chain because they are random samples that depend only on the current state and not on any other previous states. The Markov chain is ergodic if the transition probabilities do not change.

Reversibility of the Transition Function

The M-H algorithm leads to the *detailed balance* or *reversibility* condition (Gilks et al., 1996)

$$f(x)T(x \rightarrow u) = f(u)T(u \rightarrow x). \tag{6.24}$$

To show why this reversibility property occurs, combine Eqs. (6.21) and (6.23) to obtain

$$f(x)T(x \rightarrow u) = f(x)q(u|x)\alpha(x, u) = f(x)q(u|x)\min[1, R(x, u)]. \tag{6.25}$$

With Eq. (6.20), this result can be rewritten as

$$f(x)T(x \to u) = \min[f(x)q(u|x), R(x,u)f(x)q(u|x)]$$

$$= \min[f(x)q(u|x), f(u)q(x|u)]. \tag{6.26}$$

Because $\min[a,b] = \min[b,a]$ and from Eq. (6.20), the right-hand side of this result can be developed as

$$f(x)T(x \to u) = \min[f(u)q(x|u), f(x)q(u|x)]$$

$$= \min[f(u)q(x|u), f(u)q(x|u)/R(x,u)]. \tag{6.27}$$

Finally, because $1/R(x,u) = R(u,x)$ it follows that $\alpha(u,x) = \min[1, R(u,x)]$ so that Eqs. (6.27) and (6.23) give

$$f(x)T(x \to u) = f(u)q(x|u)\min[1, 1/R(x,u)]$$

$$= f(u)q(x|u)\alpha(u,x) = f(u)T(u \to x), \tag{6.28}$$

which validates Eq. (6.24).

Distribution of Metropolis–Hastings Samples

Now consider why samples formed according to the M-H algorithm are samples from the stationary PDF $f(x)$. As before, assume the PDF $f(x)$ is defined on the domain $D = [a,b]$ and further let D^+ specify the domain over which $f(x) > 0$. Next assume that the starting point is specified within D^+. In general, the transition probability from state x to state x may be nonzero, i.e., it may be possible to transition to the same state. Thus, define the function

$$p(x,u) = \begin{cases} q(u|x)\alpha, & \text{if } u \neq x, \\ 0, & \text{if } u = x, \end{cases} \tag{6.29}$$

which, in general, is not a PDF and, thus, $\int_{D^+} p(x,u)\,du$ may not be unity. Let

$$r(x) = 1 - \int_{D^+} p(x,u)\,du, \tag{6.30}$$

which is the probability that a state x remains at x. With these definitions, the transition function $T(x \to u)$ can be written as the sum of two pieces, one giving the transition probability of moving to a different state and the other giving the probability of staying in the same state, namely,

$$T(x \to u)du = p(x,u)du + r(x)\delta_x(du), \tag{6.31}$$

where

$$\delta_x(du) = \begin{cases} 1, & \text{if } du \text{ contains } x, \\ 0, & \text{otherwise.} \end{cases} \tag{6.32}$$

Now, the probability P of moving from any state x to any state within a subset S of D^+ can be expressed as

$$P = \int_{D^+} \int_S f(x)T(x \to u)\,du\,dx$$

$$= \int_{D^+} f(x)\left\{\int_S p(x, u)\,du\right\}dx + \int_{D^+} f(x)r(x)\delta_x(S)\,du$$

$$= \int_S \left\{\int_{D^+} f(x)p(x, u)\,dx\right\}du + \int_S f(x)r(x)\,dx. \tag{6.33}$$

Because of the way $p(x, y)$ was defined, the reversibility condition implies that

$$f(x)p(x, u) = f(u)p(u, x) \tag{6.34}$$

and, thus, with the definition of Eq. (6.30),

$$P = \int_S \left\{\int_{D^+} f(u)p(u, x)\,dx\right\}du + \int_S f(x)r(x)\,dx$$

$$= \int_S [1 - r(u)]f(u)\,du + \int_S f(x)r(x)\,dx = \int_S f(u)\,du. \tag{6.35}$$

This result demonstrates that the chain states are samples from the stationary PDF f.

Hence, the reversibility condition of Eq. (6.24) is sufficient to show that the samples produced by the M-H algorithm are samples from the stationary PDF. In fact, any α for which the reversibility condition is satisfied generates samples from the stationary distribution. A more complete discussion concerning the fact that Markov chain samples are samples from the stationary PDF is given by Tierney (1994). A simple illustration of using MCMC sampling is given in Example 6.2.

6.2.5 The Myth of Burn-in

It is common practice to express the MCMC estimate of a population mean $\langle z \rangle = \int_{-\infty}^{\infty} z(\mathbf{x})f(\mathbf{x})d\mathbf{x}$ in the form

$$\bar{z} = \frac{1}{N} \sum_{i=m+1}^{m+N} z(\mathbf{x}_i), \tag{6.36}$$

Example 6.2 Evaluate a One-Dimensional Integral Using MCMC

Estimate by M-H and direct sampling, the value of the integral

$$I = \int_0^\infty (x - \alpha\beta)^2 f(x)\,dx,$$

where $f(x)$ is the gamma PDF (see Appendix A.2.3)

$$f(x) = \frac{1}{\beta\Gamma(\alpha)} \left(\frac{x}{\beta}\right)^{\alpha-1} e^{-x/\beta}, \quad x \geq 0.$$

Use $\alpha = 2$ and $\beta = 1$ for which the exact value is $I = 2$.

One M-H sampling approach would be to select the proposal PDF $q(u) = (1/2)\exp[-u/2]$, $u \geq 0$, and set $x_1 = 1$. Then sample from the proposal PDF, obtaining candidate samples $u_i = -2\ln(\rho)$. Form the Metropolis ratio

$$R = \frac{f(u_i)q(x_i)}{f(x_i)q(u_i)} = \left(\frac{u_i}{x_i}\right)^{\alpha-1} e^{(u_i-x_i)/\beta} e^{-0.5(x_i-u_i)}$$

and update the samples as follows:

$$x_{i+1} = \begin{cases} u_i, & \text{if } \rho \leq \min[1, R], \\ x_i, & \text{if } \rho > \min[1, R]. \end{cases}$$

These samples are then used to form the estimate

$$\overline{I}_N = \frac{1}{N} \sum_{i=1}^N (x_i - \alpha\beta)^2 = \frac{1}{N} \sum_{i=1}^N (x_i - 2)^2.$$

Alternatively, one can sample the gamma distribution using the multistep process described in Appendix A.1.3. Results for four values of N are shown in Table 6.1 for both approaches. Both results appear to converge to the correct value of $I = 2$, suggesting that the M-H algorithm is indeed a valid way to obtain Monte Carlo samples.

Table 6.1 Estimates of I and their standard deviation using M-H and direct sampling of $f(x)$. Here $\alpha = 2$, $\beta = 1$. The exact value is $I = 2$.

N	M-H sampling $\overline{I}_N \pm \sigma(\overline{I}_N)$	Direct sampling $\overline{I}_N \pm \sigma(\overline{I}_N)$
10^4	1.946 ± 0.051	2.060 ± 0.047
10^5	2.004 ± 0.014	1.985 ± 0.014
10^6	1.997 ± 0.0047	2.003 ± 0.0045
10^7	1.999 ± 0.0014	1.999 ± 0.0014

where the $\mathbf{x}_i$ are sampled from $f(\mathbf{x})$ and m, called the "burn-in" length, is the number of samples x_i, $i = 1, 2, \ldots, m$, at the beginning of the chain that are discarded. For a process governed by a stable PDF that does not change while samples are being drawn, no burn-in is necessary, i.e., $m = 0$ is sufficient. Metropolis et al. (1953) used burn-in only because they started from a highly unlikely state and they reasoned that several wasted Markov steps would result in a more typical state from which to start accumulating scores. In essence, burn-in is a form of variance reduction, and a poor one at best.

However, burn-in is not strictly required because a long chain will generate samples having the proper distribution of the governing PDF. It is true that starting from states that are highly unlikely, as Metropolis and colleagues originally did, may lead to repeated samples of that state. However, samples will be distributed with the correct PDF, without burn-in, *if* a sufficiently large number of histories is run.

So what should one do instead of burn-in? Two rules-of-thumb are: (1) *start with any state you would not mind having in a sample* and (2) *start with a state known to have reasonably high probability, such as that at a mode.*

If MCMC is used to sample from a fixed distribution, then the samples obtained by application of the M-H algorithm will generate an ergodic Markov chain and, as just explained, no burn-in is required. Only if the underlying PDF changes during sampling is the chain of random samples generated by MCMC not ergodic. But even in this case, burn-in may not be necessary. For instance, consider that a process is being modeled in which a chemical change of phase occurs and the probabilities of certain outcomes change as a result of the phase change. Samples generated by the M-H algorithm yield valid samples for the entire process. Only if samples are desired from the PDF that results after the phase change would burn-in be useful. Whenever it is easy to start from an initial state that is highly likely, then this should be done and no burn-in is required. It is more important that N be large enough so that the samples assume their proper frequency than it is that burn-in be used. Nevertheless, the myth that burn-in is required, even for ergodic Markov chains, somehow survives as folklore.

6.2.6 The M-H Independence Sampler

There is a presumed advantage to using a proposal PDF $q(u|x)$ that is conditioned on the current value of x over using an unconditional one, namely, the number of duplicate samples is reduced. However, the proposal distribution does not have to depend on x and can have the simple form $q(u)$. Application of MCMC with a proposal distribution that is independent of x is discussed by Tierney (1994) and is commonly called the *independence sampler*. Suppose that $q(u|x) \to q(u)$ and there is some constant C such that the invariant distribution $f(x) = \pi(x)$, from which samples are to be generated, is such that $f(x) \le Cq(x)$. In such a case, samples could be generated using the acceptance/rejection algorithm that would produce independent samples from $f(x)$

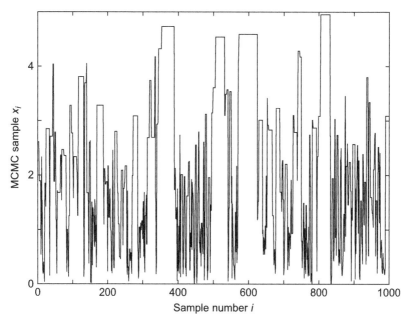

Figure 6.2 The first 1000 samples of x_i for the MCMC sampling from the proposal $q(u) = e^{-u}$. Note that large values of x_i often are successively repeated while small values of x_i are seldom repeated.

at an acceptance rate of $1/C$. By contrast, the independence sampler generates dependent samples from $f(x)$ (Kennedy, 2016).

Examples of Different Proposal Distribution

In principle, the choice of the nonconditional proposal distribution $q(u)$ is quite flexible. In fact, $q(u)$ can even be a constant, independent of u, and such a choice is often made for quick-and-dirty one-dimensional calculations, i.e., $q(u) \equiv 1/(b - a)$, where $x \in (a, b)$ is the finite domain of $f(x)$. But in practice one wants a proposal distribution whose shape is similar to that of $f(x)$.

Consider, for purposes of illustration, the case in which one seeks to generate samples from the simple PDF $f(x) = 0.2$, $0 \leq x \leq 5$, using the M-H algorithm. Suppose, for unknown reasons, a decaying exponential function is chosen as the proposal distribution, i.e.,

$$q(u) = Ce^{-\beta u} \qquad \text{with} \qquad C = \beta/[1 - e^{-5\beta}]. \qquad (6.37)$$

The constant C is a normalization constant needed to make $q(u)$ a proper PDF over the interval $u \in [0, 5]$ and β is a *scaling parameter* that determines how rapidly $h(u)$ changes with u. Then, with the inverse CDF method (see Sec-

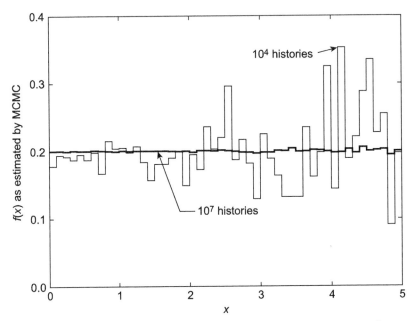

Figure 6.3 The distribution of the $\{x_i\}$ obtained by MCMC using the proposal $q(u) = e^{-u}$. A large number of samples is required to reproduce the distribution $f(x) = 1/D = 0.2$, especially for large values of x. This slow convergence for large x is a consequence of the low probability of obtaining a large sample value from the proposal distribution. Such a proposal choice would normally be a poor one but is useful for demonstration purposes.

tion 4.1.1), the ith sample is found from

$$u_i = -\frac{1}{\beta} \ln \left(1 - \frac{\beta}{C} \rho_i \right),$$

where ρ_i is a random number from $\mathcal{U}(0,1)$. To simplify the sampling, it was decided to pick $\beta = 1$ so that $C = 1.00678\ldots \simeq 1$. The first 1000 sampled values of x_i generated by the M-H algorithm for this case are shown in Fig. 6.2. Here, large values of x_i tend to repeat, often many times, because samples of u_i are likely to be smaller than a large x_i so that $q(x_i)/q(u_i) < 1$. Such repeats of large values of x_i are necessary, however, in order for the distribution of $\{x_i\}$ to approach f. The distribution of the x_i samples is shown in Fig. 6.3 for two values of N. For $N = 10^4$, there is significant variation from the true PDF $f(x) = 0.2$, especially for large x. However, the samples approximate the true PDF remarkably well for $N = 10^7$ histories. Clearly, if β had been chosen to be less than unity, recovery of $f(x)$ would require far fewer samples.

It seems obvious that duplication of the sampled points should be minimized. The term *mixing* is used to indicate the rate at which the sample space is sampled; good or rapid mixing indicates that the space of possible states is quickly sampled and few states are successively repeated. A good proposal function exhibits good mixing. This can be achieved by choosing a proposal distribution h that approximates well the shape of the stationary distribution f, so that R is generally near unity. Application of MCMC where $q = f$ results in perfect mixing because $R = 1$ always. However, if f can be sampled easily by other means, then MCMC is unnecessary.

Samples Over Infinite Domains

There are a number of distributions $f(x)$ that are defined over an infinite interval. Methods for obtaining samples from some of these distributions exist (see Appendix A), but MCMC also can be used. For instance, the proposal PDF distribution given by

$$q(u) = (\lambda/2)e^{-\lambda|u-u_o|} \tag{6.38}$$

can be used to obtain candidate values of x drawn from any PDF defined on $(-\infty, \infty)$ that has a single mode at u_o. The corresponding CDF distribution is

$$Q(u) = \frac{1}{2} \begin{cases} \exp[\lambda(u - u_o)], & u < u_o, \\ \{1 - \exp[-\lambda(u - u_o)]\}, & u > u_o. \end{cases} \tag{6.39}$$

One can sample a pseudorandom number ρ_i uniformly from the unit interval and then with the use of the inverse CDF method generate a sample u_i from $q(u)$ as

$$u_i = \begin{cases} u_o + \ln(2\rho_i)/\lambda, & u < u_o, \\ u_o - \ln(1 - 2\rho_i)/\lambda, & u > u_o. \end{cases} \tag{6.40}$$

Alternatively, a normal distribution

$$q(u) = \frac{1}{\sqrt{2\pi}\sigma} \exp\left[-\frac{(\mu - u)^2}{2\sigma^2}\right] \tag{6.41}$$

could be used as a proposal distribution. Here the parameters μ and σ determine the peak location and spread of $q(u)$ and should be chosen to best approximate $f(x)$. How to sample Eq. (6.41) to obtain a sample u_i is described in Appendix A.2.6. The use of these two proposal distributions is illustrated next.

Assume that samples x_i are needed from the stationary PDF $f(x)$ that is normally distributed with mean zero and standard deviation 2.5, i.e., from

$$f(x) = \mathcal{N}(0, 2.5) = \frac{e^{-x^2/[2(2.5)^2]}}{2.5\sqrt{2\pi}}. \tag{6.42}$$

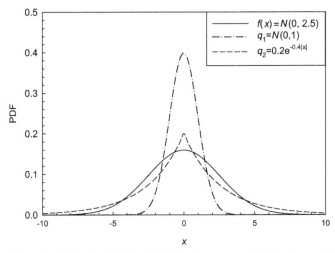

Figure 6.4 The PDF $f(x) = \mathcal{N}(0, 2.5)$ (solid line) and two proposal distributions q_1 (dashes) and q_2 (dots).

Consider the two proposal distributions,

$$q_1(u) = \mathcal{N}(0, 1) = \frac{e^{-u^2/2}}{\sqrt{2\pi}} \qquad \text{and} \qquad q_2(u) = 0.2e^{-0.4|u|}. \tag{6.43}$$

The PDFs f, q_1, and q_2 are shown in Fig. 6.4. Proposal distribution q_1 is of the same form as f but is more peaked; proposal distribution q_2 has longer tails than either f or q_1 but conforms more closely to the shape of f than does h_2 over the important range $-6 \leq x \leq 6$, the range where most of the samples are generated. The M-H algorithm was applied to generate samples from f using each proposal distribution and a starting value of $x_1 = -3.0$. These samples were then collected into 80 bins from -10.125 to 10.125, for both the cases $N = 10^3$ and $N = 10^5$. The results are shown, with the true PDF f, in Figs. 6.5 and 6.6. Note that use of $N = 10^3$ samples is insufficient for either proposal function to reproduce the desired distribution very well. On the other hand, after 10^5 histories, samples picked using both proposal functions generate samples distributed approximately according to the desired PDF. However, of the two proposals, the one that more closely resembles the stationary distribution, i.e., q_2, does a considerably better job, especially for the tails of $f(x)$. These results verify that it is best to use a proposal distribution that resembles the stationary distribution. But, if *enough* histories are drawn, almost any proposal distribution, defined on the proper interval, will suffice. But there is no way to estimate a priori what constitutes enough histories. In practice, N usually is increased until fluctuations in $\overline{z}$ are within acceptable limits.

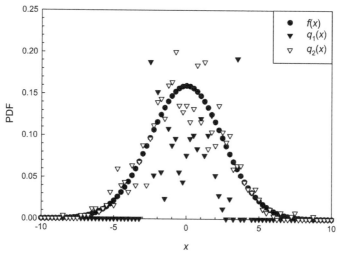

Figure 6.5 Histograms of 1000 samples from f using the two proposal distributions q_1 (open circles) and q_2 (triangles), with $m = 0$. Filled circles are used to show the actual PDF.

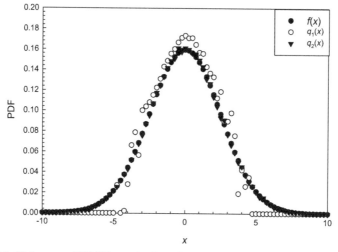

Figure 6.6 Histograms of 100,000 samples from f using the two proposal distributions q_1 (open circles) and q_2 (triangles), with $m = 0$. Filled circles are used to show the actual PDF.

6.2.7 Selecting a Proposal Distribution

To use the M-H algorithm, a proposal distribution $q(u)$, or in the more general case $q(u|x)$, must be chosen. Almost any choice, provided x and u are confined to the same domain, leads to the Markov chain having the correct stationary distribution $\pi(x) = f(x)$. However, the convergence properties of the chain can strongly be affected by the choice.

Most potential functions have scaling and shape parameters such as λ and u_o in Eq. (6.38). The values picked for these parameters in a proposal distribution greatly affect the size of the steps in the random walk over the domain or interval for x and the rate at which samples are obtained from important regions of the interval, i.e., regions in which $f(x)$ is large. If the proposal distribution of Eq. (6.38) (with $u_o = 0$ and $\lambda = 1$) were used for a function $f(x)$ that becomes large as $x \to 5$, it would take many more samples for MCMC to converge than if, for instance, $u_o = 4$ and $\lambda = .1$ had been chosen.

For the multidimensional case (discussed next), the sampling can be from conditional proposal distributions for each random variable (or subgroup of random variables) or from a global joint distribution of all the variables. In any event, all these proposal distributions generally have scale parameters that determine how rapidly the domain in phase space is explored and samples generated from important regions, i.e., where the joint distribution is large. The difficulty of the selection process is exacerbated when the dimension (number of random variables) increases and when some variables are strongly correlated and others only weakly so.

It is often the responsibility of the user of an M-H sampler not only to choose appropriate proposal distributions but also to select values for the parameters in them. Robert and Casella (2013) discuss the techniques that are often used for selecting proposal distributions and their parameters.

6.2.8 Multidimensional Sampling

In highly multidimensional cases, sampling by traditional methods can be problematic. However, the extension of MCMC to many variables is conceptually straightforward. Let $\mathbf{x} = [x_1, x_2, \ldots, x_n]$ be a collection of n random variables with joint PDF f. Often MCMC is applied in such cases to obtain random samples $\mathbf{x}_i$, $i = 1, 2, \ldots, N$, where N is the number of histories. Two basic methods are commonly employed. The first is to sample all components of the multidimensional PDF $f(\mathbf{x})$ on every history, using a multidimensional candidate distribution $q(\mathbf{u})$ and applying the M-H algorithm. This approach is sometimes called the *global* approach (Neal, 1993). Thus, all n random variables are sampled on each history, so that $\mathbf{x}_i = [(x_1)_i, (x_2)_i, \ldots, (x_n)_i]$.

The second (*local*) method is to select one component at a time and sample it from a candidate distribution using the M-H method. In this case, the proposal distribution is a one-dimensional PDF. For instance, to sample the jth component one would sample from a proposal PDF of the form $q_{x_j}(u_j)$. In this case, if variable j is sampled on history i, then

$$\mathbf{x}_{(i)} = [(x_1)_{(i-1)}, (x_2)_{(i-1)}, \ldots, $$
$$(x_{j-1})_{(i-1)}, (x_j)_i, (x_{j+1})_{(i-1)}, \ldots, (x_{n-1})_{(i-1)}, (x_n)_{(i-1)}]. \quad (6.44)$$

Of course, the local approach requires that a proposal $q_{x_j}(u_j)$ be given for each component. It is interesting that the determination of which component or variable to sample can be determined either sequentially or at random. This procedure gives rise to the concept of a "random walk," because a move is made in one direction only in a multidimensional space at each sampling point. The step size is determined by the chosen values of the parameters in the proposal distributions selected. Twelve steps of a random walk starting at x_i in a three-dimensional space are depicted in Fig. 6.7 in which the component direction is selected sequentially (x_1 first, then x_2, and so forth). Figure 6.8 depicts 12 steps in which the component direction is selected at random. Either method is valid in that the entire space eventually will be sampled if enough steps are taken. Figure 6.9 shows 12 steps in which all three components are updated during each step.

The great attractiveness of MCMC is that multidimensional PDFs need not be sampled directly; it is only necessary to evaluate these PDFs at candidate values of the random vector. The M-H algorithm for an n-dimensional PDF $f(\mathbf{x})$ proceeds as follows.

1. Choose a proposal distribution, $q(\mathbf{u}|\mathbf{x})$, which can be easy to sample. Recall that this distribution need not be conditioned on $\mathbf{x}$ but the case of a conditional PDF is treated here.

2. Choose an initial value $\mathbf{x}_1$. All components of $\mathbf{x}$ should have values that are not highly unlikely, so as to avoid successively repeating samples early in the process.

3. Sample from q to obtain a new sample $\mathbf{u}_i$. In global sampling, all components u_j, $j = 1, 2, \ldots, n$, are sampled from h on each history i, $i = 1, 2, \ldots, N$, and $\mathbf{u}_i = [(u_1)_i, (u_2)_i, \ldots, (u_n)_i]$. In local sampling, one component at a time is updated, either sequentially or at random. Thus, if the jth component is sampled, $\mathbf{u}_i = [(x_1)_i, x_2, \ldots, (x_{j-1})_i, u_j, (x_{j+1})_i, \ldots, (x_n)_i]$.

4. Evaluate the Metropolis ratio

$$R = \frac{f(\mathbf{u}_i)}{f(\mathbf{x}_i)} \frac{q(\mathbf{x}_i|\mathbf{u}_i)}{h(\mathbf{u}_i|\mathbf{x}_i)}. \tag{6.45}$$

5. Pick a pseudorandom number ρ from the unit rectangular distribution and update the vector $\mathbf{x}$ as follows:

$$\mathbf{x}_{i+1} = \begin{cases} \mathbf{u}_i, & \text{if } \rho \leq \min[1, R], \\ \mathbf{x}_i, & \text{if } \rho > \min[1, R]. \end{cases} \tag{6.46}$$

6. Use this vector to score this history, increment i by one, and repeat steps 2 through 6 until all N histories have been completed.

Results generated by global MCMC sampling from a multidimensional distribution are shown in Figs. 6.10 to 6.12, in which a bivariate normal PDF (see

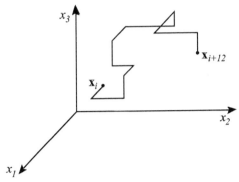

Figure 6.7 Twelve example random walk steps taken by sequentially sampling x_1, then x_2, then x_3, then x_1, and so forth.

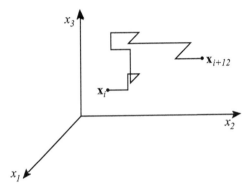

Figure 6.8 Twelve example random walk steps taken by sampling components of **x** at random.

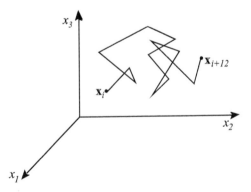

Figure 6.9 Twelve example random walk steps taken by global sampling.

Appendix A.3.1) is sampled by MCMC for three different values of N. The large spikes in the tails evident for 10^3 histories are "softened" by 10^5 histories

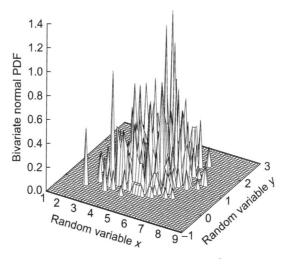

Figure 6.10 Bivariate normal PDF sampled using MCMC after 10^3 histories.

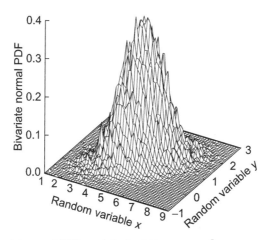

Figure 6.11 Bivariate normal PDF sampled using MCMC after 10^5 histories.

and almost completely removed by $N = 10^7$ histories. A two-dimensional case based on the bivariate normal PDF is considered in Example 6.3.

6.2.9 The Gibbs Sampler

The Gibbs sampler is named after the physicist Josiah Willard Gibbs (1839–1903), whose ideas from his pioneering studies on statistical mechanics were instrumental in the development of what today is called the Gibbs sampler. This sampling algorithm was described first by brothers Stuart and Donald Geman in 1984, some eight decades after the death of Gibbs (Geman and Geman, 1984).

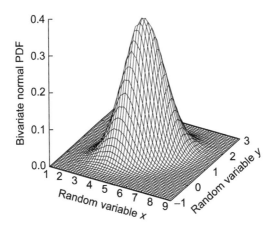

Figure 6.12 Bivariate normal PDF sampled using MCMC after 10^7 histories.

The name of the algorithm comes from their paper which describes the algorithm as a particular case of the Gibbs distribution which they then used in a Bayesian framework to reconstruct a noisy image. The Gibbs sampler became very popular after the publication of the seminal paper by Gelfand and Smith (1990). In particular, this sampler, over the next 20 years, revitalized the Bayesian community which came to realize it and MCMC were powerful tools that enabled Bayesian inference studies theretofore intractable.

The Gibbs sampler is an MCMC algorithm for obtaining a sequence of samples x_i from a *multivariate* joint probability distribution when the joint distribution is unknown or prohibitively difficult to sample. It is useful when the conditional distribution of each variable, given values for the other variables, is known and is relatively easy to sample. The Gibbs algorithm constructs a sample from the conditional distribution of each variable in turn (or in some cases, each group of variables) based on the current values of the other variables. In other words, the Gibbs sampler is a special case of the M-H algorithm in which the single-component proposal distribution is the conditional PDF for that component given values for the other components.

Example 6.3 A Modified Buffon's Needle Problem

Consider a modified Buffon's needle game in which needles of length L are dropped by a biased machine. In the original Buffon's needle problem, the PDF for the position of the needle endpoints is $f(x) = 1/D$, $0 \leq x < D$, the PDF for the orientation angle is $g(\theta) = 1/\pi$, $0 \leq \theta < \pi$, and P_{cut} is given by Eq. (1.7). The biased machine, however, drops needles such that the endpoints x and the orientation angles θ are distributed, within a unit cell, according to a bivariate normal distribution (see Appendix A.3.1), i.e.,

$$f(x, \theta) = \frac{K}{2\pi \sigma_1 \sigma_2 \sqrt{1 - \rho^2}} \exp\left[-\frac{z(x, \theta)}{2(1 - \rho^2)}\right], \quad 0 \leq x < D, \quad 0 \leq \theta < \pi,$$

where K is a normalization constant,

$$z(x, \theta) = \frac{(x - \mu_1)^2}{\sigma_1^2} - \frac{2\rho(x - \mu_1)(\theta - \mu_2)}{\sigma_1 \sigma_2} + \frac{(\theta - \mu_2)^2}{\sigma_2^2},$$

and

$$\rho = \frac{\sigma_{12}^2}{\sigma_1 \sigma_2}.$$

Employing an analysis similar to that taken in Example 1.1, the cut probability for this machine dropping needles on a grid with $D \geq L$ can be expressed as

$$P_{cut} = \int_0^\pi \int_0^D \frac{L\sin\theta}{D} f(x, \theta) \, dx \, d\theta.$$

How might one estimate this cut probability for a given set of parameters?

Consider the case where $D = 5$, $L = 2$, $\mu_1 = 3$, $\mu_2 = 1$, $\sigma_1 = 1$, $\sigma_2 = 1$, and $\sigma_{12} = 0.1$. One could drop n needles into the machine and observe k crossings on the grid below, and approximate P_{cut} as k/n. However, it is much more efficient to run an experiment on a computer using MCMC.

Choose as proposal distributions functions that are easy to sample, say $q_1(u) = 1/D$ and $q_2(v) = 1/\pi$. Then, sample u and v from these proposal functions. In this case, the Metropolis ratio is simply

$$R = \frac{f(u, v)}{f(x, \theta)},$$

because the proposals are both constants. Running 10^6 histories, an estimate of $P_{cut} = 0.2837 \pm 0.0001$ was obtained (a value statistically slightly higher than for uniform drops). Note that the two-dimensional integral for P_{cut} is evaluated without having to sample from the bivariate normal distribution or having to determine the normalization constant K. All that is required is that the ratio R of bivariate normal distributions be evaluated at both current and candidate points sampled from the simple proposal functions.

The sequence of samples generated by the Gibbs sampler constitute a Markov chain whose stationary distribution is just the sought-after joint distribution $f(\mathbf{x})$. For the Gibbs sampler, it is always true that the Hastings ratio $\alpha = 1$ and, thus, no samples are ever rejected (Geyer, 2011). Also because the Gibbs sampler is an MCMC method, i.e., a special case of the M-H method, the probability distribution of $\mathbf{x}_i$ tends to $f(\mathbf{x})$ as $i \to \infty$, provided $f(\mathbf{x})$ is not some pathological function (MacKay, 2007).

The Gibbs Algorithm

In this brief description of the Gibbs sampler, the notation $\mathbf{x}_{[-j]}$ is used to represent the vector $\mathbf{x}$ with the jth component missing, i.e.,

$$\mathbf{x}_{[-j]} = (x_1, x_2, \ldots, x_{j-1}, x_{j+1}, x_{j+2}, \ldots, x_n). \tag{6.47}$$

Then $f(x_j|\mathbf{x}_{[-j]})$ is the conditional PDF for x_j given $\mathbf{x}_{[-j]}$.

Consider an n-dimensional random variable $\mathbf{x}$, for which the conditional PDFs $f(x_j|\mathbf{x}_{[-j]})$, $j = 1, 2, \ldots, n$, are known. Local Gibbs sampling after i histories would proceed by selecting a j, either sequentially or at random, and then sampling the jth component of $\mathbf{x}$ for history $i + 1$, i.e., $x_{j,(i+1)}$, from $f(x_j|\mathbf{x}_{[-j],i})$. However, global sampling is preferred because it is straightforward to sample all n components, each from its conditional PDF.

One could proceed by sampling $x_{1,(i+1)}$ from $f(x_1|\mathbf{x}_{[-1],i})$, then $x_{2,(i+1)}$ from $f(x_2|\mathbf{x}_{[-2],i})$, and so on for $j = 3, 4, \ldots, n$. However, it is better to use the most recent sampled value of each variable, so that sampling would proceed as follows:

1. Sample $x_{1,(i+1)}$, from $f(x_1|\mathbf{x}_{[-1],i})$.
2. Next, sample $x_{2,(i+1)}$ from $f(x_2|x_{1,(i+1)}, x_{3,i}, x_{4,i}, \ldots, x_{n,i})$.
3. Then, sample $x_{3,(i+1)}$ from $f(x_3|x_{1,(i+1)}, x_{2,(i+1)}, x_{4,i}, \ldots, x_{n,i})$.
4. Continue in a similar manner for each $x_j > 3$ until $x_{n,(i+1)}$ is sampled from $f(x_n|\mathbf{x}_{[-n],(i+1)})$.

This procedure then provides an updated sample for $\mathbf{x}_{i+1}$, which is used in the MCMC estimator.

A refinement to updating each component separately, as just described above, is to partition $\mathbf{x}$ into subvectors and updating the components in each subvector (or subgroup) all at the same time. As an example, suppose $\mathbf{x}$ has three components (x_1, x_2, x_3), i.e., $n = 3$. There are four ways $\mathbf{x}_i$ can be updated to get $\mathbf{x}_{i+1}$: (1) each component is updated separately (as done with the above algorithm), (2) $x_{1,i}$ and $x_{2,i}$ are updated together and $x_{3,i}$ is updated by itself, (3) $x_{1,i}$ and $x_{3,i}$ are updated together and $x_{1,i}$ is updated separately, or (4) $x_{2,i}$ and $x_{3,i}$ are updated together and $x_{1,i}$ is updated separately. The manner in which $\mathbf{x}$ is partitioned can significantly affect the convergence rate of the Markov chain, so it is usually worthwhile to consider various groupings. Although there are no specific guidelines on how to optimize groupings, there is some evidence that strongly correlated variables should be grouped together (Ekvall and Jones, 2019).

Limitations of the Gibbs Sampler

The use of the Gibbs sampler skyrocketed after the appearance of the paper by Gelfand and Smith (1990). However, its limitations quickly became apparent (Clifford, 1993). For example, to use a Gibbs sampler, one has to become adept at manipulating conditional distributions. Also often a user has to deal with very slow convergence rates, a consequence of the sampler being unable to make diagonal moves in parameter space because each variable is updated separately. Highly correlated variables are particularly problematic.

Proponents of the Gibbs sampler are attracted by how "automatic" it appears. With M-H sampling choices of proposal distributions must be made as well as values of the parameters in them. With Gibbs sampling, no such choices are needed. But this simplicity is somewhat illusionary. Gibbs samplers have an unusual property not shared by other M-H updating methods. They are *idempotent*, meaning multiple updates of, say, x_j have the same effect as a single update. This property arises because updating x_j is based on the conditional distribution $f(x_j|\mathbf{x}_{[-j]})$, which is independent of x_j. So a single update of x_j or multiple updates do not change $\mathbf{x}_{[-j]}$. Thus, for Gibbs elementary updates to be useful, they must be combined somehow with other updates to change $f(x_j|\mathbf{x}_{[-j]})$ before x_j is updated again. The algorithm above sequentially updates the variables in the order $x_1, x_2, \ldots, x_n$ and achieves the needed change in $f(x_j|\mathbf{x}_{[-j]})$ before x_j is updated again. Alternatively, one could update the variables in random order. Many other updating orders have been used (Geyer, 2011). A "black-box" code for Gibbs sampling has already made the mixing decision, so the user is lulled into thinking Gibbs sampling is entirely free of the need to specify any sampler options.

Whether this options-free feature of Gibbs sampling is good or bad depends on how well it performs in a particular situation. It is a good feature if it works well and bad if it does not.

6.2.10 The Problem of High Dimensionality

Although the M-H algorithm is effectively used for generating samples from a joint PDF in a low-dimensional ($\lesssim 5$) state space, it does not work well in high-dimensional spaces. The reason is that most neighborhoods in the space have very low probability, and because such neighborhoods are so common the M-H algorithms end up not exploring the few regions where the joint PDF is large. The region(s) in which most probability mass is concentrated (e.g., the center of multivariate Gaussian distributions) is (are) usually small compared to the much larger hypervolume that has low probability densities (e.g., the tails of multivariate Gaussian distributions). This inability of Metropolis methods to ferret out, with their relatively slow random walks, important regions in the hypervolume V is exacerbated by the difficulty of specifying a multidimensional proposal distribution that is easily sampled and yet mimics $f(\mathbf{x})$ in shape. A very poor choice is a uniform proposal PDF, i.e., $q(\mathbf{x}, \mathbf{u}) = 1/V$. Although easy to sample, it mostly proposes candidate locations in unimportant regions. Generally, as the dimensionality d of the sampling space increases, the ratio of unimportant to important hypervolumes increases exponentially. This observation is known as the "curse of dimensionality."

To ameliorate the deficiencies of the Metropolis sampling (and its variants M-H and Gibbs sampling), several alternative sampling techniques have been proposed to reduce the time required to obtain effectively independent samples. Detailed descriptions of them are beyond the scope of this text. Here only a few are briefly mentioned; the interested reader can obtain details from the excellent report by Neal (1993) and the text by MacKay (2007).

The *Hamiltonian Monte Carlo method* is a Metropolis method, applicable to continuous state spaces, and employs gradient information to reduce Metropolis random walk behavior. This method augments the state space **x** by adding a fictitious momentum variable **p**, from which the gradients are estimated. Then an alternation between two types of proposal distributions is used to search the augmented space. The *method of overrelaxation* is used to reduce the random walk behavior in Gibbs sampling based on Gaussian conditional distributions. This method was generalized by Neal (1995) in his *ordered overrelaxation method* to any system where Gibbs sampling is used. A third method for speeding convergence is *simulated annealing*. This method introduces a "temperature" parameter which, when large, allows the system to make transitions in state space that would be improbable at lower temperatures. By initially starting with a high temperature and gradually reducing it, the random walk or simulation is less likely to get stuck in an unimportant region of state space. A fourth method is the *multistate leapfrog method*, which is used in combination with some traditional sampling procedure. It resembles, in spirit, the overrelaxation method because it moves the random walk beyond the next step made by the traditional random walk. It also works better in higher dimensions than does the overrelaxation method.

6.3   Brief Review of Probability Concepts

MCMC often is applied in statistical inference. Statistical inference in recent times has become associated with Bayesian analysis. Thus, a brief review of probability theory and Bayes' theorem[9] is given here and the potential use of MCMC in statistical inference and related applications, such as decision theory, is briefly discussed.

Bayesian analysis is in stark contrast to the frequentist approach. Frequentists interpret probability in terms of the frequency with which an event occurs among a number of trials. To a frequentist, the probability of an event is the relative frequency of occurrence. For example, if a die is thrown 100 times and a three is obtained 20 times, a frequentist is apt to say that the best estimate of the probability of throwing a three is 20/100 or 20%. Proponents of the Bayesian approach interpret probability as a measure of *likelihood* of an outcome in the absence of complete knowledge. To a Bayesian, the probability of an event is

---

[9] Thomas Bayes (1701–1761) was an English statistician, philosopher, and Presbyterian minister who is known for formulating a specific case of the theorem that bears his name. Bayes described his findings on probability in his "Essay Towards Solving a Problem in the Doctrine of Chances," which was published posthumously (1763) in the Philosophical Transactions of the Royal Society (London).

an inherent quantity, which may not be known but can be estimated. Bayesians often talk of probability as a measure of their degree of certainty about their knowledge, with 0 being completely uncertain and 1 being completely certain. A Bayesian can take the same data (e.g., 20 threes out of 100 throws of the die) and produce a different result than the frequentist. The debate between frequentists and Bayesians rages in the literature and in public and private forums. Although both approaches are widely used, it appears that the cadre of Bayesians is growing at the expense of the ranks of the frequentists.

Let Prob{A} represent the probability of A, where A may be an outcome (such as "heads or tails"), a conjecture (such as "it will rain today in Kansas City"), or a hypothesis (such as "the moon resulted from the impact over 4 billion years ago of a body about one-third the size of the current earth with a larger body that was the proto-earth"). It is useful to express the fact that probability models generally are based on some assumptions or other information, denoted here by the symbol O (denoting "other"). Thus, the expression for the probability of A more generally can be written in the conditional form Prob{$A|O$}, which in words reads "the probability of event A given other information O." It is important to recognize that models and hypotheses are based on assumptions or other information and the notation "$|O$" formally reminds one of this.

The symbol $\bar{A}$ is interpreted as "not A," i.e., the complement of A such that Prob{$\bar{A}|O$} $= 1 -$ Prob{$A|O$} or, as it is more often presented,

$$\text{Prob}\{A|O\} + \text{Prob}\{\bar{A}|O\} = 1. \tag{6.48}$$

Further, Prob{$A + B|O$} is the probability of either A or B given O, and Prob{$AB|O$} is the probability of both A and B given O. If A and B are mutually exclusive and collectively exhaustive events (e.g., A stands for "heads," B for "tails," and O assumes an unbiased throw of a fair coin), then Prob{$A + B$} $=$ 1 (the result must be one or the other) and Prob{AB} $= 0$ (you can not get both a head and a tail on one throw).

Now consider two basic tenets of probability theory. First, the product rule

$$\text{Prob}\{AB|O\} = \text{Prob}\{A|BO\}\text{Prob}\{B|O\} \tag{6.49}$$

states that the probability that both A and B are true, given O, equals the product of (1) the *conditional probability* that A is true given both B and O and (2) the *marginal probability* that B is true given O.

The second tenet is the commutativity property

$$\text{Prob}\{AB|O\} = \text{Prob}\{BA|O\}, \tag{6.50}$$

which holds because AB and BA are identical results (AB means both A and B and BA means both B and A).

6.3.1 Bayes' Theorem

An equivalent form of Eq. (6.49) for $\text{Prob}\{BA|O\}$ is

$$\text{Prob}\{BA|O\} = \text{Prob}\{B|AO\}\text{Prob}\{A|O\}. \tag{6.51}$$

It follows from Eqs. (6.49), (6.50), and (6.51) that

$$\text{Prob}\{A|BO\}\text{Prob}\{B|O\} = \text{Prob}\{B|AO\}\text{Prob}\{A|O\}, \tag{6.52}$$

which can be rearranged into the form known as Bayes' theorem, namely,

$$\boxed{\text{Prob}\{A|BO\} = \text{Prob}\{A|O\}\frac{\text{Prob}\{B|AO\}}{\text{Prob}\{B|O\}}.} \tag{6.53}$$

Thus, Bayes' theorem results from the basic commutativity and product rules of probability theory.

Bayes' theorem often is written in terms of PDFs. The derivation is slightly different, because PDFs are not probabilities but rather probability densities. Consider the case of two random variables x and y. The conditional PDF for x given y was given in Chapter 2 (see Eq. (2.42)) as

$$f(x|y) = \frac{f(x, y)}{f_y(y)}, \tag{6.54}$$

where $f(x, y)$ is the joint PDF and $f_2(y)$ is the conditional PDF of y. But Eq. (2.42) indicates that the joint PDF can be expressed as

$$f(x, y) = f(y|x)f_x(x). \tag{6.55}$$

Substitution of this result into Eq. (6.54) results in a form of Bayes' theorem for two random variables, namely,

$$f(x|y) = f_x(x)\frac{f(y|x)}{f_y(y)}. \tag{6.56}$$

In this form, Bayes' theorem relates the marginal and conditional distributions of $f(x, y)$, a result embarrassing to frequentists. To Bayesians, however, x and y need not be repeatable outcomes but may be propositions or hypotheses, and Bayes' theorem becomes a tool for making consistent inferences. Often $f_x(x)$, also called the *prior* distribution, can be based on experience, hypothesis, or subjective opinion about some property of a system. For example, let x represent the probability that a particular diesel-fired emergency electric generator will start when needed, and let y represent the observation that the generator in question has started as required in all N previous tests. The prior distribution $f_x(x)$ represents the probability that similar generators start upon demand, and may be based on data from many similar generators or even upon informed subjective opinion. Then, Bayes' theorem is used to combine this prior "guess" with the observed data for the generator in question to obtain a refined estimate of the probability $f(x|y)$ that this generator will start the next time when needed.

More generally, consider a model of a process or system with any number n of parameters $\boldsymbol{\theta}$ and a limited set of observations or "data" $\mathbf{D}$. Now, replace the variable x in Eq. (6.56) with the parameter vector $\boldsymbol{\theta}$ and the variable y with the observed data $\mathbf{D}$. Then Bayes' theorem can be expressed in the general form

$$f(\boldsymbol{\theta}|\mathbf{D}) = \frac{g(\boldsymbol{\theta})\ell(\mathbf{D}|\boldsymbol{\theta})}{M}, \qquad (6.57)$$

where the PDF $f(\boldsymbol{\theta}|D)$ is called the *posterior distribution*, the PDF $g(\boldsymbol{\theta})$ is called the *prior distribution*, and $\ell(D|\boldsymbol{\theta})$ is called the *likelihood function*. The quantity M has been called the *marginal distribution* but also the *evidence*. The global likelihood is just the probability of obtaining the observed data and is a constant. This constant, which may or may not be known, can be expressed as

$$M = \int_{\Theta} \ell(\mathbf{D}|\boldsymbol{\theta})g(\boldsymbol{\theta})\,d\boldsymbol{\theta}, \qquad (6.58)$$

where Θ is the range of possible values of $\boldsymbol{\theta}$. Thus, Bayes' theorem states that the posterior distribution (the conditional probability of $\boldsymbol{\theta}$ given the data) is proportional to the product of the prior distribution (the PDF $g(\boldsymbol{\theta})$) and a likelihood function (the probability of obtaining the data, given $\boldsymbol{\theta}$). Bayes' theorem is a result of classical statistics and agrees with frequentist conclusions whenever the prior distribution is known, as is demonstrated in Example 6.4.

Example 6.4 A Classical Statistical Analysis

Consider three bowls containing 30 candies each. The first bowl has 10 red, 10 yellow, and 10 blue candies, the second bowl contains 5 red, 15 yellow, and 10 blue candies, and the third bowl contains 10 red, 15 yellow, and 5 blue candies. Thus, there are 25 red, 40 yellow, and 25 blue candies distributed among the three bowls. A young child goes into the room with the three bowls and comes back with a yellow candy. An observer asks the simple question "what are the probabilities that the child got the candy from bowl one, from bowl two, and from bowl three?" To simplify the analysis use the following shortcut notation: (1) Prob$\{i|c\}$ denotes the conditional probability that bowl i is chosen given that color c was selected, (2) Prob$\{c|i\}$ is the conditional probability that color c was selected given bowl i, and (3) Prob$\{i\}$ is interpreted as the probability that bowl i is selected.

A *frequentist* might divide the number of yellow candies in each bowl by the total of all yellow candies to obtain

$$\text{Prob}\{1|\text{yellow}\} = \frac{10}{40} = 0.25,$$

$$\text{Prob}\{2|\text{yellow}\} = \frac{15}{40} = 0.375,$$

$$\text{Prob}\{3|\text{yellow}\} = \frac{15}{40} = 0.375.$$

Bayes' theorem provides another way to make this estimate, based on the datum of a single candy. A *Bayesian* would reason, like the frequentist, that the probability of randomly selecting each bowl is 1/3, i.e., the prior distribution $P(i)$ (discrete in this example) is known for each bowl. Here Prob$\{1\}$ = Prob$\{2\}$ = Prob$\{3\}$ = 1/3, i.e., a uniform prior distribution. Further, the normalized likelihood for bowl i is Prob$\{$yellow$|i\}/M$, where M is the global likelihood

$$M = \text{Prob}\{\text{yellow}|1\}/3 + \text{Prob}\{\text{yellow}|2\}/3 + \text{Prob}\{\text{yellow}|3\}/3$$

$$= \frac{10/30 + 15/30 + 15/30}{3} = 0.4444.$$

The Bayesian estimate of the probability that the child retrieved the yellow candy from bowl i is the (discrete) posterior distribution, namely,

$$\text{Prob}\{1|\text{yellow}\} = \frac{\text{Prob}\{1\}\text{Prob}\{\text{yellow}|1\}}{M} = \frac{(1/3)(10/30)}{0.4444} = 0.25,$$

$$\text{Prob}\{2|\text{yellow}\} = \frac{\text{Prob}\{2\}\text{Prob}\{\text{yellow}|2\}}{M} = \frac{(1/3)(15/30)}{0.4444} = 0.375,$$

$$\text{Prob}\{3|\text{yellow}\} = \frac{\text{Prob}\{3\}\text{Prob}\{\text{yellow}|3\}}{M} = \frac{(1/3)(15/30)}{0.4444} = 0.375.$$

These results, obtained from Bayes' theorem in its classical interpretation, agree with those of the strict frequentist approach because the prior distribution is known, here a uniform distribution $P(i) = 1/3$. Indeed, the use of a uniform or *uninformative* prior distribution often, but not always, produces the same results as would be obtained by a frequentist.

Implicit in both the frequentists' and Bayesian approaches used in Example 6.4 is the assumption that the child is equally likely to select any of the three bowls. But experience has revealed that a child is more likely to choose the bowl closest to the entrance door. Frequentists have great difficulty using this vague information because they must use only observed data in their analyses. A Bayesian is under no such stricture and is free to incorporate other data, personal intuition, biases, anecdotal information, etc., into the selection of a prior distribution. Before the Bayesian approach is discussed in more detail, several features about Bayes' theorem should be noted.

- Both the parameters and the data are treated as random variables; the posterior f is the PDF for the parameters and the likelihood ℓ is proportional to the PDF for the data.
- Equation (6.57) holds for any number of random variables. It also holds if the random variables are discrete, continuous, or a mixture of both.
- The "data" can be values either of a single measured quantity (such as temperature measured at different times) or of many quantities (such as temperature, pressure, and humidity measured at different times).
- Equation (6.58) is in the form of an expected value and expresses the fact that the probability a certain data set is obtained is the population mean of obtaining the data, given the model parameters, averaged over the range of the parameters.

Thus, estimation of the posterior distribution can be reduced to estimation from a prior distribution and a likelihood function.

6.3.2 Extending the Application of Bayes' Theorem

Because data about some particular rare phenomenon are often sparse and classical, (frequentists') statistical analysis usually yields results with low confidence. For example, initial drug efficacy or side effect studies involving few subjects, the probability of successfully landing robots on Mars or asteroids, the ability of viruses to mutate and transfer to other species, predicting the occurrence and strengths of earthquakes along a geological fault, and many other low-probability phenomena produce sparse data for analysis. Here is where Bayesian techniques shine because they can use not only the sparse data but also results of similar studies, expert opinion, and, alas, even political bias to construct the prior distribution.

The term "Bayesian statistics" has thus come to mean the use of Bayes' theorem in a manner different from that discussed in Example 6.4. Rather, experience, hypothesis, or some other means is used to assume a prior distribution when one is not known. This Bayesian approach is illustrated in Example 6.5. This example is deserving of the reader's attention as it used to illustrate later many features of Bayesian statistics.

Example 6.5 A Bayesian Statistical Analysis

A different application of Bayes theorem occurs when an educated guess is made at what an appropriate prior might be. To illustrate such an application, consider the problem faced by two statisticians, one a frequentist and the other a Bayesian, who are asked by a Buffon needle dropper to interpret the results from three experiments, each using a different needle length L and grid spacing D. In particular, the statisticians are asked to estimate the probability a dropped needle cuts a grid line.

In the first experiment, 682 of 1000 dropped needles intersect a grid line. The frequentist immediately says the probability that the next dropped needle will cut a grid line is $682/1000 = 0.682$, because probability is interpreted as being solely based on observed frequency.

The Bayesian's estimate depends on what prior is assumed. The Bayesian might argue that experience indicates that most Buffon needle droppers use nearly equal needle lengths and grid spacings, i.e., $L \simeq D$, but there is some chance $L \neq D$. Let p be the probability a needle cuts a grid line. One might assume the prior distribution is a beta distribution (see Appendix A.1.4), namely,

$$g(p|\alpha, \beta) = \mathcal{B}_e(\alpha, \beta) = \frac{\Gamma(\alpha + \beta)}{\Gamma(\alpha)\Gamma(\beta)} p^{\alpha-1}(1 - p)^{\beta-1}, \tag{6.59}$$

where α and β are parameters. The mean and variance of this distribution are

$$\mu(g) = \frac{\alpha}{\alpha + \beta} \quad \text{and} \quad \sigma^2(g) = \frac{\alpha\beta}{(\alpha + \beta)^2(\alpha + \beta + 1)}. \tag{6.60}$$

The Bayesian decides it is reasonable to assume $\mu(g) = 0.636620$ (P_{cut} for $L = D$) and $\sigma^2(g) = 0.01$ (to allow some variation in L/D). Solving Eq. (6.60) for α and β, the Bayesian estimates the parameters of the prior distribution as

$$\alpha = \frac{\mu(g)^2}{\sigma^2(g)}[1 - \mu(g)] - \mu(g) = 14.0906 \quad \text{and}$$

$$\beta = \frac{\mu(g)}{\sigma^2(g)}[1 - \mu(g)]^2 + \mu(g) - 1 = 8.0429.$$

Now, to apply Bayes' theorem, one must consider the likelihood ℓ and the global likelihood M. If one conducts n trials of an experiment that has constant probability p of success (such as a needle cutting a grid line), then the PDF that k successes are obtained is given by the binomial distribution $\mathcal{B}(n, p)$ (see Appendix A.2.2)

$$\ell(k|n, p) = \frac{n!}{k!(n-k)!} p^k (1 - p)^{n-k}, \qquad k = 0, 1, 2, \ldots, n. \tag{6.61}$$

The global likelihood is given by integrating the likelihood function weighted by the prior distribution (see Eq. (6.58)), and in this case assumes the form

$$M = \int_0^1 \ell(k|n, p) g(p|\alpha\beta) \, dp, \tag{6.62}$$

which can be shown to yield (Johnson et al., 1994)

$$M = \frac{n!}{k!(n-k)!} \frac{\Gamma(\alpha + \beta + n)}{\Gamma(\alpha + n)\Gamma(\beta + n - k)} \frac{\Gamma(\alpha + \beta)}{\Gamma(\alpha)\Gamma(\beta)}. \tag{6.63}$$

It can be shown that the posterior distribution is also a beta distribution[10]:

$$f(p|k, n, \alpha, \beta) = g(p|\alpha, \beta) \frac{\ell(k|n, p)}{M}$$

$$= \frac{\Gamma(\alpha + \beta + n)}{\Gamma(\alpha + k)\Gamma(\beta + n - k)} p^{\alpha+k-1}(1 - p)^{\beta+n-k-1} = \mathcal{B}_e(\alpha + k, \beta + n - k). \tag{6.64}$$

---

[10] This is true because the beta and binomial distributions are so-called *conjugate* distributions.

The Bayesian statistician, thus, estimates the probability p_o that a dropped needle cuts a grid line, given the observed data from the first experiment, as the expected value of $f(p|k, n, \alpha, \beta)$. This expected value is just the mean value of the beta distribution $g(p|\alpha + k, \beta + n)$, namely,

$$p_o = \mu(f) = \frac{\alpha + k}{\alpha + \beta + n} = \frac{14.0906 + 682}{14.0906 + 8.0429 + 1000} = 0.68102,$$

and is very close to the frequentist's estimate of $k/n = 682/1000 = 0.68200$. This excellent agreement occurs because the data are "reasonable," in the sense that the results obtained are within one standard deviation of the expected results if $L = D$ (i.e., 637 cuts in 1000 needle drops).

However, in the second experiment, astonishingly, none of the 1000 dropped needles cuts a grid line. The frequentist, therefore, would estimate the probability of a dropped needle intersecting a grid line as $p_o = 0/1000 = 0.0$. However, the Bayesian would estimate

$$p_o = \mu(f) = \frac{\alpha + k}{\alpha + \beta + n} = \frac{14.0906 + 0}{14.0906 + 8.0429 + 1000} = 0.013786.$$

The Bayesian would, thus, agree that it is very unlikely to have a grid intersection from a needle drop, but it would still be possible with a probability $p = 0.0138$, or about 1 in 72.

So a Bayesian approach appears to give results very similar to the classical frequentist approach for the first two experiments. But this is not always the case. Consider now that the needle dropper begins the third experiment, drops three needles, recording 0 cuts, and then promptly leaves for the local tavern. A frequentist analyzing these data would again assume $p_o = 0/3 = 0.0$. However, the Bayesian, who uses the same beta prior, would obtain

$$p_o = \mu(f) = \frac{\alpha + k}{\alpha + \beta + n} = \frac{14.0906 + 0}{14.0906 + 8.0429 + 3} = 0.56063.$$

The Bayesian incorporates experience that Buffon experiments frequently use $L \simeq D$ and obtains a result that is not too different from 64% (the mean of the beta distribution if it is assumed that $D = L$), because only three trials were performed.

Note in the above example that the relatively large number of needle drops, $n = 1000$, compared to the size of α and β dominate the Bayesian estimate of the cut probability in the first and second experiment. By contrast, in the third experiment, the very small number of needle drops, $n = 3$, allow the prior distribution to dominate the estimate of the cut probability.

6.3.3 Probability and Confidence Intervals

In a Bayesian analysis, the vector of model parameters $\boldsymbol{\theta}$, or in Example 6.5 the single parameter p, is treated as a random variable with an assumed PDF (devised somehow by the analyst) called the prior distribution. The posterior

PDF $g(\boldsymbol{\theta})$ is a refined PDF of $\boldsymbol{\theta}$ given both the prior PDF and the observed data. The best estimate of the "true value of $\boldsymbol{\theta}$," denoted by $\boldsymbol{\theta}_o$, is usually taken as the posterior mean, although the median or maximum could also be used. Because the posterior PDF is a probability PDF, one can also make a probability statement that the true parameter value lies in a *credible region* at the $\widehat{\alpha}\%$ level. For a single parameter p, the credible interval $p_{low} < p_o < p_{high}$ contains $\widehat{\alpha}\%$ of the area under the posterior distribution.

However, there is no unique prescription for calculating a credible interval on the posterior PDF. Methods include:

- Select p_{low} so the PDF area below it equals the area above p_{high}, i.e., the two tails each have an area $\Delta\widehat{\alpha} \equiv (1 - \widehat{\alpha})/2$. This method is widely used by Bayesians and frequentists.
- Select the narrowest interval. For a unimodal posterior PDF, one starts at the maximum and descends to lower probabilities until $\widehat{\alpha}\%$ of the PDF area is contained in the interval.
- Select the interval for which the mean of the posterior is the central point and over which the PDF area is $\widehat{\alpha}\%$.

Confidence Intervals

A frequentist, who does not believe p has a distribution, generates a closely related quantity called a *confidence interval*. The calculation of confidence and credible intervals is now illustrated for the Buffon's needle experiments of Example 6.5. Classical statisticians as well as Bayesians would both agree that the probability of observing k events (here a needle cutting a grid line) in n tries (needle drops) is described by the binomial PDF of Eq. (6.61) with $p_o = P_{cut}$ with a fixed, but unknown value. The frequentists would take the mean of this PDF, which is k/n, as the best estimate of p_o. They would also want to quantify the precision of this estimate. Specifically, they seek the maximum and minimum reasonable values of p for which k events in n tries could be expected at some confidence level $\widehat{\alpha}$. The maximum reasonable value of p at the $\widehat{\alpha}$-*level* is the value p_{high}, for which one would observe, with probability $\Delta\widehat{\alpha}$, k *or fewer* events in n tries. Similarly, the minimum reasonable value of p at the $\widehat{\alpha}$-*level* is the value p_{low}, for which one would observe, with probability $\Delta\widehat{\alpha}$, k *or more* events in n tries. It can be shown that these limiting probabilities are given by (Shultis et al., 1979)

$$I_{p_{low}}(k, n - k + 1) = \Delta\widehat{\alpha} \quad \text{and} \quad I_{p_{high}}(k + 1, n - k) = 1 - \Delta\widehat{\alpha}, \qquad (6.65)$$

where I_p in the incomplete beta function is defined as

$$I_p(a, b) = \frac{\Gamma(a + b)}{\Gamma(a)\Gamma(b)} \int_0^p t^{a-1}(1 - t)^{b-1} \, dt. \qquad (6.66)$$

At this point, numerical techniques must to used to solve for p_{low} and p_{high}.

As an illustration, consider the first experiment in Example 6.5, in which $k = 682$ grid cuts were obtained in $n = 1000$ needle drops. A frequentist says the best estimate of $p_o = 682/1000 = 0.68200$. Numerical solution of Eq. (6.65), at the 95% confidence level, gives the narrow confidence interval of $(0.65213, 0.710799)$. The narrowness of this interval is a direct result of the large values of n and k.

Probability Intervals

The Bayesians, unlike frequentists, treat p as a random variable. It may have some unknown fixed value, but the best they can do is describe it in probabilistic terms. Their best estimate of what values p could have are described by the posterior PDF $f(p|n, k, \alpha, \beta)$ of Eq. (6.64). The probability interval (p_{low}, p_{high}) about the mean

$$\mu(f) = \frac{\alpha + k}{(\alpha + k) + (\beta + n - k)}$$

is readily computed from the posterior. Explicitly, the probability that the true event probability is greater than an upper bound p_{high}, at the $\widehat{\alpha}/2$ level, is

$$\text{Prob}\{p > p_{high}\} = \Delta\widehat{\alpha} = \int_{p_{high}}^{1} f(p|k, n, \alpha, \beta)\, dp. \qquad (6.67)$$

Likewise, the probability that the true event probability is less than a lower bound p_{low}, at the $\widehat{\alpha}/2$ level, is

$$\text{Prob}\{p < p_{low}\} = \Delta\widehat{\alpha} = \int_{0}^{p_{low}} f(p|k, n, \alpha, \beta)\, dp. \qquad (6.68)$$

Substitution of Eq. (6.64) into the above two equations yields

$$I_{p_{low}}(\alpha + k, n + \beta - k) = \Delta\widehat{\alpha} \quad \text{and} \quad I_{p_{high}}(\alpha + k, n + \beta - k) = 1 - \Delta\widehat{\alpha}. \qquad (6.69)$$

Note these equations have exactly the same form as those for the classical case of Eq. (6.65).

Again consider the first experiment of Example 6.5. With the Bayesian's assumed prior distribution and the observed data, the mean value of the posterior PDF gives $p_o = 0.681017$. The solution of Eq. (6.69) yields the 95% level probability interval $(0.65213, 0.70923)$, an interval even smaller than the frequentist's confidence interval because of the extra information used by the Bayesian, namely, the prior PDF.

The Distinction

Note a credible interval (or region for a multivariate problem) makes a probability statement about the true value, namely, it has a $\widehat{\alpha}\%$ chance of being in the interval. Credible intervals are thus sometimes called *probability intervals*.

This probability statement is in contrast to *confidence intervals* of frequentist statistics. In frequentist terms, a model parameter is fixed and cannot have a distribution of possible values. Instead it is the confidence interval that is random because it depends on a random sample. In Section 2.5, the CLT is used to create $\widehat{\alpha}\%$ confidence intervals for approximating the expected value $\langle z \rangle$ of some random variable z by the arithmetic average $\overline{z}$ of a set of N samples z_i of the random variable. The interpretation of the confidence interval is *not* that $\langle z \rangle$ has an $\widehat{\alpha}\%$ chance of being in the confidence interval. Rather, it means if the same calculation was performed many times, each with a new set of N samples z_i, then $\widehat{\alpha}\%$ of the confidence intervals would contain $\langle z \rangle$. This is a subtle but important distinction between Bayesians and frequentists in describing how good their estimations of random variables are.

6.3.4 Conjugate Prior Distributions

To undertake a Bayesian statistical analysis, both the likelihood PDF $\ell(\mathbf{D}|\boldsymbol{\theta})$ and prior PDF $g(\boldsymbol{\theta})$ must be chosen by the analyst. In principle, any functions may be used; but in practice, the selection is rather limited if the posterior PDF is to be obtained explicitly. The difficulty is the analytic evaluation of the integral of Eq. (6.58). In Example 6.5, the analytic evaluation of Eq. (6.62) to obtain Eq. (6.63) is not easy.

For a given likelihood PDF, sometimes there is a family of prior PDFs for which M can be found analytically and for which the prior and posterior PDFs belong to the same family, albeit with different parameters. Such is the case in Example 6.5 where the prior PDF is a beta distribution with parameters α and β and the posterior PDF is also a beta function but with so-called *hyperparameters* $\alpha' = \alpha + k$ and $\beta' = \beta + n - k$. When the prior and posterior PDFs belong to the same family they are called *conjugate distributions*, and the prior is called a *conjugate prior* for the likelihood function.

Of the hundreds or even thousands of possible pairings of different prior and likelihood PDFs, there are relatively few conjugate prior distributions. Table 6.2 lists a few of more important conjugate pairings. An exhaustive list can be found on Wikipedia (Wikipedia Contributors, 2020) which shows only six conjugate priors for discrete likelihood PDFs and only 20 for continuous likelihood PDFs.

This paucity of analytically tractable posterior distributions severely limited the use of Bayesian methods for most of the twentieth century. However, the need for safety analyses of rare events with high consequences increased the need to bypass the analytic posterior bottleneck. Efforts were begun to use mainframe computers and standard numerical methods to calculate probability or credibility intervals for the occurrence of rare events (see, for example, (Shultis et al., 1981a, 1986)). With such brute-force efforts, the effect of using different nonconjugate priors for rare-event analyses could finally be addressed.

But such mainframe computer approaches were short-lived due to the introduction of a new technology (the personal computer) and a new way to evaluate

Table 6.2 Some example conjugate prior distributions for discrete and continuous likelihood functions. Here n is the number of observations (data) x_i, which are random vectors in the multivariate case. An exhaustive list is available on the Internet (Wikipedia Contributors, 2020).

Likelihood	Model parameters	Conjugate prior distribution	Prior parameters	Posterior hyperparameters
Likelihood function is discrete				
Binomial $\binom{n}{x} p^x (1-p)^{n-x}$ $\quad x = 0, 1, 2, \ldots$	p (prob.)	Beta $\frac{\Gamma(\alpha+\beta)}{\Gamma(\alpha)\Gamma(\beta)} x^{\alpha-1}(1-x)^{\beta-1}$	α, β	$\alpha' = \alpha + \sum_{i=1}^n x_i$ $\beta' = \beta + n + \sum_{i=1}^n x_i$
Poisson $mu^x \exp[-\mu]/x!$ $\quad x = 0, 1, 2, \ldots$	μ (rate)	Gamma $\beta^\alpha x^{\alpha-1} e^{-\beta x}/\Gamma(\alpha) \quad x>0$	$\alpha, \beta > 0$	$\alpha' = \alpha + \sum_{i=1}^n x_i$ $\beta' = \beta + n$
Negative binomial (s is fixed) $\frac{(s+x-1)!}{x!(s-1)!} p^s (1-p)^x$	p (prob.)	Beta $\frac{\Gamma(\alpha+\beta)}{\Gamma(\alpha)\Gamma(\beta)} x^{\alpha-1}(1-x)^{\beta-1}$	$\alpha, \beta > 0$	$\alpha' = \alpha + \sum_{i=1}^n x_i$ $\beta' = \beta + sn$
Likelihood function is continuous				
Normal (σ^2 is fixed) $\exp[-(x-\mu)^2/(2\sigma^2)]/\sqrt{2\pi\sigma^2}$ $\quad \infty < x < \infty$	μ (mean)	Normal $\exp[-(x-\mu_o)^2/(2\sigma_o^2)]/\sqrt{2\pi\sigma_o^2}$	μ_o, σ_o^2	$\sigma'^2 = (1/\sigma_o^2) + (n/\sigma^2)$ $\mu' = ((\mu_o/\sigma_o^2) + (\sum_{i=1}^n x_i/\sigma^2))\sigma'^2$
Pareto (β is fixed) $\alpha\beta^\alpha x^{-\alpha-1}$	α (shape) $\alpha > 0, x \geq \beta > 0$	Gamma $\beta_o^{\alpha_o} x^{\alpha_o-1} e^{-\beta_o x}/\Gamma(\alpha_o)$	α_o, β_o	$\alpha' = \alpha + n$ $\beta' = \beta_o \sum_{i=1}^n \ln(x_i/\beta)$
Weibel α is fixed $\frac{\alpha}{\beta}\left(\frac{x}{\beta}\right)^{\alpha-1}$, $x>0$	β (scale) > 0	Inverse Gamma $\frac{b^a}{\Gamma(a)}(1/x)^{a+1}\exp[-b/x]$	$a, b > 0$	$a' = a + n$ $b' = b + \sum_{i=1}^n x_i^b$
Inverse Gamma (a is fixed) $b^a x^{-(a+1)}\exp[-b/x]/\Gamma(a) \quad x > 0$	b (scale)	Gamma $\beta_o^{\alpha_o} x^{\alpha_o-1} e^{-\beta_o x}/\Gamma(\alpha_o)$	α_o, β_o	$\alpha' = \alpha_o + na$ $\beta' = \beta_o + \sum_{i=1}^n x_i$

the posterior PDF (MCMC). The revolution caused by the confluence of these two powerful events is discussed next.

6.4   Use of MCMC in Bayesian Analysis

Although Bayesian techniques were known for many decades before 1990, its use was mostly limited to analytic studies using conjugate prior distributions, which led to posterior distributions of the same form (but with different parameters). The paper by Geman and Geman (1984) introduced the Gibbs sampler to spatial statisticians because it was devoted to the optimization of the posterior PDF in the brothers' image reconstructing study. However, it took a few years before most spatial statisticians realized that the Gibbs sampler was a powerful approach for simulating the posterior PDF, thereby enabling Bayesian inference for all sorts of studies.

Then the monumental paper by Gelfand and Smith (1990) made the wider Bayesian community aware of the Gibbs sampler. This seminal paper revitalized the Bayesians who began to use MCMC for hundreds of inference analyses that could be done in no other way. MCMC has now made it possible to simulate a wide range of posterior distributions to find their parameters numerically. Further, as Bayesians began to properly understand MCMC and its sampling techniques, they began to use it for more than Bayesian inference studies. For example, it has been used in likelihood inference to compensate for missing data or to study complex interdependencies (Geyer, 1992a, 1992b, 2011, and cited reference therein).

This rapid adoption of MCMC techniques by Bayesians occurred at a time when new calculational tools became widely available to statisticians. In 1981, IBM introduced its first personal computer. By 1990, the personal computer had evolved into a machine capable of performing serious statistical calculations. Further, unlike brute-force numerical Bayesian approaches, MCMC algorithms were extremely easy to program. So, in hindsight, it is not surprising that MCMC and its sampling techniques became popular so rapidly.

6.4.1   Using MCMC to Calculate the Posterior PDF

Now consider how MCMC sampling techniques can be used to find the posterior PDF $f(\theta|\mathbf{D})$ of Eq. (6.57). The M-H algorithm (or Gibbs sampler) is used to generate samples θ_i from a function of θ that has the same *shape* as the right side of Eq. (6.57). All normalizing constants and M can be dropped from the right side of Eq. (6.57).[11]

In this example, $\theta \to p$, a random variable with $p \in (0, 1)$. The M-H algorithm of Section 6.2.4 can be used to generate samples p_i from a distribution $f(p)$ when only the shape of $f(p)$ is known. For Example 6.5,

---

[11] Recall that Bayesians treat θ as a random variable, whereas frequentists treat θ as an (unknown) constant.

$f(p) = Cp^{\alpha+k-1}(1-p)^{\beta+n-k-1}$, and for sampling with the M-H algorithm, the constant C may be taken as unity. This particular case is considered in Example 6.6.

One special precaution needs to be incorporated into the M-H algorithm of Section 6.2.4. In general, $f(x)$ and $q(x|u)$ can often assume inordinately small values such as in low-probability tails. For example, if for $x = 0.1$ and $m = 100$, $x^m = 1 \times 10^{-100}$, a value that causes the Metropolis ratio R to produce an over/underflow on most computers. Similarly, exponential functions can cause over/underflow conditions. The remedy is simple. Rather than calculate R as

$$R(x_i, u_i) = \frac{f(u_i)}{f(x_i)} \frac{q(x_i|u_i)}{q(u_i|x_i)},$$

it is better to calculate

$$\ln R(x_i, u_i) = \ln f(u_i) - \ln f(x_i) + \ln q(x_i|u_i) - \ln q(u_i|x_i).$$

Then if $\ln R \lesssim -30$ set $R = \alpha(x_i, u_i) = 0$ or, better yet, just keep the old state x_i. The number 30 can be replaced by the smallest number x_{max} such that $\exp[-x_{max}]$ does not cause an underflow. On the other hand, if $\ln R \gtrsim 30$, set $R = \infty$ or, more simply, just accept the proposed state u_i.

The last step in creating the posterior PDF is to bin the N samples x_i and then divide the number in each bin by $N\Delta x$ to obtain the bin probabilities f_i. Here Δx is the bin width for the case of contiguous equiwidth bins.

Example 6.6 MCMC Estimation of a Posterior PDF

Consider the Buffon needle drop analysis of Example 6.5 and estimate the posterior PDF as described above. Here $x \to p \in (0, 1)$. Specifically consider Experiment 1 in which $k = 682$ grid intersections would be observed in $n = 1000$ needle drops. The exact posterior PDF is given by Eq. (6.64) with $\alpha = 14.0906$ and $\beta = 8.04287$. Compare the MCMC estimated posterior PDF with the exact analytic result.

To simplify the calculations, the proposal distribution $q(u|x)$ was taken to be $\mathcal{U}(0, 1)$, i.e., $q = 1$ and the proposed sample u_i is just a pseudorandom number ρ_i uniformly distributed on $(0, 1)$. With the procedure just described to avoid under/overflow errors, a large number N of samples p_i, $i = 1, 2, \ldots, N$, were obtained using the M-H algorithm. These samples were then sorted into K equiwidth probability bins to record the total number of samples whose values fell within each bin's p-range. The midpoint probability of the ith bin is denoted by $\bar{p}_i$. The binned results are shown in Table 6.3.

With these data the posterior, averaged over bin i, is calculated as $f_i = c_i/(N\Delta p)$, where c_i is the number of counts or samples in bin i. The exact beta posterior PDF is given by Eq. (6.64) with $\alpha = 14.0906$ and $\beta = 8.0429$. The histogram of the MCMC calculated posterior is compared to the exact beta posterior in Fig. 6.13. As can be seen, the agreement is amazingly good.

Table 6.3 Computer output for the number c_i of counts per probability bin of the MCMC Bayesian simulation of the posterior distribution of Example 6.5. Sample size $= 10^9$, number of bins $K = 100$, bin width $\Delta_p = 0.01$. The values of $\bar{p}_i$ are the midpoints of each bin.

$\bar{p}_i$	c_i	$\bar{p}_i$	c_i	$\bar{p}_i$	c_i
0.515	0	0.615	21,994	0.715	18,928,254
0.525	0	0.625	288,547	0.725	2,903,507
0.535	0	0.635	2,536,567	0.735	258,885
0.545	0	0.645	14,927,120	0.745	14,018
0.555	0	0.655	58,070,649	0.755	374
0.565	0	0.665	148,101,178	0.765	18
0.575	0	0.675	245,017,669	0.775	0
0.585	8	0.685	260,324,904	0.785	0
0.595	116	0.695	175,116,332	0.795	0
0.605	1218	0.705	73,488,642	0.805	0

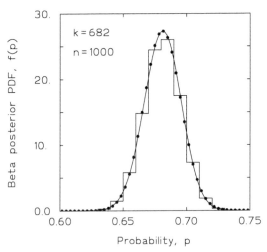

Figure 6.13 Posterior PDF $f(p)$ for Example 6.5. Exact (smooth line) and MCMC (histogram). The data points are a normal distribution whose mean and variance are those of the exact beta posterior PDF.

Finally with the MCMC discrete posterior, one can estimate the mean μ and variance σ^2 of the posterior. Exact values are given by Eq. (6.60) with the prior parameters α and β replaced by the posterior hyperparameters $\alpha' = \alpha + k$ and $\beta' = \beta + n - k$. The mean and variance of the posterior are estimated from the binned data as

$$\mu(f) \simeq \frac{1}{N} \sum_{i=1}^{K} \bar{p}_i c_i \quad \text{and} \quad \sigma^2(f) \simeq \frac{1}{N} \sum_{i=1}^{K} \bar{p}_i^2 c_i^2 - \left(\frac{1}{N} \sum_{i=1}^{K} \bar{p}_i c_i \right)^2 . \tag{6.70}$$

Alternatively, μ and σ of the posterior PDF could be estimated directly from the samples p_i, $i = 1, 2, \ldots, N$, as

$$\mu(f) \simeq \frac{1}{N} \sum_{i=1}^{N} f_i \quad \text{and} \quad \sigma^2(f) \simeq \frac{1}{N} \sum_{i=1}^{N} f_i^2 - \left(\frac{1}{N} \sum_{i=1}^{N} f_i \right)^2 , \tag{6.71}$$

where $f_i \equiv f(p_i)$. The estimated mean and standard deviation together with absolute percentage difference between the MCMC simulation and the exact results are shown in Tables 6.4 and 6.5.

The symmetry and famous bell shape of the beta function that is the posterior PDF for the $k/n = 682/1000$ case of Example 6.5 is apparent from even a cursory look at Fig. 6.13. The circular data points on Fig. 6.13 are values of a Gaussian PDF with mean μ_o and σ_o given by Eq. (6.70) or (6.71). The Gaussian PDF is seen to agree very closely with the posterior PDF. Indeed this seems to be the case whenever there is such a large number n of observed data it overwhelms whatever the prior PDF contributes to the posterior PDF. Only in the very low-probability tails does the beta posterior PDF differ from the normal approximation. Moreover, this approximation is to be expected. It is well known that if $n \gtrsim 20$ and $np < 5$ (or $(1 - p)n < 5$) the binomial likelihood PDF is well approximated by a Poisson PDF which, in turn, can be approximated by a normal PDF if $np \gtrsim 10$. This asymptotic normality behavior of the posterior PDF occurs even if the prior PDF is a nonconjugate distribution as seen in the next section.

Table 6.4 Variation of sample mean μ and standard deviation σ of the beta posterior PDF with sample size N. Also shown are the percent absolute differences between the sample-derived quantity and the exact values, $\mu = 0.681017$ and $\sigma = 0.014571$. Results are for Example 6.5 with $k/n = 682/1000$ and are based on Eq. (6.71).

N	μ	error(%)	σ	error(%)
exact	0.68102		0.01457	
10^3	0.68087	0.18524	0.01459	0.13719
10^4	0.68004	0.14381	0.01392	4.46260
10^5	0.68116	0.02165	0.01472	0.99375
10^6	0.68087	0.02200	0.01470	0.85818
10^7	0.68104	0.00357	0.01459	0.14396
10^8	0.68102	0.00032	0.01458	0.04109

Table 6.5 Variation of sample mean μ and standard deviation σ of the beta posterior PDF with sample size N and are based on Eq. (6.70). Exact values are $\mu = 0.681017$ and $\sigma = 0.014571$.

N	μ	error(%)	σ	error(%)
		100 bins		
10^4	0.68003	0.14572	0.01471	0.89670
10^5	0.68111	0.01377	0.01493	2.43689
10^6	0.68086	0.02318	0.01499	2.88474
10^7	0.68104	0.00337	0.01489	2.18707
10^8	0.68101	0.00069	0.01486	1.98277
		400 bins		
10^4	0.67994	0.15896	0.01455	0.17057
10^5	0.68121	0.02880	0.01472	1.00794
10^6	0.68087	0.02202	0.01472	1.04318
10^7	0.68104	0.00372	0.01461	0.28683
10^8	0.68102	0.00028	0.01460	0.19044

6.4.2 MCMC and Nonconjugate Prior PDFs

As already explained, MCMC with the Gibbs or M-H samplers allows Bayesian analyses based on nonconjugate prior distributions. So it is appropriate that our explorations in Monte Carlo applications in Bayesian analyses end with some examples of nonconjugate prior PDFs and when it is important to pick the prior carefully and when almost any old prior will suffice. For illustration, we mention three nonconjugate priors for the binomial likelihood function used in this chapter. These are:

1. **Tent Function:**

$$g(x|a, b, x_o) = C \begin{cases} (x-a)/(x_o - a), & a < x \le x_o, \\ (b-x)/(b-x_o), & x_o < x < b, \\ 0, & \text{otherwise,} \end{cases} \quad (6.72)$$

where C is a normalization constant to make g a PDF over the interval $[0, 1]$. However, one can set $C = 1$ since the M-H sampler requires only the shape of the prior times the likelihood to be known. For the tent function shown in Fig. 6.14(left), $a = -0.1$, $b = 1.1$, and $x_o = 0.63662$ ($= P_{cut}$ for Buffon's problem with $L = D$).

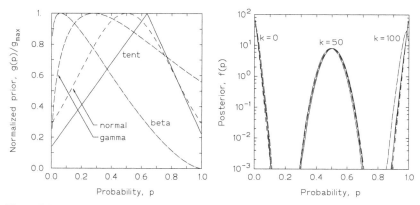

Figure 6.14 The left plot shows four different priors used in an MCMC analysis of the Buffon's needle drop problem. To the right are plots of the posterior obtained with the four priors for three cases each involving $n = 100$ drops: $k = 0$, $k = 50$, and $k = 100$. For each case 10^7 samples were used. Note the estimated posterior is independent of the prior used (except for beta prior with $k = 100$, a result of the smallness of the beta prior near $p = 1$).

2. **Normal Function:**

$$g(x|\mu, \sigma^2) = C \exp[-(x - \mu)^2/(2\sigma^2)]. \tag{6.73}$$

Again the normalization constant can be taken as unity. For the normal distribution shown in Fig. 6.14(left), $\mu = 0.5$ and $\sigma^2 = 0.1$.

3. **Gamma Function:**

$$g(x|\alpha, \beta) = C\left(\frac{x}{\beta}\right)^{\alpha-1} e^{-x/\beta}. \tag{6.74}$$

For the gamma function shown in Fig. 6.14(left), $\alpha = 1.45$ and $\beta = 0.62069$.

With the M-H sampler, it is just as easy to use these three prior distributions as it is to use the beta conjugate prior PDF. Shown in Fig. 6.14 are the four posterior density functions obtained with $n = 100$ Buffon needle drops with three outcomes $k = 0$, $k = 50$, and $k = 100$. Note that all four prior PDFs produce almost the same posterior distributions, the one exception being the beta prior PDF for the extreme case of $k = 100$ where the support of the posterior PDF is near $p = 1$ and where the assumed beta prior is very small. In this example, the large value of $n = 100$ overwhelms the prior contribution so that all four prior densities yield almost the same posterior PDFs. Again, for $k = 50$, the posterior is well approximated by a normal distribution. But for the two extreme cases $k = 0$ and $k = 100$, the binomial distribution is not at all approximated by a normal distribution.

In Fig. 6.15, posterior PDFs are shown for three cases with $n = 10$. With such a small n, the choice of the prior PDF affects the estimated posterior. Interestingly the three nonconjugate prior PDFs give similar results for the posterior

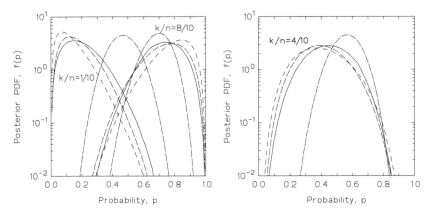

Figure 6.15 Unlike the case shown to the right in Fig. 6.14, in which n is so large that the influence of the prior contribution on the posterior PDF is lost, in the two cases here n is so small that the prior PDF has a large influence on the posterior PDF. The line styles used in Fig. 6.14(left) identify the prior PDFs on these two figures. Note that the conjugate beta prior PDF produces a posterior PDF that is noticeably different from the others, another reason to use MCMC in Bayesian analyses with nonconjugate prior PDFs.

PDF. Here it is the conjugate beta prior PDF that stands apart with a mean value quite different from the frequentist's estimate of $p_o = k/n$. In this case, the conjugate beta PDF would appear to be a poor choice.

Finally, let us revisit Example 6.5 and use the nonconjugate prior PDFs as well as the beta conjugate PDF. Results are shown in Fig. 6.16. Again for the two cases with $n = 1000$ all four prior PDFs give almost the same reasonable posterior PDF. But for the $k/n = 0/3$ case, the choice of the prior makes a great difference. Again the three nonconjugate priors give very similar results, but the beta prior yields quite a different result. One wonders if the many Bayesian studies before 1990, when MCMC methods revolutionized Bayesian analysis, were similarly biased because of the need to use the conjugate beta distribution. See Table 6.6.

6.4.3   Estimating the Prior Distribution

There is no prescription for choosing a particular family of functions unless calculations are to be done analytically, in which case a conjugate prior must be selected. Nor is there a prescription to choose values of the prior parameters other than the resulting prior should appear "reasonable." Indeed, parameters can be chosen by using only the Bayesian's intuition as was done in Example 6.5, where the analyst's past experience suggested that most Buffon needle drop experiments used needle lengths L that were close to the grid spacing D.

A less subjective way, and one widely used, is to use data obtained in similar studies. For example, in estimating the efficacy of a proposed drug, results of studies on similar drugs could be used. Likewise, to predict the size of meteors

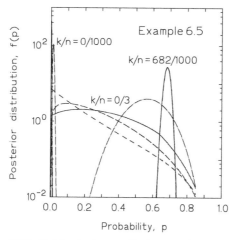

Figure 6.16 Posterior PDFs $f(p)$ for the three needle-drop experiments of Example 6.5 using the four prior PDFs shown in Fig. 6.14.

Table 6.6 Sample mean μ, standard deviation σ, and proposal acceptance percentage (PAP) of the posterior PDF to the left, in case 1 $(k/m = 682/1000)$, case 2 $(k/n = 0/1000)$, and case 3 $(k/n = 0/3)$, all with $N = 10^7$.

Prior	Case	μ	σ	PAP
tent	1	0.68115	0.01472	4.69
	2	0.00100	0.00104	0.20
	3	0.28361	0.17850	55.13
normal	1	0.68206	0.01473	4.70
	2	0.00100	0.00111	0.20
	3	0.14606	0.14457	29.23
gamma	1	0.68146	0.01470	4.69
	2	0.00120	0.00120	0.28
	3	0.22210	0.15578	42.91
beta	1	0.68103	0.01457	4.64
	2	0.01379	0.00363	1.13
	3	0.56058	0.09712	31.24

that might hit earth in the future, the distribution of crater sizes on earth or the moon should guide the choice of a prior PDF. A common use of Bayesian analysis is the description of low-probability high-consequence events such as the failure of a critical safety system to operate when activated, the so-called *failure-on-demand* problem. By design, these systems are used very infrequently so failure data are sparse but still available, usually through periodic testing of the systems. Suppose at plant i, the safety system has been observed to fail to operate k_i times in n_i demands. Typically $k_i \lesssim 5$, and often zero, in at most 100 or 200 demands. Suppose there are N plants with such safety systems, or similar ones, i.e., ones from a different manufacturer than the one being studied.

There are several methods that can be used to estimate values of the parameters of a given prior PDF from such failure-on-demand data (Shultis et al., 1979). The simplest is to equate the lowest moments of the data to those of the prior. For example, the sample mean and variance of the observed data from similar systems are

$$\widehat{\mu}_{ob} = \frac{1}{N} \sum_{i=1}^{N} \frac{k_i}{n_i} \quad \text{and} \quad \widehat{\sigma}_{ob}^2 = \frac{1}{N-1} \sum_{i=1}^{N} \left(\frac{k_i}{n_i} - \widehat{\mu}_{ob} \right)^2. \quad (6.75)$$

For a beta prior PDF, equate these values to the expressions for the mean and variance of the beta function, i.e., to Eq. (6.60), and solve for the beta parameters

$$\alpha = \frac{\widehat{\mu}_{ob}^2}{\widehat{\sigma}_{ob}^2}[1 - \widehat{\mu}_{ob}] - \widehat{\mu}_{ob} \quad \text{and} \quad \beta = \frac{\widehat{\mu}_{ob}}{\widehat{\sigma}_{ob}^2}[1 - \widehat{\mu}_{ob}]^2 + \widehat{\mu}_{ob} - 1. \quad (6.76)$$

6.4.4 Failure Rate Analysis

So far in this section, the emphasis has been on problems for which the likelihood function $\ell(\mathbf{D}|\boldsymbol{\theta})$ is a binomial distribution that describes the probability of obtaining k events (needles cutting a grid line or a component not starting) in n attempts (needle drops or tries to start the component). This type of problem is called the *failure-on-demand* problem.

There is a second important type of failure problem that describes the rate of failure (failures per unit time λ) of a component that is normally active, such as a water recirculation pump in a nuclear power plant. This problem is known as the *failure rate* problem. For such a problem the appropriate likelihood auction is the Poisson distribution

$$\ell(F|\lambda, T) = \frac{(\lambda T)^F}{\Gamma(F + 1)} e^{-\lambda}, \quad (6.77)$$

where F is the number of failures in operation time T. In the "homogeneous," or frequentist's model, it is assumed that the unknown constant being sought, λ_o, is unchanged if the component is repaired or replaced. Often several similar components are treated as a single generic class whose components all have the same λ_o. Then $T = \sum_{i=1}^{n} T_i$ is the number of component-hours of operation completed by the n components and $F = \sum_{i=1}^{n} F_i$ is the number of observed failures. The frequentist would estimate the failure rate as

$$\lambda_o = \sum_{i=1}^{n} F_i \Big/ \sum_{i=1}^{n} T_i. \qquad (6.78)$$

The Bayesian analyst would say the value of λ may vary among the components in the same class. Any reasonable prior distribution $g(\lambda, \boldsymbol{\theta})$ may be used, and with MCMC techniques, the posterior PDF $f(\lambda|\boldsymbol{\theta}, \mathbf{D})$ may be calculated. Most often the prior PDF is chosen as the gamma distribution, which is the conjugate distribution to the Poisson,

$$g(\lambda|\alpha, \beta) = \frac{\lambda^{\alpha-1} e^{-\lambda/\beta}}{\beta^{\alpha} \Gamma(\alpha)}, \qquad \alpha, \beta > 0, \quad 0 \le \lambda < \infty, \qquad (6.79)$$

or, with $\tau = 1/\beta$,

$$g(\lambda|\alpha, \tau) = \frac{\lambda^{\alpha-1} \tau^{\alpha}}{\Gamma(\alpha)} e^{-\lambda\tau}. \qquad (6.80)$$

The mean and variance of the gamma prior are

$$\mu = \alpha\beta = \alpha/\tau \qquad \text{and} \qquad \sigma^2 = \alpha\beta^2 = \alpha/\tau^2 = \mu/\tau. \qquad (6.81)$$

Because the gamma function is the conjugate PDF to the Poisson PDF, the posterior PDF is also a gamma function but with hyperparameters $\alpha' = \alpha + F$ and $\tau' = \tau + T$. Then from Eq. (6.81) the mean and variance of the posterior are

$$\langle \lambda \rangle = \frac{F + \alpha}{T + \tau} \qquad \text{and} \qquad \text{var}[\lambda] = \frac{F + \alpha}{(T + \tau)^2} = \frac{\langle \lambda \rangle}{T + \tau}. \qquad (6.82)$$

Theory and methods for estimating α and β from failure data, probability intervals from the posterior, and various goodness-of-fit statistical tests are described by Shultis et al. (1981b).

6.5   Inference and Decision Applications

The Bayesian formulation provides a powerful way to draw inferences. The goal of Bayesian inference often is to estimate some quantity that depends on an unknown distribution. For instance, one might want to estimate the expected value of a function $z(\boldsymbol{\theta})$ with respect to a distribution whose parameters are $\boldsymbol{\theta}$. Bayes' theorem allows us to express the unknown distribution as a posterior and then the expected value of z can be written as

$$\langle z \rangle = \frac{1}{M} \int_{\Theta} z(\boldsymbol{\theta}) f(\boldsymbol{\theta}|\mathbf{D}) \, d\boldsymbol{\theta} = \int_{\Theta} z(\boldsymbol{\theta}) g(\boldsymbol{\theta}) \frac{\ell(\mathbf{D}|\boldsymbol{\theta})}{M} \, d\boldsymbol{\theta}, \qquad (6.83)$$

where Θ is the domain of $\boldsymbol{\theta}$ and $\mathbf{D}$ are the data. The true PDF, call it $f_T(\boldsymbol{\theta})$, is approximated by the posterior PDF f, which is written in terms of a prior (which the user can specify) and a likelihood function (which depends on the data). In some cases, burn-in may be required in order to reach an ergodic chain.

Consider a complex situation, such as modeling the spread of a disease. The infection rate among a population may be dependent on many variables, such as the population of carriers, the proximity of carriers to transportation portals, the behavior characteristics of carriers, the age distributions among carrier and infected populations, and even ambient conditions such as temperature and humidity. These variables form a set of parameters $\boldsymbol{\theta}$, whose distribution is not fully known. Now consider that an investigator forms a complex model $z(\boldsymbol{\theta})$ for the number of cases of the disease in the US at some future time. The expected number of cases then assume a form (at least conceptually) of the expectation shown in Eq. (6.83). The investigator can then infer the expected number of cases, subject to the assumptions of the model, by MCMC. The formulation of Eq. (6.83) can be only notional, i.e., the model may not be a specific mathematical function but rather is a complex physical process. In attempting to construct the estimate of $\langle z \rangle$, the investigator averages a number of Monte Carlo histories. In each history, values of $\boldsymbol{\theta}$ are selected from a proposal distribution and accepted or rejected based on the Metropolis ratio, which depends on the assumed proposal, the assumed prior, and the likelihood function, which depends on the given data. The histories then progress through the hypothesized populations and conditions according to the proposed model z.

Another form of inference occurs when one uses observations of one set of variables to infer the values of another set of variables. For instance, Mosegaard and Sambridge (2002) discuss an interesting case in which seismometer data were collected on the moon. The seismic record was generated by moonquakes, meteor impacts, and impacts of spacecraft. The seismic data were reduced to the first arrival times of P- and S-waves at the various seismometer locations. Given these data, a complex model was used that depended on the assumed locations and epicenter times of the various events. The model was not an analytic one, in which a functional form for arrival times was assumed. Rather, a complex ray-tracing algorithm that depended on moon crust characteristics was simulated and a posterior distribution of P- and S-wave velocities was generated. These velocity distributions are clearly inferred from the data and are subject to considerable uncertainty. However, they do give plausible information regarding, for instance, lunar stratigraphy.

The MCMC formalism also has been applied in decision theory. Many decisions must be made in the presence of significant uncertainty. How does one decide among the alternatives when so much is uncertain? One way is to employ MCMC. Let $\boldsymbol{\theta}$ now represent a decision and $\boldsymbol{\theta}^*$ represent the "true scenario," i.e., the decision that should be made if there were no uncertainty. Further, let A represent an action that results from a decision. Then, it is possible to define a "loss function" $L(\boldsymbol{\theta}, A)$ that is a measure of how much is lost when a decision $\boldsymbol{\theta}$ leads to action A. In this case, the best decision can be thought of as the decision that leads to an action that minimizes the loss function L. Mathematically the value of θ is sought that minimizes

$$\Omega(\theta) = \int_{\mathcal{A}} L(\theta, A) p(A|\theta) dA, \qquad (6.84)$$

where $p(A|\theta)$ is the PDF for all actions that might result from decision θ and $\mathcal{A}$ is the domain of A. This equation is in the form of a definite integral and thus is amenable to solution by Monte Carlo techniques, including MCMC.

Bayes' theorem indicates that the posterior PDF f can be expressed in terms of a prior PDF g, a likelihood function ℓ, and a constant M. The M-H algorithm involves ratios of the posterior and proposal PDFs and hence the constant term M (when expressing the posterior in terms of the prior) cancels. This leads to the important result that the global likelihood M need not be known. This implementation extends the reach of Monte Carlo into areas where conventional Monte Carlo methods are difficult to apply.

6.5.1   Implementing MCMC with Data

Assume one is given a data vector $\mathbf{D} = \{d_j, j = 1, 2, \ldots, J\}$ of measured values of a random variable $\mathbf{x}$ and seeks to estimate the quantity

$$\langle z \rangle = \int_V z(\mathbf{x}) f_T(\mathbf{x}) d\mathbf{x}, \qquad (6.85)$$

where $z(\mathbf{x})$ is some specified function and $f_T(\mathbf{x})$ is the true (but unknown) PDF underlying the distribution of $\mathbf{x}$. In principle, one can estimate $\langle z \rangle$ by using Bayes' theorem to form the posterior

$$f(\mathbf{x}|\mathbf{D}) = \frac{1}{M} g(\mathbf{x}) \ell(\mathbf{D}|\mathbf{x}), \qquad (6.86)$$

where $g(\mathbf{x})$ is an assumed prior distribution, $\ell(\mathbf{D}|\mathbf{x})$ is the likelihood of the data given a sample of $\mathbf{x}$, and M is the global likelihood (a normalizing constant). The likelihood function ℓ uses the data $\mathbf{D}$ to correct the prior g so that the posterior f better approximates the true PDF f_T than does the prior. One can find an estimate of $\langle z \rangle$ from

$$\bar{z} = \frac{1}{N} \sum_{i=1}^{N} z(\mathbf{x}_i), \qquad (6.87)$$

where N is the number of histories on which the estimate is based and the x_i are sampled from Eq. (6.86) by MCMC. Note that in order to apply MCMC, the value of M need not be known and, thus, can be treated as unity because it will always cancel out when the Metropolis ratio is formed.

The prior should be a good approximation to the true PDF. In fact, of course, the best prior *is* the true PDF. Thus, investigators can and should use judgment and experience to select reasonable priors. It is interesting that the prior need not be a function that is easily sampled, because in MCMC candidate samples are drawn from the proposal distribution h, not from g. In fact, the prior need not even be a normalized PDF, it only need have the shape of the desired distribution.

After first selecting a prior distribution g, one then selects a proposal distribution $q(\mathbf{x}_{i+1}|\mathbf{x}_i)$, which is defined on the volume over which the random variable $\mathbf{x}$ is assumed to be defined. Note that the proposal does not have to be conditional. However, it should be a distribution that is easily sampled and better mixing is achieved if it is reasonably close in shape to the true PDF $f_T(\mathbf{x})$. The M-H algorithm described in Section 6.2 and, perhaps, modified as in Section 6.4 to prevent potential over/underflows is used to generate samples $\mathbf{x}_i$ from the posterior PDF. The resulting samples $\mathbf{x}_i$ can then be used in Eq. (6.87) to provide an estimate of the desired quantity.

6.5.2   The Likelihood Function

So far in this discussion of Bayesian applications, the "data" $\mathbf{D}$ have consisted of k events in n tries and its incorporation into the Bayesian model was easy because the likelihood $\ell(k|n, p)$ was taken as the binomial distribution whose two parameters are k and n. But what if the data consist of many (perhaps hundreds or thousands) of observed values? None of the standard PDFs can accommodate such a data set.

There are several ways to incorporate data through the likelihood function. Some rely on the intuition and guile of the investigator, while others are somewhat pedantic. Four ways are outlined as follows and two examples are given for illustration:

1. For discrete random variables, one can use a binomial distribution as the likelihood function if the data are derived from successive trials with constant probability of success, as was assumed in Example 6.5. More generally, a multinomial distribution (see Appendix A.3.2) may apply in multidimensional cases.

2. One can attempt to use the data to construct a histogram of ℓ; this works better as the number of data points increase. Then, any time a sample is drawn from within a given histogram interval, the appropriate histogram value of ℓ is used to "weight" the prior. This is demonstrated in Example 6.7.

3. The likelihood is specified according to the "physics" of the problem. With knowledge of something about the specific problem at hand, a likelihood function often can be inferred.

4. Another way to incorporate the data is to specify a likelihood function of the form

$$\ell(\mathbf{D}|\mathbf{x}) = Ce^{-S(\mathbf{x})}, \tag{6.88}$$

where C is a constant and S is the "misfit function" given, on the ith history, by

$$S(\mathbf{x}_i) = \sum_{j=1}^{J} \frac{|d_j - x_{j,i}|}{\sigma_j}, \tag{6.89}$$

with σ_j the standard deviation of d_j and $x_{j,i}$ the value of x_j on the ith history. If the MCMC samples from f are drawn from the true distribution, then the misfit function should tend toward unity and ℓ should approach the constant C/e. The more the samples differ from the data, the larger is S and thus the smaller is ℓ. This approach is discussed, for instance, by Mosegaard and Sambridge (2002).

Example 6.7 A One-Dimensional Problem with Data

Consider that the time intervals between events (such as tornado impacts within a state) are measured and used to form a data vector $\mathbf{D} = \{d_j, j = 1, 2, \dots, J\}$. You are asked to estimate the average time interval (Eq. (6.85) with $z(x) = x$). Of course, the average time interval could be estimated by simply averaging the data, but it is instructive to ignore this, for the time being, and demonstrate that the average interval can also be estimated from Eq. (6.85). Assume $J = 1000$ time intervals have been measured, where the maximum time interval was 15.6513 weeks. For simplicity, select a uniform prior

$$g(x) = \frac{1}{T}, \quad 0 \le x \le T,$$

where T is any time greater than or equal to the maximum of the measured times. Consider a proposal distribution with the simple form

$$q(u) = 0.5e^{-0.5u},$$

which is easy to sample and provides samples on the interval $[0, \infty)$. Thus, for each history,

$$u = -\frac{1}{0.5} \ln(\rho), \tag{6.90}$$

where ρ is a pseudorandom number drawn uniformly from the unit interval. Grouping the measured data into K equal bins, each of width $\delta = T/K$, a histogram likelihood function can be formed:

$$\ell(\mathbf{D}|x_j) = \begin{cases} \ell_1, & 0 < x_j \le \delta, \\ \ell_2, & \delta < x_j \le 2\delta, \\ \vdots \\ \ell_K, & (K-1)\delta < x_j \le K\delta, \end{cases} \tag{6.91}$$

where ℓ_k is the number of data points that are within the interval $[(k-1)\delta, k\delta]$. It obviously is undesirable to have the likelihood function be zero over any interval that can contain samples, and hence any of the ℓ_k that are zero are arbitrarily set to $\ell_k = 1$. Further, samples can be selected on the interval $x > T$, so the histogram contains another entry ℓ_{K+1}, which in this example is arbitrarily set to $\ell_{k+1} = 0.001$. This likelihood function clearly is not normalized, but the M-H sampling procedure can still be used.

It is easy to see that for this simple model, the Metropolis ratio on the ith history takes the form

$$R = \frac{\ell_{ku}}{\ell_{kx}} e^{-0.5(x_i - u_i)}.$$

Here ℓ_{ku} is the value of the likelihood function in the interval in which u_i is located and ℓ_{kx} is the value of the likelihood function in the interval in which x_i is located.

The results of implementing the procedure described above for $N = 10^6$ histories and $K = 50$ are as follows:

$$\bar{x} = 2.602 \pm 0.007, \quad \text{with } m = 0$$

and

$$\bar{x} = 2.616 \pm 0.007, \quad \text{with } m = 1000,$$

where m is the number of burn-in samples calculated and discarded before data are accumulated. The average of the data is $\bar{d} = 2.506$, and it turns out that the data were derived by sampling an exponential distribution of the form $f_T(x) = 0.4 e^{-0.4x}$, for which the mean value is $\mu(x) = 1/0.4 = 2.5$. This very approximate procedure wherein the data are used to generate a histogram leads to an estimate of the mean that is within 5% of the true mean. Note that the estimate of $\bar{x}$ obtained with no burn-in ($m = 0$) is just as good as the estimate with burn-in ($m = 1000$), indicating, though not proving, that burn-in is not required.

Example 6.8 Another One-Dimensional Problem with Data

For similar time-interval data as in the previous example, select a value Δt. For each d_j find the n_j for which $(n_j - 1)\Delta t \le d_j < n_j \Delta t$. This creates a new data vector $\mathbf{D_n} = (n_1, n_2, \ldots, n_J)$. Here, n_j is the number of the time interval within which d_j is located. The radioactive decay process is known to be governed by an exponential distribution. Thus, the probability p_j that a single observation will be within the jth interval, i.e., $p_j = \text{Prob}\{(n_j - 1)\Delta t \le d_j < n_j \Delta t)\}$, is given by

$$p_j = \int_{(n_j-1)\Delta t}^{n_j \Delta t} \lambda e^{-\lambda t} dt = e^{-\lambda(n_j-1)\Delta t} - e^{-\lambda n_j \Delta t}.$$

For J observations, the probability of observing $\mathbf{D}$ is the likelihood, which is simply

$$\ell(\mathbf{D}|\lambda, \Delta t) = \prod p_j = \prod_{j=1}^{J} e^{-n_j \lambda \Delta t} \left[e^{\lambda \Delta t} - 1 \right]. \tag{6.92}$$

Now, the problem is to find the "decay constant" λ of the true exponential distribution that produced the observed data. The true distribution of λ is, of course, a delta function, i.e., $f_T(\lambda) = \delta(\lambda - \lambda_T)$, where λ_T is the desired but unknown decay constant.

An estimate of the true decay constant can be obtained, using the data, by forming the posterior

$$f(\lambda|\mathbf{D}) = \frac{1}{M} g(\lambda)\ell(\mathbf{D}|\lambda),$$

where g is an assumed prior, ℓ is given by Eq. (6.92), and M is the global likelihood (which need not be known). Then, the estimate of λ_T is given by

$$\bar{\lambda} = \int_0^\infty \lambda f(\lambda|\mathbf{D}) \, d\lambda.$$

The MCMC procedure converts the prior into a posterior that approximates the true delta function. Assume a prior in the form of a gamma distribution (see Appendix A.1.3) with parameters α and β, i.e.,

$$g(\lambda) = \frac{1}{\beta\Gamma(\alpha)} \left(\frac{\lambda}{\beta}\right)^{\alpha-1} e^{-\lambda/\beta}. \tag{6.93}$$

Also, assume a proposal distribution of the simple form

$$q(u) = 0.5e^{-0.5u},$$

as in the previous example. Then, specify an initial value of λ_i, for $i = 1$, sample a candidate new value u_i from the proposal, and form the Metropolis ratio

$$R = \frac{y_i}{\lambda_i} e^{-(y_i - \lambda_i)/\beta} \frac{1 - e^{-y_i \Delta t} \prod_{j=1}^J e^{-n_j y_i \delta t}}{1 - e^{-\lambda_i \Delta t} \prod_{j=1}^J e^{-n_j \lambda_i \delta t}} e^{-0.5(\lambda_i - y_i)}. \tag{6.94}$$

The candidate samples are accepted or rejected using the criterion of Eq. (6.22) and the procedure is repeated until all desired histories have been run.

Two hundred samples were generated from an exponential with parameter $\lambda = 0.4$. Using the first 40 of these samples as given data, the results of implementing the procedure described above for $N = 10^6$ histories, with no burn-in, yielded $\bar{\lambda} = 0.4507 \pm 0.00001$. Using the first 120 of these samples as given data and again running $N = 10^6$ histories with no burn-in, the estimate $\bar{\lambda} = 0.3991 \pm 0.000003$ was obtained. Using all 200 samples as given data, the estimate $\bar{\lambda} = 0.3967 \pm 0.000002$ was obtained. The posteriors, scored as histograms over 100 intervals between zero and one, are shown in Fig. 6.17. It is obvious that, as the number of data points is increased, the posterior more closely resembles a delta function and its average more closely approximates the true value.

The MCMC method has been applied in very many situations. For example, Gilks et al. (1996) describe the use of MCMC in many medical applications (including the study of disease risks, imaging, genetics, and medical modeling), Hugtenburg (1999) applies MCMC to radiation treatment planning, and Litton and Buck (1996) consider application of MCMC to radiocarbon dating in archeological samples. Yet another application of MCMC includes the additive white

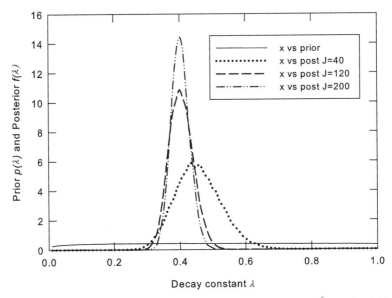

Figure 6.17 The posterior distributions for three values of J after $N = 10^6$ histories (with no burn-in) compared to the prior distribution, which is a gamma PDF with $\alpha = 1.2$ and $\beta = 1.4$. The nonprior distributions were scaled by a factor of 5.

Gaussian noise (AWGN) problem, discussed by Wang (2005). This problem considers an attempt to determine a transmitted signal from the received signal, in the presence of noise. Wang treats this by constructing an MCMC model in which samples are accepted and rejected on the basis of the M-H algorithm. Another way to look at this particular problem is as an inverse problem in which one wishes to invert a set of measurements to obtain the signal that led to those measurements. It is this class of problems that are considered in the next chapter on Monte Carlo applications to inverse problems. The reader also is referred to Barbu and Zhu (2020) and Besag and Geen (1993), who provide additional information on MCMC.

6.6 Summary

MCMC is, in essence, a particular way to obtain random samples from a PDF. Thus, it is simply a method of sampling. The method relies on using properties of Markov chains, which are sequences of random samples in which each sample depends only on the previous sample. Some chains of random samples form ergodic Markov chains from any initial state and, thus, require no burn-in. Ergodic Markov chains have transition probabilities that are constant and produce samples from a single stationary PDF.

Some of the principal advantages of MCMC based on the M-H sampling algorithm include the following:

- it is relatively simple to implement,
- it can be applied when only the shape of the distribution is specified, i.e., the normalization constants required to make functions into PDFs need not be known,
- it can be applied to problems for which Bayes' theorem is employed, allowing one to sample from a PDF f (the posterior) using another PDF g (the prior) and a likelihood function, and
- it can be applied to very complex and highly multidimensional sampling problems.

These advantages must be weighed against some disadvantages, which include the following:

- the acceptance-rejection criterion employed in the M-H algorithm results in retaining the same random sample on some successive samplings; because of this, the sample standard deviation is artificially low, until N becomes sufficiently large, and
- The production of a new sample value requires that at least two pseudorandom samples be generated, a requirement that can be somewhat inefficient.

Nevertheless, MCMC is a powerful form of Monte Carlo that has gained widespread use, especially in sampling highly multidimensional distributions or in situations where other Monte Carlo methods encounter difficulties.

Today MCMC is used in almost every quantitative discipline and even in some not normally thought to be quantitative, such as music, in which MCMC is used to generate synthetic music. The 36-page Wikipedia article on Markov chains (https://en.wikipedia.org/wiki/Markov_chain) lists dozens of areas (with citations) in which MCMC is being used.

The use of Monte Carlo methods is today increasing rapidly. The authors opine that such rapid growth results from the confluence of two factors. First the use of Monte Carlo in many applications requires little advanced mathematical background on the part of the user. Often simple arithmetic suffices. Consequently, such users treat Monte Carlo algorithms in a "black-box" manner (often a dangerous practice!) and they have little need for the sophisticated mathematics used in the proofs of the many theorems and lemmas behind the Monte Carlo methods. Using a technique and understanding it are two separate things. Second, the ubiquitous availability of inexpensive personal computers with the power of supercomputers 10 years earlier now allows anyone to use the computationally intense Monte Carlo approach. No longer is Monte Carlo the purview of researchers at the few institutions and laboratories that could afford the supercomputers of a few decades ago. Further, Monte Carlo calculations are easily done in parallel, so clusters of inexpensive personal computers, now found at most research facilities, can be used for computationally demanding Monte Carlo problems.

Problems

6.1 Consider the Markov matrix

$$T = \begin{pmatrix} 0.4 & 0.6 \\ 0.7 & 0.3 \end{pmatrix}.$$

(a) Find the stationary PDF, to five decimal places, associated with this
matrix and (b) tabulate the eigenvalues of T^n as a function of n.

6.2 Use MCMC to estimate the value of the definite integral

$$I = \int_0^1 x^2 f(x|\alpha, \beta)\, dx,$$

where $f(x|\alpha, \beta)$ is the beta distribution (see Appendix A.3.1). Consider
the case where $\alpha = 1.5$ and $\beta = 1.2$. (a) Start at $x_1 = 0.5$ and employ
burn-in lengths of $m = 0$, $m = 1000$, and $m = 10{,}000$ to estimate I using
10^6 histories. What do you conclude from your results? (b) What are the
standard deviations of your estimates?

6.3 Use MCMC to estimate the integral

$$I = \int_{-\infty}^{\infty} \int_{-\infty}^{\infty} (x + y) f(x, y)\, dx dy,$$

where $f(x, y)$ is the bivariate normal distribution (see Appendix A) with
parameters $\langle x \rangle = \mu_1 = 5$, $\sigma(x) = \sigma_1 = 1$, $\langle y \rangle = \mu_2 = 1$, $\sigma(y) = \sigma_2 = 0.5$, and $\sigma_{12} = 0.25$. Sample both α and y on each history and use pro-
posal functions $g(u) = 2e^{-4|u|}$ for α and $q(v) = 0.4e^{-0.5|v|}$ for y.

6.4 Use MCMC to estimate the value of the definite integral

$$I = \int_{-\infty}^{\infty} \int_0^{\infty} \int_2^8 (x/y)^z f(x)g(y)q(z)\, dx\, dy\, dz,$$

where $f(x) = 4/9 - x/18$, $g(y) = 4e^{-4y}$, and

$$q(z) = \frac{1}{\sqrt{2\pi}\sigma} \exp\left[-\frac{(z - \mu)^2}{2\sigma^2} \right],$$

with $\mu = 1$, $\sigma = 0.4$.

6.5 With $\mu = 2.0$ and $\sigma = \sqrt{2}$, solve the problem of Example 6.2 using the
proposal function

$$q(y|x) = xe^{-xy}$$

and compare your results with those given in Example 6.2.

6.6 Estimate the triple integral

$$I = \int_0^1 \int_0^{10} \int_0^\infty \cos(xy)\frac{3y}{\sqrt{z}}e^{-4x}\,dx\,dy\,dz$$

by MCMC using proposals for x, y, and z given by

$$q_1(u) = 3e^{-3u}, \quad u > 0,$$

$$q_2(v) = 1/10, \quad 0 \le x \le 10,$$

and

$$q_3(w) = 2w, \quad 0 \le w \le 1,$$

respectively. Compare results for $N = 10^6$ histories for the following cases. (a) Update x, y, and z in order on successive histories. (b) Next, update x, y, and z randomly on any given history. (c) Then, update x, y, and z globally on each history. Compare your results.

6.7 Repeat the analysis of Example 6.4, when it is specified that the child is twice as likely to choose bowl 2, which is closest to the door, than either of the other two bowls.

6.8 Use the M-H samplings technique to obtain samples x_i that are distributed according to the following PDFs, both defined on the interval $x \in (0, 1)$. Sort these samples into 100 equiwidth bins as in Example 6.6. From these binned results determine (1) the mean and variance of each PDF and (2) the value of the normalizing constant C to make each function a properly normalized PDF. We have

$$f_1(x) = C\frac{e^x\left[1 - \sin^2\left(\frac{\pi x}{2}\right)\right]}{1 - x^2} \quad \text{and} \quad f_2(x) = C\exp[\tan x].$$

6.9 Use the M-H samplings technique to obtain samples x_i that are distributed according to the following PDFs. Sort these samples into 100 equiwidth bins as in Example 6.6. From these binned results determine (1) the mean and variance of each PDF and (2) the value of the normalizing constant C to make each function a properly normalized PDF. We have

$$f_3(x) = Cx^4\cos^2(x), \; x \in (0,\pi/2), \quad \text{and} \quad f_4(x) = Ce^x\cos x, \; x \in (-1,1).$$

6.10 A critical safety system in a nuclear power plant is the dual emergency diesel generators needed, for example, to keep cooling pumps running should the plant lose off-site power. Normally, these generators are inactive, but when called upon they must supply power within 10 seconds after they are called upon to start. In Table 6.7, the history of testing such

Table 6.7 Failure-on-demand data for emergency diesel generators at 25 nuclear power plants collected before 1979. Here k is the number failures to start in n demands to start.

k	n	k	n	k	n	k	n	k	n
6	100	0	23	2	47	1	35	2	35
1	392	2	12	1	87	1	37	7	17
11	230	0	99	2	71	0	13	4	335
5	68	3	33	3	656	2	95	9	206
4	23	9	126	5	73	2	51	1	76

emergency generators is shown for 25 different nuclear plants up to early 1979.

In anticipation of using a beta-binomial compound, failure-on-demand, Bayesian model in safety studies, use these data to estimate values of the parameters α and β in the beta prior PDF $\mathcal{B}_e(p|\alpha, \beta)$ using the matching moments method described in Section 6.4.3.

6.11 To accept a new (or reconditioned) emergency diesel generator for service at a nuclear power plant, it must pass some number of start-on-demands from a cold standby condition without a single failure-to-start event. Such acceptance testing demonstrates the sometimes significant differences between the frequentists and Bayesians. How many successful start-on-demands in a row must a generator demonstrate before it is deemed to be fit for service? A frequentist might require that the upper limit p_{high} of the confidence interval for the failure-on-demand probability p_o be less that 0.01 at the 95% confidence level. A Bayesian might also adopt such an acceptance criterion, except p_{high} is the upper limit of the credible or probability interval for p_o.

Write a computer program, based on the results in Section 6.3.3, to estimate the minimum number n_{min} of starts without a failure, i.e., $k = 0$ so $p_{high} = 0.01$ at the $\hat{\alpha}\% = 95\%$ level. For the Bayesian analysis use the beta prior PDF determined in the previous problem. Show the details on how you obtained your estimates. You should find the Bayesian estimate of n_{min} is seven and a half times smaller than that needed by a frequentist estimation.

References

Bapat, R.B., Raghavan, T.E.S., 1997. Nonnegative Matrices and Applications. Cambridge Univ. Press, Cambridge, UK.

Barbu, A., Zhu, S-C., 2020. Monte Carlo Methods. Springer, Singapore.

Bernstein, S.N., 1926. Sur l'Extension du Théoréme Limite du Calcul des Probabilités aux Sommes de quantités dépendentes. Math. Ann. 97, 1–59.

Besag, J., Geen, P.J., 1993. Spatial statistics and Bayesian computation (with discussion). J. R. Stat. Soc. B 55, 25–37.

Clifford, P., 1993. Discussion of Smith and Roberts (1993), Besag and Green (1993), and Gilks et al. (1993). J. R. Stat. Soc. B 55, 53–54.

Ekvall, K.O., Jones, G.L., 2019. Markov chain Monte Carlo. Univ. Minnesota. http://users.stat.umn.edu/~galin/intro_mcmc.pdf.

Flegal, J.M., Jones, G.L., 2010. Batch means and spectral variance estimators in Markov chain Monte Carlo. Ann. Stat. 38 (2), 1034–1070.

Gelfand, A.E., Smith, A.F.M., 1990. Sampling-based approaches to calculating marginal densities. J. Am. Stat. Assoc. 85, 398–409.

Geman, S., Geman, D., 1984. Stochastic relaxation, Gibbs distributions, and the Bayesian restoration of images. IEEE Trans. Pattern Anal. Mach. Intell. 6, 721–741.

Geyer, C.J., 1992a. Practical Markov chain Monte Carlo. Stat. Sci. 7 (4), 473–483.

Geyer, C.J., 1992b. Likelihood inference for spatial point processes. In: Barndorff, O.E., Kendall, W.S., van Lieshout, M.N.M. (Eds.), Stochastic Geometry: Likelihood and Computation. Chapman & Hall, CRC Press, Boca Raton, FL, pp. 79–140.

Geyer, C.J., 2011. Introduction to Markov chain Monte Carlo. Ch. 1. In: Brooks, S., Gelman, A., Jones, G.L., Meng, X-L. (Eds.), Handbook of Markov Chain Monte Carlo. Chapman & Hall, CRC Press, Boca Raton, FL.

Gilks, W.R., Richardson, S., Spiegelhalter, D.J., 1996. Markov Chain Monte Carlo in Practice. Chapman & Hall, New York, NY.

Hastings, W.K., 1970. Monte Carlo sampling methods using Markov chains and their applications. Biometrika 57 (1), 97–109.

Hugtenburg, R.P., 1999. A perturbation based Monte Carlo method for radiotherapy treatment planning. Radiother. Oncol. 51, S30.

Jones, G.L., 2004. On the Markov chain central limit theorem. Probab. Survey 1, 299–320.

Johnson, N.L., Kotz, S., Balakrishnan, N., 1994. Continuous Univariate Distributions, vol. 1, 2nd ed. Wiley, New York.

Kennedy, T.G., 2016. Markov chain background. In: Monte Carlo Methods. Univ. Arizona. Ch. 8, Lecture Notes. https://www.math.arizona.edu/~tgk/mc/book_chap8.pdf.

Litton, C., Buck, C., 1996. An archaeological example: radiocarbon dating. In: Gilks, W.R., Richarson, S., Spiegelhalter, D.J. (Eds.), Markov Chain Monte Carlo in Practice. Chapman & Hall, CRC Press, Boca Raton, FL.

MacKay, D.J.C., 2007. Information Theory, Inference, and Learning Algorithms. Cambridge University Press, Cambridge, UK (Sixth Printing).

Markov, A.A., 1906. Rasprostranenie zakona bol'shih chisel na velichiny, zavisyaschie drug ot druga (Extension of the Law of Large Numbers to Dependent Quantities). Izv. Fiz.-Matem. Obsch. Kazan Univ., (2nd Ser.) 15 (94), 135–156.

Metropolis, N., Ulam, S., 1949. The Monte Carlo method. J. Am. Stat. Assoc. (Special Issue), 125–130.

Metropolis, N., Rosenbluth, A.W., Rosenbluth, M.N., Teller, A.H., Teller, E., 1953. Equation of state calculations by fast computing machines. J. Chem. Phys. 21 (6), 1087–1092.

Mosegaard, K., Sambridge, M., 2002. Monte Carlo analysis of inverse problems. Inverse Probl. 18, R29–R54.

Meyn, S.P., Tweedie, R.L., 1993. Markov Chains and Stochastic Stability. Springer, London.

Neal, R.M., 1993. Probabilistic Inference Using Markov Chain Monte Carlo Methods. Report CRG-TR-93-1. Dept. of Comp. Sci., Univ. Toronto.

Neal, R.M., 1995. Suppressing Random Walks in Markov Chain Monte Carlo Using Ordered Over-relaxation. Report 9508. Dept. of Stat., Univ. Toronto.

Popp, J.G., Wilson, D.B., 1996. Exact sampling with coupled Markov chains and applications to statistical mechanics. Random Struct. Algorithms 9 (1–2), 223–252.

Robert, C.P., Casella, G., 2013. Monte Carlo Statistical Methods. Springer, New York.

Shultis, J.K., Tillman, F.A., Eckhoff, N.D., Grosh, D., 1979. Bayesian Analysis of Component Failure Data. NUREG/CR-1110. U.S. Nuc. Reg. Com. 207 pp.

Shultis, J.K., Johnson, D.E., Milliken, G.A., Eckhoff, N.D., 1981a. Use of Non-Conjugate Prior Distributions in Compound Failure Models. NUREG/CR-2374. U.S. Nuc. Reg. Com. 135 pp.

Shultis, J.K., Johnson, D.E., Milliken, G.A., Eckhoff, N.D., 1981b. GAMMA: A Code for the Analysis of Component Failure Rates with a Compound Poisson-Gamma Model. NUREG/CR-2373. U.S. Nuc. Reg. Com. 30 pp.

Shultis, J.K., Johnson, D.E., Milliken, G.A., 1986. Non-conjugate prior distributions and their estimation for use in compound failure models. Commun. Stat., Theory Methods 15 (9), 2835–2865.

Tierney, L., 1994. Markov chains for exploring posterior distributions (with discussion). Ann. Stat. 22 (4), 1701–1762.

Wang, X., 2005. Monte Carlo signal processing for digital communications: principles and applications. Ch. 26. In: Ibnkahia, M. (Ed.), Signal Processing for Mobile Communications Handbook. CRC Press, New York, NY.

Wikipedia Contributors, 2020. Conjugate prior. In: Wikipedia, The Free Encyclopedia. https://en.wikipedia.org/wiki/Conjugate_prior. (Accessed 2 June 2020).

Chapter 7

Inverse Monte Carlo

Karen had the right answer, you see;
but wondered what the question might be.
After looking around
she finally found.
she could use symbolic IMC.

Most of this book has focused on "forward" Monte Carlo (MC) methods that can be applied to solving "direct" problems. In direct problems, the response of a specified process is sought, i.e., one seeks to determine the output of a given system for a certain input. A definite integral poses a direct problem, but so does a model of a complex system, whether or not the model is in the form of a single mathematical formula or is expressed as a complicated algorithm. Monte Carlo solves direct problems by using samples of independent variables in order to estimate values of dependent variables. The independent variables can be discrete, continuous, or a combination of the two. Often, the independent variables are stochastic but this need not be the case. In direct problems, Monte Carlo can be considered to be a form of quadrature or, alternatively, a means to simulate a process in which the "physics" (the underlying processes) of the problem are known or specified.

"Inverse" problems, by contrast, arise when the outcome or response is given and the conditions that might have led to these results are sought. Meteorologists measure microwave intensity distributions from which they seek to determine atmospheric concentration profiles in order to make weather predictions. Doctors measure temperature and conduct blood tests in order to assess a patient's health. Cooks use a variety of measures (some rather subjective) to ascertain when a dish is ready to serve. Our eyes are sensors that detect light, allowing us to recognize each other. Our brains perform inverse analyses to identify words from sounds produced by others or to recognize specific symphonies from the sounds produced by musical instruments. People are detectors who make other detectors to enhance their ability to perceive the world. The responses of these detectors provide data that are turned into information. Thus, inverse problems differ from direct problems in that the values of the independent variables that could have led to a set of observations are sought.

In many direct engineering problems, initial and/or boundary conditions are specified and some conditions at later times and/or at internal positions are

Exploring Monte Carlo Methods. https://doi.org/10.1016/B978-0-12-819739-4.00015-9

desired in systems with specified properties. For example, a direct problem might be stated as: given an initial surface temperature distribution on an object of known composition, what is the internal temperature distribution at future times? There are many forms of inverse problems. In some engineering inverse problems, system responses are measured and one wants to determine either the properties of the system or the initial and/or boundary conditions that could have led to these responses. In other inverse problems, a desired response to a given input is specified and one seeks the properties of a system that would produce this desired response. Unlike direct problems, however, inverse problems are plagued by questions of *existence* and *uniqueness*.

The existence of solutions to some inverse problems may not always be guaranteed in advance. It is certainly possible, in some cases, to specify a set of final or internal conditions that cannot be achieved by any set of system properties and, thus, no solution exists. For example, one may want the surface temperature of an object to be constant under some specified heating conditions, but no geometry may exist that will achieve this result for the specified material. In direct problems, the system is responding, according to natural laws, to a set of imposed conditions. In inverse problems, the parameters of some hypothetical system are sought that would produce some specified conditions. It is certainly possible to specify conditions that cannot be realized physically (and perhaps not even mathematically).

Even if a solution does exist, that solution may not be unique. There may be several sets of parameters that could produce the same specified output. Alternatively, some solutions may be allowed mathematically that are not physically realizable. For instance, negative mass densities have no physical meaning and temperatures that are expressed as complex numbers may solve an equation but make no physical sense. Thus, uniqueness also is an issue in inverse problems. Moreover, it often is difficult to prove existence and uniqueness. However, for many practical problems, it is highly likely that at least one solution exists, and it is important to devise methods to find the physically reasonable solutions, from which an optimum solution generally can be determined.

Consider, for example, the backscatter response of a gamma ray density gauge, which is a function of the density of the interrogated medium of which the composition is known. A possible model for the response R in terms of density ρ for a medium with known composition is of the form

$$R(\rho) = \rho e^{\alpha - \beta \rho} \qquad (7.1)$$

and behaves as shown in Fig. 7.1. A direct problem is to calculate R, given α and β, for a specified density. An inverse problem is to calculate ρ, given α and β, for a measured response. The uniqueness problem is concerned first with whether any solution for ρ exists. For instance, the response R_1 shown in Fig. 7.1 does not correspond to any density. Second, if one solution exists, do others? For instance, the response R_2 leads to two positive densities. One must determine which of the two is most physically reasonable.

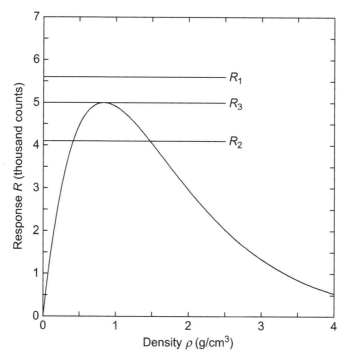

Figure 7.1 Plot of backscatter response versus density.

The situation in practice is even more complicated. Measurements are imprecise and the uncertainty in measurements leads to uncertainty in the recovered density. The response R_3 in Fig. 7.1 corresponds to a single density. However, an actual response from a sample with that density is likely either slightly above R_3, for which there is no inverse solution, or slightly below R_3, for which there are two solutions. Also, responses generally are functions of several variables, some of which are sought and others of which "interfere" with the measurement. Suffice it to say that existence and uniqueness are generic issues for inverse problems with which all solution techniques must deal. MC approaches to solving inverse problems usually just assume that one or more solutions exist and then try to deduce the most physically reasonable solution among the solutions obtained.

A "well-posed" inverse problem provides sufficient information to determine if at least one solution exists, even if that solution is unknown and not unique (perhaps several sets of parameters could lead to the observed result). An "ill-posed" inverse problem is one for which not enough information is given to specify whether or not a solution exits, for instance, because there are more variables to determine than there are equations relating these variables. In this chapter, it is assumed that the inverse problems are well posed and, thus, that at least one solution exists. However, multiple solutions may exist, within the uncertainties of the data provided, and hence uniqueness is not guaranteed.

7.1 Formulation of the Inverse Problem

Two general approaches to formulating inverse problems are briefly reviewed. The first is the formal mathematical approach, which parallels the integral formulation of direct Monte Carlo. The other is the practical or physical approach, whose analog is the simulation approach of direct Monte Carlo. Neither approach is inherently better than the other. Some problems are easily specified in equation form while others are more easily posed in terms of the physical processes. This does not mean that there are two classes of inverse problems, only that there are two ways to specify these problems. The situation is rather like wave–particle duality. Photons can be considered waves for some experiments (the two-slit experiment) and particles for other experiments (photoelectric absorption). So far as is known, both models are correct and just illuminate different aspects of the behavior of photons. The two general approaches for inverse problems are outlined in the following two subsections. This is followed by a brief section that identifies certain optimization problems as inverse problems.

7.1.1 Integral Formulation

Consider the expected value of some kernel z expressed in the form

$$\langle z_k \rangle = \langle z(\mathbf{y}_k) \rangle = \int_V z(\mathbf{x}, \mathbf{y}_k, \boldsymbol{\theta}) f(\mathbf{x}, \boldsymbol{\theta}) d\mathbf{x}, \quad k = 1, 2, \ldots, K, \tag{7.2}$$

where $\mathbf{x}$ is an n-dimensional random variable distributed according to the joint probability density function (PDF) f, $\mathbf{y}$ is a vector of independent parameters that can take on specified values $\mathbf{y}_k$, with $k = 1, 2, \ldots, K$, V is the domain of $\mathbf{x}$, and $\boldsymbol{\theta}$ is a set of $J \geq 1$ parameters of the integrand. In direct problems, the $\boldsymbol{\theta}$ are assumed known or specified and the object is to estimate the expectations of Eq. (7.2) for $K \geq 1$ values of $\mathbf{y}_k$. Straightforward MC does this by using the sample mean defined by

$$\bar{z}(\mathbf{y}_k) = \frac{1}{N} \sum_{i=1}^{N} z(\mathbf{x}_i, \mathbf{y}_k, \boldsymbol{\theta}), \tag{7.3}$$

where the $\mathbf{x}_i$ are sampled from $f(\mathbf{x}, \boldsymbol{\theta})$.

A similar formulation exists for inverse problems. Let $R(\mathbf{y}_k, \boldsymbol{\theta})$ be the response model for a system when an independent variable $\mathbf{y}$ has the specific value $\mathbf{y}_k$ and let $\boldsymbol{\theta}$ be a vector of $J \geq 1$ unknown parameters of the system. Then, the equations

$$R(\mathbf{y}_k, \boldsymbol{\theta}) = \int_V z(\mathbf{x}, \mathbf{y}_k, \boldsymbol{\theta}) f(\mathbf{x}, \boldsymbol{\theta}) d\mathbf{x}, \quad k = 1, 2, \ldots, K, \tag{7.4}$$

where the "kernel" z is determined by the physics of the problem, pose a classic inverse problem in the form of a Fredholm integral equation of the first kind to find the parameters θ. This chapter addresses how MC offers a means to solve inverse problems such as those posed in the form of Eq. (7.4).

For instance, it is known that the response of a radiation detector can be expressed as a function of energy in the form (e.g., see Dunn and Shultis (2009))

$$R(E) = Q_o \int_0^\infty \chi(E_o) K(E_o \rightarrow E) \, dE_o, \tag{7.5}$$

where Q_o is the number of radiation particles incident on the detector during some counting time, $\chi(E_o)$ is the PDF defining the energy distribution of source particles incident on the detector, $K(E_o \rightarrow E)$ is the response of the detector at energy E due to a unit source at energy E_o, and $R(E)$ is the response of the detector per unit energy at energy E. In practice, this formulation is usually discretized and the response in a finite number of contiguous energy bins or "channels" is recorded. In radiation detection, it is a direct problem to estimate the detector response for a known source and a detector with specified response kernel K. However, Eq. (7.5) is an inverse problem if the measured response $R(E)$ is given and one seeks the energy spectrum $\chi(E_o)$ of a source of known intensity Q_o, or the intensity of a source having a known spectrum, or both the intensity and the energy spectrum.

7.1.2 Practical Formulation

Many inverse problems are not easily cast into the form of Fredholm integral equations. Direct Monte Carlo can be viewed and implemented as a simulation of a process, such as estimating the neutron density at some position in a re-actor core, given the core composition and size, without resorting to writing a definite integral. Similarly, inverse Monte Carlo (IMC) can be viewed as using simulation to solve inverse problems for which no integral equation is formally considered. One can in fact use MC to find the size of a reactor core given its composition or, alternatively, the composition of a reactor given its size, without ever considering Fredholm integral equations.

It is easy to imagine many other examples. Say that you find a metal artifact in the shape of a rectangular plate, whose thickness L is much smaller than the other two dimensions, and want to determine the thermal conductivity, in an effort to identify the metal. You introduce a constant heat flux q'' on one face and measure the temperature difference ΔT between that face and the opposite face in steady state. You know that in this simple case, Fourier's law of heat conduction can be written approximately as

$$q'' = k \frac{\Delta T}{L}, \tag{7.6}$$

where k is the thermal conductivity of the metal. You can then use the measured temperature difference and thickness to estimate the thermal conductivity, and thus infer the material out of which the plate is made. This problem differs from the direct problem of estimating the temperature difference for a plate of known material and thickness. The inverse problem in this case takes measured temperature values and seeks a property of the medium under consideration.

Many other inverse problems are posed as physical or conceptual inverse problems for which a process can be modeled. The retrieval of P- and S-wave velocities from lunar seismometer data discussed by Mosegaard and Sambridge (2002) is one example. The first arrival times at dispersed seismometers on the lunar surface were used with a ray-tracing model for wave propagation through unknown lunar strata to reconstruct possible wave velocities, which also demonstrated approximate lunar stratigraphy. Similarly, assume a known radiation source is placed on one side of a slab shield and n radiation detectors are placed on the other side of the shield. The thickness of the shield is known but its composition is not. One may, in principle, determine the unknown shield composition, expressed as weight fractions of $m \leq n$ elements, given the responses from the n detectors. This is an inverse problem that is stated without writing an integral equation.

7.1.3   Optimization Formulation

Inverse problems also arise from attempts to design or optimize systems. A design optimization problem might be posed in a form such as "what system design produces a device for which the error in results using this device is minimized when measuring a set of samples?" Optimization problems typically seek to find extrema of objective functions, which are identified by locations where the partial derivatives have the desired value of zero. Thus, quite often, it is possible to consider an optimization problem as an inverse problem in which the measurements are replaced by the constraints that derivatives vanish.

Consider a system for which the location of an extremum of the objective function given by

$$\Omega(\boldsymbol{\theta}) = \int_V z(\mathbf{x}, \boldsymbol{\theta}) f(\mathbf{x}, \boldsymbol{\theta}) d\mathbf{x} \tag{7.7}$$

is desired, where the variables are as defined in Section 7.1.1. Then, the values of θ_j at which the partial derivatives of the objective function vanish, i.e.,

$$\frac{\partial \Omega(\boldsymbol{\theta})}{\partial \theta_j} = 0, \quad j = 1, 2, \dots, J, \tag{7.8}$$

identify the position of the extremum (maximum or minimum). In the remainder of this chapter, optimization problems are treated as inverse problems in which the "response models" are objective functions of the form of Eq. (7.8) and the "measured responses" are all zero.

7.1.4   Monte Carlo Approaches to Solving Inverse Problems

The term "inverse Monte Carlo" (IMC) has been used (e.g., see Dunn (1981)) to identify a specific approach, which is noniterative in the MC simulations, to solving a class of inverse problems. Here, a broader perspective is taken and IMC refers to any MC method applied to inverse or optimization problems. Certain implementations of Markov chain Monte Carlo (MCMC) that involve data to infer parameters of models to complex problems can be viewed as using MC to solve inverse problems. There are two other ways to proceed. One involves IMC by iterative simulation. This approach exploits the fact that computer simulation is relatively inexpensive and, thus, many scenarios can be considered and the one whose simulated results match the observed results most closely can be selected. The other approach, which may be advantageous when the cost of doing a simulation is expensive or the number of simulations is large, converts the inverse problem into a system of algebraic equations that are then solved by standard numerical methods. The conversion involves scoring histories with symbols and numbers rather than only numbers and, hence, is called *symbolic Monte Carlo* (SMC). The SMC approach often avoids iterative simulations.

7.2   Inverse Monte Carlo by Iteration

Consider the problem of obtaining solutions to Eq. (7.4) assuming K responses R_k, $k = 1, 2, \ldots, K$, are known. In general, a "brute-force" MC approach to solving such inverse problems can be implemented. This approach involves iterative simulations, as unknown parameter values are varied. Only when simulated responses match measured or specified responses to an acceptable degree do the simulations cease. The procedure might be implemented as follows:

1. Choose an initial estimate of θ and call it θ_m with $m = 1$.
2. Construct estimates $\overline{R}_k$, $k = 1, 2, \ldots, K$, by direct MC methods that generally require sampling from a PDF using the assumed θ_m.
3. Compare the estimates $\overline{R}_k$ with the known values $\mathcal{R}_k$.
4. Choose a new estimate of θ (call it θ_{m+1}) by any of several techniques, such as by using a fixed grid superimposed over the volume V, a Fibonacci search, the simplex method (Press et al., 1996), or some other search procedure.
5. Use direct MC to construct new estimates of $\overline{R}_k$, $k = 1, 2, \ldots, K$, and go to step 3.

The procedure is repeated until the Monte Carlo estimates $\overline{R}_k$ agree with the given responses $\mathcal{R}_k$ to within an acceptable tolerance, for all k. This process requires that a direct Monte Carlo simulation be run for each assumed value of θ and, thus, can become cumbersome and "expensive" in terms of simulation resources as the complexity of the problem or the dimensionality of θ increases. Nevertheless, many inverse problems can be solved by this iterative IMC approach, as the following example demonstrates.

Example 7.1 A Two-Parameter Inversion Using Iterative IMC

Assume that the expected response of a process involves an exponentially distributed stochastic variable and can be written as follows:

$$R(y) = \int_0^\infty \frac{x}{\sqrt{\pi y + ax}} be^{-bx} dx.$$

Assume further that you are given values of $\mathcal{R}_1 \simeq R(y_1 = 1.0) = 0.2973 \pm 0.0002$ and $\mathcal{R}_2 \simeq R(y_2 = 5.0) = 0.1781 \pm 0.0002$ and are asked to determine the values of a and b. One way to proceed is to select initial values a and b and apply the iterative procedure described above. Starting with $a = b = 1.0$ and estimating the above integral for both $y_1 = 1$ and $y_2 = 5$ using $N = 100{,}000$ histories, one might obtain the values given in the first row of Table 7.1. One of the parameters, say b, is then incremented to obtain the results shown in rows 2 through 4. The results in row 3 are closest to the given responses, so one might next increment a for $b = 1.5$, generating the results in rows 5 through 10. Proceeding along similar lines generates results for various cases until the values in the final row of Table 7.1 are obtained. The parameter values are acceptably close to the given responses. Note that one must either evaluate the output of each run and decide on new values for the parameters or establish a close-spaced grid of very many parameter pair values.

Table 7.1 Iteration toward a solution to a two-parameter inverse problem. The value of N was 10^5 in all cases. The correct values are $a = 3.23$ and $b = 1.24$.

j	a_j	b_j	$\overline{R}_{1j} \pm \sigma(\overline{R}_{1j})$	$\overline{R}_{2j} \pm \sigma(\overline{R}_{2j})$
1	1.00	0.50	0.7877 ± 0.0019	0.4540 ± 0.0013
2	1.00	1.00	0.4510 ± 0.0012	0.2377 ± 0.0007
3	1.00	1.50	0.3198 ± 0.0009	0.1617 ± 0.0005
4	1.00	2.00	0.2482 ± 0.0007	0.1224 ± 0.0004
5	0.50	1.50	0.3418 ± 0.0010	0.1640 ± 0.0005
6	1.50	1.50	0.3003 ± 0.0008	0.1582 ± 0.0005
7	2.00	1.50	0.2872 ± 0.0007	0.1567 ± 0.0005
8	2.50	1.50	0.2722 ± 0.0007	0.1529 ± 0.0004
9	3.00	1.50	0.2620 ± 0.0006	0.1509 ± 0.0004
10	3.50	1.50	0.2531 ± 0.0006	0.1492 ± 0.0004
11	3.00	1.25	0.3011 ± 0.0007	0.1782 ± 0.0005
12	3.25	1.25	0.2949 ± 0.0007	0.1767 ± 0.0005
13	3.30	1.25	0.2937 ± 0.0007	0.1764 ± 0.0005
14	3.20	1.25	0.2953 ± 0.0007	0.1764 ± 0.0005
15	3.20	1.24	0.2974 ± 0.0007	0.1779 ± 0.0005

A few things are apparent from Example 7.1. First, the retrieved parameter values may not be very precise. In fact, the best one can do is determined by (a) the uncertainty inherent in the measured values $\mathcal{R}_j$ and (b) the precision of the MC simulation, which is dependent on the number of histories N. Second, a relatively small number of histories ($N = 100,000$ in this case) often is sufficient to obtain results of practical relevance. Finally, however, many iterations might be required, especially as the number of unknown parameters increases.

The example above is in the form of an integral. However, it is just as likely that a random process, such as X-ray attenuation by an unknown shield, could produce one or more responses. Understanding the processes for X-ray scattering and absorption (see Chapter 9), one can in principle simulate the process for assumed sample variables (such as shield composition and thickness) until the simulated responses match the measured responses sufficiently well. Thus, it is possible to employ iterative IMC to solve very many inverse problems. Many of these applications require a knowledge of the specific processes that are involved.

7.3 Symbolic Monte Carlo

An alternative approach, previously referred to as IMC (Dunn, 1981, 2006) but here and elsewhere (Dunn and Shultis, 2009) called SMC, proceeds formally as follows. Instead of sampling from the PDF $f(\mathbf{x}, \boldsymbol{\theta})$ for the value of $\boldsymbol{\theta}$ selected on a given iteration, sample from an assumed PDF $f^*(\mathbf{x})$, also defined on V, and weight the histories appropriately, as in importance sampling. The weight factor for history i is of the form $f(\mathbf{x}_i, \boldsymbol{\theta})/f^*(\mathbf{x}_i)$ and thus contains the unknown parameters $\boldsymbol{\theta}$ as symbols. Proceeding formally, one obtains

$$\overline{R}(\mathbf{y}_k, \boldsymbol{\theta}) = \frac{1}{N} \sum_{i=1}^{N} z(\mathbf{x}_i, \mathbf{y}_k, \boldsymbol{\theta}) \frac{f(\mathbf{x}_i, \boldsymbol{\theta})}{f^*(\mathbf{x}_i)}, \quad k = 1, 2, \ldots, K. \tag{7.9}$$

Substitution of measured or specified responses $\mathcal{R}_k$ for $\overline{R}(\mathbf{y}_k, \boldsymbol{\theta})$ produces a system of equations in which all quantities are known except $\boldsymbol{\theta}$. Note that the assumed PDF f^* is independent of $\boldsymbol{\theta}$ and, thus, can be sampled without regard to the unknown $\boldsymbol{\theta}$. In principle, Eq. (7.9) forms a system of K algebraic equations for the J unknown parameters of $\boldsymbol{\theta}$.

In the underdetermined case, where $K < J$, there are an infinite number of solutions and additional constraints must be applied to obtain a single solution. If $K = J$, then, in principle, these equations can be solved for the unknown $\boldsymbol{\theta}$. In the overdetermined case, where $K > J$, methods such as least-squares can be applied (see Press et al. (1996)) to obtain a solution that best fits all the responses. Although the system of K equations given by Eq. (7.9) may require that a numerical procedure, such as trial and error or least-squares, be employed in order to find the unknown $\boldsymbol{\theta}$, in SMC only *one* simulation needs to be performed. When simulations are lengthy, such as in highly multidimensional cases, the SMC approach offers an attractive alternative to the iterative IMC approach.

At first, it may seem odd that symbols are inserted into history scores. However, there is no reason that this cannot be done. Of course, the insertion of the symbols $\theta_1, \theta_2, \ldots, \theta_J$ into the Monte Carlo scores can mean that the system of Eq. (7.9) is quite complex. Nevertheless, as is to be seen, there are cases where this procedure can be successfully implemented.

7.3.1 Uncertainties in Retrieved Values

The SMC method can be used not only to generate a system of equations for the desired variables but also to estimate, in principle, the uncertainties in the estimated values of the parameters (Dunn, 1981). The system of equations used to determine values of the parameters θ_j in the SMC process are obtained by equating Eq. (7.9) to measured or specified responses $\mathcal{R}_k$, i.e.,

$$\overline{R}(\mathbf{y}_k, \boldsymbol{\theta}) = \mathcal{R}_k, \quad k = 1, 2, \ldots, K. \tag{7.10}$$

The individual estimates of θ_j, $j = 1, 2, \ldots, J$, can then be written, at least conceptually, as

$$\theta_j = G_j(\mathbf{y}, \overline{\mathbf{R}}, \mathcal{R}), \quad j = 1, 2, \ldots, J, \tag{7.11}$$

where G_j is the function, often unknown, that defines the solution of Eq. (7.10) for the parameter θ_j.

Even if it is assumed that $\mathbf{y}_k$ have no uncertainty, the measurements $\mathcal{R}_k$ presumably have some uncertainty, quantified by the standard deviation $\sigma(\mathcal{R}_k)$, that affects the uncertainty in the retrieved values of θ_j. Likewise, the MC evaluation of $\overline{R}_k$ has an inherent uncertainty that affects the uncertainty in the calculated values of θ_j. Recall that the central limit theorem indicates that, for sufficiently large N, a direct MC estimate $\overline{R}_k$ of the integral of Eq. (7.4), with $\boldsymbol{\theta}$ given, has a standard deviation $\sigma(z)/\sqrt{N}$, where σ is the population standard deviation of $z(\mathbf{x}, \mathbf{y}_k, \boldsymbol{\theta})$ with respect to the PDF f. Substitution of the sample standard deviation for the population standard deviation leads to an expression for the standard deviation of the MC estimate of the form

$$\sigma(\overline{R}_k) = \frac{\sigma}{\sqrt{N}} \simeq \left[\frac{1}{N-1} (\overline{R_k^2} - \overline{R}_k^2) \right]^{1/2}, \quad k = 1, 2, \ldots, K. \tag{7.12}$$

With these uncertainties for $\mathcal{R}_k$ and $\overline{R}_k$, the uncertainty in θ_j can be estimated. If the K measurements $\mathcal{R}_k$ and the K values of $\overline{R}_k$ are all independent, then one can use the standard propagation of errors formula (e.g., see Bevington (1969)), to estimate the standard deviations of the individual recovered parameters due to uncertainties in the measured responses and $\overline{R}_k$ by

$$\sigma(\theta_j) = \left\{ \sum_{k=1}^{K} \left(\frac{\partial G_j}{\partial \mathcal{R}_k} \right)^2 \sigma^2(\mathcal{R}_k) + \sum_{k=1}^{K} \left(\frac{\partial G_j}{\partial \overline{R}_k} \right)^2 \sigma^2(\overline{R}_k) \right\}^{1/2}, \quad j = 1, 2, \ldots, J.$$

$$(7.13)$$

However, in inverse problems, θ is not known until the inverse problem has been solved, making estimation of $\sigma(\overline{R}_k)$ problematic. Further, the function G_j often is quite complex or it may be impossible even to formulate it, except through the algorithmic process used to solve Eq. (7.10) numerically. In these cases, the evaluation of $\partial G_j/\partial \mathcal{R}_k$ and $\partial G_j/\partial \overline{R}_k$ must rely on approximations. Only for the simplest cases can G_j be explicitly determined (e.g., see Examples 7.2 and 7.7). Thus, the above formal treatment for estimating the uncertainties in the retrieved parameters is generally difficult to implement fully. Instead, approximate methods must be employed. For instance, in some cases it may be easy to evaluate $\partial G_j/\partial \mathcal{R}_k$, but difficult or impossible to evaluate $\partial G_j/\partial \overline{R}_k$. In such cases, the errors due only to uncertainties in the measurements $\mathcal{R}_k$, $k = 1, 2, \ldots, K$, can be estimated as lower bounds to the σ_j, $j = 1, 2, \ldots, J$, i.e.,

$$\sigma(\theta_j) \geq \sum_{k=1}^{K} \frac{\partial G_j}{\partial \mathcal{R}_k} \sigma(\mathcal{R}_k), \quad j = 1, 2, \ldots, J. \tag{7.14}$$

Of course, this estimate will be largely unaffected by the number of histories run because the uncertainties in the MC summations are ignored. Nevertheless, lower bounds on the uncertainties in the retrieved parameters can be estimated and these estimates may be fairly good if N is large.

An obvious alternative approach to determine how the parameter values θ_j depend on uncertainties in $\mathcal{R}_k$ and $\overline{R}_k$ is to perform a sensitivity analysis. This approach requires that the appropriate summations used to form the function $\overline{R}_k$ in the SMC simulation be saved. Then, one determines how the resulting values of θ_j, $j = 1, 2, \ldots, J$, vary as $\mathcal{R}_k$ and these summations are varied by small amounts (say, 1 or 2%). Such a procedure provides an indication of how sensitive the retrieved parameters are to variations in uncertainties in the response values and the Monte Carlo model terms.

7.3.2 The PDF Is Fully Known

The particular implementation of SMC depends on where and how the unknown parameters are located in the kernel and the PDF. If all of the unknown parameters are contained only in the kernel, the implementation is inherently straightforward. If, on the other hand, the PDF is dependent on some or all of the unknown parameters, then the implementation is more complex; this situation is considered in Section 7.3.3.

When the PDF is fully known, i.e., f is independent of $\boldsymbol{\theta}$ and only the kernel z in Eq. (7.4) is dependent on the unknown parameters, the inverse problem simplifies considerably. In fact, any of a number of quadrature schemes, including Monte Carlo, can be applied to such inverse problems. If the PDF is easy to sample, then one might as well sample it directly, in which case Eq. (7.9) simplifies to

$$\overline{R}_k(\boldsymbol{\theta}) = \frac{1}{N} \sum_{i=1}^{N} z(\mathbf{x}_i, \mathbf{y}_k, \boldsymbol{\theta}), \quad k = 1, 2, \ldots, K. \tag{7.15}$$

In those cases where $f(\mathbf{x})$ is difficult to sample directly, MCMC can be employed or one can use another PDF $f^*(\mathbf{x})$ that is more easily sampled and Eq. (7.9) reduces to the form

$$\overline{R}_k(\boldsymbol{\theta}) = \frac{1}{N} \sum_{i=1}^{N} z(\mathbf{x}_i, \mathbf{y}_k, \boldsymbol{\theta}) \frac{f(\mathbf{x}_i)}{f^*(\mathbf{x}_i)}, \quad k = 1, 2, \ldots, K. \tag{7.16}$$

In some cases, the system of equations given by Eq. (7.15) or Eq. (7.16) can be solved by some direct procedure, such as matrix inversion. More often, an iterative numerical procedure must be employed. However, the iterations do not involve further simulations.

A Simple Linear Kernel

For the purpose of illustration, consider the simple case where the random variable $\mathbf{x}$ is a single variable x, the vector of unknowns is a single parameter θ, and z has the simple linear form

$$z(\mathbf{x}, \mathbf{y}_k, \boldsymbol{\theta}) = z(x, \theta) = \theta x, \tag{7.17}$$

which indicates that z has no dependence on $\mathbf{y}$. Further, assume the PDF $f(x)$ is uniform over $[a, b]$. Then Eq. (7.4) reduces to

$$R(\theta) = \int_a^b \theta x \frac{1}{b-a} \, dx. \tag{7.18}$$

It is trivial to sample from $f(x) = 1/(b - a)$ and each history is scored with the product θx. Then, Eq. (7.15) takes the simple form

$$\overline{R}(\theta) = \frac{\theta}{N} \sum_{i=1}^{N} x_i = \theta \overline{x}. \qquad (7.19)$$

The known or measured $\mathcal{R}$ can be substituted for $\overline{R}(\theta)$ and the resulting equation can be solved for θ, namely,

$$\theta = \frac{\mathcal{R}}{\overline{x}}. \qquad (7.20)$$

In this trivial case, the estimation of $\sigma(\theta)$ by Eq. (7.13) is straightforward and leads to

$$\sigma(\theta) = \frac{1}{\overline{x}} [\sigma(\overline{R}) + \sigma^2(\mathcal{R})]^{1/2}, \qquad (7.21)$$

where

$$\sigma(\overline{R}) = \left[\frac{1}{N-1} (\overline{\theta x^2} - \theta \overline{x}^2) \right]^{1/2}. \qquad (7.22)$$

Numerical results for this simple case are given in Example 7.2.

Example 7.2 A One-Dimensional Inversion

Consider the particular case where $a = 1$, $b = 3$, and $\theta = 4$. Then, Eq. (7.18) can be used to show that $R(\theta) = 8$. However, assume a measured value of $\mathcal{R} = 7.98 \pm 0.05$. Then, a single forward Monte Carlo simulation can be used to estimate $\overline{x}$ and $\overline{x^2}$ and it is possible to solve Eq. (7.20) for θ and Eq. (7.22) for $s(\theta)$. This problem was simulated and the results are shown in Table 7.2 for three different values of N. It is seen that the solution approaches the correct value of $\theta = 4$ and the uncertainties decrease roughly as $1/\sqrt{N}$.

Table 7.2 A symbolic Monte Carlo solution for θ.

N	$\overline{x}$	θ	$s(\theta)$	N	$\overline{x}$	θ	$s(\theta)$	N	$\overline{x}$	θ	$s(\theta)$
10^2	1.987	4.015	0.115	10^4	2.001	3.989	0.012	10^6	2.000	3.990	0.003

The above procedure can be extended, generally, to any kernel that is linear in the unknown parameters $\boldsymbol{\theta}$. In most cases, estimation of $\sigma(\theta_j)$ is more difficult, but estimation of the unknown parameters is numerically straightforward, as Example 7.3 illustrates.

Example 7.3 A Three-Parameter Linear Kernel

Consider a system whose response can be modeled as

$$R(y, \theta_1, \theta_2, \theta_3) = \int_1^4 \int_0^2 z(x_1, x_2, y, \theta_1, \theta_2, \theta_3) f(x_1, x_2)\, dx_2\, dx_1,$$

where the kernel has the form

$$z(x_1, x_2, y, \theta_1, \theta_2, \theta_3) = \theta_1 x_1 / y + \theta_2 y e^{-2x_2} + \theta_3 x_1 x_2 y^2.$$

The joint PDF is the bivariate normal (see Appendix A)

$$f(x_1, x_2) = \frac{1}{2\pi \sigma_1 \sigma_2 \sqrt{1 - \rho^2}} \exp\left\{ -\frac{\xi}{2(1 - \rho^2)} \right\},$$

with

$$\xi = \frac{(x_1 - \mu_1)^2}{\sigma_1^2} - \frac{2\rho(x_1 - \mu_1)(x_2 - \mu_2)}{\sigma_1 \sigma_2} + \frac{(x_2 - \mu_2)^2}{\sigma_2^2},$$

and ρ is the correlation coefficient. Here it is assumed $\mu_1 = 0$, $\mu_2 = 1$, $\sigma_1 = 1$, $\sigma_2 = 2.5$, and $\rho = 0.5$. How might one use SMC to obtain θ_1, θ_2, and θ_3 given measurements $\mathcal{R}_1 = 0.17407$, $\mathcal{R}_2 = 0.36406$, and $\mathcal{R}_3 = 0.73845$ for $y_1 = 1$, $y_2 = 2$, and $y_3 = 3$, respectively?

The SMC formulation of Eq. (7.9) leads to

$$\overline{R}(y_k | \theta_1, \theta_2, \theta_3) = \frac{1}{N} \sum_{i=1}^N z(x_1, x_2, y_k | \theta_1, \theta_2, \theta_3) \frac{f(x_{1i}, x_{2i})}{f^*(x_{1i}, x_{2i})}, \quad k = 1, 2, 3.$$

If, for simplicity, x_1 and x_2 are sampled uniformly over their respective intervals, then $f^*(x_1, x_2) = (1/3)(1/2) = 1/6$ and these equations can be written in terms of the measurements as

$$\mathcal{R}_k = \frac{6}{N}\left\{ \frac{\theta_1}{y_k} \mathcal{S}_1 + \theta_2 y_k \mathcal{S}_2 + \theta_3 y_k^2 \mathcal{S}_3 \right\}, \quad k = 1, 2, 3,$$

where the summations $\mathcal{S}_1 = \sum_{i=1}^N x_{1i} f(x_{1i}, x_{2i})$, $\mathcal{S}_2 = \sum_{i=1}^N e^{-2x_{2i}} f(x_{1i}, x_{2i})$, and $\mathcal{S}_3 = \sum_{i=1}^N x_{1i} x_{2i} f(x_{1i}, x_{2i})$ are easily scored. A single simulation can be run in which the sums $\mathcal{S}_1$, $\mathcal{S}_2$, and $\mathcal{S}_3$ are accumulated and the solution can be obtained by matrix inversion of $\mathcal{R} = \mathbf{AS}$. The SMC solutions for $N = 10^6$ histories are compared to the actual values in Table 7.3.

Table 7.3 Comparison of actual and SMC parameters.

	θ_1	θ_2	θ_3
Actual values	1.500	0.800	1.000
SMC values	1.505	0.802	1.004

Thus, the SMC procedure obtained results good to within 0.5%. This problem demonstrates that, if the kernel is linear *in the unknown parameters* $\boldsymbol{\theta}$ and the PDF is known, the SMC procedure can be reduced to a matrix inversion.

It is possible to solve the equations in Examples 7.2 and 7.3 directly or by matrix inversion because the kernels z are linear in the unknown parameters and f is completely known. However, in many cases the kernel is nonlinear in $\boldsymbol{\theta}$ and the SMC approach leads to a system of equations that must be solved by some numerical technique. An extreme case is one in which each history leads to a score whose value is unique and thus the SMC model produces a system of equations each of which has N terms. Such a case is illustrated in Example 7.4.

Example 7.4 A Nonlinear Kernel

Consider the definite integral

$$R(y, \theta_1, \theta_2) = \frac{1}{\sqrt{2\pi}\theta_2} \int_a^b e^{-[(x - y\theta_1)/(\sqrt{2}\theta_2)]^2} \lambda e^{-\lambda x} dx,$$

which can be converted to the form

$$R(y, \theta_1, \theta_2) = \frac{\lambda}{\sqrt{2\pi}A\theta_2} \int_a^b e^{-[(x - y\theta_1)/(\sqrt{2}\theta_2)]^2} A e^{-\lambda x} dx,$$

where λ is known and $A = 1/(e^{-\lambda a} - e^{-\lambda b})$ is a normalization constant. Samples from the PDF $f(x) = A e^{-\lambda x}$ are of the form

$$x_i = -\frac{1}{\lambda} \ln[e^{-a\lambda} - \rho_i(e^{-a\lambda} - e^{-b\lambda})],$$

where ρ_i is a pseudorandom number uniformly distributed on the unit interval. Next, form the SMC estimates, from Eq. (7.15), to obtain

$$\overline{R}_k(\theta_1, \theta_2) = \frac{\lambda}{\sqrt{2\pi}A\theta_2} \frac{1}{N} \sum_{i=1}^N e^{[(x_i - y_k\theta_1)/(\sqrt{2}\theta_2)]^2}, \quad k = 1, 2, \ldots, K.$$

Substitution of the known $\mathcal{R}_k$ for $\overline{R}_k$ yields K equations in two unknowns. The model is not linear in the unknowns θ_1 and θ_2, and x_i assume arbitrary values over the interval $[a, b]$. Thus, each term in the summations has a unique value. If the x_i values are stored, then the above equations, each with N terms on the right-hand side, can be solved numerically for θ_1 and θ_2. This solution requires numerical iteration, but only one MC simulation need to be implemented.

For the particular case that $a = 2$, $b = 10$, $\theta_1 = 3$, $\theta_2 = 2$, $\lambda = 1$, $K = 2$, $y_1 = 1$, and $y_2 = 2$, calculated responses are $\mathcal{R}_1 = 0.02458$ and $\mathcal{R}_2 = 0.009157$. Given these response values, an SMC model was run and the above two equations were solved for θ_1 and θ_2. The results shown in Table 7.4 were obtained by specifying values of θ_1 and θ_2 and comparing the calculated $\overline{R}_k$ to the specified $\mathcal{R}_k$. The search was stopped when $[(\overline{R}_1 - \mathcal{R}_1)^2 + (\overline{R}_2 - \mathcal{R}_2)^2]^{1/2}/(\mathcal{R}_1 + \mathcal{R}_2) < 0.01$. Of course, other quadrature formulas could be used to obtain similar results. However, this problem demonstrates that even if each history leads to a unique term in the SMC model, it still may be possible to utilize SMC to obtain solutions to inverse problems utilizing only a single simulation.

Table 7.4 The retrieved parameters θ_1 and θ_2 for various values of N.

N	θ_1	θ_2	N	θ_1	θ_2	N	θ_1	θ_2
10^2	2.98	2.01	10^4	3.00	2.00	10^6	3.00	2.00

7.3.3 The PDF Is Unknown

The SMC solution for the general case in which both the kernel and the PDF depend on some set of unknown parameters $\boldsymbol{\theta}$ is treated by Eq. (7.9). Section 7.3.2 considered the simpler case where the PDF has no dependence on the unknown parameters. The case in which the kernel is independent of $\boldsymbol{\theta}$ but the PDF is not is now considered. When some parameters of the PDF are unknown or the dependence of the PDF on the parameters is unknown, implementation of SMC becomes more involved. The complexity depends on the specifics of the problem. For instance, a situation in which the PDF is linear in the unknown parameters (not necessarily in the random variables) may be simpler than a situation in which the dependence on the parameters is nonlinear. Some approaches that can be applied in certain cases are considered next.

Approximating an Unknown PDF as a Histogram

In some cases, it is possible to linearize the search for an unknown PDF by using a histogram approach. For instance, consider a response model of the form

$$R(y) = \int_a^b z(x, y) f(x) \, dx, \tag{7.23}$$

where the PDF and its dependence on hidden parameters are unknown but responses are known at K values of y. In this inverse problem, one seeks to infer the shape of the PDF from the K responses. If K is large enough, then one can divide the interval $[a, b]$ into $J \leq K$ subintervals and express the PDF in histogram form. For instance, if each subinterval of a one-dimensional PDF is of width $\Delta = (b - a)/J$, then the PDF can be expressed as

$$f(x) = f_1, \quad a \leq x < a + \Delta$$
$$= f_2, \quad a + \Delta \leq x < a + 2\Delta$$
$$\vdots$$
$$= f_J, \quad a + (J - 1)\Delta \leq x \leq b. \tag{7.24}$$

Sampling from a known PDF f^*, an SMC solution can be expressed in the form

$$\overline{R}_k = \frac{1}{N} \sum_{i=1}^{N} z(x_i, y_k) \frac{f(x_i)}{f^*(x_i)}. \tag{7.25}$$

On any given history, x_i is in only one subinterval; if this subinterval is called j, then the contribution to the kth response for that history is of the form

$$\delta R_k = z(x_i, y_k) \frac{f_j}{f^*(x_i)}. \tag{7.26}$$

Everything in this equation is known except f_j. Thus, a system of linear equations of the form

$$\mathbf{Af} = \mathcal{R} \tag{7.27}$$

results, where $\mathbf{f} = \{f_j, \ j = 1, 2, \ldots, J\}$ and

$$\mathbf{A} = \{\mathcal{S}_{kj}\}, \quad k = 1, 2, \ldots, K, \ j = 1, 2, \ldots, J,$$

with

$$\mathcal{S}_{kj} = \frac{1}{N} \sum_{i=1}^{N} \frac{z(x_i, y_k)}{f^*(x_i)}, \quad a + (j-1)\Delta \le x_i < a + j\Delta. \tag{7.28}$$

For the case $J = K$, Eq. (7.27) can be solved, in principle, by direct matrix inversion. However, it may be the case that many measurements are available (K is large) and fewer histogram values ($J < K$) are adequate to characterize the PDF. In such cases, it may be possible to find values of the $J < K$ parameters of the PDF by a numerical procedure such as least-squares (Press et al., 1996). In practice, the uncertainties in the data and the uncertainties inherent in the MC summations often lead to a system of equations whose solution by these standard techniques fits the data fairly well but is not physically reasonable. For instance, some PDF values may be negative.

In Example 7.5, it was found that for $K = 10$, standard numerical procedures employed to find $4 < J < 10$ PDF histogram values did not give results better than those shown in Fig. 7.2. It is obvious that solutions that contain physically unreasonable values should be ignored and other solutions should be sought until all the PDF values appear to be reasonable. Even in cases where all of the PDF parameters are positive, it is possible, because of uncertainties in the data and the MC summations, that the retrieved parameters are not very accurate. It may be useful to try several values of J and see if the trend in the histogram of the PDF remains consistent. It also deserves mention that various variance reduction methods can, and often should, be employed. For instance, rather than sampling over $[a, b]$ and then determining the subinterval within which x_i exists, it is better to allocate $1/J$ of the histories to each subinterval, in the systematic sampling manner.

Many quantities have a PDF-like character in that they can be decomposed into a magnitude times a shape function whose integral is normalized to unity. For instance, in spectroscopy, a spectrum (often consisting of $K > 10^3$ response values) can be written in terms of the source strength, the PDF on source energy,

Example 7.5 A Linearized Quadratic PDF

For purposes of illustration, consider Eq. (7.23) with $a = 0$, $b = \pi$, and $z(x, y) = 1000/[y + 0.5\cos(x)]$. If the PDF has the quadratic form

$$f(x) = 0.640819 - 0.025633(x - 5)^2,$$

then it can be shown, by numerical integration, that for $J = K = 4$ this model leads to the following responses: $\mathcal{R}_1(y_1 = 1.0) = 1{,}367.25$, $\mathcal{R}_2(y_2 = 1.5) = 791.007$, $\mathcal{R}_3(y_3 = 2.0) = 561.927$, and $\mathcal{R}_4 = \overline{R}(y_4 = 2.5) = 436.943$.

Assume one is unaware of how these responses were obtained and one is asked to estimate f_j, $j = 1, 2, 3, 4$. One might choose to sample from the uniform PDF $f^*(x) = 1/\pi$, $0 \le x \le \pi$, and then obtain a system of equations of the form

$$\mathcal{S}_{k1} f_1 + \mathcal{S}_{k2} f_2 + \mathcal{S}_{k3} f_3 + \mathcal{S}_{k4} f_4 = \mathcal{R}_k, \quad k = 1, 2, 3, 4,$$

where

$$\mathcal{S}_{kj} = \frac{\pi}{N} \sum_{i=1}^{N} \frac{1000}{y_k + 0.5\cos(x_i)}, \quad (j - 1)\pi/4 \le x_i \le j\pi/4.$$

With $N = 10^7$ histories sampled from the uniform PDF on $[0, \pi]$ and evaluation of the summations $\mathcal{S}_{ki}$, $i = 1, \ldots, 4$, the following results were obtained: $f_1 = 0.07104$, $f_2 = 0.2848$, $f_3 = 0.3939$, and $f_4 = 0.5234$. These values are plotted, along with the actual PDF, in Fig. 7.2. This example demonstrates that a reasonable approximation can be obtained by this approach.

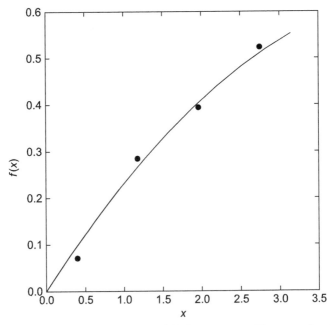

Figure 7.2 The true quadratic PDF (line) and the four-component histogram values (points) recovered using SMC.

and the detector response function, as in Eq. (7.5). In cases such as these, one can treat the magnitude (e.g., Q_0 in Eq. (7.5)) as an additional unknown and solve the system of K equations for the magnitude and $J - 2$ PDF histogram values (there are $J - 1$ of these values but one is determined by the PDF normalization condition). Alternatively, rather than perform matrix inversion, one can solve an overdetermined system, where $K > J + 1$ by methods such as least-squares or maximum entropy (Press et al., 1996).

The PDF Is Separable

Sometimes $f(\mathbf{x}, \boldsymbol{\theta})$ can be approximated by a finite series whose terms are products of separable terms, i.e., one term contains $\mathbf{x}$ and the other $\boldsymbol{\theta}$. In such cases, the PDF can be written as

$$f(\mathbf{x}, \boldsymbol{\theta}) \simeq \sum_{m=0}^{M} h_m(\boldsymbol{\theta}) g_m(\mathbf{x}), \tag{7.29}$$

where the functions $g_j(\mathbf{x})$ and $h_j(\boldsymbol{\theta})$ are known and $\boldsymbol{\theta}$ has J unknown components. Generally, the larger M, the better is the approximation.

In such cases, a general inverse problem for $\boldsymbol{\theta}$ can be written as

$$R(\mathbf{y}_k, \boldsymbol{\theta}) = \int_V z(\mathbf{x}, \mathbf{y}_k) \sum_{m=0}^{M} h_m(\boldsymbol{\theta}) g_m(\mathbf{x}) \, d\mathbf{x}. \tag{7.30}$$

An SMC solution could, in general, take the form

$$\overline{R}_k = \frac{1}{N} \sum_{i=1}^{N} z(\mathbf{x}_i, \mathbf{y}_k) \frac{f(\mathbf{x}_i, \boldsymbol{\theta})}{f^*(\mathbf{x}_i)} = \sum_{m=0}^{M} h_m(\boldsymbol{\theta}) S_{km}, \quad k = 1, 2, \dots, K, \tag{7.31}$$

where

$$S_{km} = \frac{1}{N} \sum_{i=1}^{N} z(\mathbf{x}_i, \mathbf{y}_k) \frac{g_m(\mathbf{x}_i)}{f^*(\mathbf{x}_i)}, \quad k = 1, 2, \dots, K, \tag{7.32}$$

and where $f^*(x)$ is the PDF used to sample values of x_i.

Upon setting Eq. (7.31) to measured or specified responses $\mathcal{R}_k$, a set of K equations in the J unknown values of θ_j are obtained, namely,

$$\mathcal{R}_k = \sum_{m=0}^{M} h_m(\boldsymbol{\theta}) S_{km}, \quad k = 1, 2, \dots, K. \tag{7.33}$$

To avoid over- and underdetermined sets of equations, usually K is chosen to equal J. However, even in this case nonlinear numerical techniques generally must be used to obtain a solution.

A PDF of One Random Variable

A PDF with a single random variable x can always be cast into a separable PDF by using a truncated Taylor series expansion about some point x_o, namely,

$$f(x, \boldsymbol{\theta}) \simeq \sum_{m=0}^{M} \frac{(x - x_o)^m}{m!} \frac{\partial^m f(x, \boldsymbol{\theta})}{\partial x^m}\bigg|_{x=x_o}. \tag{7.34}$$

This has the form of Eq. (7.29) with

$$h_m(\boldsymbol{\theta}) = \frac{\partial^m f(x, \boldsymbol{\theta})}{\partial x^m}\bigg|_{x=x_o} \quad \text{and} \quad g_m(x) = \frac{x^m}{m!}. \tag{7.35}$$

Exponential PDF

To illustrate the above separable PDF technique, consider the exponential PDF $f(x, \theta) = \theta e^{-\theta x}$ for $x \in [0, \infty)$. Exponential PDFs are encountered in many physical processes. For instance, this PDF, which is nonlinear in the parameter θ, applies to many physical phenomena, such as radioactive decay, neutral particle transport, and growth and decay of biological systems.

With an M-term Taylor series expansion of Eq. (7.34) about $x_o = 0$, this PDF can be approximated as

$$f(x, \theta) \simeq \sum_{m=0}^{M} h_m(\theta) g_m(x), \tag{7.36}$$

where

$$h_m(\theta) = \frac{d^m}{dx^m}(\theta s^{-\theta x})\bigg|_{x=0} = (-1)^m \theta^{m+1} \quad \text{and} \quad g_m(x) = \frac{x^m}{m!}. \tag{7.37}$$

Now consider a model of the form

$$R(y, \theta) = \int_0^\infty z(x, y) e^{-\theta x} \, dx, \tag{7.38}$$

where z is a kernel for which the integral exists. Assume $\mathcal{R}_1$ is known or can be determined for some value y_1. How can SMC be used to obtain the unknown parameter θ in a single simulation? One way to proceed is to rewrite the above equation as

$$R_1(\theta) = \frac{1}{\theta} \int_0^\infty z(x, y_1) \theta e^{-\theta x} \, dx. \tag{7.39}$$

Because θ is unknown, it is not possible to sample from this PDF directly, but it is possible to sample from $f^*(x) = \theta_o e^{-\theta_o x}$, where θ_o is known. Doing so, an estimate of $\overline{R}_1(\theta)$ can be constructed as

$$\overline{R}_1(\theta) = \frac{1}{\theta} \sum_{m=0}^{M} h_m(\theta) \left[\frac{1}{N} \sum_{i=1}^{N} z(x_i, y_1) \frac{g_m(x_i)}{f^*(x_i)} \right], \qquad (7.40)$$

where the x_i are sampled from $f^*(x)$. Substitution for f^*, h_m, and g_m then gives an equation from which θ can be calculated, namely,

$$\mathcal{R}_1 \simeq \overline{R}_1 = \frac{1}{\theta_o N} \sum_{m=0}^{M} (\theta_o - \theta)^m \left[\sum_{i=1}^{N} \frac{z(x_i, y_1) x_i^m}{m! e^{-\theta_o x_i}} \right]. \qquad (7.41)$$

This equation is a polynomial equation of degree M for the unknown θ and generally has M solutions. Only solutions that are real and positive are of interest. If more than one real positive solution is found (the nonuniqueness of inverse problems), then the one that is most physically realistic is chosen.

Example 7.6 A Specific Exponential PDF

Consider the particular case

$$R(\theta) = \int_0^\infty x^2 e^{-\theta x} \, dx,$$

where it can be argued on physical grounds that $0.5 \le \theta \le 1.5$. This integral has a known value, for a given θ, of $R(\theta) = 2/\theta^3$. However, for purposes of illustration, this fact is ignored and one is asked to use SMC to determine the value of θ corresponding to a given measured value $\mathcal{R} = 3.90625$. A five-term expansion of the exponential function can be used with $\theta_0 = 1$. Because

$$\frac{f(x_i, \theta)}{f^*(x_i)} = \theta e^{-(\theta-1)x_i},$$

the Taylor series expansion of this exponential can be written as

$$e^{-(\theta-1)x_i} \simeq 1 + \sum_{j=1}^{5} \frac{(-(\theta-1)x_i)^j}{j!}.$$

Then, the SMC solution can be written as

$$3.90625 = \frac{1}{N} \left\{ \sum_{i=1}^{N} x_i^2 - (\theta-1) \sum_{i=1}^{N} x_i^3 + \frac{(\theta-1)^2}{2} \sum_{i=1}^{N} x_i^4 - \frac{(\theta-1)^3}{6} \sum_{i=1}^{N} x_i^5 \right.$$
$$\left. + \frac{(\theta-1)^4}{24} \sum_{i=1}^{N} x_i^6 - \frac{(\theta-1)^5}{120} \sum_{i=1}^{N} x_i^7 \right\}. \qquad (7.42)$$

A single simulation allows estimations of the six summations on the right side of the above equation. This equation for θ can then be solved by any of several numerical techniques. For example, $N = 10^6$ histories were used to estimate the six summations, and then a simple numerical search with θ varying from 0.5 to 1.5 in increments of 0.01 was conducted. Linear interpolation between the two closest points yielded $\theta = 0.798$ (the correct value is $\theta = 0.800$). Note that a five-term expansion was sufficient because $\theta - 1$ is small. In general, the larger the difference between θ and θ_o, the more terms may be required.

The PDF Is Linear in the Unknown Parameters

The simplest linear PDF defined on a finite interval $[a, b]$ is the uniform PDF. Such a PDF has no adjustable parameters because its value is constrained to be $f(x) = 1/(b - a)$. More generally, the specific form of Eq. (7.29) in which $h_j(\boldsymbol{\theta}) = \theta_j$, $j = 0, 1, 2, \ldots, J$, leads to a PDF of the form

$$f(\mathbf{x}, \boldsymbol{\theta}) = \sum_{j=0}^{J} \theta_j g_j(\mathbf{x}), \quad \mathbf{x} \in V, \tag{7.43}$$

which is linear in the parameters θ_j, $j = 0, 1, 2, \ldots, J$. The normalization condition can be used to relate one of the unknown parameters, say θ_0, to the others, yielding

$$\theta_0 = \sum_{j=1}^{J} \theta_j \int_V g_j(\mathbf{x}) \, d\mathbf{x} \Big/ \int_V g_0(\mathbf{x}) \, d\mathbf{x}.$$

Thus, in principle, only $K = J$ responses are needed from a response model of the form of Eq. (7.4) in order to construct an SMC model to estimate the $J + 1$ unknown parameters.

For such a PDF, Eq. (7.4) assumes the form

$$\mathcal{R}_k \simeq R(\mathbf{y}_k, \boldsymbol{\theta}) = \int_V z(\mathbf{x}_i, \mathbf{y}_k) \sum_{j=0}^{J} \theta_j g_j(\mathbf{x}) \, d\mathbf{x}, \quad k = 1, 2, \ldots, K, \tag{7.44}$$

which is an inverse problem for $\boldsymbol{\theta}$ if at least one $\mathcal{R}_k$ is known. The general SMC estimate of R can be written as

$$\overline{R}_k = \frac{1}{N} \sum_{i=1}^{N} z(\mathbf{x}_i, \mathbf{y}_k) \frac{f(\mathbf{x}, \boldsymbol{\theta})}{f^*(\mathbf{x}_i)}$$

$$= \sum_{j=0}^{J} \theta_j \mathcal{S}_{kj}, \quad k = 1, 2, \ldots, K, \tag{7.45}$$

where

$$S_{kj} = \frac{1}{N} \sum_{i=1}^{N} z(\mathbf{x}_i, \mathbf{y}_k) \frac{g_j(\mathbf{x}_i)}{f^*(\mathbf{x}_i)}, \quad k = 1, 2, \dots, K. \qquad (7.46)$$

A One-Dimensional Linear PDF

For instance, consider the linear PDF, $f(x) = \alpha + \beta x$, for which the normalization condition $\int_a^b (\alpha + \beta x) dx = 1$ can be used to show that

$$\alpha = \frac{1}{b-a} - \beta \frac{b+a}{2}. \qquad (7.47)$$

Then the integral

$$R(y_k, \alpha, \beta) = \int_a^b z(x, y_k)(\alpha + \beta x) dx, \quad k = 1, 2, \dots, K, \qquad (7.48)$$

poses an inverse problem for α and β if $\mathcal{R}_k$ is known for at least one value of k. An SMC solution, from Eq. (7.9), is given by

$$\begin{aligned}
\mathcal{R}_k &= \frac{1}{N} \sum_{i=1}^{N} z(x_i, y_k) \left[\frac{f(x_i)}{f^*(x_i)} \right] \\
&= \frac{1}{b-a} S_{1k} + \beta S_{2k} - \beta \frac{b+a}{2} S_{1k}, \quad k = 1, 2, \dots, K,
\end{aligned} \qquad (7.49)$$

where

$$S_{1k} = \frac{1}{N} \sum_{i=1}^{N} z(x_i, y_k) \frac{1}{f^*(x_i)} \quad \text{and} \quad S_{2k} = \frac{1}{N} \sum_{i=1}^{N} z(x_i, y_k) \frac{x_i}{f^*(x_i)},$$

and the x_i are sampled from the PDF $f^*(x)$ of one's choice. The solution for β, for arbitrary k, can be written as

$$\beta_k = \left\{ \mathcal{R}_k - \frac{1}{b-a} S_{1k} \right\} \bigg/ \left\{ S_{2k} - \frac{b+a}{2} S_{1k} \right\}, \quad k = 1, 2, \dots, K. \qquad (7.50)$$

Clearly, only one response is needed to obtain an estimate of β but if $K > 1$, then β can be estimated as an average of K estimates, i.e.,

$$\beta = \frac{1}{K} \sum_{k=1}^{K} \beta_k. \qquad (7.51)$$

An estimate of $s(\beta_k)$, due only to uncertainties in the responses, is easily obtained by differentiating Eq. (7.50) with respect to $\mathcal{R}_k$, which yields

$$\frac{\partial \beta_k}{\partial \mathcal{R}_k} = \frac{1}{S_{2k} - (b + a/2)S_{1k}}, \quad k = 1, 2, \ldots, K,$$

and then using Eq. (7.14) to calculate

$$s(\beta_k) \simeq \sum_{k=1}^{K} \left(\frac{\partial \beta_k}{\partial \mathcal{R}_k} \right)^2 \sigma^2(\mathcal{R}_k) = \sum_{k=1}^{K} \left(S_{2k} - \frac{b+a}{2} S_{1k} \right)^{-2} \sigma^2(\mathcal{R}_k), \quad k = 1, 2, \ldots, K.$$

$$(7.52)$$

Example 7.7 Analysis of an Unknown Linear PDF

Consider the specific case of Eq. (7.49) for which $z(x, y_k) = e^{-\lambda x/y_k}$ for $K = 2$, where $a = 0$, $b = 5$, $\lambda = 0.1$, $y_1 = 0.5$, and $y_2 = 2.5$. Quadrature can be used to show that $\mathcal{R}_1 = 0.619166$ and $\mathcal{R}_2 = 0.902527$ (for $\alpha = 0.1750$ and $\beta = 0.0100$).

But suppose that measurements yield $\mathcal{R}_1 = 0.6210 \pm 0.0031$ and $\mathcal{R}_2 = 0.9018 \pm 0.0045$. Also assume it is known that $\beta \geq 0$, i.e., the PDF is nondecreasing as x increases. Then, it is straightforward to show that $0 \leq \beta \leq 0.08$, in order for $f(x) = \alpha + \beta x$ to be a PDF. A single SMC simulation is run, sampling from the uniform PDF $f^*(x) = 1/(b - a) = 1/5$ to provide estimates of β and $\sigma(\beta)$. The results in Table 7.5 are obtained for three values of N. The corresponding value of α can be obtained from Eq. (7.47). Of course, the three values of $s(\beta) \approx \sigma(\beta)$ are underestimated because Eq. (7.14) ignores the contributions from the N histories.

Table 7.5 The parameter β obtained for different values of N. The correct value is $\beta = 0.0100$.

N	β	$s(\beta)$	N	β	$s(\beta)$	N	β	$s(\beta)$
10^3	0.0109	0.0086	10^5	0.0103	0.0121	10^7	0.0103	0.0122

7.3.4 Unknown Parameter in Domain of x

Sometimes the domain V of the random variable $\mathbf{x}$ may be unknown. For such problems, the best one can do is to use the iterative method to try different domains until agreement with $\mathcal{R}_k$ is obtained. To illustrate a systematic approach to this iterative procedure, the following example is given.

Example 7.8 A Buffon Needle Inverse Problem

Consider the problem of finding the grid spacing D, for Buffon's needle problem, when $P_{cut}(=\mathcal{R})$ and needle length L are known. This is a somewhat contrived inverse problem because if the needle length could be measured, then surely the grid could also be measured. Nevertheless, the problem is a valid inverse problem and is illustrative of an unknown domain. In Chapter 1, it was shown that

$$P_{cut} = \frac{2}{\pi D} \int_0^U \cos^{-1}\left(\frac{x}{L}\right) dx, \tag{7.53}$$

where

$$U = \begin{cases} D, & D < L, \\ \\ L, & D \geq L. \end{cases}$$

It is not clear how to evaluate the integral because D is unknown. One way to proceed is to rewrite Eq. (7.53) in the form

$$P_{cut} = \frac{2L}{\pi D} \int_0^L [H(x) - H(D - x)] \cos^{-1}\left(\frac{x}{L}\right) \frac{1}{L} dx, \tag{7.54}$$

where H is the unit step function. Then, in principle, one can sample uniformly over the interval $[0, L]$, because $1/L$ is a properly normalized PDF over this interval, to obtain the result

$$P_{cut} \simeq \frac{2L}{\pi DN} \sum_{i=1}^N h_i \cos^{-1}\left(\frac{x_i}{L}\right),$$

where

$$h_i = \begin{cases} 1, & \text{if } x_i \leq D, \\ \\ 0, & \text{if } x_i > D. \end{cases}$$

This may not seem promising, because D is unknown and so it is unknown when h_i is unity or zero.

One can, however, proceed by dividing the interval $[0, L]$ into J contiguous subintervals of equal size $\Delta = L/J$. Then Eq. (7.54) can be rewritten as

$$P_{cut} \simeq \frac{2L}{\pi D} \sum_{j=1}^J \delta_j S_j, \tag{7.55}$$

where

$$S_j = \frac{1}{N} \sum_{i=1}^{N_j} \cos^{-1}\left(\frac{x_{ij}}{L}\right), \tag{7.56}$$

N_j is the number of samples in the jth interval, x_{ij} is the ith sample in the jth interval, and

$$\delta_j = \begin{cases} 1, & \text{if } D \leq j\Delta, \\ 0, & \text{if } D > j\Delta. \end{cases}$$

A possible IMC solution proceeds as follows. Conduct a single Monte Carlo simulation sampling x uniformly over $[0, L]$ and score the S_j, $j = 1, 2, \ldots, J$. Then, start at $k = 1$, set $\overline{D}_k = \overline{D}_1 = \Delta/2$ (the midpoint of the first subinterval), and calculate the right-hand side of Eq. (7.56), setting $\delta_j = 0$ for all $j > k = 1$. Incrementing k, one can perform similar operations for each $\overline{D}_k = \overline{D}_{k-1} + \Delta$, where δ_j is set to 0 for all $j > k$. Because the calculated P_{cut} is larger than the correct value if $\overline{D}_k < D$, the incrementing of k is stopped when the calculated P_{cut} is less than the given P_{cut}. The estimated value of D is interpolated as

$$\overline{D} = \overline{D}_k - \frac{P_{cut} - \overline{P}_k}{\overline{P}_{k-1} - \overline{P}_k} \Delta,$$

where $\overline{P}_k$ is the right-hand side of Eq. (7.55) when the value $\overline{D}_k$ is used and $\overline{P}_{k-1}$ is the right-hand side of Eq. (7.55) when the value $\overline{D}_{k-1}$ is used. Thus, the SMC procedure requires only a single simulation to form the S_j of Eq. (7.56) and an iterative algebraic solution of Eq. (7.55).

It is clear that the values of P_{cut} range from 0 (for infinite grid spacing) to 1 (for zero grid spacing), because it is a probability. For any case in which $D \geq L$, all δ_j are unity and

$$P_{cut} = \frac{2}{\pi D} \int_0^L \cos^{-1}\left(\frac{x}{L}\right) dx.$$

For any case in which $D < L$, some of the δ_j are zero (for large enough J) and hence some of the computation is "wasted" because some of the S_j are unused in the inversion process. It is also clear that the accuracy of the process depends on the accuracy with which L and P_{cut} are known and the precision of estimated grid spacing can be improved by increasing N and J.

The SMC process outlined above was implemented using $N = 10^6$ histories and $J = 10^3$. The results shown in Table 7.6 were obtained.

Table 7.6 Recovery of grid spacing D by SMC.

L	Actual D	P_{cut}	Estimated D
4.00	1.00	0.920000	1.010
4.00	3.00	0.747487	2.998
4.00	4.00	0.636620	3.998
4.00	5.00	0.509296	4.996
4.50	10.0	0.254648	9.993

7.4   Inverse Monte Carlo by Simulation

The previous discussion has considered IMC problems when the response can be written in the form of the integral of Eq. (7.4). However, there are many inverse problems (perhaps the majority) that cannot be put into the form of an integral. In such problems, multiple simulations are performed by varying the problem parameters until the difference between measured responses and the simulation results is minimized as discussed in Section 7.1.3. Such an IMC approach is a "brute-force" approach and may be computationally expensive if each simulation requires a large computational effort. The use of SMC methods may be difficult in many simulation problems and requires that the tallies can be analytically written in terms of the unknown parameters. It is also true that construction of an SMC model may be time consuming and is "worth the effort" primarily in cases where it saves considerable computation time. The simulation approach to SMC has been employed in various ionizing radiation transport problems (e.g., (Dunn and O'Foghludha, 1982; Floyd et al., 1986; Yacout and Dunn, 1987; Michael, 1991; and Dunn and Womersley, 1994)). This approach is illustrated next by considering a simple hypothetical case.

7.4.1   A Simple Two-Dimensional World

To illustrate the simulation approach to SMC, consider a two-dimensional world in which light scatters only at angles that are multiples of $\pi/2$. In this world, a rectangular medium is encountered that is 5 units wide and 10 units tall. Light particles are projected at the center of the left wall and detectors are mounted on all four sides that count all of the particles that exit each of the sides. It is stipulated that the medium is composed of

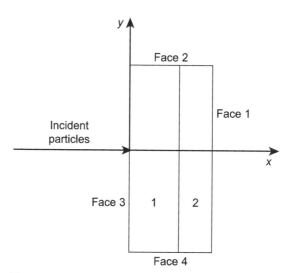

Figure 7.3 A two-dimensional scattering medium consisting of two materials, with particles entering normally from the left. The particles exiting each face are counted.

at most two materials, whose scattering and absorption probabilities are known. The probability of an interaction per unit differential distance in material 1 is

μ_1 and in material 2 it is μ_2. In the kth material, a photon is scattered left with probability p_l^k, is scattered backward with probability p_b^k, is scattered right with probability p_r^k, and is absorbed with probability p_a^k, for $k = 1$ and 2.

An experiment is run in which one photon per second is introduced normally on the center of the front face, as shown in Fig. 7.3, and the detectors record the number of particles exiting each of the four sides per second $\mathcal{R}_j$, $j = 1, ..., 4$. The problem is to determine the thicknesses of each of the two materials that compose the medium given the four measured responses. This inverse problem is not posed in integral form. Nevertheless, an SMC solution can be obtained.

A possible approach is outlined below. Specify the particle direction of travel by a variable id such that

$$id = \begin{cases} 1, & \text{if moving in the } x\text{-direction toward face 1,} \\ 2, & \text{if moving in the } y\text{-direction toward face 2,} \\ 3, & \text{if moving in the } -x\text{-direction toward face 3,} \\ 4, & \text{if moving in the } -y\text{-direction toward face 4.} \end{cases}$$

Then, construct a simulation model that proceeds as follows.

1. Introduce particles at $x = 0$ and $y = 0$ moving in the x-direction ($id = 1$).

2. First, presume the medium is all material 1. Thus, set $k = 1$ and $t_k = 5$, where t_k is the thickness of material k.

3. Sample a distance d to the next interaction. This requires sampling d from the PDF $f(d) = \mu_k e^{-\mu_k d}$, which leads to $d = -\ln(\rho)/\mu_k$, where ρ is sampled uniformly from $(0, 1)$.

4. Move the particle to a new position. This is accomplished by moving the particle d units in the direction indicated by id. This yields a new position (x, y).

5. Check to see if the particle has exited the medium. For instance, if $x > 5$, the particle has exited the right face and the detector response for the right detector is incremented by unity. A similar procedure is used for the other three faces. After scoring the appropriate response, start a new history. If the particle is still in the medium, proceed to the next step.

6. Sample a new direction by picking a pseudorandom number ρ. If $\rho \leq p_l^k$, then the particle scatters to the left and we set $id = id + 1$. If $p_l^k < \rho \leq p_l^k + p_b^k$, then the particle scatters backward and we set $id = id + 2$. If $p_l^k + p_b^k \leq \rho < p_l^k + p_b^k + p_r^k$, then the particle scatters to the right and we set $id = id + 3$. Otherwise, the particle is absorbed and this history stops. It is necessary to renormalize id to values from 1 to 4, so if $id > 4$, then $id = id \bmod 4$.

7. After all histories have been completed, save the four responses $\overline{R}_j$ =(no. leaving face j)/(no. particles tracked)= (no. escaping through face j per second).

8. Next assume the medium is all material 2, so that $k = 2$ and $t_k = 5$, where t_k is now the thickness of material 2. Again start particles at $x = 0$ and $y = 0$ with $id = 1$. Repeat steps 3 through 6 and output the responses for this case.

9. Now, assume the medium is part material 1 and part material 2. Start with some thickness for t_1 and let $t_2 = 5 - t_1$. Again, introduce particles at $x = 0$ and $y = 0$ in direction $id = 1$. Repeat steps 3 through 6, except this time, note crossings between the two materials and sample distances using the appropriate μ_k and sample new directions according to the probabilities for the medium in which the interactions occur.

10. Step through some set of discrete thicknesses of material 1, with corresponding thicknesses of material 2, and score the responses.

11. Finally, compare values of the objective function $\Omega^i = \sum_{j=1}^{4} |\mathcal{R}_j - \overline{R}_j^i|$, where $\overline{R}_j^i$ is the response of the jth detector for the ith simulation. The minimum value of Ω^i gives the best estimate of the thicknesses the two materials.

This procedure may seem a little involved, but in fact is easily programmed and implemented.

7.5   General Applications of IMC

The examples given to this point have been somewhat contrived in order to illustrate applications of IMC and SMC. In this section, several actual applications are characterized and reference is made to the published literature, which provides more details about these applications.

Iterative IMC has been used in countless inverse problem applications, some published and many others not publicly shared. For instance, Mosegaard and Tarantola (1995) consider IMC methods applied to geophysics inverse problems. As another example, the paper by Hammer et al. (1995) describes a detailed iterative IMC model for determining properties of ocular fundus tissues from transmittance and reflectance measurements.

The symbolic version of IMC has seen application in several fields. For instance, Floyd et al. (1985) constructed an algorithm for application to emission computed tomography (ECT) and Floyd et al. (1985) applied SMC to single-photon emission computed tomography (SPECT). At the time, the term in use was IMC, but since then the distinction between IMC, which is iterative in the simulations, and SMC, which is noniterative in the simulations, has been made. Several other applications of SMC are briefly discussed below. Each application requires a thorough understanding of the underlying processes and allows the user much freedom in implementation.

Example 7.9 A Simplified Photon Scattering Inverse Problem

Assume a medium for which $t_1 = 3$ and $t_2 = 2$. Given that $\mu_1 = 0.5$ (in appropriate units) and $\mu_2 = 1.0$ (in the same units) and $p_l^1 = 0.25$, $p_b^1 = 0.25$, $p_r^1 = 0.25$, $p_a^1 = 0.25$, $p_l^2 = 0.25$, $p_b^2 = 0.5$, $p_r^2 = 0.25$, and $p_a^2 = 0$, one can run a forward Monte Carlo model to estimate the four responses. For 10^6 histories, the following responses were obtained: $\mathcal{R}_1 = 0.2520 \pm 0.0004$ s^{-1}, $\mathcal{R}_2 = 0.0554 \pm 0.0002$ s^{-1}, $\mathcal{R}_3 = 0.2240 \pm 0.0004$ s^{-1}, and $\mathcal{R}_4 = 0.0551 \pm 0.0002$ s^{-1}. The results of conducting the SMC procedure outlined above, given these data, are shown in Fig. 7.4. It is obvious that Ω is minimized near the correct value of $t_1 = 3$ and thus the SMC procedure can be used to solve inverse problems that are in the form of simulation models and not expressed explicitly as integral equations.

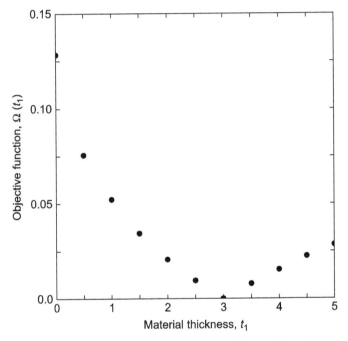

Figure 7.4 The figure of merit is minimized at the correct thickness of material 1, i.e., three units.

7.5.1 Radiative Transfer

Radiative transfer (RT) concerns the transport of electromagnetic radiation through media. Of particular interest is the transport of the sun's radiation through the earth's atmosphere. In his original paper on IMC, Dunn (1981) considered an idealized problem in radiative transfer. Next, Dunn (1983) applied SMC (then called IMC) to three one-dimensional RT problems.[1] Each of the three RT problems assumed that measurements of atmospheric transmitted and/or reflected radiation were known and the single-scatter albedo α (a property of the medium) was to be determined. The single-scatter albedo is a ratio of the photon scattering probability to its total (scattering plus absorption) interaction probability. Thus, the single-scatter albedo is a relative measure of atmospheric scattering and may depend on depth in the medium.

One of the problems Dunn considered involved a finite slab with two regions both with unknown single-scatter albedos α_1 and α_2. Given the total reflectance from the upper irradiated surface and the total transmission through the bottom surface at the earth's surface (values calculated by Shouman and Ozisik (1981)), it was shown that only $N = 40,000$ Monte Carlo histories were sufficient to recover α_1 and α_2 by SMC to within 3% in all cases considered. Subramaniam and Menguc (1991) also used SMC to address similar RT problems.

In short- and intermediate-wavelength RT, media are often characterized by their scattering and absorption properties. Even though a medium is homogeneous, these properties can change with depth into the medium because they are dependent on the wavelength (or alternatively, the energy or frequency) of the radiation. In some media, absorbed energy is re-emitted; thus RT problems can be highly nonlinear. Galtier et al. (2017) combined SMC with the idea of "null collisions," described earlier by Galtier et al. (2013), to construct solutions for scattering and absorption coefficients in media of unknown optical thickness. In their approach, a null collision term κ_n is added to the extinction coefficient κ, so that $\kappa = \kappa_a + \kappa_s + \kappa_n$, where κ_a is the absorption coefficient, κ_s is the scattering coefficient, and κ_n is the null collision coefficient. Null collisions occur within the medium and their MC scores are incremented with terms involving the unknown coefficients. Thus, the unknown problem parameters are carried along in the scores. In this manner, the parameters are contained as symbols in polynomial equations whose solutions are readily obtained by standard numerical methods.

Roger et al. (2018) extended these solutions to RT problems in high-temperature media. Maanane et al. (2020) applied SMC to estimating the RT properties of a heterogeneous semitransparent material at room temperature. They used directional-hemispherical transmittance and reflectance responses as known. Their implementation of SMC is similar to that of Roger et al. (2018) and again results in polynomial equations that can be readily solved for the desired quantities.

---

[1] The simplification of a planetary atmosphere to one dimension may seem extreme. But because the average distance visible light travels in the atmosphere is small compared to the radius of the earth, at any location on the earth's surface the atmosphere is effectively infinite in two dimensions and the sun's radiation may be considered to be monodirectional and uniformly distributed over the top of the atmosphere.

7.5.2 Photon Beam Modifier Design

In medical physics applications, it is sometimes desirable to introduce beam modifiers into radiation beams in order to create beams that produce desired *dose* or energy-absorption distributions in tissue (water) phantoms or patients. One such simple situation arises when a wedge-shaped filter is inserted into the beam to produce isodose curves in a water phantom that has desired characteristics. The filter can be characterized by a single parameter, namely, the wedge angle ω. The purpose of such a filter is to create isodose curves, a portion of which have constant dose along lines that are at an angle θ to a given axis. This situation is depicted in Fig. 7.5. From actual dose measurements in a water phantom produced by two wedge filters (whose wedge angles were known), SMC was applied in order to estimate the wedge angles. The wedge angles were retrieved to within 5% (see Dunn and Womersley (1994)).

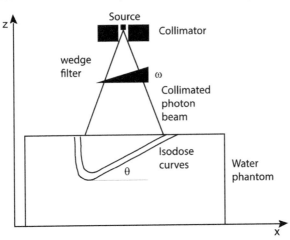

In another case, described by Dunn and Shultis (2009), the wedge filters were divided into $J = 30$ sections and the thicknesses of these sections were determined by solving an SMC model using $K = 60$ detector measurements in a phantom. By fitting a straight line to the 30 thicknesses, results to within 10% were obtained. This is significant because it indicates

Figure 7.5 A collimated source irradiates a water phantom with an intervening wedge filter. The wedge angle ω is desired that produces isodose curves that have an approximately linear behavior at an angle of θ from the horizontal.

that SMC can be used to find estimates of many parameters. One can assume that an SMC solution can be used to start an experimental process that hones in on the wedge angle that produces a desired dose distribution.

7.5.3 Energy-Dispersive X-Ray Fluorescence

X-ray fluorescence is produced when incident radiation removes an inner-shell electron from its position around a nucleus to produce a free electron and an ion. A characteristic X ray is then emitted as the empty electron shell is filled by an electron from an outer shell producing yet another empty shell. This process is repeated as X rays of lower and lower energies are emitted by electrons filling the vacant shells until a neutral atom is achieved.

Energy-dispersive X-ray fluorescence is concerned with X-ray spectroscopy of the emitted X rays. For a sample containing several elements, absorption and enhancement can occur. Absorption occurs when an X ray from element "A" is absorbed by another element and thus the measured spectrum contains fewer characteristic X rays from element A. Enhancement occurs when the absorbing element emits additional X rays due to absorption of some other element's emissions. In a material such as stainless steel, which may contain several elements of similar atomic number, absorption and enhancement can complicate the analysis of the X-ray spectra.

SMC was applied by Yacout and Dunn (1987) to account for secondary X rays, i.e., those emitted from an element that absorbed X rays from element A. Then Michael (1991) extended this SMC model to account for tertiary X rays, i.e., those emitted by element A being absorbed by element B, some of whose X rays are then absorbed by element C, which enhances the characteristic X-ray production from element C.

7.6 Summary

Inverse problems are a generic class of problems for which outcomes are either measured or specified and the values of parameters that could lead to the outcomes are sought. The existence and uniqueness of solutions to inverse problems are usually unknown. As a practical matter, however, inverse problems often are approached by presuming at least one solution exists and then attempting to find one or more solutions. If more than one solution is found, other considerations often must be used to select the most plausible solution. The term IMC is sometimes used to describe the use of Monte Carlo methods to attempt to solve inverse problems.

The most direct form of IMC is the approach of specifying values of the unknown parameters, performing direct MC simulations to estimate the outcomes, comparing the estimated outcomes to the given ones, and iterating until the estimated and given outcomes converge to an acceptable difference. This iterative approach requires that a full direct MC simulation be run for each set of candidate parameter values. Multiple simulations may not be a significant problem if each simulation is simple, but can become problematic if the simulations are complex (requiring long run-times) or highly multidimensional. Also, this "brute-force" approach to IMC requires that some sort of parameter update procedure be implemented. The "art" in using this approach is in how the parameter values are updated from iteration to iteration.

Another form of IMC, called SMC, is motivated by the desire to eliminate the need to perform multiple simulations. If the PDF, from which the Monte Carlo samples are being drawn, is dependent on some or all of the unknown parameters, then one can proceed by sampling from an assumed PDF that is independent of the unknown parameters. The histories must be weighted, in the importance sampling manner, by the ratio of actual to assumed PDFs. The actual PDF may contain the unknown parameters and, thus, the weight factors contain these parameters. Equating the Monte Carlo estimators to the measured or specified outcomes produces a system of equations that, in principle, can be solved for values of the unknown parameter.

In IMC, as in direct MC, it is important to quantify the uncertainties in the estimates obtained. This is less straightforward in IMC than in direct MC, but can be accomplished using an approximate procedure involving the uncertainties in the measured outcomes, the uncertainties in the direct MC estimates, and the sensitivities of the responses to the unknown parameters.

Problems

7.1 It can be shown that, for a particular value of θ, $0 < \theta \le 2$, the definite integral

$$R = \int_0^{\pi/2} \sin^\theta (x)\, dx$$

assumes the value $\mathcal{R} = 0.814183$. Use iterative IMC to estimate the value of θ to two decimal places.

7.2 The definite integral

$$R(y) = \int_0^5 e^{-\theta xy}\, dx$$

assumes the value 1.567 for $y = 2$ and a certain value of θ, $0 < \theta \le 2$. Use iterative IMC to estimate the value of θ to two decimal places.

7.3 Consider the model

$$R(y) = \int_a^b (\theta_1 + \theta_2 xy)\frac{1}{b-a}\, dx.$$

Assume that, for $a = 0$ and $b = 1$, the model gives $R(1) = 2$ and $R(3) = 5$. Use SMC to find θ_1 and θ_2.

7.4 Consider the model

$$R(y) = \int_a^b (\theta_1 + \theta_2 x^2 y)\, dx.$$

Assume that, for $a = 0$ and $b = 5$, the model gives $R(1) = 156.6667$ and $R(4) = 656.6667$. Use SMC to find θ_1 and θ_2.

7.5 Consider the definite integral

$$R = \int_0^\infty e^{-bx}\, dx,$$

which can be rewritten as

$$R = \frac{1}{b} \int_0^\infty b e^{-bx} \, dx.$$

Sample from $f^*(x) = e^{-x}$, $0 \leq x < \infty$, to find the value of the parameter b if you are told that $R = 0.8333$.

7.6 A process involving a stochastic variable, which is exponentially distributed on the interval $[0, \infty)$, is known to have a linear kernel and expected response of the form

$$R(y) = \int_0^\infty (\theta_1 y + \theta_2 x) 4 e^{-4x} \, dx.$$

Assume the values $\mathcal{R}_1 = R(y_1 = 1) = 2.0$ and $\mathcal{R}_2 = R(y_2 = 3) = 10.0$ are given and the parameters θ_1 and θ_2 are unknown. Use SMC to estimate the values of θ_1 and θ_2.

7.7 Use a five-term Taylor series expansion of the exponential function to find the value of b if

$$R(b) = \int_0^\infty x^2 e^{-bx} \, dx = 1.9223.$$

7.8 Generalize the example in Section 7.4 to the case of isotropic scattering in the 2-D world, i.e., probability of scattering through an angle θ is uniform in $[0, 2\pi]$. The total probabilities of scattering in the two media are $p_s^1 = 0.8$ and $p_s^2 = 1.0$ and the absorption probabilities are $p_a^1 = 0.2$ and $P_a^2 = 0.0$.

7.9 Consider Example 7.5, where $a = 0$ and $b = \pi$. (a) Calculate by Monte Carlo the values of $\mathcal{R}_1(y_1 = 0.20)$, $\mathcal{R}_2(y_2 = 0.40)$, $\mathcal{R}_3(y_3 = 0.60)$, and $\mathcal{R}_4 = \overline{R}(y_4 = 0.80)$. (b) Use these values as known and determine the values of f_1, f_2, f_3, and f_4 by SMC. (c) Plot the quadratic PDF and the values you obtained from part (b), as in Fig. 7.2.

References

Bevington, P.R., 1969. Data Reduction and Error Analysis for the Physical Sciences. McGraw-Hill Book Company, New York, NY.

Dunn, W.L., 1981. Inverse Monte Carlo analysis. J. Comput. Phys. 41 (1), 154–166.

Dunn, W.L., 1983. Inverse Monte Carlo solutions for radiative transfer in inhomogeneous media. J. Quant. Spectrosc. Radiat. Transf. 29 (1), 19–26.

Dunn, W.L., 2006. The inverse Monte Carlo formalism for solving inverse radiation problems. Trans. Am. Nucl. Soc. 45, 532–533.

Dunn, W.L., O'Foghludha, F., 1982. Feasibility study for design of photon beam modifiers using Monte Carlo methods. In: Proc. World Congress on Medical Physics and Biomedical Engineering. Hamburg.

Dunn, W.L., Shultis, J.K., 2009. Monte Carlo methods for design and analysis of radiation detectors. Radiat. Phys. Chem. 78, 852–858.

Dunn, W.L., Womersley, W.J., 1994. Solutions of inverse and optimization problems in high energy and nuclear physics using inverse Monte Carlo. In: Dragovitsch, P., Linn, S.L., Burbank, M. (Eds.), MC93 International Conference Monte Carlo Simulation in High Energy and Nuclear Physics. World Scientific, Singapore, pp. 78–87.

Floyd, C.E., Jaszczak, R.J., Coleman, R.E., 1985. Inverse Monte Carlo: a unified reconstruction algorithm for SPECT. IEEE Trans. Nucl. Sci. NS-32, 779–785.

Floyd, C.E., Jaszczak, R.J., Greer, K.L., Coleman, R.E., 1986. Inverse Monte Carlo as a unified reconstruction algorithm for ECT. J. Nucl. Med. 27 (10), 1577–1585.

Galtier, M., Blanco, S., Caliot, C., Coustet, C., Dauchet, J., et al., 2013. Integral formulation of null-collision Monte Carlo algorithms. J. Quant. Spectrosc. Radiat. Transf. 125, 57–68.

Galtier, M., Roger, M., Andre, F., Delmas, A., 2017. A symbolic approach for the identification of radiative properties. J. Quant. Spectrosc. Radiat. Transf. 196, 130–141.

Hammer, M., Roggan, A., Muller, G., 1995. Optical properties of ocular fundus tissues—an in vitro study using the double-integrating-sphere technique and inverse Monte Carlo simulation. Phys. Med. Biol. 40, 963–978.

Maanane, Y., Roger, M., Dalmas, A., Galtier, M., André, F., 2020. Symbolic Monte Carlo method applied to the identification of radiative properties of a heterogeneous material. J. Quant. Spectrosc. Radiat. Transf. 249.

Michael, M., 1991. A complete inverse Monte Carlo model for energy-dispersive X-ray fluorescence analysis. Nucl. Instrum. Methods Phys. Res., Sect. A 301 (3), 523–542.

Mosegaard, K., Sambridge, M., 2002. Monte Carlo analysis of inverse problems. Inverse Probl. 18, R29–R54.

Mosegaard, K., Tarantola, A., 1995. Monte Carlo sampling of solutions to inverse problems. J. Geophys. Res. 100 (B7), 12,431–12,447.

Press, W.H., Teukolsky, S.A., Vetterling, W.T., Flannery, B.P., 1996. Numerical Recipes in Fortran 77, 2nd ed. Cambridge University Press, New York, NY.

Roger, M., Galtier, M., André, F., Delmas, A., 2018. Symbolic Monte Carlo methods: an analysis tool for the experimental identification of radiative properties at high temperature. In: European Seminar 110, Computational Radiation in Participating Media VI. April 11–13, 2018, Cascals, Portugal.

Shouman, S.M., Özişik, N., 1981. Radiative transfer in an isotropically scattering two-region slab with reflecting boundaries. J. Quant. Spectrosc. Radiat. Transf. 26 (1), 1–9..

Subramaniam, S., Menguc, M.P., 1991. Solution of the inverse radiation problem for inhomogeneous and anisotropically scattering media using a Monte Carlo technique. Int. J. Heat Mass Transf. 34 (1), 253–266.

Yacout, A.M., Dunn, W.L., 1987. Application of the inverse Monte Carlo method to energy-dispersive X-ray fluorescence. Adv. X-ray Anal. 30, 113–120.

Chapter 8

Linear Operator Equations

The equations gave Susan great fears,
and solving them drove her to tears.
But try as she might
the answers weren't right
until she tried dancing on spheres.

One interesting application of Monte Carlo methods is to solve numerically mathematical equations of different types. Such equations include linear algebraic equations, integral equations, ordinary differential equations (ODEs), and partial differential equations (PDEs) that describe boundary value or initial value problems. It should be noted, however, that many of these equations usually can be solved more efficiently and accurately using standard numerical analysis techniques than they can with Monte Carlo calculations. Nevertheless, there are situations where only approximate values of the solution are needed or where standard methods perform badly, such as when the problem is very large or very intricate. For example, multidimensional integral equations or boundary value problems with complicated boundaries are often better treated using Monte Carlo methods. Also Monte Carlo approaches are usually much easier to code than are the more accurate but more complex state-of-the-art numerical algorithms.

In this chapter, some basic Monte Carlo methods are introduced for solving different types of mathematical equations. All of these equations involve a *linear operator*. Appendix D presents the different types of linear operators most frequently encountered and the important properties that all of these linear operators have in common. Here, no attempt is made to be comprehensive in the application of Monte Carlo; rather, the methods discussed are chosen to illustrate the power of Monte Carlo techniques and the ease with which these methods are implemented. For more complex methods and for details and extensions of the methods presented here, the reader is referred to Chandrasekhar (1943), Brown (1956), Hammersley and Handscomb (1964), Rubinstein (1981), Sabelfeld (1991), and Mikhailov (1995).

8.1 Linear Algebraic Equations

Although many sets of linear algebraic equations can be solved using Monte Carlo methods, such equations are almost always more effectively solved using standard deterministic techniques of numerical analysis. This "conventional wisdom" arises from early studies, such as those by Curtiss (1956), all made with serial computers. However, with modern massively parallel computers or with distributed cloud and network computing, for which Monte Carlo methods are well-suited, this wisdom needs to be re-examined. More on the practical utility of Monte Carlo linear equation solvers is discussed later. Also, Monte Carlo methods for linear algebraic equations are discussed here because analogous methods can be applied to integral equations, for which standard numerical methods often are not as effective, particularly for highly multidimensional integral equations.

Consider the set of n linear algebraic equations $\mathbf{A}\mathbf{x} = \mathbf{f}$ or, alternatively,

$$\mathbf{x} = \mathbf{f} + \mathbf{B}\mathbf{x}, \tag{8.1}$$

where the $n \times n$ matrix $\mathbf{B} \equiv \mathbf{I} - \mathbf{A}$ and has elements b_{ij}. Equations of this type often can be solved by the following iteration scheme:

$$\mathbf{x}^{(k+1)} = \mathbf{f} + \mathbf{B}\mathbf{x}^{(k)}, \quad k = 0, 1, 2, \ldots \quad \text{with} \quad \mathbf{x}^{(0)} = \mathbf{f}, \tag{8.2}$$

$$= (\mathbf{I} + \mathbf{B} + \mathbf{B}^2 + \cdots + \mathbf{B}^k)\mathbf{f}. \tag{8.3}$$

If the sequence $\mathbf{x}^{(1)}, \mathbf{x}^{(2)}, \mathbf{x}^{(3)}, \ldots$ converges, then the converged vector $\mathbf{x}$ satisfies Eq. (8.2), which is equivalent to Eq. (8.1), so that the converged $\mathbf{x}$ is the desired solution. It can be shown (Wachspress, 1966) that a necessary condition for convergence is that the spectral radius $\mu(\mathbf{B})$ of the matrix $\mathbf{B}$ be less than unity, i.e.,

$$\mu(\mathbf{B}) \equiv \max_i |\lambda_i(\mathbf{B})| < 1, \tag{8.4}$$

where $\lambda_i(\mathbf{B})$ are the eigenvalues of $\mathbf{B}$. The smaller the spectral radius, the more rapid is the convergence. The so-called L_∞ norm

$$||\mathbf{B}|| \equiv \max_i \sum_{j=1}^{n} |b_{i,j}| \tag{8.5}$$

gives an upper bound on $\mu(\mathbf{B})$, namely, $\mu(\mathbf{B}) \leq ||\mathbf{B}||$.

From Eq. (8.3) the ith component of $\mathbf{x}^{(k+1)}$ is

$$x_i^{(k+1)} = f_i + \sum_{j_1} b_{j_0 j_1} f_{j_1} + \sum_{j_1 j_2} b_{i j_1} b_{j_1 j_2} f_{j_2} + \ldots + \sum_{j_1 j_2 \cdots j_k} b_{i j_1} b_{j_1 j_2} \ldots b_{j_{k-1} j_k} f_{j_k}$$

$$= \sum_{m=0}^{k} \sum_{j_1=1}^{n} \sum_{j_2=1}^{n} \cdots \sum_{j_m=1}^{n} b_{i j_1} b_{j_1 j_2} \cdots b_{j_{m-1} j_m} f_{j_m}. \tag{8.6}$$

Often it is not the solution $\mathbf{x}$ that is of interest but, rather, some inner product of $\mathbf{x}$, namely,

$$H \equiv (\mathbf{h}, \mathbf{x}) \equiv \sum_{i=1}^{n} h_i x_i, \tag{8.7}$$

where $\mathbf{h}$ is a specified weighting vector. In particular, if $h_i = \delta_{ij}$, then $(\mathbf{h}, \mathbf{x}) = x_j$, the solution for the jth component of $\mathbf{x}$.

8.1.1 Solution of Linear Equations by Random Walks

To solve Eq. (8.1) by Monte Carlo techniques, an $n \times n$ *transition* matrix $\mathbf{P}$ and an n-component *starting* vector $\mathbf{p}$ are constructed whose nonnegative elements have the properties (Rubinstein, 1981)

$$\sum_{j=1}^{n} P_{ij} = 1, \quad i = 1, \dots, n, \quad \text{and} \quad \sum_{i=1}^{n} p_i = 1. \tag{8.8}$$

Moreover $p_i > 0$ if $f_i \neq 0$ or $p_i = 0$ if $f_i = 0$, and $P_{ij} > 0$ if $b_{ij} \neq 0$ or $P_{ij} = 0$ if $b_{ij} = 0$. The $\mathbf{P}$ and $\mathbf{p}$ can be chosen more or less arbitrarily (subject to the above constraints), although better results are obtained if the elements are approximately proportional to the magnitudes of the elements in $\mathbf{B}$ and $\mathbf{f}$, respectively. One choice for the transition matrix is

$$P_{ij} = |b_{ij}| \bigg/ \sum_{m=1}^{n} |b_{im}|, \tag{8.9}$$

which leads to what is called *almost optimal* Monte Carlo (Dimov et al., 1998, 2008).

With these auxiliary quantities, a series of *random walks* (RWs) or Markov chains is made through a set of states consisting of the integers 1 through n. Here p_i determines the probability of starting a random walk at state i and P_{ij} determines the probability, at each step in the random walk, of moving from state i to state j. An integer k is initially chosen, corresponding to the value of k in Eq. (8.2). Thus each random walk consists of a sequence of $k + 1$ states $i_0 \to i_1 \to i_2 \to \dots \to i_k$, where i_j is a natural (integer) number in $[1, n]$.

At each step in a random walk, the quantity

$$U_m = \frac{b_{i_0 i_1} b_{i_1 i_2} \cdots b_{i_{m-1} i_m}}{P_{i_0 i_1} P_{i_1 i_2} \cdots P_{i_{m-1} i_m}} \tag{8.10}$$

is accumulated, which can be constructed recursively during the random walk as

$$U_m = U_{m-1} \frac{b_{i_{m-1} i_m}}{P_{i_{m-1} i_m}}, \quad m = 1, 2, \dots, k \quad \text{with} \quad U_0 = 1. \tag{8.11}$$

Also at each step in the random walk, the sum $\sum_m U_m f_m$ is constructed, so that, at the end of the sth random walk, a value of the random variable

$$Z_k^{(s)}(\mathbf{h}) = \frac{h_{i_0}}{p_{i_0}} \sum_{m=0}^{k} U_m f_{i_m} \tag{8.12}$$

is obtained. A large number N of random walks is performed, each producing a value of $Z_k^{(s)}(\mathbf{h})$, $s = 1, \ldots, N$. The desired inner product is then estimated as

$$(\mathbf{h}, \mathbf{x}^{(k+1)}) \simeq \frac{1}{N} \sum_{s=1}^{N} Z_k^{(s)}(\mathbf{h}). \tag{8.13}$$

Proof of Eq. (8.13)

The path $i_0 \rightarrow i_1 \rightarrow i_2 \rightarrow \ldots \rightarrow i_k$ has a probability of being realized of $p_{i_0} P_{i_0 i_1} P_{i_1 i_2} \ldots P_{i_{k-1} i_k}$. Thus, the expected value of the random variable $Z_k(\mathbf{h})$ is

$$\langle Z_k(\mathbf{h}) \rangle = \sum_{i_0=1}^{n} \sum_{i_1=1}^{n} \ldots \sum_{i_k=1}^{n} Z_k(\mathbf{h}) p_{i_0} P_{i_0 i_1} P_{i_1 i_2} \ldots P_{i_{k-1} i_k}. \tag{8.14}$$

Substitution of Eqs. (8.12) and (8.10) into this expression yields

$$\langle Z_k(\mathbf{h}) \rangle = \sum_{i_0=1}^{n} \ldots \sum_{i_k=1}^{n} h_{i_0} \sum_{m=0}^{k} \frac{b_{i_0 i_1} b_{i_1 i_2} \cdots b_{i_{m-1} i_m}}{P_{i_0 i_1} P_{i_1 i_2} \cdots P_{i_{m-1} i_m}} f_{i_m} P_{i_0 i_1} P_{i_1 i_2} \ldots P_{i_{k-1} i_k}$$

$$= \sum_{i_0=1}^{n} \ldots \sum_{i_k=1}^{n} h_{i_0} \sum_{m=0}^{k} b_{i_0 i_1} b_{i_1 i_2} \cdots b_{i_{m-1} i_m} f_{i_m} P_{i_m i_{m+1}} \cdots P_{i_{k-1} i_k}. \tag{8.15}$$

Now, use the fact that $\sum_{j=1}^{n} P_{ij} = 1$ to reduce Eq. (8.15) as follows:

$$\langle Z_k(\mathbf{h}) \rangle = \sum_{i_0=1}^{n} h_{i_0} f_{i_0} + \sum_{i_0=1}^{n} \sum_{i_1=1}^{n} h_{i_0} b_{i_0 i_1} f_{i_1} + \cdots + \sum_{i_0=1}^{n} \ldots \sum_{i_k=1}^{n} h_{i_0} b_{i_0 i_1} \cdots b_{i_{k-1} i_k} f_{i_k}$$

$$= \sum_{m=0}^{k} \sum_{i_0=1}^{n} \ldots \sum_{i_m=1}^{n} h_{i_0} b_{i_0 i_1} b_{i_1 i_2} \cdots b_{i_{m-1} i_m} f_{i_m}$$

$$= \sum_{m=0}^{k} \sum_{i_0=1}^{n} h_{i_0} [\mathbf{B}^m \mathbf{f}]_{i_0} = \left(\mathbf{h}, \sum_{m=0}^{k} \mathbf{B}^m \mathbf{f} \right) = (\mathbf{h}, \mathbf{x}^{(k+1)}), \tag{8.16}$$

where the last line follows from Eq. (8.6). Thus, Eq. (8.13) is proven.

Special Cases

The limit of Eq. (8.16) then gives

$$\lim_{k \to \infty} \langle Z_k(\mathbf{h}) \rangle = \lim_{k \to \infty} (\mathbf{h}, \mathbf{x}^{(k+1)}) = (\mathbf{h}, \mathbf{x}) = H. \tag{8.17}$$

Hence, *provided* the sequence $\mathbf{x}^{(1)}, \mathbf{x}^{(2)}, \mathbf{x}^{(3)}, \ldots$ converges and the random walk path $i_0 \to i_1 \to \cdots \to i_k \to \cdots$ is infinitely long, an unbiased estimator of $(\mathbf{h}, \mathbf{x})$ is obtained. In practice, good estimates of $(\mathbf{h}, \mathbf{x})$ can be obtained from Eq. (8.13) for small values of k. The faster the iterative solution converges, i.e., the smaller the spectral radius $\mu(\mathbf{B})$, the smaller can be k.

Three features of the above Monte Carlo procedure for evaluation of $(\mathbf{h}, \mathbf{x}^{(k+1)})$ should be noted. The same random paths can be used to evaluate this inner product for different values of $\mathbf{f}$. Also, if the $\mathbf{B}$ and $\mathbf{P}$ matrices are such that $\mathbf{B} = \alpha\mathbf{P}$, where $0 < \alpha < 1$, then Eq. (8.11) reduces to $U_m = \alpha^m$ so that

$$Z_k(\mathbf{h}) = \frac{h_{i_0}}{p_{i_0}} \sum_{m=0}^{k} \alpha^m f_{i_m}. \tag{8.18}$$

Finally, if $h_i = \delta_{ij}$ (the Kronecker delta function), the random walks always start at $i_0 = j$ with $h_{i_0}/p_{i_0} = 1$ so that

$$\lim_{k \to \infty} \langle Z_k(\mathbf{h}) \rangle = \lim_{k \to \infty} x_j^{(k+1)} = x_j, \tag{8.19}$$

the jth component of $\mathbf{x}$. An example of the above method is shown in Example 8.1.

Example 8.1 Solution of Linear Algebraic Equations x = f + Bx

As an example of the Monte Carlo method described in Section 8.1.1, consider the 4×4 set of linear algebraic equations

$$\begin{pmatrix} x_1 \\ x_2 \\ x_3 \\ x_4 \end{pmatrix} = \frac{1}{10} \begin{pmatrix} -7 \\ 18 \\ 33 \\ 28 \end{pmatrix} + \frac{1}{10} \begin{pmatrix} 4 & 1 & 1 & 2 \\ 1 & -5 & 1 & 2 \\ 1 & 2 & -4 & 1 \\ 1 & 2 & 1 & 1 \end{pmatrix} \begin{pmatrix} x_1 \\ x_2 \\ x_3 \\ x_4 \end{pmatrix}.$$

The exact solution of this equation is $\mathbf{x} = (1\ 2\ 3\ 4)$ and for $\mathbf{h} = (1\ 1\ 1\ 1)$, the exact value of $(\mathbf{h}, \mathbf{x}) = 10$. The spectral radius for the above matrix is $\mu(\mathbf{B}) = 0.616\,375\,428\,606\ldots$ and $||\mathbf{B}|| = 0.9$. Hence there is a transition matrix $\mathbf{P}$ for which the Neumann-series Monte Carlo algorithm converges. For the above Monte Carlo method, the starting probability vector was taken as $\boldsymbol{p} = (.25\ .25\ .25\ .25)$ and the transition matrix **P** was chosen as

$$P = \begin{pmatrix} 0.55 & 0.15 & 0.15 & 0.15 \\ 0.15 & 0.55 & 0.15 & 0.15 \\ 0.20 & 0.20 & 0.40 & 0.20 \\ 0.25 & 0.25 & 0.25 & 0.25 \end{pmatrix}.$$

The results of the Monte Carlo solution are shown in Fig. 8.1. The left-hand figure shows how the error in $(\mathbf{h}, \mathbf{x}^{(k)})$ decreases as the number of histories N increase for the case of 10 steps in the random walk. As seen from the right-hand figure, the maximum number of steps k used to terminate each random walk for this example should be about 10 or greater to avoid errors from random walks that are too short. However, using very long random walks affords little, if any, increase in accuracy while wasting computing time.

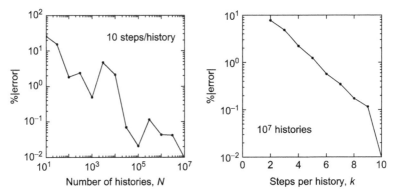

Figure 8.1 The percent error in $(\mathbf{h}, \mathbf{x})$ as a function of the number of histories N (left) and for the number of steps k in each history (right).

Convergence of the Random Walk Algorithm

As mentioned earlier, the necessary condition for the Neumann series $(\mathbf{I} + \mathbf{B} + \mathbf{B}^2 + \cdots + \mathbf{B}^k)\mathbf{f}$ to converge to $(\mathbf{I} - \mathbf{B})^{-1}\mathbf{f}$, as $k \to \infty$, is that the spectral radius $\mu(\mathbf{B}) < 1$. However, this condition is not sufficient for Eq. (8.13) to converge to H as $k \to \infty$. The convergence of the RW algorithm depends not only on **B** but also on the transition matrix **P** used. Ji et al. (2013) have studied the convergence problem and found the following.

1. If $||\mathbf{B}|| < 1$ (so then $\mu(\mathbf{B}) < 1$), there are easy-to-find transition matrices **P** with which the Monte Carlo linear solver is guaranteed to converge. However, there are some **P** matrices that result in divergence.
2. If $||\mathbf{B}|| \geq 1$ and $\mu(\mathbf{B}) < 1$, there are transition matrices that result in convergence provided $\mu(\mathbf{B}^+) < 1$, where $b_{ij}^+ = |b_{ij}|$. However, finding transition matrices that produce convergence is generally difficult.
3. All other cases, even though $\mu(\mathbf{B}) < 1$, produce divergence.

4. The *necessary and sufficient* condition for convergence of the Monte Carlo linear solver based on a Neumann series is that $\mu(\mathbf{B}^*) < 1$, where $b_{ij}^* = b_{ij}^2/Pij$.

8.1.2 Solving the Adjoint Linear Equations by Random Walks

Associated with the set of linear equations of Eq. (8.1) is the set of equations

$$\mathbf{x}^\dagger = \mathbf{h} + \mathbf{B}^T\mathbf{x}^\dagger, \tag{8.20}$$

where $\mathbf{B}^T$ is the transpose or *adjoint* of $\mathbf{B}$. Again this adjoint set of equations can be solved by an iterative solution

$$\mathbf{x}^{\dagger(k+1)} = \mathbf{h} + \mathbf{B}^T\mathbf{x}^{\dagger(k)}, \quad k = 1, 2, \ldots \quad \text{with } \mathbf{x}^{\dagger(0)} = 0. \tag{8.21}$$

Because $\mathbf{B}$ and $\mathbf{B}^T$ have the same eigenvalues, $\mu(\mathbf{B})$ equals $\mu(\mathbf{B}^T)$ and this iterative solution converges whenever that of Eq. (8.2) converges.

Multiplication of Eq. (8.1) by the transpose of $\mathbf{x}^\dagger$ yields[1]

$$(\mathbf{x}^\dagger, \mathbf{x}) = (\mathbf{x}^\dagger, \mathbf{f}) + (\mathbf{x}^\dagger, \mathbf{B}\mathbf{x}). \tag{8.22}$$

Similarly, from Eq. (8.20), it is found that

$$(\mathbf{x}, \mathbf{x}^\dagger) = (\mathbf{x}, \mathbf{h}) + (\mathbf{x}, \mathbf{B}^T\mathbf{x}^\dagger). \tag{8.23}$$

It is readily shown that $(\mathbf{x}, \mathbf{x}^\dagger) = (\mathbf{x}^\dagger, \mathbf{x})$ and that $(\mathbf{x}^\dagger, \mathbf{B}\mathbf{x}) = (\mathbf{x}, \mathbf{B}^T\mathbf{x}^\dagger)$. Subtraction of Eq. (8.22) from Eq. (8.23) then yields

$$(\mathbf{h}, \mathbf{x}) = (\mathbf{x}^\dagger, \mathbf{f}). \tag{8.24}$$

Thus, by solving Eq. (8.20) for $\mathbf{x}^\dagger$ by a random walk algorithm, another independent estimator of the inner product $(\mathbf{h}, \mathbf{x})$ can be obtained.

To find $(\mathbf{x}^\dagger, \mathbf{f})$ using Markov chains, a state-transition matrix $\mathbf{P}^\dagger$ and initial state probability vector $\mathbf{p}^\dagger$ are defined such that (Rubinstein, 1981)

$$\sum_{j=1}^{n} P_{ij}^\dagger = 1, \quad i = 1, \ldots, n, \quad \text{and} \quad \sum_{i=1}^{n} p_i^\dagger = 1, \tag{8.25}$$

where $P_{ij}^\dagger > 0$ if $b_{ij}^T = b_{ji} \neq 0$ or $P_{ij}^\dagger = 0$ if $b_{ij}^T = b_{ji} = 0$, and $p_i^\dagger > 0$ if $h_i \neq 0$ or $p_i^\dagger = 0$ if $h_i = 0$. Then, to estimate $(\mathbf{x}^\dagger, \mathbf{f})$, first select an integer k (again

---

[1] The *inner product* of two n-component real vectors $\mathbf{x}$ and $\mathbf{y}$ is defined as

$$(\mathbf{x}, \mathbf{y}) \equiv \mathbf{x}^T \bullet \mathbf{y} = \sum_{i=1}^{n} x_i y_i.$$

corresponding to the k in Eq. (8.3)) and simulate a large number N of random walks, each of k steps. The initial state i_0 is selected with probability $p_{i_0}^\dagger$. For each random walk, calculate the random variable

$$Z_k^\dagger(\mathbf{f}) = \frac{f_{i_0}}{p_{i_0}^\dagger} \sum_{m=0}^{k} U_m^\dagger h_{i_m}, \tag{8.26}$$

where $U_m^\dagger$ is accumulated at each step of the random walk according to

$$U_m^\dagger = U_{m-1}^\dagger \frac{b_{i_{m-1}i_m}^T}{P_{i_{m-1}i_m}^\dagger} = U_{m-1}^\dagger \frac{b_{i_m i_{m-1}}}{P_{i_{m-1}i_m}^\dagger}, \quad \text{with} \quad U_0^\dagger = 1. \tag{8.27}$$

The inner product $(\mathbf{x}^{\dagger(k+1)}, \mathbf{f})$ then is estimated as

$$\boxed{(\mathbf{x}^{\dagger(k+1)}, \mathbf{f}) \simeq \frac{1}{N} \sum_{s=1}^{N} Z_k^{\dagger(s)}(\mathbf{f}),} \tag{8.28}$$

where $Z_k^{\dagger(s)}(\mathbf{f})$ is the value of Eq. (8.26) for random walk s. Finally, as $k \to \infty$,

$$\lim_{k\to\infty} \langle Z_k^\dagger(\mathbf{f}) \rangle = \lim_{k\to\infty} (\mathbf{f}, \mathbf{x}^{\dagger(k+1)}) = (\mathbf{f}, \mathbf{x}^\dagger) = (\mathbf{h}, \mathbf{x}) = H. \tag{8.29}$$

Note that the same random walk can be used to evaluate different $\mathbf{h}$ vectors by simply changing h_{i_m} in Eq. (8.26). In particular, if $h_i = \delta_{ij}$, $i = 1, 2, \ldots n$, then

$$x_j^{(k+1)} = \langle Z_k^\dagger(\mathbf{f}) \rangle = \langle (f_{i_0}/p_{i_0}^\dagger) \sum_{m=1}^{k} U_m^\dagger \delta_{i_m j} \rangle, \quad j = 1, 2, \ldots, n. \tag{8.30}$$

Thus, the adjoint method is more suitable for finding the general shape of $\mathbf{x}$ because the same random walk can be used to build estimates of all x_j, whereas the direct method of Section 8.1.1 is best used for a single $\mathbf{f}$ vector.

8.1.3 Solution of Linear Equations by Finite Random Walks

A difficulty with the previous two methods for solving linear algebraic equations is that the Markov chain is stopped after k steps in the random walk. The value of k is chosen rather arbitrarily and one hopes that the result for $(\mathbf{h}, \mathbf{x}^{(k+1)}) \simeq (\mathbf{h}, \mathbf{x})$. To verify this, estimates based on different values of k need to be performed to ensure convergence has been achieved.

There is another approach that avoids this difficulty and produces estimates of $(\mathbf{h}, \mathbf{x})$ directly. In fact this method, which introduces *absorption states* into the random walk, predates the previous methods. This alternative method was first proposed by Neumann and Ulam in the 1940s and was later published and expanded by Forsythe and Liebler (1950).

As before, a vector p with $\sum_{i=1}^{n} p_i = 1$ is used to define the starting state of each random walk. But in this method, the state transition matrix $\mathbf{P}$, which has elements $P_{ij} \geq 0$ such that $P_{ij} > 0$ if $b_{ij} \neq 0$, has rows that do not always sum to unity. The *kill vector*, which has components

$$g_i^{kill} \equiv 1 - \sum_{j=1}^{n} P_{ij} \geq 0, \quad i = 1, \ldots, n, \tag{8.31}$$

is used to terminate each random walk. A random walk is begun at state i_0 with probability p_{i_0}. At each step in the walk, the particle goes from state i_{j-1} to state i_j with probability P_{ij} *or* the walk is terminated with probability $g_{i_{j-1}}^{kill}$. Thus, each random walk is finite in length, $i_0 \to i_1 \to \ldots \to i_k$, where the number of steps k may be different for each random walk. For each random walk calculate the random variable

$$Z_k(\mathbf{h}) = \frac{h_{i_0}}{p_{i_0}} U_k \frac{f_{i_k}}{g_{i_k}^{kill}}, \tag{8.32}$$

where U_k is given by Eq. (8.10). The inner product $(\mathbf{h}, \mathbf{x})$ is then estimated by averaging Z_k over N random walks, i.e.,

$$(\mathbf{h}, \mathbf{x}) \simeq \frac{1}{N} \sum_{s=1}^{N} Z_k^{(s)}(\mathbf{h}). \tag{8.33}$$

Proof of Eq. (8.33)

The probability of obtaining the path $i_0 \to i_1 \to \ldots \to i_k$ is $p_{i_0} P_{i_0 i_1} P_{i_1 i_2} \cdots P_{i_{k-1} i_k} g_{i_k}^{kill}$. Hence, the expected value of $Z_k(\mathbf{h})$ is

$$\langle Z_k(\mathbf{h}) \rangle = \sum_{k=0}^{\infty} \sum_{i_0=1}^{n} \cdots \sum_{i_k=1}^{n} Z_k(\mathbf{h}) p_{i_0} P_{i_0 i_1} P_{i_1 i_2} \cdots P_{i_{k-1} i_k} g_{i_k}^{kill}. \tag{8.34}$$

Substitution of Eqs. (8.32) and (8.10) into this result then gives

$$\langle Z_k(\mathbf{h}) \rangle = \sum_{k=0}^{\infty} \sum_{i_0=1}^{n} \cdots \sum_{i_k=1}^{n} h_{i_0} b_{i_0 i_1} b_{i_1 i_2} \ldots b_{i_{k-1} i_k} f_{i_k}$$

$$= \sum_{i_0=1}^{n} h_{i_0} [f_{i_0} + (\mathbf{Bf})_{i_0} + (\mathbf{B}^2 \mathbf{f})_{i_0} + (\mathbf{B}^3 \mathbf{f})_{i_0} + \cdots]$$

$$= \sum_{i_0=1}^{n} h_{i_0} x_{i_0} = (\mathbf{h}, \mathbf{x}), \qquad (8.35)$$

where the last line follows from Eq. (8.3) and, thus, Eq. (8.33) is justified.

8.1.4 Recent Interest in Monte Carlo Linear Equation Solvers

Because modern deterministic linear equation solvers (Press et al., 1996) are generally more efficient and accurate for solving $\mathbf{Ax} = \mathbf{f}$ for $\mathbf{x}$, Monte Carlo solvers have historically been mostly neglected and were primarily of academic interest. Only if the matrix size was very large ($\gtrsim 10^4$), very sparse, and only if a rough or approximate solution was required would Monte Carlo methods have been considered. Similarly, if $(\mathbf{h}, \mathbf{x})$ was to be estimated for many different $\mathbf{h}$ and $\mathbf{f}$, Monte Carlo methods might have been used because the same random walks could be used to obtain estimates of the different inner products.

However, recent emphasis on "big data" problems and the emergence of new computer paradigms such as network and cloud computing, massively parallel machines, and general-purpose graphics processing units have renewed interest in Monte Carlo methods. Big data, characterized by a huge volume and rapid growth as more data are accumulated, have turned to Monte Carlo sampling methods for effective handling of various operations involving enormous matrices, matrices so large they cannot even be stored in any computer's memory. Matrices with 10^{200} elements and greater have been proposed. Simultaneously, Monte Carlo linear equation solvers have also attracted renewed interest because of their unique advantages over deterministic solvers for treating very large matrices (Ji et al., 2013). First, because Monte Carlo solvers are based on sampling, all the elements of $\mathbf{A}$ need not be accessed during an RW. This feature is attractive for large-scale sensor networks for which obtaining every element, while possible, is costly or impractical. Likewise, this feature is very useful when some data are missing or imperfect. Second, the RWs can be performed independently in a distributed manner using parallel platforms such as network and cloud computing or massively parallel machines. Third, Monte Carlo linear solvers can very quickly obtain low-accuracy approximate solutions, which often are sufficient for many big data applications, or the rough solutions can be refined to obtain higher accuracy for particular components of the solution vector. Fourth, the algorithms for Monte Carlo linear solvers have modest memory requirements and scale well with the size of the matrix. Finally, with adjoint Monte Carlo linear solvers, computational effort can be greatly reduced by solving for only the components of the solution vector that are of interest.

8.2 Linear Integral Equations

In contrast to the limited utility of Monte Carlo methods for linear algebraic equations, Monte Carlo is often the best method for numerically solving linear integral equations, particularly if the integrals are multidimensional. Moreover, many important differential equations can be formally converted into integral equations, so, in principle, Monte Carlo again can be used. In fact, one of the first applications of Monte Carlo analysis was to solve the integral form of the Boltzmann equation for neutral particle transport (see Chapter 10).

8.2.1 Frequently Encountered Integral Equations

A general integral equation for an unknown function $y(x)$ can be written as

$$g(x)y(x) = f(x) + \int_a^b K(x, x')y(x')\,dx', \tag{8.36}$$

where $f(x)$ and $g(x)$ are specified functions and $K(x, x')$ is a specified *kernel*. If $g(x) = 0$, Eq. (8.36) is called a Fredholm equation of the *first kind*; otherwise it is a Fredholm equation[2] of the *second kind*. The equation is *homogeneous* if $f(x) = 0$. The integral $\int_a^b$ can be cast into a standard form $\int_0^1$ by a simple change of variables $\tilde{x} = (x - a)/(b - a)$ and $\tilde{x}' = (x' - a)/(b - a)$. Closely related to Fredholm integral equations are Volterra integral equations[3] of the first and second kind given, respectively, by

$$f(x) = \int_a^x K(x, x')y(x')\,dx' \quad \text{and} \quad y(x) = f(x) + \int_a^x K(x, x')y(x')\,dx'. \tag{8.37}$$

These equations are a special case of Fredholm equations with the kernel defined on the rectangle $a \le x \le b$, $a \le x' \le b$ and vanishing in the triangle $a \le x < x' \le b$. If f and K are differentiable, then a Volterra equation of the first kind is equivalent to one of the second kind. To see this, differentiate the first-kind equation (see the footnote on page 308) and rearrange the result to obtain

$$y(x) = \frac{f'(x)}{K(x, x)} - \int_a^x \frac{K_x(x, x')}{K(x, x)} y(x')\,dx'. \tag{8.38}$$

These one-dimensional equations can also be extended to multidimensional situations, so that, for example, the Fredholm equation of the second kind becomes

$$y(\mathbf{x}) = f(\mathbf{x}) + \int_V K(\mathbf{x}, \mathbf{x}')y(\mathbf{x}')\,d\mathbf{x}', \quad \mathbf{x} \in V. \tag{8.39}$$

In this generalization from one dimension, some properties of the one-dimensional case are lost. For example, Volterra equations of the first kind

---

[2] Named after the Swedish mathematician Erik Ivar Fredholm (1866–1927), whose research on the theory of integral operators foreshadowed the theory of Hilbert spaces.
[3] Named after the Italian mathematician Vito Volterra (1860–1940) known for his research in mathematical biology and integral equations and for being one of the founders of functional analysis.

generally cannot be reduced to equations of the second kind. Finally there are nonlinear forms of these integral equations in which the kernel is also a function of the dependent variable, i.e., $K(x, x', y)$. In these cases Monte Carlo methods are particularly useful (Vajargah and Moradi, 2007).

8.2.2 Approximating Integral Equations by Algebraic Equations

Equation (8.36) (with $g(x) = 1$) can be approximated by a set of linear algebraic equations in several ways. Application of some M-point deterministic quadrature formula immediately gives

$$y(x_i) = f(x_i) + \sum_{j=1}^{M} w_j K(x_i, x_j) y(x_j), \quad i = 1, \ldots, M, \tag{8.40}$$

where x_j and w_j are the quadrature abscissas and weights, respectively. Equations (8.40) are M linear algebraic equations that can be written in the standard form $\mathbf{y} = \mathbf{f} + \mathbf{A}\mathbf{y}$, where $y_i \equiv y(x_i)$.

Alternatively, a complete set of basis functions $\phi_1(x), \phi_2(x), \ldots$ can be used to expand $f(x)$ as

$$f(x) = f_1 \phi_1(x) + f_2 \phi_2(x) + \ldots. \tag{8.41}$$

If the expansions

$$\int_a^b K(x, x') \phi_j(x') \, dx' = a_{1j} \phi_1(x) + a_{2j} \phi_2(x) + \ldots \tag{8.42}$$

are known, then $y(x)$ is approximated by an M-term expansion as

$$y(x) \simeq y_1 \phi_1(x) + y_2 \phi_2(x) + \ldots + y_M \phi_M(x), \tag{8.43}$$

where the unknown expansion coefficients y_j are the solution of the M linear algebraic equations $\mathbf{y} = \mathbf{f} + \mathbf{A}\mathbf{y}$.

For multidimensional problems, the region V can be partitioned into M contiguous (mutually disjoint) subdomains V_i, $i = 1, \ldots M$, so that $V = \cup_{i=1}^{M} V_i$. The subdomains are chosen so that $f(\mathbf{x})$ and $K(\mathbf{x}, \mathbf{x}')$ are approximately constant in each subdomain, i.e., $f(\mathbf{x}) \simeq f_i$ for $\mathbf{x} \in V_i$ and $K(\mathbf{x}, \mathbf{x}') \simeq a_{ij}$ for $\mathbf{x} \in V_i$ and $\mathbf{x}' \in V_j$. Equation (8.39) can, thus, be approximated, for $\mathbf{x} \in V_i$, as

$$y_i = f_i + \sum_{j=1}^{M} \int_{V_j} K(\mathbf{x}, \mathbf{x}') y(\mathbf{x}') \, d\mathbf{x}' \simeq f_i + \sum_{j=1}^{M} a_{ij} \int_{V_j} y(\mathbf{x}') \, d\mathbf{x}'$$

$$\simeq f_i + \sum_{j=1}^{M} (V_j a_{ij}) y_j, \tag{8.44}$$

or, with $b_{ij} = V_j a_{ij}$,

$$y_i \simeq f_i + \sum_{j=1}^{M} b_{ij} y_j, \quad i = 1, 2, \ldots, M, \tag{8.45}$$

a set of M linear algebraic equations.

However, as the dimensionality of the integral in Eq. (8.39) increases, converting the integral equation to linear algebraic equations becomes increasingly complex, resulting in very large $\mathbf{A}$ matrices. For dimensions greater than 3 or 4, Monte Carlo techniques become increasingly attractive.

8.2.3 Monte Carlo Solution of a Simple Integral Equation

In this section, Monte Carlo techniques are combined with a Neumann expansion to solve an integral equation whose kernel and inhomogeneous term have special properties in order to show the general idea behind the method. Later a more robust procedure is introduced to handle equations whose kernels do not have these special properties.

Suppose the value of some inner product of $y(\mathbf{x})$, i.e.,

$$H = (h(\mathbf{x}), y(\mathbf{x})) \equiv \int_V h(\mathbf{x}) f(\mathbf{x}) \, d\mathbf{x},$$

is sought, where $h(\mathbf{x})$ is specified and $y(\mathbf{x})$ is the solution of the integral equation

$$y(\mathbf{x}) = \int_V K(\mathbf{x}, \mathbf{x}') y(\mathbf{x}') \, d\mathbf{x}' + f(\mathbf{x}), \quad \mathbf{x} \in V. \tag{8.46}$$

Here it is assumed $f(\mathbf{x}) \geq 0$ and $K(\mathbf{x}, \mathbf{x}') \geq 0$ for $\mathbf{x}', \mathbf{x} \in V$. For simplicity, it is also assumed that the source term and kernel are normalized such that

$$\int_V f(\mathbf{x}) \, d\mathbf{x} = 1 \quad \text{and} \quad \int_V K(\mathbf{x}, \mathbf{x}') \, d\mathbf{x}' = 1. \tag{8.47}$$

To gain insight into what sampling scheme should be used, first solve the integral equation for $y(\mathbf{x})$ using the Neumann expansion method and then evaluate H. Thus, one seeks a solution of the form

$$y(\mathbf{x}) = \sum_{n=0}^{\infty} y_n(\mathbf{x}). \tag{8.48}$$

Here

$$y_0(\mathbf{x}) = f(\mathbf{x}), \quad y_1(\mathbf{x}) = \int_V K(\mathbf{x}, \mathbf{x}') y_0(\mathbf{x}') \, d\mathbf{x}', \quad y_2(\mathbf{x}) = \int_V K(\mathbf{x}, \mathbf{x}') y_1(\mathbf{x}') \, d\mathbf{x}', \ldots,$$

or, in general,

$$y_n(\mathbf{x}) = \int_V K(\mathbf{x}, \mathbf{x}') y_{n-1}(\mathbf{x}')\, d\mathbf{x}', \quad n = 1, 2, 3, \ldots. \qquad (8.49)$$

It is assumed that the kernel is sufficiently well behaved so that all of these integrals exist and the series converges, i.e., $\lim_{n \to \infty} y_n(\mathbf{x}) = y(\mathbf{x})$. With this *formal* solution for $y(\mathbf{x})$, the functional H is given by

$$H = \int_V h(\mathbf{x}) \sum_{n=0}^{\infty} y_n(\mathbf{x})\, d\mathbf{x}. \qquad (8.50)$$

Note that, because of the nonnegativity of $K(\mathbf{x}, \mathbf{x}')$, $y_n(\mathbf{x})$ are nonnegative and that, because of the normalization of Eq. (8.47), $y_n(\mathbf{x})$ have unit (hyper)area. Thus, $f(\mathbf{x})$ is a probability density function (PDF) characterizing a source, $K(\mathbf{x}, \mathbf{x}')$ is a PDF describing the transition of a random walk going from $\mathbf{x}$ to a unit differential volume about $\mathbf{x}'$, and $y_n(\mathbf{x})$ is the PDF of $\mathbf{x}$ after n events or transitions. Further, H can be interpreted as the sum of the expected values of H after each transition, namely,

$$H = \sum_{n=0}^{\infty} \int_V h(\mathbf{x}) y_n(\mathbf{x})\, d\mathbf{x} \equiv \sum_{n=0}^{\infty} \langle h(\mathbf{x}) \rangle_n.$$

But how does one estimate H by using random sampling techniques? Each random walk (or *history*) consists of a series of points (or *events*) in phase space V, namely, $\mathbf{x}_0, \mathbf{x}_1, \mathbf{x}_2, \ldots, \mathbf{x}_n, \ldots$. For each random walk s, $\mathbf{x}_0$ is randomly sampled from $f(\mathbf{x})$, $\mathbf{x}_1$ is sampled from $K(\mathbf{x}_0, \mathbf{x})$, $\mathbf{x}_2$ is sampled from $K(\mathbf{x}_1, \mathbf{x})$, $\ldots$ $\mathbf{x}_n$ is sampled from $K(\mathbf{x}_{n-1}, \mathbf{x})$, and so on. Thus, after N histories, the contribution to H from event n is obtained by averaging overall histories, and the expected value of H is then found by summing overall events, i.e.,

$$H \simeq \sum_{n=0}^{\infty} \left[\frac{1}{N} \sum_{s=1}^{N} h(\mathbf{x}_n^{(s)}) \right].$$

In practice, the sth history is truncated after M_s events when $h(\mathbf{x}_{M_i})$ becomes negligibly small. Hence, the estimate of H is

$$H \simeq \frac{1}{N} \sum_{i=1}^{N} \sum_{n=0}^{M_s} h(\mathbf{x}_n^{(s)}). \qquad (8.51)$$

This Monte Carlo approach is now generalized to treat integral equations whose kernels and inhomogeneous terms do not have the special properties of Eq. (8.47).

8.2.4 A More General Procedure for Integral Equations

Equation (8.39) can be written as $y(\mathbf{x}) = f(\mathbf{x}) + \mathcal{K}y(\mathbf{x})$, where the integral operator $\mathcal{K}$ is defined as

$$\mathcal{K}\psi(\mathbf{x}) \equiv \int_V K(\mathbf{x}, \mathbf{x}')\psi(\mathbf{x}')\,d\mathbf{x}'. \tag{8.52}$$

Rather than use a Neumann expansion, as is done in Section 8.2.3, an iterative solution, analogous to the iterative solution of linear algebraic equations given by Eq. (8.2), is used, namely,

$$y^{(k+1)}(\mathbf{x}) = f(\mathbf{x}) + \mathcal{K}y^{(k)}(\mathbf{x}) = \sum_{m=0}^{k} \mathcal{K}^m f(\mathbf{x}), \tag{8.53}$$

with $\mathcal{K}^{(0)} = 1$ and $y^{(0)} = 0$. If this Neumann series $y^{(0)}, y^{(1)}, y^{(2)}, \ldots$ converges, then[4]

$$y(\mathbf{x}) = \lim_{k \to \infty} y^{(k)}(\mathbf{x}) = \sum_{m=0}^{\infty} \mathcal{K}^m f(\mathbf{x}). \tag{8.54}$$

Suppose that what is sought is not $y(\mathbf{x})$, but the inner product

$$H \equiv (h(\mathbf{x}), y(\mathbf{x})) \equiv \int_V h(\mathbf{x})y(\mathbf{x})\,d\mathbf{x}. \tag{8.55}$$

This inner product can be estimated by a procedure analogous to that used in Section 8.1.1 by using a *continuous* random walk. A PDF $p(\mathbf{x})$ is devised to sample the starting position $\mathbf{x}_0$ of each random walk in V. Next, devise a transition function $P(\mathbf{x}, \mathbf{x}')$ such that $P(\mathbf{x}, \mathbf{x}') > 0$ if $K(\mathbf{x}, \mathbf{x}') \neq 0$ and

$$\int_V P(\mathbf{x}, \mathbf{x}')\,d\mathbf{x}' = 1. \tag{8.56}$$

Ideally, $p(\mathbf{x})$ and $P(\mathbf{x}, \mathbf{x}')$ should be proportional to the magnitudes of $f(\mathbf{x})$ and $K(\mathbf{x}, \mathbf{x}')$, respectively. But uniform distributions often give satisfactory results. The inner product (h, y) can then be estimated by the following procedure.

1. Choose a value of k corresponding to k in Eq. (8.53).
2. Select a starting location $\mathbf{x}_0$ in V by sampling from $p(\mathbf{x})$.
3. Create a continuous random walk $\mathbf{x}_0 \to \mathbf{x}_1 \to \ldots \to \mathbf{x}_k$ where the probability of moving from state $\mathbf{x}_i$ to another state or location in phase space $\mathbf{x}$ is determined by sampling from $P(\mathbf{x}_i, \mathbf{x})$.

---

[4] Here it is assumed that $f(\mathbf{x}) \in L_2(V)$ (i.e., f is square integrable) and $K(\mathbf{x}, \mathbf{x}') \in L_2(V \times V)$. A sufficient (but not necessary) condition for Eq. (8.54) to converge is

$$\sup_{\mathbf{x} \in V} \int_V |K(\mathbf{x}, \mathbf{x}')|\,d\mathbf{x}' < 1.$$

4. For each random walk $s = 1, \ldots, N$, the random variable

$$Z_k^{(s)}(h) \equiv \frac{h(\mathbf{x}_0)}{p(\mathbf{x}_0)} \sum_{m=0}^{k} U_m f(\mathbf{x}_m) \tag{8.57}$$

is calculated. Here U_m is

$$U_m = \frac{K(\mathbf{x}_0, \mathbf{x}_1)}{P(\mathbf{x}_0, \mathbf{x}_1)} \frac{K(\mathbf{x}_1, \mathbf{x}_2)}{P(\mathbf{x}_1, \mathbf{x}_2)} \cdots \frac{K(\mathbf{x}_{m-1}, \mathbf{x}_m)}{P(\mathbf{x}_{m-1}, \mathbf{x}_m)}, \quad \text{with} \quad U_0 = 1.$$

5. After N such random walks, (h, y) is estimated as

$$H = (h(\mathbf{x}), y(\mathbf{x})) \simeq \frac{1}{N} \sum_{s=1}^{N} Z_k^{(s)}(h). \tag{8.58}$$

Should the solution $y(\mathbf{x}_0)$ be sought, set $h(\mathbf{x}) = p(\mathbf{x}) = \delta(\mathbf{x} - \mathbf{x}_0)$ so that every random walk begins at $\mathbf{x}_0$ in phase space V. Also in Eq. (8.57), the term $h(\mathbf{x}_0)/p(\mathbf{x}_0)$ is set to unity. An example of the above procedure is shown in Example 8.2.

Example 8.2 Monte Carlo Solution of an Integral Equation

Consider the integral equation

$$y(x) = x + \lambda \int_0^1 e^{x-x'} y(x') \, dx'.$$

The analytical solution of this equation, for any real value of $\lambda \neq 1$, is

$$y(x) = x + \frac{\lambda}{1-\lambda} \left[1 - \frac{2}{e} \right] e^x,$$

which is easily verified by substitution into the integral equation.

It can also be shown that $\lambda = 1$ is the only eigenvalue with the corresponding eigenfunction (unnormalized) $y(x) = e^x$. Thus, the spectral radius of the integral operator $\mu(\mathcal{K}) = 1$ and the Neumann series of Eq. (8.53) converges for $\lambda < 1$. Thus, the Monte Carlo solution of this integral equation can be realized only for $\lambda < 1$. Moreover, as $\lambda \to 1$, the length of the random walk must become ever larger to obtain an estimate of $(h(x), y^{(k)}(x))$ that approximates the true value of the inner product $(h(x), y(x))$, as can be seen from the Monte Carlo results of Fig. 8.2. In this example, $h(x) = 1$ so that the analytical value of the inner product is

$$(h, y) = \int_0^1 h(x) y(x) \, dx = \int_0^1 \left[x + \frac{\lambda(1 - 2e^{-1})}{1 - \lambda} e^x \right] dx$$

$$= \frac{1}{2} + \frac{\lambda}{1-\lambda} \left[e + \frac{2}{e} - 3 \right].$$

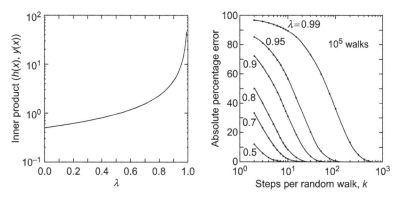

Figure 8.2 The inner product $(h(x), y(x))$ (left) and the percent error (right) for the length of the continuous random walk (right) for different values of λ. In these results $h(x) = 1$. For simplicity, the PDFs used were $p(x) = 1$ and $P(x, x') = 1$.

8.3 Ordinary Differential Equations

There is an intimate relation between integral equations (IEs) and ODEs. One can often be converted into the other and vice versa. Consider the initial value problem

$$y'(x) = f(x, y) \qquad \text{with} \qquad y(x_o) = y_o \text{ specified.} \qquad (8.59)$$

Integration of this equation from x_o to x and use of the initial values converts this equation into a Volterra equation, namely,

$$y(x) = y_o + \int_{x_o}^{x} f(t, y(t)) \, dt. \qquad (8.60)$$

Note that if Eq. (8.60) is evaluated at $x = x_o$, the initial condition is recovered, i.e., $y(x_o) = y_o$. If Eq. (8.60) is differentiated, with the help of Leibniz' rule,[5] the original ODE is recovered. Thus, Eqs. (8.59) and (8.60) are equivalent problems. In general, initial value ODEs and Volterra equations can be converted from one to the other. Likewise, boundary value ODEs and Fredholm equations can be converted from one to the other as illustrated by the following two examples.

---

[5] Gottfried Wilhelm Leibniz (1648–1716) developed the following formula for differentiating an integral:

$$\frac{d}{dx}\left(\int_{a(x)}^{b(x)} f(x, t) \, dt\right) = \int_{a(x)}^{b(x)} \frac{\partial f(x, t)}{\partial x} \, dt + f(x, a(x)) \frac{da(x)}{dx} - f(x, b(x)) \frac{db(x)}{dx},$$

under the assumption of continuity of $a(x)$, $b(x)$, and $f(x, t)$ and their derivatives.

Example 8.3 Fredholm Equation to Differential Equation

Consider the Fredholm integral equation

$$y(x) = \lambda \int_0^1 K(x, x') y(x), \tag{8.61}$$

in which λ is a constant and the symmetric kernel is

$$K(x, x') = \begin{cases} x(1 - x'), & 0 \le x < x' \le 1, \\ x'(1 - x), & 0 \le x' < x \le 1. \end{cases} \tag{8.62}$$

Substitution of Eq. (8.62) into Eq. (8.61) gives

$$y(x) = \lambda \int_0^x x'(1 - x) y(x') \, dx' + \lambda \int_x^1 x(1 - x') y(x') \, dx'. \tag{8.63}$$

Differentiation of this form of the Fredholm equation with respect to x gives

$$y'(x) = \lambda \int_0^x (-x') y(x') \, dx' + \lambda x(1 - x) y(x)$$

$$+ \lambda \int_x^1 (1 - x') y(x') \, dx' - \lambda x(1 - x) y(x)$$

$$= -\lambda \int_0^x x' y(x') \, dx' + \lambda \int_x^1 (1 - x') y(x') \, dx'. \tag{8.64}$$

Differentiate this result with respect to x to obtain

$$y''(x) = -\lambda x y(x) - \lambda (1 - x) y(x) = -\lambda y(x).$$

Hence $y(x)$, which is the solution of the Fredholm Eq. (8.61) is also the solution of the ODE

$$y''(x) + \lambda y(x) = 0, \tag{8.65}$$

subject to the boundary conditions $y(0) = y(1) = 0$, which are obtained by evaluating Eq. (8.63) at $x = 0$ and $x = 1$.

Example 8.4 Differential Equation to Integral Equation

Consider the ODE obtained in the previous example and convert it back to its equivalent Fredholm integral equation. Begin by integrating Eq. (8.65) over $(0, x)$ to obtain

$$y'(x) - y'(0) + \lambda \int_0^x y(x') \, dx' = 0, \tag{8.66}$$

where $y'(0)$ is yet to be found. Integration of this result, again over $(0, x)$, then yields

$$y(x) - y(0) - xy'(0) + \lambda \int_0^x \int_0^{x'} y(x'') \, dx'' \, dx' = 0. \tag{8.67}$$

Use the boundary condition $y(0) = 0$ and the identity[6]

$$\int_a^x \int_a^{x'} f(x'') \, dx'' = \int_a^x (x - x') f(x') \, dx', \tag{8.68}$$

with $a = 0$, to write Eq. (8.67) as

$$y(x) - xy'(0) = \lambda \int_0^x (x - x') y(x') \, dx' = 0. \tag{8.69}$$

To find $y'(0)$, set $x = 1$ in Eq. (8.69) and use the boundary condition $y(1) = 0$. In this manner $y'(0)$ is found to be

$$y'(0) = \lambda \int_0^1 (1 - x') y(x') \, dx'.$$

Finally, substitute this value for $y'(0)$ into Eq. (8.67):

$$\begin{aligned} y(x) &= x\lambda \int_0^1 (1 - x') y(x') \, dx' - \lambda \int_0^x (x - x') y(x') \, dx' \\ &= \lambda \int_x^1 x(1 - x') y(x') \, dx' - \lambda \int_0^x x'(1 - x) y(x') \, dx' \\ &= \lambda \int_0^1 K(x, x') y(x') \, dx', \end{aligned} \tag{8.70}$$

which is the Fredholm equation of the previous example. The ODE considered in this example is a very simple one; however, *any* second-order, boundary value, differential equation can be converted to a Fredholm integral equation (see Problem 8.6).

---

[6] To prove this identity, assume $f(x)$ is any continuous function for $x \geq a$. Let $F(x) = \int_a^x f(x') \, dx'$ or, equivalently, $f(x) = F'(x)$. Then

$$\int_a^x \int_a^{x'} f(x'') \, dx'' \, dx' = \int_a^x F(x') \, dx' = \int_a^x 1 \cdot F(x') \, dx'$$

$$\underset{\text{by parts}}{=} \left. x' F(x') \right|_a^x - \int_a^x x' F'(x') \, dx' = xF(x) - aF(a) - \int_a^x x' f(x') \, dx'$$

$$= x \int_a^x f(x') \, dx' - 0 - \int_a^x x' f(x') \, dx' = \int_a^x (x - x') f(x') \, dx'.$$

8.3.1 Reduction to First-Order Equations

Any first degree ODE of order n, either linear or nonlinear, can be written as

$$y^{(n)}(x) = F\left(x, y, y^{(1)}, y^{(2)}, \ldots, y^{(n-1)}\right), \tag{8.71}$$

where $y^{(m)}(x) \equiv d^m y(x)/dx^m$. Such an equation can be expressed as a set of first-order coupled ODEs by introducing $n-1$ auxiliary functions:

$$\begin{cases} y_0(x) \equiv y(x), \\[2mm] y_1(x) \equiv \dfrac{dy}{dx} = \dfrac{dy_0}{dx}, \\[2mm] y_2(x) \equiv \dfrac{d^2 y}{dx^2} = \dfrac{dy_1}{dx}, \\[2mm] \quad\vdots \quad \equiv \quad \vdots \\[2mm] y_{n-1}(x) \equiv \dfrac{d^{n-1} y}{dx^{n-1}} = \dfrac{dy_{n-2}}{dx}, \\[2mm] y_n(x) \equiv \dfrac{d^n y}{dx^n} = \dfrac{dy_{n-1}}{dx} F\left(x, y, \dfrac{dy}{dx}, \ldots, \dfrac{d^{n-2} y}{dx^{n-2}}, \dfrac{d^{n-1} y}{dx^{n-1}}\right), \end{cases} \tag{8.72}$$

or

$$\frac{d}{dx} \begin{pmatrix} y_0(x) \\ y_1(x) \\ \vdots \\ y_{n-2}(x) \\ y_{n-1}(x) \end{pmatrix} = \begin{pmatrix} y_1(x) \\ y_2(x) \\ \vdots \\ y_{n-1}(x) \\ F(x, y_0, y_1, \ldots, y_{n-2}, y_{n-1}) \end{pmatrix}. \tag{8.73}$$

Equation (8.73) can then be written more compactly as

$$\boxed{\frac{d\mathbf{y}(x)}{dx} = \mathbf{f}(x, \mathbf{y}).} \tag{8.74}$$

A set of coupled differential equations, each of any order, can likewise be transformed into a set of first-order ODEs. Consequently, any numerical method that can solve Eq. (8.74) can be used to solve almost any ODE provided a unique solution exists, a condition that may not be true for nonlinear equations.

8.3.2 Monte Carlo Solution of Initial Value Problems

Consider Eq. (8.74), in which values of $y(x)$ and its $n-1$ derivatives are specified at x_0, i.e., $\mathbf{y}(x_0) = \mathbf{y}_0$ is specified. Then one seeks $\mathbf{y}(x)$ for some $x > x_0$.

Integration of Eq. (8.74) from x_0 to x gives

$$\mathbf{y}(x) = \mathbf{y}_0 + \int_{x_0}^{x} \mathbf{f}(x', \mathbf{y}(x')) \, dx'. \tag{8.75}$$

Thus finding the solution is reduced to a problem of simply evaluating an integral. An integral, you say! Why not use Monte Carlo?

Begin by dividing the interval (x_0, x) into M equiwidth contiguous subintervals (x_i, x_{i+1}) in which $x_M = x$ and $\Delta x = x_{i+1} - x_i$. Thus the integral in Eq. (8.75) is broken into M subintegrals so that

$$\mathbf{y}(x_i) = \mathbf{y}_0 + \sum_{i=1}^{M} \int_{x_{i-1}}^{x_i} \mathbf{f}(x', \mathbf{y}(x')) \, dx'. \tag{8.76}$$

Equivalently,

$$\mathbf{y}(x_i) = \mathbf{y}(x_{i-1}) + \int_{x_{i-1}}^{x_i} \mathbf{f}(x', \mathbf{y}(x')) \, dx', \quad i = 0, 1, \dots, (M-1). \tag{8.77}$$

Now evaluate the integral over the interval (x_{i-1}, x_i) by Monte Carlo using values of x uniformly distributed over the interval, i.e., K values of $x_{ik} = x_{i-1} + \Delta x \, \rho_k$. Then Eq. (8.77) becomes

$$\mathbf{y}(x_i) = \mathbf{y}(x_{i-1}) + \frac{\Delta x}{K} \sum_{k=1}^{K} \mathbf{f}(x_{ik}, \mathbf{y}(x_{ik})). \tag{8.78}$$

The only difficulty with this formulation is that $\mathbf{y}(x)$ is not known in the integration interval. However, one could use a truncated Taylor series to approximate $\mathbf{y}(x)$ in the integration interval, i.e.,

$$\mathbf{y}(x) = \mathbf{y}(x_i) + (x - x_i) \left. \frac{d\mathbf{y}}{dx} \right|_{x_i} + \frac{(x - x_i)^2}{2!} \left. \frac{d^2\mathbf{y}}{dx^2} \right|_{x_i} + \cdots$$

$$= \mathbf{y}(x_i) + (x - x_i)\mathbf{f}(x_i, \mathbf{y}_i) + \frac{(x - x_i)^2}{2!} \left. \frac{d\mathbf{f}}{dx} \right|_{x_i} + \cdots. \tag{8.79}$$

The crudest approximation ("zeroth order") is to keep just the first term so that in Eq. (8.78) $\mathbf{f}(x_{ik}, \mathbf{y}(x_{ik})) \simeq \mathbf{f}(x_{ik}, \mathbf{y}(x_i))$. The first-order approximation keeps the first two terms in Eq. (8.79) so that

$$\mathbf{f}(x_{ik}, \mathbf{y}(x_{ik})) \simeq \mathbf{f}(x_{ik}, \mathbf{y}(x_i)) + (x_{ik} - x_i)\mathbf{f}(x_i, \mathbf{y}(x_i)). \tag{8.80}$$

Higher-order approximations are possible, thereby allowing use of a coarser grid, but the calculation of derivatives of $\mathbf{f}$ generally becomes messy and they are seldom used. For example,

$$\frac{d\mathbf{f}}{dx}\bigg|_{x_i} = \left[\frac{\partial \mathbf{f}}{\partial x} + \sum_{j=1}^{N} \frac{\partial \mathbf{f}}{\partial y_j} f_j\right]_{x_i}.$$

The following two examples demonstrate this Monte Carlo approach for solving ODEs, one linear and the other nonlinear.

Example 8.5

Consider the initial value problem

$$\frac{dy(x)}{dx} = 4x - 2xy \equiv f(y, x) \quad \text{with} \quad y(0) = 5. \tag{8.81}$$

Plot the solution for $x \in (0, 3)$.

The exact solution is $y(x) = C \exp(-x^2) + 2$, where C is an arbitrary constant. Verify by substituting this expression into the ODE. Application of the initial condition gives $C = 3$. This exact solution is plotted as a solid line in the left figure of Fig. 8.3.

The Monte Carlo values were obtained with Eq. (8.80) (first-order approximation) and are shown by the data points in the left figure of Fig. 8.3 for two grid sizes. In the shown calculation, $K = 1000$ random points per subinterval; however, almost the same results are obtained with $K = 100$. The error in the Monte Carlo calculations for the number M of subintervals used for $x \in (0, 3)$ is shown to the right in Fig. 8.3.

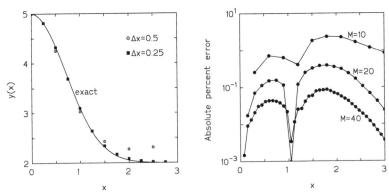

Figure 8.3 (left) The exact and Monte Carlo calculated values for Eq. (8.81). (right) Error in the Monte Carlo calculations for three different grid sizes. The dip in the error around $x \simeq 1$ occurs when the Monte Carlo calculations change from underpredicting $(x < 1)$ to overpredicting $(x > 1)$.

Example 8.6

In nature populations of predators and their prey, e.g., lions and zebras, lynx and hares, foxes and rabbits, etc., are observed to oscillate in time. Surprisingly perhaps, such periodic behavior is predicted by the following two simple coupled ODEs:

$$\frac{dy_1(x)}{dx} = \alpha y_1(x) - \beta y_1(x)y_2(x),$$
$$\frac{dy_2(x)}{dx} = \delta y_1(x)y_2(x) - \gamma y_2(x)$$
$$\text{or} \quad \frac{\mathbf{y}(x)}{dx} = \mathbf{f}(x, \mathbf{y}).$$

Here y_1 is the population (in thousands) of rabbits whose reproduction rate αy_1 is exponential (assuming unlimited food supply) and whose predation rate $\beta y_1 y_2$ is proportional to the frequency they meet foxes. The fox population y_2 (in thousands) increases as $\delta y_1 y_2$ in proportion to the frequency they find rabbits. The term γy_2 describes the loss of foxes through natural death or migration from the study area.

From the results in Fig. 8.4, it is seen that the Monte Carlo solution method discussed in this section is quite capable of producing results very comparable to those obtained with a much more complex (but accurate) numerical technique.

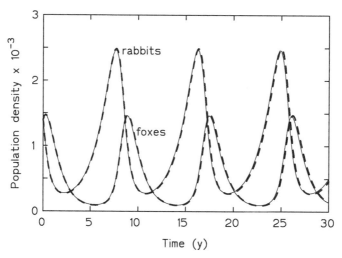

Figure 8.4 The predator–prey model for $\alpha = 2/3$, $\beta = 4/3$, and $\delta = \gamma = 1$ with the initial conditions $y_1(0) = y_2(0) = 1.4$. The light lines are results from a sophisticated Runge–Kutta differential equation solver (Press et al., 1996) and the heavy dashed lines are results from a first-order Monte Carlo analysis using Eqs. (8.78) and (8.80).

8.3.3 Monte Carlo Solution of Partial Differential Equations

The application of Monte Carlo techniques to evaluate solutions of various important PDEs received much early attention (Curtiss, 1949; Wasow, 1951;

Muller, 1956). Even today, work continues on developing new Monte Carlo methods for particular differential equations as a search using Google Scholar quickly reveals.

The Monte Carlo solution of linear differential equations can be evaluated by using three different approaches. The first is to use some form of discretization such as finite differences or finite elements to *approximate* the differential equation by a set of linear algebraic equations. These algebraic equations can then be treated using the methods discussed in Section 8.1. Such an approach, which uses random walks over the finite discretization grid, produces results that not only have the statistical uncertainties associated with Monte Carlo estimation but also truncation (or systematic) errors arising from the discretization process itself. An example of this approach is given in Example 2.4 in which the one-dimensional heat conduction equation is solved by random walks over a discrete grid.

The second approach is to replace the differential equation by an integral equation. This can always be done for a second-order Hermitian differential operator by finding the Green's functions for the specified geometry and boundary conditions. Once an equivalent integral equation is found, the Monte Carlo methods of Section 8.2 can be used to estimate the desired solution.

The third approach is to use Monte Carlo methods in which a large number of random walks or histories are constructed in the domain of the PDE for which a solution is desired. The steps in each random walk form a Markov chain because the ending position of each step of the walk depends only on the ending position of the previous step. From the scores accumulated by each random walk, estimates of the dependent variable can be obtained. The random walks can be performed with each step being of a predetermined length (so-called *grid-based* Markov chains) or can be continuously variable (so-called *continuous Markov chains*). Both of these Markov chain approaches are explored in the following sections.

Monte Carlo methods have numerous advantages over those used for the finite difference or finite element approaches. First, for complex geometries these later approaches require a small spatial grid size to model accurately the boundary. Further, if the PDE is a time-dependent problem, stability issues force small time steps for small spatial grids. The result is a very large matrix that must be assembled and then inverted. The Monte Carlo approach avoids the bookkeeping effort to assemble a large matrix and the computation expense of the matrix inversion. Second, Monte Carlo allows solving just for a region of interest, instead of solving for the entire domain. This feature is particularly beneficial in inverse transient problems in which information in the interior of the domain is used to predict values at the boundaries. Third, the Monte Carlo approach is stable and very well suited to parallel computing, a consideration of increasing importance with today's powerful computers based on large multicore and multithread architectures. Fourth, the Monte Carlo approach by and large avoids the need for sophisticated mathematical analyses often needed by other methods.

Fifth, Monte Carlo solvers automatically estimate uncertainties in results, a feature often lacking in other numerical approaches. Finally, Monte Carlo methods make it a powerful method to treat problems with inherent stochastic behavior or properties, such as the simulation of fluid flow in a porous medium.

In this section, two important differential equations are considered. First considered is Poisson's equation

$$\nabla^2 u(\mathbf{r}) = -s(\mathbf{r}), \quad \mathbf{r} \in D, \tag{8.82}$$

which reduces to the Laplace equation when the source term $s(\mathbf{r}) = 0$, and is routinely encountered, for example, in steady-state heat conduction problems, electric potential calculations, and absorption-free diffusion analyses. For diffusion problems with absorption, the Helmholtz equation

$$\nabla^2 u(\mathbf{r}) - \beta^2 u(\mathbf{r}) = -s(\mathbf{r}), \quad \mathbf{r} \in D, \tag{8.83}$$

is needed. Both equations are usually called boundary value problems because conditions on $u(\mathbf{r})$ on the boundary Γ of D are needed in order to obtain a unique solution.

Analytic solutions of these equations generally can be obtained only for homogeneous media with simple boundaries such as planes, spheres, or cylinders. For heterogeneous media with irregular boundaries, numerical methods must be employed. Even though most boundary value problems described by differential equations are usually far more efficiently evaluated using standard numerical techniques designed for these equations, the Monte Carlo approach is discussed here both for completeness and for application to cases with irregular boundaries where the standard methods often have difficulty.

8.3.4   Discretization of Poisson's Equation

By using first-order finite differences, Poisson's equation can be approximated by a set of linear algebraic equations whose solutions can be estimated by discrete Markov random walks through the domain D as was done in Section 8.1. However, in this case the walks are considerably simplified because the transition probabilities are known. First the finite difference equations are derived somewhat succinctly for Dirichlet boundary conditions, i.e., the value of $u(\mathbf{r})$ is specified on the boundary Γ of the domain D. More detailed derivations and other boundary conditions can be found in many texts such as those by Özişik et al. (2017), Nakamura (1977), and Patankar (1980). Once the equations are derived, the use of Monte Carlo to estimate solutions of the equations is addressed.

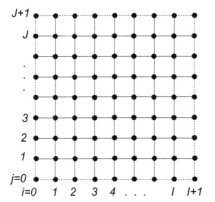

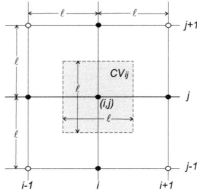

Figure 8.5 A Cartesian grid for 2-D problems. The boundary Γ is shown by the dotted lines and interior nodes have position indices of $i = 1, \ldots, I$ and $j = 1, \ldots, J$.

Figure 8.6 The control volume CV_{ij} about node ij and its four nearest neighbor nodes are shown as solid circles.

The Finite Difference Approximation

First consider an equimesh discrete Cartesian grid, such as that in Fig. 8.5, with adjacent nodes separated by a small distance ℓ.[7] Integration of Eq. (8.82) over the "control volume" about node $\mathbf{r}_o$ (see Fig. 8.6) and use of first-order finite differences to evaluate the normal derivatives at the control-volume surfaces, which are located halfway between $\mathbf{r}_o$ and its adjacent nodes, yields the following finite difference equations:

$$u(\mathbf{r}_l) = \sum_{nn} p_{nn} u(\mathbf{r}_{nn}) + Q(\mathbf{r}_l), \quad l = 1, 2, \ldots L, \qquad (8.84)$$

where $Q(\mathbf{r}_l) = s(\mathbf{r}_l)\ell^2/m$ and $\mathbf{r}_l$ is any of the L "interior nodes," i.e., the nodes not on Γ. For example, $\mathbf{r}_l$ is the node with position indices ij in Fig. 8.6. The $\mathbf{r}_{nn}$ are the location of the m "nearest neighbor" nodes all of which are a distance ℓ from $\mathbf{r}_l$. The transition probability that in a random walk a particle goes from $\mathbf{r}_l$ to an adjacent node is $p_{nn} = 1/m$. In 1-D, $\mathbf{r}_l = x_i$, $m = 2$, and $\mathbf{r}_{nn} = x_{i-1}$ and x_{i+1}. In 2-D, $\mathbf{r}_l = (x_i, y_j)$, $m = 4$, and $\mathbf{r}_{nn} = (x_{i+1}, y_j)$, (x_{x-1}, y_j), (x_i, y_{j+1}), and (x_i, y_{j-1}). In 3-D, $\mathbf{r}_l = (x_i, y_j, z_k)$, $m = 6$, and $\mathbf{r}_{nn} = (x_{i+1}, y_j, z_k)$, (x_{i-1}, y_j, z_k), (x_i, y_{j+1}, z_k), (x_i, y_{j-1}, z_k), (x_i, y_j, z_{k+1}), and (x_i, y_j, z_{k-1}).

For interior nodes adjacent to the boundary Γ, one of the nearest nodes in Eq. (8.84) is a node on Γ at which u is known (Dirichlet boundary condition), its normal derivative is given (Neumann boundary condition), or some other

---

[7] Although square or cubical grids are widely used for irregularly shaped 2-D or 3-D objects, the boundary can be, at best, be modeled in a stair-step fashion with grid lines having varying numbers of nodes.

condition on u is imposed. In any event, the term involving the boundary node is removed from Eq. (8.84) by including it in the source term Q. Likewise the source in the smaller-than-normal control volume around the boundary node is moved into the source term for the adjacent interior node. To illustrate this simplification, consider a 1-D problem with a constant source s_0. The left-most interior nodes are shown in Fig. 8.7. At the boundary $u_0 \equiv u(x_0)$ is known. The source emission from the half-size control volume around x_0 is $s(x_0)\ell^2/(2m)$. Half ($p_{nn} = 1/m = 1/2$) of this emission goes to the left and is lost and half goes toward x_1 and augments the emissions around this node. Thus Eq. (8.84) for node x_1 can be written as

$$u_1 - \frac{1}{2}u_2 = \frac{u_0}{2} + p_{nn}u_0 + \frac{s_0\ell^2}{2}\left(1 + \frac{1}{2}p_{nn}\right) = \frac{1}{2}u_0 + \frac{5s_0\ell^2}{8} \equiv Q(\mathbf{r}_1). \quad (8.85)$$

A similar treatment holds for the rightmost interior node I.

The same simplification for interior nodes adjacent to the boundaries holds for higher dimensions. In 2-D, yet another special treatment of sources in the control volumes at corner boundary nodes must be used. Similarly, in 3-D nodes adjacent to boundary edges and corners must be afforded special treatment.

The finite difference equations can be written in matrix form as $\mathbf{Au} = \mathbf{Q}$, where the components of $\mathbf{Q}$ for nodes adjacent to Γ have been adjusted $\overline{Q}$ as was done for u_1 in Eq. (8.85). In 1-D, the $\mathbf{A}$ matrix is tridiagonal with 1s on the diagonal and $(-1/2)$s above and below the main diagonal. The matrix is symmetric and diagonally dominant and the equation $\mathbf{Au} = \mathbf{Q}$ is easily solved with the so-called tridiagonal algorithm (Nakamura, 1977). In 2-D or 3-D geometries, $\mathbf{A}$ has five or seven diagonals, respectively, provided "consistent ordering of the mesh" is used; the resulting $\mathbf{A}$, while not tridiagonal, is block tridiagonal.

8.3.5 Discrete Random Walks for Poisson's Equation

The solution of Eq. (8.84) at each of the L interior nodes $\mathbf{r}_l$ can be estimated by making a large number N of discrete random walks (RWs) through the discrete grid placed over D (see Fig. 8.8). The following algorithm is a generalization of that used in Example 2.6.

1. Choose a starting node l' with probability $p_{l'} = Q_{l'}/Q_{tot}$, where Q_{tot} is the total source strength, i.e., $Q_{tot} = \sum_{l=1}^{L} Q_l$.
2. At each step in the walk, move the particle to one of the nearest nodes, each with a probability $p_{nn} = 1/m$.
3. Each time that a particle lands on node l increase a counter $c_l = c_l + 1$ in order to record the cumulative number of visits to node l.
4. When the particle reaches a node on the boundary, kill it and start a new RW as in step 1.
5. Perform steps 1 through 4 for a large number N of particles.

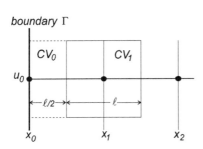

Figure 8.7 Nodes near the left-hand boundary for a 1-D problem. Note the control volume is half the size of those around interior nodes.

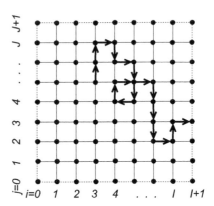

Figure 8.8 A possible random walk through a discrete 2-D grid. The walk ends when the boundary Γ is reached.

After N RWs, the value of $u(\mathbf{r}_l)$ is estimated as

$$u_l \equiv u(\mathbf{r}_l) \simeq \frac{Q_{tot}}{N} c_l. \tag{8.86}$$

An Adjoint Approach

Associated with the forward problem, $\mathbf{A}\mathbf{u} = \mathbf{Q}$ is an adjoint problem

$$\mathbf{A}^{\dagger} \mathbf{u}^{\dagger} = \mathbf{Q}^{\dagger}, \tag{8.87}$$

where $\mathbf{Q}^{\dagger}$ is the adjoint source. Because $\mathbf{A}$ is real $\mathbf{A}^{\dagger} = \mathbf{A}^{T}$, and because $\mathbf{A}$ is symmetric $\mathbf{A}^{T} = \mathbf{A}$. Thus one can use the same RWs used for the forward problem to solve the associated adjoint problem

$$u^{\dagger}(\mathbf{r}_l) = \sum_{nn} p_{nn} u^{\dagger}(\mathbf{r}_{nn}) + Q^{\dagger}(\mathbf{r}_l), \quad l = 1, 2, \ldots L. \tag{8.88}$$

Suppose one seeks some inner product $H = (\mathbf{h}|\mathbf{u})$. As shown in Appendix D, the same inner product can be obtained with the adjoint solution of Eq. (8.87) when the adjoint source $\mathbf{Q}^{\dagger} = \mathbf{h}$, i.e.,

$$H = (\mathbf{h}|\mathbf{u}) = (\mathbf{Q}|\mathbf{u}^{\dagger}). \tag{8.89}$$

In particular, if $h_l = \delta_{ll'}$, then

$$u_l = \sum_{l'=1}^{L} Q_{l'} u_{l'}^{\dagger}. \tag{8.90}$$

To find $u_{l'}$, perform the same RWs as for the forward problem, but now every RW begins at node l and, as before, the number of visits $c_{l'}$ to node l' are recorded. After a large number N of RWs, $u^{\dagger}$ is estimated as $u_{l'}^{\dagger} = c_{l'}/N$. Then from Eq. (8.90) one has

$$u_l \simeq \frac{1}{N} \sum_{l'=1}^{L} Q_{l'} c_{l'}. \tag{8.91}$$

The advantage of this adjoint approach over the direct approach is that far fewer RWs are needed to find u at a specified node. However, if u is wanted at all interior nodes, this adjoint calculation would have to be repeated a large number L of times.

One drawback of tracking the number of visits to each node is that one cannot directly estimate a standard deviation for u_l other than run the same problem M times, each with the same number of random walks but with a different sequence of random numbers. If $u_l^{(m)}$ is the estimate obtained for u_l for run m, then

$$u_l \simeq \overline{u_l^{(m)}} = \frac{1}{N_m} \sum_{m=1}^{M} u_l^{(m)} \quad \text{and} \quad s^2(u_l) \simeq \frac{1}{M-1} \sum_{m=1}^{M} \left(u_l^{(m)} - \overline{u_l^{(m)}} \right)^2. \tag{8.92}$$

An Alternative Adjoint Approach

To avoid running multiple batches, an alternative approach using a single batch of random walks can be used. From a probabilistic interpretation of Eq. (8.84), it is seen that a particle at a node $\mathbf{r}$ has an equal probability to going to any of the nearest neighbors on the next step of its RW. However, the presence of $Q(\mathbf{r})$ in Eq. (8.84) indicates that this quantity should be added to the score of the particle. Then the tally for the RW that ends at boundary node $\mathbf{r}_b$ on Γ is

$$\text{tally} = u(\mathbf{r}_b) + \sum_{k=1}^{K} Q(\mathbf{r}_k), \tag{8.93}$$

where K is the number of steps in the RW. This tally is added to the tally accumulator S_1 and its square is added to S_2. From the results presented in Section 4.2, $\overline{u_l}$ and $s(\overline{u_l}) \simeq \sigma(\overline{u_l})$ are readily estimated. An example of this alternative adjoint approach is given in Example 8.7.

Example 8.7 Adjoint Monte Carlo Solution of the 2-D Poisson's Equation

Poisson's equation, given by Eq. (8.82), can be written in 2-D Cartesian geometry as

$$\frac{\partial^2 u(x, y)}{\partial x^2} + \frac{\partial^2 u(x, y)}{\partial y^2} = -s(x, y). \tag{8.94}$$

Consider the case of a constant source s_o in a square plate $D: -L/2 \le x \le L/2$ and $-L/2 \le y \le L/2$. In this problem, edges of the medium are kept at a constant value $u_o = 100$, $s_o = 20$, and $L = 10$. What is the value of $u(x, y)$ at the center of the medium, i.e., at $x = y = 0$?

Analytic Solution
Method 1:
The solution is symmetric in both x and y about the origin. Hence a double Fourier series in the even cosine functions is appropriate:

$$u(x, y) = u_o + \sum_{m \text{ odd}} \sum_{n \text{ odd}} A_{mn} \cos(\gamma_m x) \cos(\gamma_n y), \tag{8.95}$$

where $\gamma_n \equiv n\pi/L$. Application of the boundary condition and the orthogonality property of the $\cos(\gamma_n x)$ functions to this general solution then yield the expansion coefficients

$$A_{mn} = \frac{16L^2 s_o}{\pi^4 mn(n^2 + m^2)} (-1)^{(n+m-2)/2}. \tag{8.96}$$

Method 2:
The problem with the above solution is that the double summation in Eq. (8.95) is difficult to evaluate numerically. A simpler form can be obtained as follows. Let

$$u(x, y) - u_o = \sum_{n \text{ odd}} A_n(y) \cos(\gamma_n x). \tag{8.97}$$

Substitution of this expansion into Eq. (8.94) gives

$$-\sum_{n \text{ odd}} A_n(y)\gamma_n^2 \cos(\gamma_n x) + \sum_{n \text{ odd}} \frac{d^2 A_n(y)}{dy^2} \cos(\gamma_n x) = -s_o. \tag{8.98}$$

Multiply this result by $\cos(\gamma_n x)$ and use the orthogonality of the cosine functions to obtain

$$\frac{d^2 A_n(y)}{dy^2} - \gamma_n^2 A_n(y) = -\frac{2s_o}{L} \int_{-L/2}^{L/2} \cos(\gamma_n x) \, dx$$

$$= -\frac{4s_o}{L\gamma_n} (-1)^{\frac{n-1}{2}} = -\frac{4s_o}{L\gamma_n} (-1)^{\frac{n+1}{2}} \equiv \alpha_n. \tag{8.99}$$

The most general solution of this differential equation for $A_n(y)$ is

$$A_n(y) = C \cosh(\gamma_n y) + D \sinh(\gamma_n y) - \alpha_n/\gamma_n^2, \tag{8.100}$$

where C and D are arbitrary constants. The boundary condition $u(x, \pm L/2) - u_o = 0$ requires that $A_n(\pm L/2) = 0$, which, when applied to the above solution, gives

$$C = \frac{\alpha_n}{\gamma_n^2} \frac{1}{\cosh(\gamma_n L/2)} \quad \text{and} \quad D = 0. \tag{8.101}$$

Substitution of this result for $A_n(y)$ into Eq. (8.97) gives the desired solution

$$u(x, y) = u_o + \sum_{n \text{ odd}} \frac{\alpha_n}{\gamma_n^2} \left[\frac{\cosh(\gamma_n y)}{\cosh(\gamma_n L/2)} - 1 \right] \cos(\gamma_n x). \tag{8.102}$$

To evaluate Eqs. (8.96) and (8.102), the infinite sums are truncated after some large value of n and m, here taken as 201. Both equations give $u_{exact}(0, 0) = 247.34272\ldots$.

Adjoint Monte Carlo Solution

Results obtained with the alternative adjoint approach described in Section 8.3.5 are summarized in Table 8.1. From these results, as would be expected, the estimated standard deviation of $u_{MC}(0, 0)$ decreases as $1/\sqrt{N}$, i.e., 100 times more histories decrease $s(\overline{u_{MC}})$ by a factor of 10. The error between the exact value and u_{MC} initially decreases with the number of RWs but beyond about 300,000 histories remains relatively constant at about 0.1%. This asymptotic error is the error introduced by using the finite difference approximation. A finer mesh would decrease this saturation error.

Table 8.1 Estimate of $u(0, 0) \equiv u_{MC}$ using adjoint Monte Carlo on a discrete grid of 41 nodes in both the x- and y-directions. An equimesh was used with a spacing between adjacent nodes of 0.23810. The exact value is $u_{exact}(0, 0) = 247.3427\ldots$.

No. of histories	u_{MC}	Standard deviation	% error
10	245.759	40.611	−0.6403
30	277.089	31.257	12.03
100	274.860	18.590	11.13
300	255.010	8.8996	3.100
1000	252.765	4.8301	2.192
3000	246.710	2.6601	−0.2557
10,000	247.944	1.4491	0.2430
30,000	247.327	0.8321	−0.0064
100,000	246.990	0.4556	−0.1426
300,000	247.073	0.2631	−0.1090
1,000,000	247.093	0.1442	−0.1009

8.3.6 Continuous Random Walks for the 2-D Laplace's Equation

Let us begin the exploration of continuous Monte Carlo by considering the simple two-dimensional heat conduction equation without any internal heat source, i.e.,[8]

$$\frac{\partial^2 u(x, y)}{\partial x^2} + \frac{\partial^2 u(x, y)}{\partial y^2} = 0, \quad (x, y) \in D \quad \text{with } u(x, y)|_\Gamma = \psi(x, y).$$

(8.103)

This can be rewritten in polar coordinates, with the origin at an arbitrary position $\mathbf{r}_0 = (x_0, y_0)$, as

$$\frac{\partial^2 u(r, \theta)}{\partial r^2} + \frac{1}{r}\frac{\partial u(r, \theta)}{\partial r} + \frac{1}{r^2}\frac{\partial^2 u(r, \theta)}{\partial \theta^2} = 0.$$

(8.104)

Form of Solution

To gain insight into how $u(r, \theta)$ varies around $\mathbf{r}_0$, examine the form of the solution to Eq. (8.104). Assume the solution is separable, i.e., $u(r, \theta) = R(r)\Theta(\theta)$, and substitute this into Eq. (8.104) to obtain

$$\frac{r^2}{R}\frac{d^2 R}{dr^2} + \frac{r}{R}\frac{dR}{dr} + \frac{1}{\Theta}\frac{d^2\Theta}{d\theta^2} = 0.$$

(8.105)

The first two terms are a function only of r and the third term is a function only of θ so that the third term must equal a constant (α^2 say) and the first two terms must, therefore, sum to $-\alpha^2$. Thus,

$$\frac{d^2\Theta}{d\theta^2} + \alpha^2\Theta(\theta) = 0,$$

(8.106)

and

$$\frac{d^2 R}{dr^2} + \frac{1}{r}\frac{dR}{dr} - \frac{\alpha^2}{r^2}R(r) = 0.$$

(8.107)

The general solution of Eq. (8.106) is

$$\Theta(\theta) = A\cos\alpha\theta + B\sin\alpha\theta,$$

(8.108)

where A and B are arbitrary constants. The cyclic requirement that $\Theta(\theta) = \Theta(\theta + 2\pi)$ requires that α be an integer $\ell = 0, 1, 2, \dots$. The solution of the radial equation, Eq. (8.107), is

$$R(r) = \begin{cases} A\ln r + B, & \ell = 0, \\ Ar^{-\ell} + Br^\ell, & \ell > 0. \end{cases}$$

(8.109)

---

[8] Here the value of $u(x, y)$ on the boundary Γ is specified as $\psi(x, y)$. This is known as a Dirichlet boundary condition, which is the most studied one for boundary value problems. However, there are other possible boundary conditions that are discussed later in this chapter.

Because $u(r, \theta)$ must be finite as $r \to 0$, the arbitrary constant A in this result must vanish. Hence, the most general, physically realistic, solution of Eq. (8.104) must have the form

$$u(r, \theta) \equiv R(r)\Theta(\theta) = a_0 + \sum_{\ell=1}^{\infty} r^{\ell} [a_{\ell} \cos \ell\theta + b_{\ell} \sin \ell\theta], \qquad (8.110)$$

where a_{ℓ} and b_{ℓ} are arbitrary constants whose values are determined by the boundary conditions imposed on Eq. (8.103).

An Important Property of the Solution

An important property of this solution can be observed by integration of Eq. (8.110) over all $\theta \in [0, 2\pi]$ to obtain

$$\int_0^{2\pi} u(r, \theta) \, d\theta = 2\pi a_0. \qquad (8.111)$$

Also, the solution at the origin $u(0, \theta) \equiv u(\mathbf{r}_0)$ is independent of θ so that, from Eq. (8.110), $u(\mathbf{r}_0) = a_0$. Combining this result with Eq. (8.111) gives

$$u(\mathbf{r}_0) = \frac{1}{2\pi} \int_0^{2\pi} u(r, \theta) \, d\theta. \qquad (8.112)$$

In words, Eq. (8.112) says that *the value of u at the center of a circle of radius r equals the average of u on the circle.*

Monte Carlo Estimation of $u(\mathbf{r}_0)$

Suppose we seek the value of u at some point $\mathbf{r}_0$. Begin by drawing a circle that is centered at $\mathbf{r}_0$ and that just touches the nearest boundary (see Fig. 8.9). Then pick a point $\mathbf{r}_1$ uniformly sampled on this circle. The value of $u(\mathbf{r}_1)$ is then one particle's estimate of $u(\mathbf{r}_0)$. Averaged over many particles, each starting from $\mathbf{r}_0$, Eq. (8.112) shows that the average $\overline{u(\mathbf{r}_1)} \simeq u(\mathbf{r}_0)$.

However, $u(\mathbf{r}_1)$ is generally unknown *unless* $\mathbf{r}_1$ lies on the boundary Γ, in which case $u(\mathbf{r}_1) = \psi(\mathbf{r}_1)$, the specified boundary value. For $\mathbf{r}_1$ not on Γ, $u(\mathbf{r}_1)$ is estimated in the same manner used to estimate $u(\mathbf{r}_0)$, namely, a point $\mathbf{r}_2$ is uniformly sampled on the largest circle wholly in D and centered at $\mathbf{r}_1$ (the *maximal* circle) and $u(\mathbf{r}_2)$ is the particle's estimate of $u(\mathbf{r}_1)$. If $\mathbf{r}_2$ does not lie on the boundary, the random walk continues on successive circles until a point $\mathbf{r}_n$ "on" the boundary is found and the history is terminated. The score or tally for this walk is then recorded as $\psi(\mathbf{r}_n)$. This procedure is repeated for a large number N of particles, all starting from $\mathbf{r}_0$. The estimate of $u(\mathbf{r}_0)$ is then found

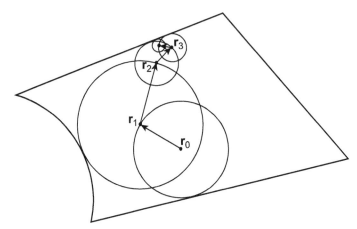

Figure 8.9 The continuous random walk used to estimate $u(\mathbf{r}_0)$. The "walking-on-circles" continues until the particle is within some small distance ϵ from the boundary.

by averaging all the scores, i.e.,

$$u(\mathbf{r}_0) = \frac{1}{N} \sum_{i=1}^{N} \psi(\mathbf{r}_{n_i}), \qquad (8.113)$$

where $\mathbf{r}_{n_i}$ is the terminating position on the boundary for the ith history.

The only difficulty with this scheme is that randomly selected points on a circle have zero probability of actually being on the boundary. Moreover, as a particle approaches the boundary, the circles get smaller and smaller, requiring more and more computing effort. One way to avoid this is to use a small parameter $\epsilon > 0$ such that, if $\mathbf{r}_n$ is within ϵ of the closest boundary point $\mathbf{r}^*$, the walk is terminated and $u(\mathbf{r}_n)$ is set to $\psi(\mathbf{r}^*)$.

The above *walking-on-circles* (WOC) procedure[9] is exact in the interior. However, it is only approximate near the boundary because it allows a particle to get only arbitrarily close to the boundary but never actually reach it. Clearly, as ϵ decreases, this approximation error decreases. Although standard numerical techniques for solving the Laplace equation are usually far more efficient, this Monte Carlo procedure can be used to good advantage when the boundary Γ assumes a complicated shape, such as that in Fig. 8.9.

---

[9] Continuous random walks have also been called *floating* random walks (Haji-Sheikh and Sparrow, 1967; Haji-Sheikh and Howell, 2006). Random walks through a discrete mesh require only integer arithmetic, whereas the WOC random walks require floating-point arithmetic, whence the alternative name.

8.3.7 Continuous Monte Carlo for 2-D Poisson Equation

The inhomogeneous form of Laplace' equation, given by Eq. (8.104) in two dimensions, is known as Poisson's equation:

$$\frac{\partial^2 u(r,\theta)}{\partial r^2} + \frac{1}{r}\frac{\partial u(r,\theta)}{\partial r} + \frac{1}{r^2}\frac{\partial^2 u(r,\theta)}{\partial \theta^2} = -s(r,\theta), \quad (x,y) \in D. \quad (8.114)$$

Again the boundary condition is $u(x,y)|_\Gamma = \psi(x,y)$. This equation can also be solved by the WOC Monte Carlo algorithm. First consider the case in which $s(r,\theta)$ is a constant s_o. Later, the more general case is considered.

The most general solution of Eq. (8.114) is the solution of the corresponding homogeneous equation, i.e., that with $s = 0$, plus any particular solution. It is verified upon substitution into Eq. (8.114) that $u_p(r) = -r^2 s_o/4$ is a particular solution. Thus, the most general, physically realistic solution of Eq. (8.114) is

$$u(r,\theta) \equiv u(\mathbf{r}) = a_0 + \sum_{\ell=1}^{\infty} r^\ell \left[a_\ell \cos \ell\theta + b_\ell \sin \ell\theta \right] - \frac{r^2 s_o}{4}. \quad (8.115)$$

Note that, as $r \to 0$,

$$u(\mathbf{r}_o) \equiv u(0,\theta) = a_0. \quad (8.116)$$

Integration of Eq. (8.115) over all polar angles gives

$$\int_0^{2\pi} u(r,\theta)\, d\theta = 2\pi a_0 - 2\pi r^2 s_o/4. \quad (8.117)$$

Combining Eqs. (8.116) and (8.117) and rearranging give the useful result

$$u(\mathbf{r}_0) = \frac{1}{2\pi} \int_0^{2\pi} u(r,\theta)\, d\theta + S(r), \quad (8.118)$$

where $S(r) \equiv r^2 s_0/4$. This result, which is very similar to Eq. (8.112) for the homogeneous problem ($s_o = 0$), says that *the value of u at the center of a circle of radius r equals the average of u on the circle plus a source contribution of $S(r)$.*

Monte Carlo Estimation of $u(\mathbf{r}_0)$

The continuous Monte Carlo WOC calculation of $u(\mathbf{r}_0)$ is almost the same as that used for Laplace' equation except at each step in a walk's history a source contribution $S(r_j) = r_j^2 s_o/4$ is added to the tally, where r_j is the radius of the jth circle. The random walk through D with $\mathbf{r}$ on a series of circles continues until $\mathbf{r}$ reaches Γ (or within a small distance ϵ from the boundary) where the ith walk terminates and the value $\psi(\mathbf{r}_{n_i}^*)$ is added to the tally of the ith random walk. Here $\mathbf{r}^*_{n_i}$ is the nearest point on Γ when $|\mathbf{r}_{n_i} - \mathbf{r}^*_{n_i}| < \epsilon$ for the ith history when $\mathbf{r}$ is on the $i$circle.

After N histories (or N WOC random walks), the value of $u(\mathbf{r}_o)$ is estimated as

$$u(\mathbf{r}_0) = \frac{1}{N} \sum_{i=1}^{N} \left[\psi(\mathbf{r}^*_{n_i}) + s_o \sum_{j=1}^{n_i} \frac{r_j^2}{4} \right], \qquad (8.119)$$

where $\mathbf{r}^*_{n_i}$ is the terminating position on the boundary for the ith history.

8.3.8 Other Boundary Conditions

So far, the WOC Monte Carlo approach has terminated a walk when the walker reaches a surface at which u is specified (a Dirichlet boundary condition). Such a boundary is called an *absorbing* boundary.

But other boundary conditions, particularly in heat conduction problems, are frequently encountered. In a Neumann boundary condition, the normal derivative of u on the boundary is specified, i.e.,

$$\frac{\partial u}{\partial n} = -q, \qquad (8.120)$$

where n is the direction of the inward-pointing normal to Γ and q is a specified normalized flow into D per unit area and time. For example, in heat conduction $q \to q''/k$, where q'' is the heat flux into D, k is the thermal conductivity, and $u(x, y)$ is the temperature. Using a second-order Taylor series expansion, the value of u on Γ is approximated by (Haji-Sheikh and Sparrow, 1967)

$$u|_\Gamma = u|_{\delta n} + q \, \delta n + \frac{S(\delta n)^2}{4}, \qquad (8.121)$$

where a source contribution is included for generality. The quantity $u|_{\delta n}$ is the value of u at a distance δn from the boundary, i.e., the reflection distance. Thus, in a WOC calculation the particle or walker, when it reaches within ϵ of such a boundary, is reflected back into D, a distance δn and the tally is increased by an amount $[q\delta n + S(\delta n)^2/4]$. Hence a Neumann boundary is a *reflecting* boundary.

A Neumann boundary condition with $q = 0$ can, in principle, be used as a pure reflecting boundary, so as to take advantage of any problem symmetries. However, as shown in Example 8.3, this use of a Neumann boundary to simulate boundaries where $\partial u/\partial n = 0$ is generally a poor way to take advantage of symmetries.

In a Robin boundary condition a linear combination of u and $\partial u/\partial n$ is specified. For example, in heat conduction a convective boundary condition has the form

$$\frac{\partial u}{\partial n} = \alpha(u_o - u|_\Gamma), \qquad (8.122)$$

where u_o is the ambient temperature and α is the heat transfer coefficient divided by the thermal conductivity of the medium. Use of a Taylor series expansion gives

$$u|_\Gamma = \frac{1}{1+\alpha\delta n} u|_{\delta n} + \frac{\alpha\delta n}{1+\alpha\delta n} u_o + \frac{S(\delta n)^2/4}{1+\alpha\delta n}. \tag{8.123}$$

This result states that when a WOC particle reaches (within ϵ of) the wall, it is reflected with probability $1/[1+\alpha\delta n]$ and absorbed with probability $[\alpha\delta n]/[1+\alpha\delta n]$. In the latter case, u_o is added to the tally. In both cases, $[S(\delta n)^2/4]/[1+\alpha\delta n]$ is added to the tally. Thus, a Robin boundary condition is a *partially reflecting and partially absorbing* boundary.

The WOC Monte Carlo method has also been used in 2-D transient calculations by Haji-Sheikh and Sparrow (1967). They also addressed nonlinear boundary conditions such as a convective and radiative boundary at which

$$q = \alpha(u|_\Gamma - u_o) + \beta(u_R^4 - u_o^4), \tag{8.124}$$

where u_R is the radiation temperature and β is the product of the radiation interchange factor and the Stefan–Boltzmann constant. Haji-Sheikh and Sparrow also used the WOC method to solve the moving boundary problem.

Example 8.8 Monte Carlo Solution of the 2-D Poisson's Equation

Consider the 2-D Poisson's equation given by Eq. (8.94) for the case considered in Example 8.7, namely, a constant source s_o in a square plate D : $-L/2 < x, y < L/2$. The edges of the medium are kept at a constant value $u_o = 100$, $s_o = 20$, and $L = 10$. What is the value of $u(x, y)$ at $x = y = 2.5$?

Analytic Solution

The analytic solution is given by Eq. (8.102), from which the exact solution is found to be $u_{exact}(2.5, 2.5) = 190.57232\ldots$.

Monte Carlo Solution Using the WOC Method

Three approaches are used. First solve this problem using the entire plate with Dirichlet boundary conditions on the four edges of the plate. Call this approach Problem (a). Because there is symmetry about the x- and y-axes, one could solve this problem for the first quadrant using Neumann boundary conditions with $q = 0$ on the axes and Dirichlet boundary conditions at $x = 5$ and $y = 5$. Call this approach Problem (b). These two equivalent problems are shown in Fig. 8.10. In Problem (c) the axes are treated as true reflecting surfaces so that, if at any time in the WOC walk, the center of a circle ventures into quadrants 2, 3, or 4, it is "reflected" back to quadrant 1, i.e., the center of the ith circle is always taken as $(|x_i|, |y_i|)$. Of course, this simple reflection prescription is possible because of the simple geometry of this problem. Generally, incorporation of a reflecting boundary is very problem-dependent.

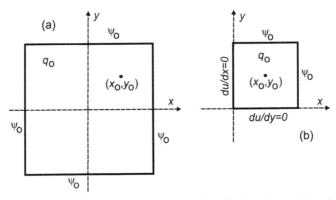

Figure 8.10 Two equivalent problems for a square plate. Problem (a) uses the entire plate plus four Dirichlet boundaries. Problem (b) uses problem symmetry and two Neumann boundary conditions. The point of interest (x_o, y_o) is the midpoint of quadrant 1.

The results obtained with the WOC method for these three problems are summarized in Table 8.2. All three problems give the correct value of u_{exact} for a sufficiently large number N of walks. The error decreases with increasing N about as quickly in all problems. However, Problem (b) is much more computationally expensive, as shown by the large number of steps per walk (SPW). This huge increase in the number of steps is a result of replacing two absorber boundaries by Neumann boundaries. At a Neumann boundary, the reflected particle or walker is still very close to the boundary ($\delta n = 0.01$) and, consequently, it is very likely to run into the same boundary again and again. Hence, it is seen that the use of a Neumann boundary condition with $q = 0$ to simulate a reflecting boundary is generally ill advised. Problem (c), by contrast, is computationally the most attractive since only the distances to the two Dirichlet boundaries are needed in order to calculate the radius of maximal circles. Moreover, Problem (c) avoids the numerical approximation of a reflection distance δ.

Table 8.2 Estimate of $u(2.5, 2.5)$ for Problems (a) and (b). N is the number of walks, %err is the percent difference between u_{exact} and the WOC estimate u, and "SPW" is the average number of steps per walk. Here $\epsilon = 0.001$ and $\delta n = 0.01$.

N	Problem (a)			Problem (b)			Problem (c)		
	$u(x_o, y_o)$	%error	SPW	$u(x_o, y_o)$	%error	SPW	$u(x_o, y_o)$	%error	SPW
10^1	198.74	-4.287	12.80	223.34	-17.196	1752.	230.14	-20.763	12.20
10^2	179.50	5.808	11.65	199.36	-4.610	1534.	183.60	3.659	11.92
10^3	187.24	1.749	11.39	194.19	-1.899	1412.	191.95	-0.724	11.80
10^4	189.67	0.475	11.91	190.72	-0.079	1300.	190.45	0.066	11.89
10^5	190.79	-0.116	11.93	190.96	-0.205	1315.	190.48	0.049	11.92
10^6	190.53	0.020	11.91	190.53	0.020	1303.	190.42	0.082	11.91
10^7	190.56	0.008	11.90	190.54	0.015	1310.	190.54	0.010	11.90

8.3.9 Extension to Three Dimensions

The above continuous WOC Monte Carlo procedure for evaluating the solution to the Laplace equation can be extended directly to arbitrary dimension K (Mikhailov, 1995). In particular, for three dimensions, points are uniformly sampled on the surfaces of maximal spheres. For a sphere centered at $\mathbf{r}$ the radius of the sphere is

$$r = \min[d_1(\mathbf{r}), d_2(\mathbf{r}), \ldots, d_M(\mathbf{r})], \tag{8.125}$$

where $\{d_i(\mathbf{r})\}$ are the shortest distances from $\mathbf{r}$ to the M elementary surfaces that form the boundary Γ of D and on which the value of u is specified. In this section, the discussion is limited to Dirichlet boundary conditions although other types of boundary conditions can be accommodated. This method of generating a random walk between points on spheres is known as the *walking-on-spheres* (WOS) algorithm or, perhaps more poetically, the *dancing on bubbles* algorithm.

To see the generalization to three dimensions, consider the Laplace equation in three-dimensional spherical coordinates (r, θ, ϕ) centered at some arbitrary point $\mathbf{r}_0 \in D$, namely,

$$\frac{1}{r^2}\frac{\partial}{\partial r}\left[r^2 \frac{\partial u}{\partial r}\right] + \frac{1}{r^2 \sin\theta}\frac{\partial}{\partial \theta}\left[\sin\theta \frac{\partial u}{\partial \theta}\right] + \frac{1}{r^2 \sin^2\theta}\frac{\partial^2 u}{\partial \phi^2} = 0, \tag{8.126}$$

with $u(r, \theta, \phi)|_\Gamma = \psi(\mathbf{r})$ prescribed on the surface. The most general nonsingular solution of this equation is (Margenau and Murphy, 1956)

$$u(r, \theta, \phi) = a_0 + \sum_{\ell=1}^{\infty}\sum_{m=1}^{\ell} r^\ell P_\ell^m(\cos\theta)[a_\ell \cos\ell\phi + b_\ell \sin\ell\phi], \tag{8.127}$$

where the functions $P_\ell^m(\cos\theta)$ are the associated Legendre functions and a_ℓ and b_ℓ are arbitrary constants. Integration of the result over all directions gives

$$\int_{4\pi} u(\mathbf{r})\, d\Omega = \int_0^\pi d(\cos\theta) \int_0^{2\pi} d\phi\, u(r, \theta, \phi) = 4\pi a_0. \tag{8.128}$$

Because $u(0, \theta, \phi) \equiv u(\mathbf{r}_0) = a_0$ and is independent of θ and ϕ, Eq. (8.128) then requires

$$u(\mathbf{r}_0) = \frac{1}{4\pi}\int_{4\pi} u(r, \theta, \phi)\, d\Omega, \tag{8.129}$$

i.e., the average value of u over the surface of a sphere equals the value of u at the center of the sphere, exactly analogous to Eq. (8.112).

To estimate $u(\mathbf{r}_0)$ by the WOS algorithm, start each particle at $\mathbf{r}_0$ and pick a point $\mathbf{r}_1$ uniformly sampled on the surface of the largest sphere contained in D. Point $\mathbf{r}_1$ is then taken as the center of the next maximal sphere in D and a point $\mathbf{r}_2$ is uniformly sampled on this sphere. The nth particle's walking-on-spheres terminate when a point $\mathbf{r}_{n_i}$ is selected that is within ϵ of the surface Γ, the nearest surface point being $\mathbf{r}_{n_i}^*$. The value of $\psi(\mathbf{r}_{n_i}^*)$ is then scored as the tally for the particle. Finally, after N histories, the estimate of $u(\mathbf{r}_o)$ is taken as

$$u(\mathbf{r}_0) \simeq \frac{1}{N} \sum_{i=1}^{N} \psi(\mathbf{r}_{n_i}^*). \qquad (8.130)$$

8.3.10 Continuous Monte Carlo for the 3-D Poisson Equation

The inhomogeneous form of Laplace' equation, also known as Poisson's equation,

$$\frac{1}{r^2}\frac{\partial}{\partial r}\left[r^2\frac{\partial u}{\partial r}\right] + \frac{1}{r^2\sin\theta}\frac{\partial}{\partial\theta}\left[\sin\theta\frac{\partial u}{\partial\theta}\right] + \frac{1}{r^2\sin^2\theta}\frac{\partial^2 u}{\partial\phi^2} = -s(\mathbf{r}), \quad (8.131)$$

can also be solved by the WOS Monte Carlo algorithm. First considered is the special case in which $s(\mathbf{r})$ is a constant, followed later by the more general case.

Case: q Is a Constant

It is readily verified that a particular solution of Eq. (8.131), with $s(\mathbf{r}) = s_o$, is $u_p = -r^2 s_o/6$ so that the most general nonsingular solution is the solution of the homogeneous equation, Eq. (8.127), plus the particular solution, i.e.,

$$u(r,\theta,\phi) = a_0 + \sum_{\ell=1}^{\infty}\sum_{m=1}^{\ell} r^\ell P_\ell^m(\cos\theta)[a_\ell\cos\ell\phi + b_\ell\sin\ell\phi] - s_o\frac{r^2}{6}. \quad (8.132)$$

As before, $u(0,\theta,\phi) \equiv u(\mathbf{r}_0) = a_0$, and integration of Eq. (8.132) over all directions yields $\int_{4\pi} u(\mathbf{r})\,d\Omega = 4\pi[a_0 - s_o(r^2/6)]$. Combining these two results yields

$$\boxed{u(\mathbf{r}_0) = \frac{1}{4\pi}\int_{4\pi} u(\mathbf{r})\,d\Omega + S(r),} \qquad (8.133)$$

where $S(r) = s_o r^2/6$. This important result says that the value of u at the center of a sphere of radius r equals the average u over the surface *plus* the function $S(r)$.

Thus, the WOS algorithm again can be used with a slight modification, namely, that at each step in a particle's random walk, the quantity $s_o r_j^2/6$ (where r_j is the radius of the jth sphere in the walk) is added to the particle's tally. When the particle on the n_ith sphere is within ϵ of Γ, the walk is, as before, terminated and the value of $\psi(\mathbf{r}_{n_i}^*)$ is also added to the tally. After N random walks, the estimate of $u(\mathbf{r}_0)$ is given by

$$u(\mathbf{r}_0) \simeq \frac{1}{N}\sum_{i=1}^{N}\left[s_o\sum_{j=1}^{n_i}\frac{r_j^2}{6} + \psi(\mathbf{r}_{n_i}^*)\right]. \qquad (8.134)$$

In Example 8.8, an example of this WOS algorithm is given. The estimate of Eq. (8.134) can also be used with the 2-D Poisson equation with $S(r) = s_o r^2/4$ (Booth, 1981).

Example 8.9 Monte Carlo Solution of Poisson's Equation

Consider Poisson's equation, Eq. (8.131), with $q = s_o$ a constant. The domain D is a cube with sides of length L parallel to the axes and with the origin in the center of the cube, i.e.,

$$D = \{x, y, z : -L/2 \le x, y, z \le L/2\}.$$

On the surface, $u = \psi_o$, a constant. For this problem, the analytic solution can be expressed as

$$u(x, y, z) = \psi_o + \sum_{\ell=1}^{\infty}{}' \sum_{m=1}^{\infty}{}' \sum_{n=1}^{\infty}{}' \frac{64 s_o L^2}{\ell m n \pi^5} \frac{(-1)^{(\ell+m+n-3)/2}}{\ell^2 + m^2 + n^2} \cos \gamma_\ell x \, \cos \gamma_m y \, \cos \gamma_n z,$$

where $\sum'$ indicates that the summations are over only the odd integers and $\gamma_i \equiv i\pi/L$. Of interest is the largest value of u in D, which occurs at $\mathbf{r}_0 = (0, 0, 0)$. In this example, the problem parameters are chosen as $L = 10$, $s_o = 10$, and $\psi = 100$ at all points on the surface. For this particular case, the exact value of $u(0, 0, 0) \simeq 156.21283\ldots$.

The WOS algorithm produced the results for this problem shown in Fig. 8.11. In the left figure, the percent error decreases roughly as $1/N^2$, where N is the number of histories for a very small value of $\epsilon = 0.0001$. Because the Monte Carlo algorithm is only approximate in the ϵ layer, the value of ϵ determines the accuracy of the "converged" Monte Carlo result. In the right graph of Fig. 8.11, the converged accuracy is seen to decrease proportionally to the fraction of the cube's volume in the ϵ layer, i.e., as $\% V_\epsilon = 100 \times (L^3 - V_\epsilon)/L^3$ decreases, where

$$V_\epsilon = L^3 - (L - 2\epsilon)^3.$$

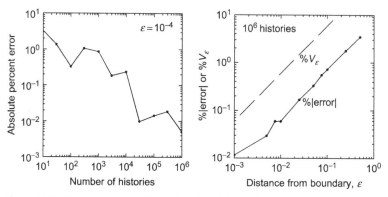

Figure 8.11 The percent error (left) as a function of the number of random walks and (right) as a function of the ϵ parameter.

Case: $s(\mathbf{r})$ Is Variable

For the case that $s(\mathbf{r})$ varies throughout D, Eq. (8.133) still is valid but $S(\mathbf{r})$ is now given by (Mikhailov, 1995)

$$S(r) = \int_{|\mathbf{r}'|<r} G(0, \mathbf{r}')s(\mathbf{r}')dV', \qquad (8.135)$$

where the integration is over all differential volume elements dV' about $\mathbf{r}'$ in a sphere of radius r centered at the origin. Here $G(\mathbf{r}, \mathbf{r}')$ is the Green's function for a sphere and gives the contribution to $u(\mathbf{r})$ caused by a unit strength source at $\mathbf{r}'$. For the case of a sphere of radius r, the Green's function is

$$G(\mathbf{r}, \mathbf{r}') = \frac{1}{4\pi}\left[\frac{1}{|\mathbf{r} - \mathbf{r}'|} - \frac{1}{r}\right]. \qquad (8.136)$$

Note that $G(0, r')$ is spherically symmetric about $\mathbf{r}' = 0$, i.e., there is no θ' or ϕ' dependence, so that dV' in Eq. (8.135) equals $4\pi r'^2 dr'$. Substitution of Eq. (8.136) into Eq. (8.135) yields

$$S(r) = \frac{1}{4\pi}\int_{|\mathbf{r}'|<r}\left[\frac{1}{|\mathbf{r}'|} - \frac{1}{r}\right]s(\mathbf{r}')dV'. \qquad (8.137)$$

Rather than evaluate this integral numerically at each step in the random walk, Monte Carlo quadrature can be used. Pick a point $\mathbf{r}'_i$ inside the sphere of radius r from a PDF proportional to $G(0, \mathbf{r}')$. Hence

$$S(r) \simeq s(\mathbf{r}'_i)\frac{1}{4\pi}\int_{|\mathbf{r}'|<r}\left[\frac{1}{|\mathbf{r}'|} - \frac{1}{r}\right]dV' = s(\mathbf{r}'_i)\int_0^r\left[\frac{1}{r'} - \frac{1}{r}\right]r'^2 dr' = s(\mathbf{r}'_i)\frac{r^2}{6}. \qquad (8.138)$$

When this approximation is averaged over many random walks, the proper value for Eq. (8.137) is obtained. The random position $\mathbf{r}'_i$ is taken as $\mathbf{r}'_i = r'_i\mathbf{\Omega}_i$, where the direction vector $\mathbf{\Omega}_i$ is taken from an isotropic distribution and r'_i is sampled from a PDF proportional to

$$\left[\frac{1}{r'} - \frac{1}{r}\right]r'^2 \propto r'(r - r').$$

From Eq. (8.138) we see that at the jth step of the WOS algorithm, the value of the particle's tally is increased by s (evaluated at some selected point $\mathbf{r}_j + \mathbf{r}'_j$ in the sphere of radius r_j centered at $\mathbf{r}_j$) multiplied by $r_j^2/6$. Thus, the tally for the ith history, which terminates on step n_i, is computed as

$$T_i = \sum_{j=1}^{n_i} s(\mathbf{r}'_j + \mathbf{r}_j)\frac{r_j^2}{6} + \psi(\mathbf{r}^*_{n_i}), \qquad (8.139)$$

where $\mathbf{r}_{n_i}^*$ is the point on the boundary closest to $\mathbf{r}_{n_i}$ that is within ϵ of the boundary.

Finally, when this tally is averaged over many histories N, the estimate of $u(\mathbf{r}_0)$ is obtained, namely,

$$u(\mathbf{r}_0) \simeq \frac{1}{N} \sum_{i=1}^{N} \left\{ \sum_{j=1}^{n_i} s(\mathbf{r}'_j + \mathbf{r}_j) \frac{r_j^2}{6} + \psi(\mathbf{r}_{n_i}^*) \right\}. \tag{8.140}$$

As would be expected, this result reduces to Eq. (8.134) when $s(\mathbf{r}) \to s_0$.

8.3.11 Continuous Monte Carlo for the 2-D Helmholtz Equation

Diffusion processes with absorption are governed by the Helmholtz equation

$$\nabla^2 u(\mathbf{r}) - \beta^2 u(\mathbf{r}) = -s(\mathbf{r}), \quad \mathbf{r} \in D, \quad \text{with } u(\mathbf{r})|_\Gamma = \psi(\mathbf{r}), \tag{8.141}$$

with $\beta^2 > 0$. In the following subsections we have relied on Booth (1981), Hwang and Mascagni (2001), and Mikhailov (1995), who present many more results than presented here.

Solution of the Homogeneous Equation

To see how the previous Monte Carlo methods must be altered for this equation, first consider the homogeneous form of Eq. (8.141) in two dimensions. In polar coordinates with respect to an arbitrary origin at $\mathbf{r}_0$, Eq. (8.141) can be written as

$$\frac{\partial^2 u(r,\theta)}{\partial r^2} + \frac{1}{r} \frac{\partial u(r,\theta)}{\partial r} + \frac{1}{r^2} \frac{\partial^2 u(r,\theta)}{\partial \theta^2} - \beta^2 u(r,\theta) = 0. \tag{8.142}$$

As was done in Section 8.3.6, first separate the dependent variable by assuming $u(r,\theta) = R(r)\Theta(\theta)$. This separation results in two ODEs (with a separation constant α), namely,

$$\frac{d^2\Theta}{d\theta^2} + \alpha^2 \Theta(\theta) = 0, \tag{8.143}$$

$$\frac{d^2 R}{dr^2} + \frac{1}{r} \frac{dR}{dr} + \left(-\beta^2 - \frac{\alpha^2}{r^2} \right) = 0. \tag{8.144}$$

The general solution of the angular equation, Eq. (8.143), subject to the requirement that Θ be cyclic in 2π is

$$\Theta(\theta) = A \cos \ell\theta + B \sin \ell\theta, \tag{8.145}$$

where $\alpha \equiv \ell = 0, 1, 2, \ldots$. The radial equation, Eq. (8.144) with α^2 replaced by ℓ^2, is the modified Bessel equation whose two independent solutions are $I_n(\beta r)$

and $K_n(\beta r)$, the modified Bessel functions of order n. Because $u(r, \theta)$ must be finite at $r = 0$, the singular $K_n(\beta r)$ functions are discarded. Thus, the most general physically realistic solution of Eq. (8.142) has the form

$$u(r, \theta) \equiv R\Theta = a_0 I_0(\beta r) + \sum_{\ell=1}^{\infty} I_\ell(\beta r) [a_\ell \cos \ell\theta + b_\ell \sin \ell\theta], \qquad (8.146)$$

where a_ℓ and b_ℓ are constants.

Integration of this solution over θ yields

$$\int_0^{2\pi} u(r, \theta) \, d\theta = 2\pi a_0 I_0(\beta r). \qquad (8.147)$$

The solution at the origin $u(0, \theta) \equiv u(\mathbf{r}_0)$ is seen from Eq. (8.146) to be simply a_0 because $I_0(0) = 1$ and $I_\ell(0) = 0$, $\ell = 1, 2, 3, \ldots$. Substitution of $a_0 = u(\mathbf{r}_0)$ into Eq. (8.147) produces the useful result

$$u(\mathbf{r}_0) = \frac{1}{I_0(\beta r)} \frac{1}{2\pi} \int_0^{2\pi} u(r, \theta) \, d\theta. \qquad (8.148)$$

This result shows that the solution of the Helmholtz equation at any point $\mathbf{r}_0$ in D equals the average of u over the circumference of a circle of radius r (that is contained in D and centered at $\mathbf{r}_0$) multiplied by the factor $1/I_0(\beta r)$.

To evaluate $u(\mathbf{r}_0)$ by continuous Monte Carlo, a random WOC procedure can again be used. At each step j in the random walk, the particle's weight W is modified by multiplying its previous weight W_{j-1} by $1/I_0(\beta r_j)$, i.e.,

$$W_j = W_{j-1}/I_0(\beta r_j), \quad j = 1, 2, 3, \ldots, n-1 \quad \text{with } W_0 = 1, \qquad (8.149)$$

where r_j are the radii of the maximal circles used in the random WOC. As before, the ith random walk is terminated when $\mathbf{r}_{n_i}$ is within ϵ of the surface Γ, and the score of the walk is set to $W_{n_i}^{(i)} \psi(\mathbf{r}_{n_i}^*)$, where $\mathbf{r}_{n_i}^*$ is the nearest point on Γ from $\mathbf{r}_{n_i}$. Finally, after N random walks, the value of $u(\mathbf{r}_0)$ is estimated as

$$u(\mathbf{r}_0) \simeq \frac{1}{N} \sum_{i=1}^{N} W_{n_i}^{(i)} \psi(r_{n_i}^*). \qquad (8.150)$$

Solution for a Constant Source Term

To obtain the form of the solution to

$$\frac{\partial^2 u(r, \theta)}{\partial r^2} + \frac{1}{r} \frac{\partial u(r, \theta)}{\partial r} + \frac{1}{r^2} \frac{\partial^2 u(r, \theta)}{\partial \theta^2} - \beta^2 u(r, \theta) = -s_o, \qquad (8.151)$$

observe that $u_p = s_o/\beta^2$ is a particular solution. Addition of u_p to the most general physically realistic solution of the corresponding homogeneous equation gives the general solution of Eq. (8.151), namely,

$$u(r, \theta) \equiv R\Theta = a_0 I_0(\beta r) + \sum_{\ell=1}^{\infty} I_\ell(\beta r) [a_\ell \cos \ell\theta + b_\ell \sin \ell\theta] + s_o/\beta^2.$$

$$(8.152)$$

To see how $u(\mathbf{r})$ at the center of a circle is related to the value on the circumference, integrate Eq. (8.152) over θ to obtain

$$\int_0^{2\pi} u(r, \theta) \, d\theta = 2\pi a_0 I_0(\beta r) + 2\pi s_o/\beta^2. \qquad (8.153)$$

The value at the circle center $u(0, \theta) = u(\mathbf{r}_0)$ is independent of θ and, from Eq. (8.152), has the value

$$u(\mathbf{r}_0) = a_0 + s_o/\beta^2. \qquad (8.154)$$

Combining these two results to eliminate a_0 gives the important result

$$\boxed{u(\mathbf{r}_0) = \frac{1}{I_0(\beta r)} \frac{1}{2\pi} \int_0^{2\pi} u(r, \theta) \, d\theta + S(r),} \qquad (8.155)$$

where

$$S(r) \equiv \frac{s_o}{\beta^2} \left(1 - \frac{1}{I_0(\beta r)} \right). \qquad (8.156)$$

From this result, it is seen that u at the center of a circle of radius r equals $S(r)$ plus the average of $u(r, \theta)/I_0(\beta r)$ on the circumference. Thus, the previous random WOC procedure is altered to score the particle's weight times $S(\mathbf{r})$ at each step, in addition to scoring the surface contribution at the end of the walk. After N random walks, the value of $u(\mathbf{r}_0)$ is then estimated as

$$u(\mathbf{r}_0) \simeq \frac{1}{N} \sum_{i=1}^{N} \left\{ \sum_{j=1}^{n_i} W_j^{(i)} S(r_j) + W_{n_i}^{(i)} \psi(r_{n_i}^*) \right\}. \qquad (8.157)$$

In Example 8.4, this WOC procedure is used to solve a particular diffusion problem.

Case for Negative β^2

Sometimes for a multiplying medium, for example neutrons diffusing in a medium containing fissionable material, the constant β^2 in Eq. (8.141) may become negative. If $-\beta^2 < \lambda^*$, where λ^* is the smallest eigenvalue of the system, a unique solution for $u(\mathbf{r})$ still exists. Let $\gamma^2 = -\beta^2 > 0$. Then repeating the above analysis produces the same result except $I_0(\beta r)$ is replaced by $J_0(\gamma r)$, the regular Bessel function of the first kind (see Problem 8.15).

Example 8.10 Monte Carlo Solution of a Diffusion Problem

Consider the two-dimensional Helmholtz equation on a square with constant source s_o and boundary value ψ_o, namely,

$$\frac{\partial^2 u(x,y)}{\partial x^2} + \frac{\partial^2 u(x,y)}{\partial y^2} - \beta^2 u(x,y) = -s_o, \quad \frac{-L}{2} \le x, y \le \frac{L}{2} \quad \text{with } u(x,y)|_\Gamma = \psi_o.$$

For this problem, the analytic solution can be expressed as

$$u(x,y) = \psi_o + \sum_{n=1}^{\infty}{}' \frac{S_n}{\gamma_n^2}\left[1 - \frac{\cosh \gamma_n x}{\cosh \gamma_n L/2}\right]\cos\frac{n\pi}{L}y,$$

where $\sum'$ indicates the summation is over only the odd integers. Here

$$\gamma_n = \left(\frac{n\pi}{L}\right)^2 + \beta^2 \quad \text{and} \quad S_n = \frac{4}{n\pi}[s_o - \beta^2\psi_o](-1)^{(n-1)/2}.$$

For the case $L = 10$, $s_o = \psi_o = 100$, and $\beta^2 = 0.2$, the numerical value of $u(0,0)$ at the center of the square is computed to be $u_{exact}(0,0) = 371.62142\ldots$.

The WOC algorithm, discussed above, produced the results for $u(0,0)$ shown in Fig. 8.12. In the left-hand figure, the percent error decreases roughly as $1/\sqrt{N}$, where N is the number of histories. For these results, a very small value of $\epsilon = 0.00001$ was used. Because the Monte Carlo algorithm is only approximate in the ϵ layer, the value of ϵ determines the accuracy of the "converged" Monte Carlo result (here assumed to be obtained after 10^7 histories). In the right-hand graph of Fig. 8.12, the converged accuracy is seen to decrease as the fraction of the square's area in the ϵ layer decreases, i.e., as $\%A_\epsilon = 100(L^2 - A_\epsilon)/L^2$ decreases, where $A_\epsilon = L^2 - (L - 2\epsilon)^2$. Although initially the decrease in the error is directly proportional to $\%A_\epsilon$, it decreases less rapidly as ϵ becomes very small because the small boundary layer has less effect on u at the center than does the internal source s_o.

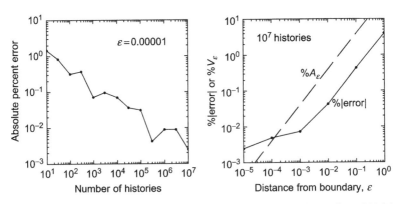

Figure 8.12 The percent error (left) as a function of the number of random walks and (right) as a function of the ϵ parameter.

8.3.12 The 3-D Helmholtz Equation

The three-dimensional diffusion (Helmholtz) equation, Eq. (8.141) with $\beta^2 > 0$, can also be solved by the WOC algorithm. As shown by Mikhailov (1995), the weight of the particle as it leaves the jth sphere, $j = 0, 1, 2, 3, \ldots$, with radius r_j, for $\beta^2 > 0$, is given by

$$W_j = \frac{\beta r_j}{\sinh(\beta r_j)} W_{j-1}, \quad j = 1, 2, 3, \ldots, n - 1 \quad \text{with } W_0 = 1. \quad (8.158)$$

Here the quantity $(\beta r_j)/\sinh(\beta r_j)$ is the probability that the particle, beginning the diffusion process from the center of the sphere, escapes absorption and reaches the surface (for the first time) (see Problem 8.10). At each step in the random walk, the tally for the particle is augmented by a source contribution $W_j S_j$, where

$$S_j = \frac{1}{4\pi} \int_{V_j} \frac{\sinh(\beta[r_j - |\mathbf{r}_j - \mathbf{r}'|])}{|\mathbf{r}_j - \mathbf{r}'| \sinh(\beta r_j)} s(\mathbf{r}') dV', \quad (8.159)$$

where integration is over the volume of the jth sphere. This integral could be evaluated by standard numerical techniques, but, more aptly, Monte Carlo methods can be used. A point $\mathbf{r}'_j$ is picked randomly and uniformly in the jth sphere. Then evaluation of the integrand of Eq. (8.159) at $\mathbf{r}'_j$ multiplied by the sphere's volume $V_j = (4/3)\pi r_j'^3$ gives the source contribution for the jth step as

$$S_j = \frac{r_j^3}{3} \frac{\sinh(\beta[r_j - |\mathbf{r}_j - \mathbf{r}'_j|])}{|\mathbf{r}_j - \mathbf{r}'_j| \sinh(\beta r_j)} s(\mathbf{r}'_j). \quad (8.160)$$

This quantity, averaged over many particle walks, then produces a Monte Carlo estimate of Eq. (8.159). For the case that $s(\mathbf{r}) = s_o$, a constant, the integral in Eq. (8.159) can be evaluated analytically as (see Problem 8.16)

$$S_j = \frac{s_o}{\beta^2} \left[1 - \frac{\beta r_j}{\sinh(\beta r_j)} \right]. \quad (8.161)$$

The tally for the ith random walk, which reaches a point within ϵ of the boundary on step n_i, is

$$T_i = \sum_{j=1}^{n_i} W_j S_j + W_{n_i} \psi(\mathbf{r}_{n_i}^*), \quad (8.162)$$

where $\mathbf{r}_{n_j}^*$ is the nearest point on Γ from $\mathbf{r}_{n_j}$. Thus, after N random walks, the estimate of $u(\mathbf{r}_0)$ is

$$u(\mathbf{r}_o) = \frac{1}{N} \sum_{i=1}^{N} T_i. \quad (8.163)$$

The Monte Carlo WOS algorithm for treating the Helmholtz equation can be extended to cases where not only $s(\mathbf{r})$ is variable, but also β^2, the parameter that determines how readily a diffusing particle is absorbed (or multiplied), varies with position in the medium (Mikhailov, 1995). The WOS method can also be extended to an arbitrary number of dimensions in which the random walks are performed on the surfaces of hyperspheres (Booth, 1981; Mikhailov, 1995).

8.4 Transient Partial Differential Equations

Parabolic PDEs are often encountered in the modeling of the transient responses of many systems. Exact solutions of the relevant equations are available only for idealized problems and numerical methods generally must be employed. Two Monte Carlo methods are among these. In one approach, Markov chains of either random walks through a discrete spatial grid or continuous random walks are used to estimate the solution at a specified time to that specified by the initial condition for the problem. In the second approach, Green's functions, appropriate to the problem geometry, can be used to perform continuous (floating) random walks (much like the WOC algorithm) to relate the solution at a fixed time to the initial condition. Alternatively, the Green's function can be convoluted with the inhomogeneous term of the equation and evaluated by Monte Carlo integration. All of these approaches are reviewed in this section.

8.4.1 Reverse-Time Monte Carlo

One approach to obtain transient solutions is to use Markov chains to obtain the solution at some specified time by walking backwards in time to reach the known solution at some initial time. To illustrate this reverse-time Monte Carlo (RMC) approach, the time-dependent heat conduction equation for an isotropic homogeneous medium without a volumetric heat source and subject to Dirichlet boundary conditions is used. This parabolic equation is

$$\frac{\partial T(\mathbf{r}, t)}{\partial t} = \alpha \nabla^2 T(\mathbf{r}, t), \quad \mathbf{r} \in D \quad \text{with} \quad T(\mathbf{r}, t)|_\Gamma = T_{BC}(\mathbf{r}_s, t), \quad (8.164)$$

with the initial condition (IC) $T(\mathbf{r}, 0) = T_{IC}(\mathbf{r})$. Here $T_{BC}(\mathbf{r}_s, t)$ and $T_{IC}(\mathbf{r})$ are specified, $\mathbf{r}_s$ is any point on the surface Γ that encloses D, and α is the thermal diffusivity of the medium.[10]

Both discrete (fixed grid) and continuous random walks through D can be used in the RMC approach. First consider an equimesh discrete Cartesian grid with adjacent nodes separated by a small distance ℓ, such as the 2-D grid shown

---

[10] Throughout this section it is assumed that D, Γ, and the medium properties are constant, i.e., independent of t and T.

in Fig. 8.6. Use of finite differences to approximate the derivatives in Eq. (8.164) yields (see Problem 8.17) an "interior" node, i.e., a node not on Γ,

$$T(\mathbf{r}_o, t) \simeq \frac{\alpha \Delta t}{\ell^2} \sum_{nn} T(\mathbf{r}_{nn}, t - \Delta t) + \left(1 - \frac{m\alpha \Delta t}{\ell^2}\right) T(\mathbf{r}_o, t - \Delta t). \quad (8.165)$$

Here $\mathbf{r}_{nn}$ are the m "nearest neighbor" nodes, all of which are a distance ℓ from $\mathbf{r}_o$. The value of m and the nearest neighbor nodes are the same as in the steady-state finite difference Eq. (8.84). To interpret Eq. (8.165) in probabilistic terms, define

$$p_{nn} = \frac{\alpha \Delta t}{\ell^2} \quad \text{and} \quad p_o = \left(1 - \frac{m\alpha \Delta t}{\ell^2}\right), \quad (8.166)$$

so that Eq. (8.165) can be written as

$$T(\mathbf{r}_o, t) = \sum_{nn} p_{nn} T(\mathbf{r}_{nn}, t - \Delta t) + p_o T(\mathbf{r}_o, t - \Delta t). \quad (8.167)$$

Equation (8.167) forms the basis of using random walks through the fixed grid with each step or transition giving T at ever earlier times. To interpret p_{nn} and p_o as transition probabilities for the RWs, they must (1) be nonnegative and (2) satisfy $p_o + \sum_{nn} p_{nn} = 1$ so that, at each time step, the particle either stays at its present node (with probability p_o) or moves to one of the neighboring nodes (with probability p_{nn}). Moreover, Eq. (8.167) says that if a particle is at $\mathbf{r}_o$ at time t, it arrives at its next node (or stays at the same node) at an earlier time $t - \Delta t$. In other words, at each step of a particle's RW, the time associated with the moving particle decreases by an amount Δt. Hence the name reverse-time Monte Carlo.

It is often tempting to choose $p_{nn} = 1/m$ so that $p_o = 0$, thereby simplifying Eq. (8.167) by eliminating the last term. With this choice, the RW always moves to a neighboring node at each time step. However, eliminating the possibility that an RW can stay at its current node at the next time step can introduce severe errors, as is discussed shortly. Also Δt (and hence p_{nn}) must be chosen so the time of interest t is an *integer* multiple of Δt. Thus, an integer number n_s of reverse time steps can reach $t = 0$, i.e., $t = n_s \Delta t$.

RMC Algorithm

Suppose the temperature at some interior node at $\mathbf{r}_o$ at time $t = n_s \Delta t$ sought in an object whose surface temperature is prescribed at a function of position and time. The RW algorithm is as follows:

1. A particle begins its RW at node $\mathbf{r}_o$ at time $n_s \Delta t$ and moves through the grid according to Eq. (8.167). Keep track of the number of steps or transitions s made by the particle.
2. If the particle reaches a point on the boundary when $(n_s - s) > 0$, then the boundary temperature is scored and a new RW is begun.
3. If the particle is still in the object after n_s steps, the initial temperature at the node reached by the particle when $(n_s - s) = 0$ is scored.
4. Repeat steps 1 through 3 for a large number N of RWs, each beginning at $\mathbf{r}_o$.
5. The scores for each RW are summed and the result is divided by N to give the average score for an RW. This average is the RMC estimate of $T(\mathbf{r}_o, t)$.

Two examples of using RMC with a discrete grid are given in Examples 8.11 and 8.12. Although these examples are both for 1-D spatial domains, they are readily adapted to 2-D and 3-D domains as was done earlier for steady-state finite difference problems (although the analytic solutions are not).

Example 8.11 One-Dimensional Quenching Problem

Consider a 1-D infinite homogeneous slab of thickness L initially at temperature $T(x, 0) = T_o$, $x \in [0, L]$. At $t = 0$, the boundaries are reduced in a stepwise manner to $T_b = T(0, t) = T(L, t) = 0$ for $t \geq 0$.

The analytical solution to this quenching problem is (Cole et al., 2011)

$$\frac{T(x,t)}{T_o} = D(0) + \sum_{n=1}^{\infty} [D(n) + D(-n)],$$

where

$$D(n) = \frac{1}{2}\left[\mathrm{erfc}\left(\frac{(2n-1)L+x}{\sqrt{4\alpha t}}\right) - 2\mathrm{erfc}\left(\frac{2nL+x}{\sqrt{4\alpha t}}\right) + \mathrm{erfc}\left(\frac{(2n+1)L+x}{\sqrt{4\alpha t}}\right) \right].$$

In practice, only the first few terms ($\simeq 5$) in the infinite sum are needed to evaluate this exact solution.

An RMC calculation can give astoundingly accurate results, as shown in Fig. 8.13, with far less effort than that required to obtain the above analytic solution. Such good agreement was obtained by using $N = 10^5$ RMC random walks starting at each node. For smaller values of N, the stochastic nature of the RMC results becomes increasingly apparent. Also this is an "easy" problem because each random walk ending at an interior node contributes a nonzero value to the tally of the starting node.

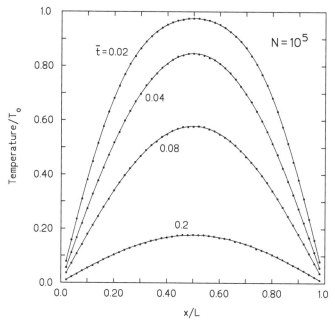

Figure 8.13 Comparison of exact solutions (lines) with RMC solutions (data points) for the quenching problem. One-sigma error bars for the RMC points are almost of the same size as the points and, hence, are not shown. Results for four dimensionless times $\bar{t} = \alpha t / L^2$ are shown. Here $p_{nn} = 1/4$ and $p_o = 1/2$ so that the number of time steps is $n_s = 200, 400, 800,$ and 2000 for the smallest to largest times $\bar{t}$, respectively.

Example 8.12 One-Dimensional Transient Green's Function

Consider a 1-D infinite homogeneous slab of thickness L initially at ambient temperature, i.e., $T(x, 0) = 0$, $x \in [0, L]$. At $t = 0$ a unit impulse of thermal energy is deposited at x_o so that the initial condition is $T(x, 0) = \delta(t)\delta(x - x_o)$ causing the node at x_o to instantaneously acquire a temperature T_o and all the other nodes remain at zero (ambient) temperature. The resulting solution $T(x, t | x_o)$ is the Green's function for a slab whose surfaces are kept at $T_b = 0$.

The exact solution for this Green's function is (Cole et al., 2011)

$$\frac{T(x, t | x_o)}{T_o} = \frac{L}{2\sqrt{\pi \alpha t}} \left\{ B(0) + \sum_{n=1}^{\infty} [B(n) + B(-n)] \right\},$$

where

$$B(n) = \exp\left(\frac{(2nL + x - x_o)^2}{4\alpha t} \right) - \exp\left(\frac{(2nL + x + x_o)^2}{4\alpha t} \right).$$

Again only the first few terms ($\simeq 10$) in the infinite sum are needed to evaluate this exact solution.

In Fig. 8.14 an RMC calculation is compared to the exact analytic solution. This is a much more difficult problem for RMC than is the quenching problem given in Example 8.11 because the only nonzero tally contribution for a node is from a random walk that ends at node x_o. Most random walks thus make no contribution. Even with 10^5 random walks for each node, the standard error of the Monte Carlo estimate is still appreciable.

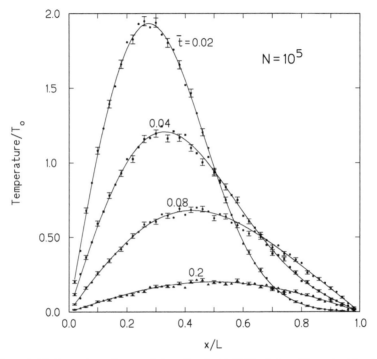

Figure 8.14 Comparison of exact solutions (lines) with RMC solutions (data points) for the 1-D slab Green's function with zero boundary temperatures. In this example, results are for $x_0/L = 0.26$. four dimensionless times $\bar{t} = \alpha t/L^2$ are shown. Here $p_{nn} = 1/4$ and $p_o = 1/2$ so that the number of time steps are $n_s = 200, 400, 800$, and 2000 for the smallest to largest times $\bar{t}$, respectively.

Consequences of Using a Discrete Grid

The use of a discrete grid and the finite difference in the RMC method approximation introduces potential difficulties in addition to those already encountered earlier for steady-state finite difference problems. Recall that these earlier difficulties include (1) difficulty in modeling nonplanar, noncylindrical, or nonspherical borders (i.e., the stair-step effect), (2) difficulty with irregularly shaped objects with boundary conditions involving normal derivatives, (3) the discretization or systematic error introduced by the first-order finite difference spatial approximation, and (4) the rapid increase in the number of RWs needed before a terminating boundary is reached when the number of nodes are increased in an effort to reduce difficulty (3).

In the transient application of finite differences, two additional difficulties are encountered, one obvious and the other more subtle. First, in order to reduce the discretization error, small values of the grid spacing ℓ must be used. But because the time step Δt is constrained by transition probability requirement $p_{nn} = \alpha \Delta t / \ell^2 \leq 1/m$ (see Eq. (8.166)), the time step must become increasingly smaller as ℓ is decreased. In other words, doubling the number of spatial nodes in any direction quadruples the number of times steps n_s needed to obtain the solution at time t.

A second difficulty, and one that is often ignored, arises in "difficult" problems in which nonzero scores are produced by initial temperatures at only a few nodes. The Green's function of Example 8.12 is a prime example in which an RW starting at any node produces a nonzero score only if the walk ends at node $\mathbf{r}_o = x_o$. For the extreme case, in an effort to make Δt as large as possible, $p_{nn} = 1/m$ and $p_o = 0$. With such a choice, a particle must go to a neighboring node at every step in the RW. In such a case and when an RW ends at an interior node, the parity of the ending node is the same as that of the starting node if n_s is even.[11] The opposite is true if n_s is odd. In either case, half of the nodes can never be ending nodes for an RW, and if $\mathbf{r}_o$ is amongst these neglected nodes, then the RMC estimate for the temperature of the starting node is always zero! To avoid this situation, it is important that Δt be chosen so that p_o is comparable to p_{nn}.

Finally, all of these difficulties associated with the finite difference approximation can be avoided by using continuous (floating) RWs (such as the WOC and WOS RWs used earlier for steady-state problems).

WOC/WOS Method for Transient Problems

A continuous random walk for the transient heat diffusion equation is very similar to the WOC/WOS method used for elliptic equations in Section 8.3.6. To illustrate the continuous random walk approach, consider a homogeneous two-dimensional region within a circle of radius r, centered at (x, y), and with an initially uniform temperature. At $t \geq 0$ the temperature on the circumference begins to vary in an arbitrary but prescribed manner $T(r, \theta, t)$ or $T(r, \omega, t)$, where $\omega \equiv \cos \theta$. The temperature at the center of the circle $T(x, y, t)$ is formally given by (Haji-Sheikh, 1965)

$$T(x, y, t) = \int_{F=0}^{1} \int_{\tau=0}^{t} T(x, y, t - \tau) \, dF \, dH^{(2)}, \qquad (8.168)$$

---

[11] The parity (evenness/oddness) of a node is the parity of the sum of its position indices. For example in 3-D, the parity of node $\mathbf{r}_{ijk} \equiv (x_i, y_j, z_k)$ is the parity of $i + j + k$.

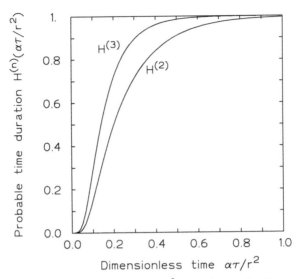

Figure 8.15 The CDFs for dimensionless time $\alpha\tau/r^2$ required to travel a distance r as given by Eqs. (8.169) and (8.171).

where

$$F = \frac{\omega}{2\pi} \qquad \text{and} \qquad H^{(2)}\left(\frac{\alpha\tau}{r^2}\right) = 1 - 2\sum_{k=1}^{\infty} \frac{\exp[-\beta_k^2 \alpha\tau/r^2]}{\beta_k J_1(\beta_k)}. \qquad (8.169)$$

Here β_k are the roots of the Bessel function $J_0(\beta)$ ordered such that $0 < \beta_1 < \beta_2 < \dots$.

Equation (8.168) can be interpreted in a probabilistic sense much as was done with Eq. (8.167); F is a PDF with the property $dF/d\omega$ is a constant so that a particle starting at (x, y) has an equal chance of reaching any angular position on the circumference. The cumulative distribution function (CDF) $H^{(2)}$ is the distribution of the time needed for a particle to step from the center to the circumference of the circle (see Fig. 8.15). So setting $\rho = H^{(2)}(\alpha\tau/r^2)$, where ρ is a sample from $\mathcal{U}(0, 1)$, one can obtain a sample of $\tau \equiv \Delta t.$[12] Because $0 < \rho < 1$, samples of τ for a given value of r vary or "float" as does the step size.

The continuous, reverse-time random walk used to estimate $T(x, y, t)$ is, as before, based on the average scores of a large number of particle histories. Consider a homogeneous, source-free domain D with a specified initial temperature $T_{IC}(x, y, 0)$ and a boundary temperature $T_{BC}(x_s, y_s, t)$, $t > 0$, for (x_s, y_s)

---

[12] In practice, one interpolates from a table of a large number of $\alpha\tau/r^2$ and corresponding $H^{(2)}(\alpha\tau/r^2)$ values to avoid the huge numerical expense of solving $\rho = H^{(2)}(\alpha\tau/r^2)$. Alternatively, empirical results fitted to such a table are available (Haji-Sheikh and Howell, 2006).

on the boundary Γ. Each history for this Dirichlet problem is generated as follows.

1. Start each history or RW at (x, y) and at time t. A maximal circle centered at (x, y) with radius equal to the shortest distance between the center and Γ is constructed. Then two random numbers are used: one to determine where on the circle (F) and the other to sample the time $(H^{(2)})$ the particle reaches the circle. Then, with the location on the circle as a center, a new maximal circle is constructed. This procedure is continued as in the WOC algorithm. Let τ_i be the time increment for the particle to travel from the center to the circumference of circle i.
2. If the particle arrives at (or within a small distance ϵ of) the boundary when $(t - \sum_i \tau_i) = t_b \geq 0$, the temperature T_{BC} at that boundary location at time t_b is scored and the history is terminated.
3. If a point inside Γ is reached when $(t - \sum_i \tau_i) \leq 0$, the initial temperature T_{IC} at that point is scored and the history is ended.

Finally, the scores for a large number N of histories are summed and divided by N to give an estimate of $T(x, y, t)$.

Extension to Three Dimensions

The analog of Eq. (8.168) in three dimensions is

$$T(x, y, z, t) = \int_0^1 \int_0^1 \int_{\tau=0}^t T(r, \omega, \phi, t - \tau) \, dF \, dG \, dH^{(3)}, \qquad (8.170)$$

where $F = \omega/2\pi$, $G = (1 - \cos\theta)/2$, and

$$H^{(3)}\left(\frac{\alpha\tau}{r^2}\right) = 1 + 2\sum_{k=1}^{\infty}(-1)^k \exp[-k^2\pi^2\alpha\tau/r^2]. \qquad (8.171)$$

The CDF $H^{(3)}(\alpha\tau/r^2)$ is shown in Fig. 8.15. To estimate $T(x, y, x, t)$ histories are generated as in the two-dimensional case except now the continuous random walk proceeds from points on maximal spherical surfaces just as in the WOS method.

Other Extensions

The above analysis applies to regions that initially have a uniform temperature. The extension of the method using continuous (floating) random walks to accommodate nonuniform initial temperature distributions is given by Haji-Sheikh (1965). He also shows how the above method used for the Dirichlet transient problem can also be used to analyze problems with a variety of other boundary conditions. Also if $T(\mathbf{r}, t)$ is sought for a sequence of times up to $t = t_{max}$, only a single simulation for $t = t_{max}$ is needed. Clever bookkeeping then can give

$T(\mathbf{r}, t)$ at all intermediate times, thereby saving considerable computational expense.

Since its introduction in 1967 (Haji-Sheikh and Sparrow, 1967), the floating random walk method has been incrementally extended. For example, recent work (Bahadori et al., 2018) has extended the floating random walk method to handle composite layered materials with temperature-dependent thermal properties.

8.4.2 Green's Function Approach

The Monte Carlo random walks (both discrete and continuous) described in the previous sections are all based on the underlying PDE describing the diffusion of heat in an object. However, an entirely different approach based on integrals involving Green's functions can also be used. Because integrals, especially multidimensional ones, are readily evaluated with Monte Carlo, this alternative approach is ideally suited to Monte Carlo methods.

Green's Function Solution Equation

The transient temperature in an object is described by a generalization of Eq. (8.164), namely,[13]

$$\nabla^2 T(\mathbf{r}, t) + \frac{1}{k} q(\mathbf{r}, t) = \frac{1}{\alpha} \frac{\partial T(\mathbf{r}, t)}{\partial t}, \qquad \mathbf{r} \in D, \quad t > 0, \qquad (8.172)$$

where $q(\mathbf{r}, t)$ is the volumetric source of heat production. For simplicity, the medium is taken as isotropic and homogeneous so the thermal conductivity k and thermal diffusivity α are constant. An initial condition (IC) and a boundary condition (BC) are imposed on Eq. (8.172) to obtain a unique solution, namely,

$$\text{IC:} \quad T(\mathbf{r}, 0) = T_{IC}(\mathbf{r}) \qquad \text{and} \qquad \text{BC:} \quad g_i \frac{\partial T}{\partial n_i} + h_i T = f_i(\mathbf{r}_{si}, t),$$

$$(8.173)$$

where T_{IC} and f_i are prescribed functions and the temperature T and the *outward* normal derivative $\partial T / \partial n_i$ are evaluated at the boundary surfaces Γ_i, with $\mathbf{r}_{si}$ being a point on the surface Γ_i. For $g_i = 0$ and $h_i = 1$, the boundary temperature $T_{BC_i}(\mathbf{r}_{si}, t)$ is specified (Dirichlet boundary condition), i.e., $f_i(\mathbf{r}_{si}, t) = T_{BC_i}(\mathbf{r}_{si}, t)$. For $h_i = 0$ the boundary heat flux is specified (Neumann boundary condition), and if $k_i \neq 0$ and $h_i \neq 0$ one has a mixed boundary condition (Robin boundary condition). Here $\partial / \partial n_i$ is the derivative along the *outward* normal to the surface Γ_i, $i = 1, \ldots, s$. The surface Γ that encloses domain D is $\Gamma = \bigcup_{i=1}^{s} \Gamma_i$. The subsurfaces Γ_i are introduced to allow for the

---

[13] In 2-D coordinates, $\mathbf{r} = (x, y)$ (Cartesian) or $\mathbf{r} = (r, \theta)$ (polar). In 3-D coordinates, $\mathbf{r} = (x, y, z)$ (Cartesian), $\mathbf{r} = (r, \theta, z)$ (cylindrical), or $\mathbf{r} = (r, \theta, \phi)$ (spherical).

possibility that different types of boundary conditions are applied to different parts of Γ, e.g., T may be fixed along one part of Γ and the heat flux may be specified along another portion of Γ.

To obtain the so-called Green's function solution equation (GFSE), begin by defining the auxiliary problem of finding the Green's function associated with the above temperature problem.[14] The associated Green's function is the solution of

$$\nabla^2 G(\mathbf{r}, t | \mathbf{r}', \tau) + \frac{1}{\alpha} \delta(\mathbf{r} - \mathbf{r}') \delta(t - \tau) = \frac{1}{\alpha} \frac{\partial G(\mathbf{r}, t | \mathbf{r}', \tau)}{\partial t}, \tag{8.174}$$

for the *same* geometry used for finding $T(\mathbf{r}, t)$ but with *homogeneous* boundary conditions and initial condition, i.e.,

$$\text{IC:} \quad G(\mathbf{r}, t | \mathbf{r}', \tau) = 0 \quad t < \tau \qquad \text{and} \qquad \text{BC:} \quad k_i \frac{\partial T}{\partial n_i} + h_i T = 0. \tag{8.175}$$

The Green's function has the following properties:

1. The Green's function obeys causality, i.e., $G \geq 0$ in domain D for $t - \tau \geq 0$ and $G = 0$ for $t - \tau < 0$.
2. The Green's function obeys the reciprocity relation (see Appendix D): $G(\mathbf{r}, t | \mathbf{r}', \tau) = G(\mathbf{r}', -\tau | \mathbf{r}, -t)$.
3. The time dependence of the Green's function is always $t - \tau$.
4. In Cartesian coordinates, $G(\mathbf{r}, t | \mathbf{r}', \tau)$ has units of m^{-1} in 1-D, m^{-2} in 2-D, and m^{-3} in 3-D.
5. $G(\mathbf{r}, t' | \mathbf{r}, t') = \delta(\mathbf{r} - \mathbf{r}')$.
6. $G(\mathbf{r}, t | \mathbf{r}', t') = 0$ for $\mathbf{r} \in \Gamma$ and for any $t \geq t'$.

The derivation of the GFSE starts with the reciprocity property $G(\mathbf{r}, t | \mathbf{r}', \tau) = G(\mathbf{r}', -\tau | \mathbf{r}, -t)$. With this relation, Eq. (8.174) can be written as

$$\nabla_{\mathbf{r}'}^2 G(\mathbf{r}', -\tau | \mathbf{r}, -t) + \frac{1}{\alpha} \delta(\mathbf{r}' - \mathbf{r}) \delta(t - \tau) = -\frac{1}{\alpha} \frac{\partial G(\mathbf{r}', -\tau | \mathbf{r}, -t)}{\partial \tau}, \tag{8.176}$$

where $\nabla_{\mathbf{r}'}^2$ is the Laplacian operator for the $\mathbf{r}'$ variable. Then a change of variables in Eq. (8.172) is made by replacing $\mathbf{r}$ by $\mathbf{r}'$ and t by τ to give

$$\nabla_{\mathbf{r}'}^2 T(\mathbf{r}', \tau) + \frac{1}{k} q(\mathbf{r}', \tau) = \frac{1}{\alpha} \frac{\partial T(\mathbf{r}', \tau)}{\partial \tau}. \tag{8.177}$$

Next, multiply Eq. (8.177) by $T(\mathbf{r}', \tau)$, multiply Eq. (8.177) by $G(\mathbf{r}', -\tau | \mathbf{r}, -t)$, subtract the results, integrate the subtracted result over all $\mathbf{r}'$ in D and over τ from 0 to $\tau + \epsilon$ (where ϵ is a small positive number), and rearrange the integrated

---

[14] The derivation given here is based on a much more detailed one by Cole et al. (2011).

result to obtain

$$
T(\mathbf{r}, t) = -\int_D G(\mathbf{r}', -\tau | \mathbf{r}, -t) T(\mathbf{r}', \tau) \Big|_{\tau=0}^{t+\epsilon} d\mathbf{r}'
$$

$$
+ \int_{\tau=0}^{t+\epsilon} \int_D \alpha [G(\mathbf{r}', -\tau | \mathbf{r}, -t) \nabla_{\mathbf{r}'} T(\mathbf{r}', \tau) - T(\mathbf{r}', \tau) \nabla_{\mathbf{r}'} G(\mathbf{r}', -\tau | \mathbf{r}, -t)] d\mathbf{r}' d\tau
$$

$$
+ \int_{\tau=0}^{t+\epsilon} \int_D \frac{\alpha}{k} G(\mathbf{r}', -\tau | \mathbf{r}, -t) q(\mathbf{r}', \tau) d\mathbf{r}' d\tau. \tag{8.178}
$$

The first and second terms on the right side of Eq. (8.178) can be simplified and expressed in terms of the initial and boundary conditions of Eq. (8.173), respectively. Consider the first term on the right-hand side of Eq. (8.178). At the upper limit at $\tau = t + \epsilon$, $G(\mathbf{r}', -(t + \epsilon) | \mathbf{r}, -t) = G(\mathbf{r}, t | \mathbf{r}', t + \epsilon)$ (by reciprocity), which is zero because of causality. Also at the lower limit $\tau = 0$, $T(\mathbf{r}', 0) = T_{IC}(\mathbf{r}')$ the initial temperature distribution. Thus, the first term is determined solely by the initial condition as

$$
I_{IC} = \int_D G(\mathbf{r}, t | \mathbf{r}', \tau) T_{IC}(\mathbf{r}') d\mathbf{r}'. \tag{8.179}
$$

Now consider the second term on the right-hand side of Eq. (8.178). From Green's second integral theorem, this term can be written as[15]

$$
I_{BC} = \int_0^{t+\epsilon} \sum_{\Gamma_i} \int_{\Gamma_i} \alpha \left[G(\mathbf{r}', -\tau | \mathbf{r}, -t) \frac{\partial T}{\partial n_i'} \Big|_{\mathbf{r}'=\mathbf{r}_{si}'} - T \frac{\partial G(\mathbf{r}', -\tau | \mathbf{r}, -t)}{\partial n_i'} \Big|_{\mathbf{r}'=\mathbf{r}_{si}'} \right] d\Gamma_i' d\tau. \tag{8.180}
$$

Finally, for simplicity, assume a Dirichlet boundary condition is applied to all of Γ, i.e., $T(\mathbf{r}_s, t) = T_{BC}(\mathbf{r}_s, t)$ is specified for $\mathbf{r}_s$ on the surface Γ. Also $G(\mathbf{r}_{si}', -\tau | \mathbf{r}, -t) = 0$ on the surface (see Eq. (8.175)) so that Eq. (8.180), after applying the reciprocity relation, becomes

$$
I_{BC} = -\alpha \int_0^{t+\epsilon} \int_\Gamma T_{BC}(\mathbf{r}_s', \tau) \frac{\partial G(\mathbf{r}, t | \mathbf{r}', \tau)}{\partial n'} \Big|_{\mathbf{r}'=\mathbf{r}_s'} d\Gamma' d\tau. \tag{8.181}
$$

Simplifications for the other types of boundary conditions are given by Cole et al. (2011).

In the third term on the right-hand side of Eq. (8.178), the reciprocity property allows $G(\mathbf{r}', -\tau | \mathbf{r}, -t)$ to be written as $G(\mathbf{r}, t | \mathbf{r}', \tau)$. Finally, taking the

---

[15] Green's second theorem says that for any continuous scalar functions $u(\mathbf{r})$ and $v(\mathbf{r})$

$$
\int_D [u \nabla^2 v - v \nabla^2 u] dV = \int_\Gamma [u \nabla v - v \nabla u] \cdot \mathbf{n} \, d\Gamma = \int_\Gamma \left[u \frac{\partial v}{\partial n} - v \frac{\partial u}{\partial n} \right] d\Gamma,
$$

where $\partial/\partial n$ is the derivative with respect to the *outward* normal.

limit as $\epsilon \to 0$ affects nothing in the simplified results, so the GFSE *for Dirich-let boundary conditions* becomes

$$T(\mathbf{r}, t) = \int_D G(\mathbf{r}, t | \mathbf{r}', \tau) T_{IC}(\mathbf{r}') \, d\mathbf{r}' \qquad \text{(IC)}$$

$$- \alpha \int_0^t \int_\Gamma T_{BC}(\mathbf{r}'_s, \tau) \left. \frac{\partial G(\mathbf{r}, t | \mathbf{r}', \tau)}{\partial n'} \right|_{\mathbf{r}' = \mathbf{r}'_s} d\Gamma' \, d\tau \qquad \text{(BC)}$$

$$+ \frac{\alpha}{k} \int_0^t \int_D G(\mathbf{r}, t | \mathbf{r}', \tau) q(\mathbf{r}', \tau) \, d\mathbf{r}' \, d\tau. \qquad \text{(heat source)}$$

$$\text{(8.182)}$$

If the Green's function $G(\mathbf{r}, t | \mathbf{r}', \tau)$ were known analytically in closed form, the three integral terms in Eq. (8.182) could be evaluated straightforwardly with numerical quadrature. In particular Monte Carlo integration, discussed in Section 2.6, is ideally suited to evaluate these multidimensional integrals. However, analytic closed-form Green's functions are available only for geometrically simple domains D such as circles, spheres, rectangles, and parallelepipeds. In the section below, Green's functions for such simple domains are applied to more complex-shaped domains.

An Important Property of the Green's Function

But before the decomposition of D into simpler subdomains is discussed, an important property of the Green's function should be noted. Consider the case in which (1) there is no heat source in D ($q = 0$), (2) the boundary condition is $T_{BC}(\mathbf{r}_s, t) = 1$ for $\mathbf{r}_s \in \Gamma$, and (3) the initial condition $T_{IC}(\mathbf{r}) = 1$ for $\mathbf{r} \in D$. Then $T(\mathbf{r}, t) = 1$ is the solution for all $\mathbf{r} \in D$ and $t > 0$. Substitute this solution into Eq. (8.182) to find

$$1 = \int_D G(\mathbf{r}, t | \mathbf{r}', 0) \, d\mathbf{r}' - \alpha \int_0^t \int_\Gamma \left. \frac{\partial G(\mathbf{r}, t | \mathbf{r}', \tau)}{\partial n'} \right|_{\mathbf{r}'_s} d\Gamma' \, d\tau. \qquad (8.183)$$

Now let $t \to \infty$ and because $\lim_{t \to \infty} G(\mathbf{r}, t | \mathbf{r}', 0) = 0$, this result reduces to

$$1 = -\alpha \int_0^\infty \int_\Gamma \left. \frac{\partial G(\mathbf{r}, t | \mathbf{r}', \tau)}{\partial n'} \right|_{\mathbf{r}'_s} d\Gamma' \, d\tau. \qquad (8.184)$$

Finally, substitution of Eq. (8.184) into Eq. (8.183) gives

$$- \alpha \int_t^\infty \int_\Gamma \left. \frac{\partial G(\mathbf{r}, t | \mathbf{r}', \tau)}{\partial n'} \right|_{\mathbf{r}'_s} d\Gamma' \, d\tau = \int_D G(\mathbf{r}, t | \mathbf{r}', 0) \, d\mathbf{r}'. \qquad (8.185)$$

Denote the first integral term in Eq. (8.183) as $w(t)$, which has the properties (1) $w(0) = 1$, (2) $0 < w(t) < 1$, and (3) $w(\infty) = 0$. From Eq. (8.182), the physical significance of $w(t)$ is seen to be the total weight or importance of the initial condition contribution to $T(\mathbf{r}, t)$. Likewise the second integral term in Eq. (8.182), or equivalently, $1 - w(t)$, is the total weight for the contribution of the boundary condition $T_{BC}(\mathbf{r}'_s, t)$ to $T(\mathbf{r}, t)$. Thus, for small t, $T(\mathbf{r}, t)$ is determined largely by the initial condition $T_{IC}(\mathbf{r})$, whereas for large t it is determined mainly by the boundary condition $T_{BC}(\mathbf{r}'_s, t)$ and the initial condition has little influence.

Use of Subdomains

To evaluate Eq. (8.182) for an arbitrarily shaped domain D, consider a sub-domain V of a simple shape that is contained in D. For example, two rectangular subdomains are shown in Fig. 8.16. The Green's function for this subdomain is the solution of Eq. (8.186), with D and Γ replaced by V and S, which can now be solved analytically, to give Eq. (8.186). Further, the analysis from Eq. (8.172) through Eq. (8.185) remains valid when D and Γ are replaced by V and S provided only that V is contained in D. In practice, one chooses V so that S and Γ coincide as much as possible.

To evaluate Eq. (8.182) for $T(\mathbf{r}_o, t)$ by Monte Carlo, pick a subdomain V containing $(\mathbf{r}_o, t)$. A Monte Carlo game is played in which a large number of histories, each starting at $(\mathbf{r}_o, t)$, is generated. The average of the history scores then estimates $T(\mathbf{r}_o, t)$. A history ends whenever the initial time $t' = 0$ reached or the particle reaches S overlapping Γ (or is within some small distance ϵ of Γ). If a point $\mathbf{r}'$ on the surface of S is reached at time $t' > 0$ that is not on Γ, then $T(\mathbf{r}', t')$ is considered a new unknown that is evaluated with a new encompassing V' whose surface S' coincides with Γ as much as possible. The history continues until a time $t' = 0$ is reached. Such an RW is shown in Fig. 8.17 for a domain that is easily decomposed into rectangular subdomains.

The algorithm for generating RWs or histories using the Green's function approach of Eq. (8.182) is as follows (Nakamura, 1977).

1. Estimate the time-volume integral involving internal heat sources (if any are present) by ordinary Monte Carlo numerical quadrature. Add this to the particle score. The first and second terms are then evaluated as follows.

2. Select a random number ρ. If $\rho < w(t)$, one samples from $T_{IC}(\mathbf{r}')$ by going to step 2(a). However, if $\rho \geq w(t)$, then a sample is taken from $T_{BC}(\mathbf{r}'_s, t)$ by going to step 2(b).

   (a) If $\rho < w(t)$, then the first term of Eq. (8.182) has to be sampled. This is done by sampling a point $\mathbf{r}'$ in V from a PDF proportional to $G(\mathbf{r}, t | \mathbf{r}', 0)$ and adding the known $T_{IC}(\mathbf{r}')$ to the particle score. The history is then terminated.

   (b) If, on the other hand, $\rho \geq w(t)$, then the second term of Eq. (8.182) is to be evaluated by sampling a point $\mathbf{r}'_s$ on S from a PDF proportional to $-\alpha \partial G(\mathbf{r}, t | \mathbf{r}'_s, t - \tau)/\partial n'$. Two possibilities arise. In one, $\mathbf{r}'_s$ lies

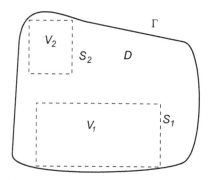

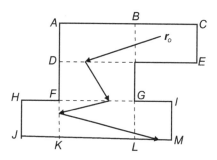

Figure 8.16 Two rectangular subdomains in domain D with boundary Γ.

Figure 8.17 An example four-legged random walk in rectangular subdomains of D. Here $V_1 = ACED$, $V_2 = ABGF$, $V_3 = ABLK$, and $V_4 = HIMJ$.

on a part of S common to (or within ϵ of) Γ and the known value of $T_{BC}(\mathbf{r}'_s, t - \tau)$ is scored and the history is terminated. In the other case, $\mathbf{r}'_s$ is on S but internal to D. This point is then surrounded by another volume V and $T(\mathbf{r}'_s, t - \tau)$ is estimated in the same manner just described to estimate $T(\mathbf{r}, t)$, i.e., go to step 1. This random walk continues until a known temperature is encountered at $t' = 0$ or $\mathbf{r}'_s$ is on (or within ϵ of) the boundary Γ. The known temperature is then added to the particle score and the history ends.

Further details of the use of the Green's function approach using subdomains with other boundary conditions besides the Dirichlet boundary condition considered here are provided by Troubetzkoy et al. (1976).

Rectangular Subdomains for V

For 2-D domains D that have many surfaces parallel to the x- and y-axes, rectangular subdomains are a natural choice because their surfaces have a significantly greater overlap with Γ than other shaped subdomains. For a rectangular V, $\mathbf{r} = (x, y)$ with $x_{min} \le x \le x_{max}$ and $y_{min} \le y \le y_{max}$, the Green's function can be written as (Cole et al., 2011; Nakamura, 1977)

$$G(\mathbf{r}, t | \mathbf{r}', t') = \sum_{k=1}^{\infty} \sum_{l=1}^{\infty} a_{kl} \sin \frac{k\pi(x_{max} - x')}{X} \sin \frac{l\pi(y_{max} - y')}{Y} \exp[-\gamma_{kl}(t - t')],$$

(8.186)

where $X = x_{max} - x_{min}$, $Y = y_{max} - y_{min}$, and

$$\gamma_{kl} = \alpha \left[\left(\frac{k\pi}{X} \right)^2 + \left(\frac{l\pi}{Y} \right)^2 \right],$$

$$a_{kl} = \frac{4}{XY} \sin \frac{k\pi(x_{max} - x)}{X} \sin \frac{l\pi(y_{max} - y)}{Y}.$$

Circular Subdomains for V

The Green's function for a circular subdomain is the same as that for an infinitely long cylinder and may be found by using a Bessel–Fourier series. The result is complex; however, the result simplifies considerably if $\mathbf{r}$ is the center of the circle. In this case

$$G(\mathbf{r}, t|\mathbf{r}', t') = \sum_{k=1}^{\infty} \frac{2}{\pi R^2 J_1^2(\beta_k)} J_0\left(\beta_k \frac{r}{R}\right) \exp\left[\frac{-\alpha \beta_k^2 (t - t')}{R^2}\right], \qquad (8.187)$$

where $r = |\mathbf{r} - \mathbf{r}'|$, R is the radius of the circle, and β_k are the roots (zeros) of $J_0(r) = 0$ ordered such that $0 < \beta_1 < \beta_2 < \dots$. This Green's function has the following properties:

1. $G(\mathbf{r}, t|\mathbf{r}', t') = 0$ if $\mathbf{r}'$ lies on the circumference S of the circle;
2. $G(\mathbf{r}, t'|\mathbf{r}', t') = \delta(\mathbf{r}' - \mathbf{r})$;

3.
$$\left.\frac{\partial G(\mathbf{r}, t|\mathbf{r}', t')}{\partial n'}\right|_{\mathbf{r}' \in S} = \left.\frac{\partial G(\mathbf{r}, t|\mathbf{r}', t')}{\partial r'}\right|_{r=R}$$

$$= -\sum_{k=1}^{\infty} \frac{2\beta_k}{\pi R^3 J_1(\beta_k)} J_0\left(\beta_k \frac{r}{R}\right) \exp\left[\frac{-\alpha \beta_k^2 (t - t')}{R^2}\right];$$

4. $\displaystyle w(\tau) = \int_V G(\mathbf{r}, t|\mathbf{r}', t')\, d\mathbf{r}' = 2\pi \int_0^R G(\mathbf{r}, t|\mathbf{r}', t')r'\, dr'$

$$= \sum_{k=1}^{\infty} \frac{4}{\beta_k J_1(\beta_k)} \exp\left[\frac{-\alpha \beta_k^2 (t - t')}{R^2}\right].$$

With circular subdomains, the Monte Carlo game to generate a heat particle history is very similar to the WOC algorithm of Section 8.3.6 used for steady-state heat conduction. Specifically:

1. Construct the maximal circle in D with $\mathbf{r} \equiv (r, \theta)$ as the center. The value of $\tau = t - t'$ is taken from $w(\tau) = \rho$, where ρ is a random number from $\mathcal{U}(0, 1)$. If $\tau \leq 0$ go to step 2; otherwise go to step 3.
2. Find $\mathbf{r}'$ in V from the following CDFs. The CDF for the angular polar coordinate with $\mathbf{r}$ as the origin is $F(\cos\theta) = \cos\theta/2\pi$. The CDF for the radial distribution is found by integrating Eq. (8.187) with respect to r to obtain

$$H(r) = \left[\sum_{k=1}^{\infty} \gamma_k r J_1(\beta_k(r/R))\right] \Bigg/ \left[\sum_{k=1}^{\infty} \gamma_k R J_1(\beta_k)\right],$$

where

$$\gamma_k = \frac{2}{\pi R^2 \beta_k [J_1(\beta_k)]^2} \exp\left[-\frac{\alpha \beta_k^2 t}{R^2}\right].$$

Give a score of $T_{IC}(\mathbf{r}')$ to this history and construct another history.

3. Find $\mathbf{r}'$ on S with the CDF $F(\cos\theta) = (\cos\theta)/(2\pi)$, where θ is the angular coordinate on the circle with center at $\mathbf{r}$. If $\mathbf{r}'$ is on Γ, set the history score to $T_{BC}(\mathbf{r}', t')$; otherwise go to step 1 and use a maximal circle with center at $\mathbf{r}'$. The values of $\mathbf{r}$ and t are reset to $\mathbf{r}'$ and t', respectively, found in this step.

8.5 Eigenvalue Problems

Monte Carlo techniques can also be used to find eigenvalues and eigenfunctions of many linear operators. However, Monte Carlo methods are not widely known or used to solve eigenvalue problems because they tend to be complex and specialized to specific applications. See, for example, Donsker and Kac (1951), who used a Monte Carlo method to find the fundamental eigenvalue for a problem in quantum mechanics. The three eigenvalue problems considered in this section are

$$\mathbf{Ax} = \lambda\mathbf{x}, \tag{8.188}$$

$$y(\mathbf{x}) = \lambda \int_V K(\mathbf{x}, \mathbf{x}') y(\mathbf{x}') \, d\mathbf{x}' \equiv \lambda \mathcal{K} y(\mathbf{x}), \tag{8.189}$$

and the Sturm–Liouville equation (see Appendix D.7)

$$\frac{d}{dx}\left[p(x)\frac{dy(x)}{dx} \right] + q(x)y(x) + w(x)\lambda y(x) = 0. \tag{8.190}$$

These homogeneous equations generally have only the trivial null solution $\mathbf{x} = 0$ or $y(x) = 0$. However, for certain values of λ (the *eigenvalues*), there may exist nontrivial solutions (the *eigenfunctions* or *eigenvectors*). Monte Carlo techniques usually can find only the extreme eigenvector and associated eigenfunction. Of most interest are *Hermitian* (self-adjoint) operators because their eigenvalues are real and the eigenfunctions form a complete set of functions that can be used to expand other functions. The familiar Fourier series is one such example. A matrix $\mathbf{A}$ is Hermitian if it is symmetric, and the integral operator is Hermitian if $K(\mathbf{x}, \mathbf{x}') = K(\mathbf{x}, \mathbf{x}')$. Equation (8.190), subject to appropriate boundary conditions, also involves a Hermitian differential operator.

8.5.1 Matrix Eigenvalue Problem

The Monte Carlo method can usually estimate only the largest or smallest eigenvalue and then only if $\mathbf{A}$ satisfies certain conditions. Here it is assumed that the components of $\mathbf{A}$ are real and the largest eigenvalue λ_1 is also real and distinct, i.e., there is no eigenvalue degeneracy in λ_1. Consider the matrix eigenvalue problem of Eq. (8.188). For almost any $\mathbf{x}^{(0)}$, the largest eigenvalue of $\mathbf{A}$ is given by

$$|\lambda|_{max} = \lim_{k \to \infty} \frac{(\mathbf{h}, \mathbf{A}^k \mathbf{x}^{(0)})}{(\mathbf{h}, \mathbf{A}^{k-1} \mathbf{x}^{(0)})}, \tag{8.191}$$

where $\mathbf{h}$ is an arbitrary weighting vector often taken simply as $h_i = 1$, $i = 1, \ldots, n$. The evaluation of the inner products is readily performed using the Monte Carlo techniques described in Section 8.1.

A Markov chain or random walk of k steps $i_0 \to i_1 \to i_2 \to \ldots i_k$ is created using a transition matrix $\mathbf{P}$ and starting vector $\boldsymbol{p}$ described by Eq. (8.8). A particular choice, used here, is

$$p_i = |h_i| \Big/ \sum_{m=1}^{n} |h_m|, \quad i = 1, \ldots, n, \tag{8.192}$$

and

$$P_{ij} = |a_{ij}| \Big/ \sum_{m=1}^{n} |a_{im}|, \quad i = 1, \ldots, n, \quad j = 1, \ldots, n. \tag{8.193}$$

Such a choice leads to *almost optimal* Monte Carlo algorithms for matrix calculations (Dimov et al., 1998).

Then following a procedure similar to that given in Section 8.1 one calculates recursively, for each random walk s,

$$U_m^{(s)} = U_{m-1}^{(s)} \frac{|a_{i_{m-1}i_m}|}{P_{i_{m-1}i_m}}, \quad \text{with} \quad U_0^{(s)} = \frac{h_{i_0}}{p_{i_0}}, \quad m = 1, \ldots, k. \tag{8.194}$$

It can be shown that (Halton, 1992)

$$(\mathbf{h}, \mathbf{A}^m \mathbf{x}^{(0)}) = \langle U_m^{(s)} x_{i_m}^{(0)} \rangle, \quad m = 1, 2 \ldots, \tag{8.195}$$

from which it follows that, for sufficiently large k,

$$|\lambda|_{max} \simeq \frac{\langle U_k x_{i_k}^{(0)} \rangle}{\langle U_{k-1} x_{i_{k-1}}^{(0)} \rangle} \simeq \sum_{s=1}^{N} U_k^{(s)} x_{i_k}^{(0)} \Big/ \sum_{s=1}^{N} U_{k-1}^{(s)} x_{i_{k-1}}^{(0)}. \tag{8.196}$$

An example of using the above method is given in Example 8.13.

The deterministic power method of Eq. (8.191) is widely used to estimate $|\lambda|_{max}$. However, for large ($n \times n$) matrices, the number of arithmetic operations required for k iterations is $\mathcal{O}(4kn^2 + 3kn)$ and, thus, this deterministic method becomes computationally very expensive. Moreover, the method is not well-suited for implementation on massively parallel machines. By contrast, the Monte Carlo approach for large sparse matrices has a computational time almost independent of n and a parallel efficiency that is superlinear (Dimov et al., 1998).

Just as there are adjoint RWs and RWs with absorption states that can be used to solve linear algebraic equations, such RWs can also be used to estimate $|\lambda|_{max}$. The interested reader is referred to Hammersley and Handscomb (1964). The power method has also been extended to find the second, third, and even higher largest eigenvalues (Booth, 2006).

Example 8.13 Monte Carlo Calculation of the Dominant Matrix Eigenvalue

As an example of the Monte Carlo method for finding the largest eigenvalue of a matrix, consider the matrix

$$\mathbf{A} = \begin{pmatrix} 0.6 & -0.3 & 0 & 0 & 0 \\ -0.3 & 0.6 & -0.3 & 0 & 0 \\ 0 & -0.3 & 0.6 & -0.3 & 0 \\ 0 & 0 & -0.30 & 0.6 & -0.3 \\ 0 & 0 & 0 & -0.3 & 0.6 \end{pmatrix}.$$

With standard numerical methods, the eigenvalues are found to be $\lambda_{max} = \lambda_1 = 1.119\ 615\ 242\ 270\ 663\ldots$, $\lambda_2 = 9/10$, $\lambda_3 = 6/10$, $\lambda_4 = 3/10$, and $\lambda_5 = 0.080\ 384\ 757\ 729\ 3368\ldots$.

Results obtained with the Monte Carlo method described above are summarized in Fig. 8.18. For these results, the weight vector $h(\mathbf{x}) = (1, 1, \ldots, 1)$. As can be seen in Fig. 8.18 (right), the *stochastic* relative error inherent in the Monte Carlo process decreases as $1/\sqrt{N}$. But there is also a *systematic* error inherent in the power method caused by using a finite number of iterations k. The systematic error is proportional to $(\lambda_2/\lambda_1)^k$. In this example for $k = 50$, the systematic error is $(0.9/1.1196)^{50} = 0.0000176 \simeq 0.002\%$. Generally, for good results the two types of errors should be roughly comparable. In this example, the slow decrease in the systematic error with increasing k interrupts the initial steady decrease in the error of λ_{max} around $k \geq 10$ as seen in Fig. 8.18 (right).

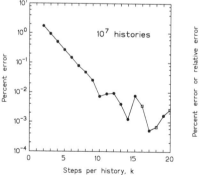 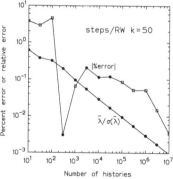

Figure 8.18 (left) The percent error in $\bar{\lambda}_{max}$ as a function of the number of steps k per history. Solid circles indicate an underestimate and open squares an overestimation. Each datum is an average of $N = 10^7$ random walks. (right) Percent error in $\bar{\lambda}_{max}$ and relative error $\bar{\lambda}_{max}/\sigma(\bar{\lambda})$ as a function of the number N of histories for $k = 50$ steps per history.

8.5.2 Eigenvalues of Integral Operators

The use of Monte Carlo to estimate the extreme eigenvalue of an integral operator dates back to a few years after the introduction of the term "Monte Carlo" by Metropolis in 1949 and well before the widespread availability of electronic computers (Curtiss, 1954; Cutkosky, 1951; Vladimirov, 1956). In fact Vladimirov estimated an eigenvalue for a particular integral equation *by hand* using 100 random walks based on a table of random numbers. Throughout this section, we follow Sobol (1973) as interpreted by Rubinstein (1981).

For simplicity, the case of only a single independent variable, i.e., $y(\mathbf{x}) \rightarrow y(x)$, is considered here. However, everything to follow is also valid for the multidimensional case with the one-dimensional integrals replaced by the multidimensional integrals as in Eq. (8.189). The eigenvalue Eq. (8.189) can then be written as[16]

$$y(x) = \lambda \int_a^b K(x, x') y(x') \, dx'. \tag{8.197}$$

The existence and properties of the eigenvalues and corresponding eigenfunctions of a linear integral operator, defined as the nontrivial solutions of Eq. (8.197), depend critically on the properties of the kernel $K(x, x')$.

Symmetric Kernels

If $K(x, x')$ is symmetric, real, well-behaved (continuous, smooth, etc.), and quadratically bounded ($\int_a^b \int_a^b K^2(x, x') \, dx' \, dx < \infty$), then $\mathcal{K}$ is a *compact Hermitian* operator (see Appendix D), for which the eigenvalues and eigenfunctions have many useful properties. These properties include:

- Equation (8.197) has an infinite number of eigenfunctions $\{y_i(x)\}$, which can be made real and orthogonal. These eigenfunctions form a complete basis set so that any well-behaved function $g(x)$, $x \in [a, b]$, can be expanded as

$$g(x) = \sum_{i=1}^{\infty} (g, y_i) y_i(x), \qquad \text{where} \quad (g, y_i) = \int_a^b g(x) y_i(x) \, dx.$$

- Equation (8.197) has at least one eigenvalue $\lambda_i \neq 0$.

---

[16] To be consistent with eigenvalue equations for other linear operators, the eigenvalue equation is sometimes written equivalently as

$$\mathcal{K} y(x) \equiv \int_a^b K(x, x') y(x') \, dx' = \tilde{\lambda} y(x),$$

where $\tilde{\lambda} = 1/\lambda$. In this formulation, $|\tilde{\lambda}|_{max} = 1/|\lambda|_{min}$.

- All eigenvalues λ_i are real, and all eigenfunctions y_i may be chosen to be real and orthonormal.
- Two or more eigenfunctions may have the same eigenvalue; however, only a finite number of linearly independent eigenfunctions can have the same *nonzero* eigenvalue. Such eigenfunctions can be made orthonormal.
- If there are an infinite number of nonzero eigenvalues, then $\lim_{i \to \infty} \lambda_i = \infty$, where the eigenvalues are ordered so that $\lambda_i < \lambda_{i+1}$.

The eigenfunctions $\{y_i(x)\}$ and the kernel $K(x, x')$ of the operator $\mathcal{K}$ are intimately related. To see this, treat $K(x, x')$ as a function of x' and expand it in terms of the eigenfunctions $\{y_i(x')\}$, namely,

$$K(x, x') = \sum_{i=1}^{\infty} (K(x, x'), y_i(x')) y_i(x'). \qquad (8.198)$$

The expansion coefficients $(K(x, x'), y_i(x'))$ are evaluated as

$$(K(x, x'), y_i(x')) = \int_a^b K(x, x') y_i(x') \, dx' = (\mathcal{K} y_i)(x) = \frac{1}{\lambda_i} y_i(x), \qquad (8.199)$$

so that

$$K(x, x') = \sum_{i=1}^{\infty} \frac{1}{\lambda_i} y_i(x) y_i(x'). \qquad (8.200)$$

Nonsymmetric Kernels

If the kernel $K(x, x') \neq K(x', x)$, then most of the properties listed above for a symmetric kernel disappear. The integral operator may have no eigenvalues at all. Nor is it always possible to know if nonzero eigenvalues exist. Even if eigenfunctions and eigenvalues exist, they may be complex. The expansion (completeness) and orthogonality properties of the eigenfunctions also vanish. Only the property that remains is $\lim_{i \to \infty} \lambda_i = \infty$.

Finding Eigenvalues

Analytic Approach:

For the remainder of this section, it is assumed that $K(x, x')$ is symmetric and the operator $\mathcal{K}$ is Hermitian and positive definite, i.e., $(\mathcal{K}\psi, \psi) > 0$ for all $\psi \neq 0$. The usual method used to find the eigenvalues of Eq. (8.197) is to convert the integral equation to an ODE (see Section 8.3) and then find the eigenvalues of the ODE. This procedure is illustrated in Example 8.14.

Example 8.14 Finding Eigenvalues of an Integral Operator

Find the eigenvalues of

$$y(x) = \lambda \int_0^1 K(x, x')y(x')\,dx,$$

in which the symmetric kernel is

$$K(x, x') = \begin{cases} x(1 - x'), & 0 \le x < x' \le 1, \\ x'(1 - x), & 0 \le x' < x \le 1. \end{cases}$$

As shown in Example 8.3, this integral equation is equivalent to the ODE

$$y''(x) + \lambda y(x) = 0 \qquad \text{subject to} \qquad y(0) = y(1) = 0.$$

For $\lambda > 0$, the most general solution of this ODE is

$$y(x) = A\cos\sqrt{\lambda}x + B\sin\sqrt{\lambda}x,$$

where the arbitrary constants A and B are found from the boundary condition. From the requirement $y(0) = 0$, it is seen that $A = 0$. Then, from $y(1) = 0$, one finds either $B = 0$ (which leads only to the trivial solution $y(x) = 0$) or the need for $\lambda = n^2\pi^2$, where n is any integer. Thus, the eigenvalues are $\lambda_n = n^2\pi^2$ and the corresponding eigenfunctions are $y_n(x) = \sin(n\pi x)$.

Monte Carlo Approach:

The solution of Eq. (8.197) by Monte Carlo techniques depends on the convergence of, for almost any initial guess $y_0(x') = f(x)$ of the eigenfunction,

$$y_m(x) = \lambda \int_a^b K(x, x')y_{m-1}(x')\,dx'. \tag{8.201}$$

To estimate $H = (h(x), y(x))$, perform N continuous random walks on the interval $x \in [a, b]$, each consisting of k steps, $x_0 \to x_1 \to \ldots \to x_k$, by the same procedure used in Section 8.2.4. These random walks are governed by an arbitrary starting PDF $p(x)$ and an arbitrary transition PDF $P(x, x')$ subject to

$$\int_a^b p(x)\,dx = 1 \qquad \text{and} \qquad \int_a^b P(x, x')\,dx' = 1. \tag{8.202}$$

During random walk s, construct the random variable

$$Z_k^{(s)}(h) = \frac{h(x_0)}{p(x_0)}U_k\,f(x_k), \tag{8.203}$$

where $f(x)$ is any positive function and

$$U_k = \frac{K(x_0, x_1)}{P(x_0, x_1)} \frac{K(x_1, x_2)}{P(x_1, x_2)} \cdots \frac{K(x_{k-1}, x_k)}{P(x_{k-1}, x_k)}. \tag{8.204}$$

In practice, U_k is calculated recursively after each step in a random walk from

$$U_m = U_{m-1} \frac{K(x_{m-1}, x_m)}{P(x_{m-1}, x_m)}, \quad m = 0, 1, 2, \ldots k \quad \text{with } U_0 = 1. \tag{8.205}$$

It can be shown (Rubinstein, 1981) that

$$H = \langle Z_k^{(s)}(h) \rangle \simeq \frac{1}{N} \sum_{s=1}^{N} Z_k^{(s)}(h), \qquad \text{provided} \qquad \sum_{m=0}^{\infty} |\mathcal{K}^m f| < \infty. \tag{8.206}$$

Now suppose the kernel is real, symmetric, and positive definite (i.e., $(\mathcal{K}\psi, \psi) > 0$ for $\psi(x) \neq 0$). Then for *any* positive functions $h(x)$ (Sobol, 1973)

$$|\lambda|_{\min} = \lim_{k \to \infty} \left\{ \sum_{s=1}^{N} Z_{k-1}^{(s)}(h) \bigg/ \sum_{s=1}^{N} Z_k^{(s)}(h) \right\}. \tag{8.207}$$

The construction of the random walks and $Z_k^{(s)}$ is considerably simplified if h, f, p, and P are chosen to be constant over the interval $[a, b]$. Specifically, choose $f(x) = 1$, $h(x) = p(x) = 1/(b-a)$ so in Eq. (8.203) $h(x_0)/p(0) = 1$. Likewise, choose the PDF $P(x, x') = 1/(b-a)$ to be a uniform distribution in x' on the interval $[a, b]$ for all x. Choose the starting location of each RW as $x_0 = a + (b-a)\rho_0$, where ρ_0 is sampled from $\mathcal{U}[a, b]$. Likewise, each step is calculated in the same manner, i.e., $x_i' = a + (b-a)\rho_i$. Then

$$Z_k^{(s)} = \prod_{m=0}^{k} \frac{K(x_{m-1}, x_m)}{(b-a)} = U_k^{(s)}, \tag{8.208}$$

and $|\lambda|_{\min}$ is then estimated from Eq. (8.207). An example using this simplification is shown in Example 8.15.

Example 8.15 Monte Carlo Eigenvalue Estimate for an Integral Operator

Consider the eigenvalue problem of Example 8.14 in which the eigenvalues were found to be $\lambda_n = n^2\pi^2$, $n = 1, 2, \ldots$. The smallest eigenvalue is thus $\lambda_{\min} = \lambda_1 = \pi^2 = 9.869\,604\,950\ldots$.

To use the Monte Carlo technique, one must first select a weight function $h(x)$, an initial guess of the eigenfunction $y_0(x)$, a PDF $p(x)$ for starting the random walk, and a transition PDF $P(x, x')$. Because in this example $a = 0$ and $b = 1$, the simplest choices for these functions are $h(x) = f(x) = p(x) = 1$ and $P(x, x') = \mathcal{U}[0, 1]$ for selecting x' for any value of x. With these simplifications, Eqs. (8.203) and (8.204) reduce to

$$Z_k^{(s)} = U_k^{(s)},$$

where $U_k^{(s)}$ is calculated recursively as

$$U_m^{(s)} = K(x_{m-1}, x_m)U_{m-1}^{(s)} \qquad \text{with} \qquad U_0^{(s)} = 1.$$

Then the smallest eigenvalue is given by

$$\lambda_{\min} = \lim_{k \to \infty} \left\{ \sum_{s=1}^{N} U_{k-1}^{(s)} \bigg/ \sum_{s=1}^{N} U_k^{(s)} \right\}.$$

Results for different vales of k (the number of steps in a random walk) are shown in Table 8.3. Of interest is that the random walks need not be particularly long. Here $k \simeq 5$ seems to give the most rapidly converging results and increasing k beyond this optimal value not only slows the convergence with increasing N but also requires more computational effort.

Table 8.3 The Monte Carlo estimate of the minimum eigenvalue $\lambda_{\min}^{MC}$ and its percentage error from the exact value of $\pi^2 = 9.869\,604\,40\ldots$ for the integral equation of Example 8.15. Shown are results for four different numbers k of steps per random walk and the number N of random walks.

N	$k = 5$		$k = 10$		$k = 15$		$k = 20$	
	$\lambda_{\min}^{MC}$	% error	$\lambda_{\min}^{MC}$	% error	$\lambda_{\min}^{MC}$	% error	$\lambda_{\min}^{MC}$	% error
1E1	13.5604	37.396	23.1455	134.510	14.5594	47.517	12.7255	28.936
3E1	9.13259	−7.468	14.1586	43.457	9.30039	−5.767	7.90533	−19.902
1E2	8.05734	−18.36	9.55904	−3.147	12.6743	28.417	8.41882	−14.700
3E2	9.52543	−3.487	9.34787	−5.286	11.7971	19.530	7.88246	−20.134
1E3	9.69967	−1.722	10.9761	11.211	8.70696	−11.780	6.83202	−30.777
3E3	9.93707	0.684	11.0049	11.503	10.2963	4.324	8.18144	−17.105
1E4	9.91787	0.489	10.1779	3.123	10.0867	2.200	11.9064	20.637
3E4	9.89731	0.281	10.4740	6.124	9.11601	−7.636	10.6290	7.695
1E5	9.84158	−0.284	9.91722	0.482	9.81657	−0.537	10.6884	8.296
3E5	9.86834	−0.013	9.91883	0.498	9.62766	−2.451	9.85616	−0.136
1E6	9.86351	−0.062	9.88994	0.206	9.77052	−1.004	9.36639	−5.099
3E6	9.86769	−0.019	9.90351	0.344	9.78913	−0.815	9.74032	−1.310
1E7	9.85922	−0.011	9.87445	0.049	9.71243	−1.593	10.1273	2.611

Other Approaches:

A slight variation of the Monte Carlo method presented above for estimating the extremal eigenvalue is given by Hammersley and Handscomb (1964). However, Fortet (1952) has proposed an entirely different approach for estimating the extremal eigenvalue. If the kernel $K(x, x')$ is real, symmetric, and positive definite, then there is a Gaussian process $X(x)$ with $K(x, x')$ as its covariance function, i.e., the random variable X is normally distributed with zero mean, variance $K(x, x)$, and covariance $\text{cov}(x, x') = K(X(x), X(x'))$. Then by generating random samples of $X(x)$, an estimate of the random variable

$$Y = \int_a^b [X(x)]^2 \, dx \qquad (8.209)$$

can be obtained by Monte Carlo integration. Once Y is known, the Fredholm determinant $D(\eta)$ can be constructed. The zeros of this determinant are the inverses of the eigenvalues. Although of academic interest, Fortet's approach does not seem to have had much use, partly because of its relative mathematical complexity as compared to most other Monte Carlo applications.

8.5.3 Eigenvalues of Differential Operators

There do not appear to be any general methods for estimating the extremal eigenvalue of a differential operator, even for Hermitian operators, for which there are an infinite number of real eigenvalues (see Appendix D). The first effort to use Monte Carlo methods to find the smallest eigenvalue of a Hermitian differential operator was that of Donsker and Kac (1951). They devised a scheme to find the smallest eigenvalue and associated eigenfunction of the one-dimensional Schrödinger equation

$$\frac{1}{2} \frac{d^2 \psi(x)}{dx^2} - V(x)\psi(x) + \lambda \psi(x) = 0. \qquad (8.210)$$

Their analysis treats x as a random variable with a distribution $X(x)$ that is a *Weiner process*, which governs, for example, *Brownian motion*. This method is closely related to the approach used by Fortet (1952) to find the principal eigenvalue of an Hermitian integral operator. In fact, Wasow commented at length in his 1952 paper about the Donsker and Kac analysis. Suffice it to say that the mathematical complexity of the theory and analysis is well beyond the scope of this book. However, it should be noted that the analysis of Donsker and Kac can be readily extended to more than one dimension by using multidimensional Wiener processes.

Another early approach for finding the eigenvalues of elliptic PDEs was that of Wasow (1951), who used a finite difference approximation of the PDE to estimate the eigenvalues. Since the pioneering work of Donsker and Kac, refined methods for finding the principal eigenvalue for various classes of PDEs have been developed. For example, Lejay and Maire (2007) consider three different schemes for the Laplace operator subject to Dirichlet boundary conditions. These methods are based on the speed of absorption of the Brownian motion by the boundary.

8.6   Summary

In this chapter, Monte Carlo has been applied to a variety of mathematical equations: algebraic, integral, and differential. Linear algebraic equations are almost always better solved by standard numerical methods using serial computers. However, multidimensional integral equations and differential equations with irregular boundaries are often best treated by Monte Carlo techniques. Because most problems in the real world can be modeled by linear equations, it is, thus, not surprising that Monte Carlo analysis can be applied to a large variety of problems. Although the analyses presented in this chapter are very analytical in nature and do not immediately suggest simulations at an underlying level, an attempt has been made to demonstrate how the Monte Carlo algorithms proposed are, fundamentally, random walks through phase space, that, with the accumulation of random variates, allow quantification of the solution to these linear equations.

The methods chosen for inclusion here are often those developed by early Monte Carlo researchers because (1) they are of historical interest and (2) they are generally much simpler to understand compared to later more sophisticated and robust methods. With the development of new computer architectures such as massively parallel machines and neural networks, there has been a resurgence of interest in Monte Carlo methods. For example, Limy and Wearez (2017) review modern Monte Carlo techniques and propose new fast algorithms capable of treating enormous linear algebra problems. These algorithms have been used in quantum Monte Carlo calculations to treat matrices as large as $10^{108} \times 10^{108}$! Recently Han et al. (2020) proposed a new method to solve eigenvalue problems for second-order differential operators in high dimensions based on deep neural networks. Finally, the new era of big data and the need to manipulate enormous data sets have brought new life to the development of Monte Carlo methods.

Problems

8.1 Repeat the analysis of Example 8.1 using the following two extreme choices for the transition matrix **P**. (a) A transition matrix in which the probability of going from one state to any other state is equally probable, i.e., $P_{ij} = 0.25$ for all i and j. (b) The components of **P** are proportional to the magnitude of the corresponding component of **B**, i.e.,

$$P_{ij} = |b_{ij}| \Big/ \sum_{j=1}^{n} |b_{ij}|, \quad i = 1, \dots, n.$$

This latter choice leads to the *almost optimal* Monte Carlo calculation (Dimov et al., 1998).

8.2 Write a program to solve the linear algebraic equations of Example 8.1 using the adjoint random walk method of Section 8.1.2. Explore how the accuracy of the Monte Carlo solution varies with the number of walks used and with the length k of each walk.

8.3 Write a program to solve the linear algebraic equations of Example 8.1 using random walks with absorption states as described in Section 8.1.3. Use the following transition matrix:

$$\mathbf{P} = \begin{pmatrix} 0.60 & 0.10 & 0.10 & 0.10 \\ 0.10 & 0.60 & 0.10 & 0.10 \\ 0.05 & 0.10 & 0.60 & 0.05 \\ 0.20 & 0.20 & 0.20 & 0.20 \end{pmatrix}.$$

Explore how the accuracy of your estimate of $(\mathbf{h}, \mathbf{x})$ varies with the number of histories. Also try different transition matrices to determine how increasing the strength of the absorption influences how rapidly the Monte Carlo converges.

8.4 Consider the integral equation in Example 8.2 with $\lambda = 0.95$. Write a program to evaluate $y(0.5)$ by the continuous random walk method discussed in Section 8.2.4. Investigate how your choice of k, the number of steps per random walk, affects the convergence and accuracy of your calculations.

8.5 Consider, once again, the integral equation of Example 8.2 with $\lambda = 0.95$. This time the convergence of the inner product $H \equiv (h, y)$ with $h(x) = 1$ is to be investigated. For ease of programming, the PDFs $p(x)$, used to choose the starting position of a random walk, and the transition function $P(x', x)$ are often taken as $\mathcal{U}(0, 1)$. Investigate how the use of nonuniform PDFs of your devising affects the Monte Carlo calculations compared to the use of uniform PDFs.

8.6 Convert the most general, linear, second-order, boundary value ODE

$$\frac{d^2 y(x)}{dx^2} + A(x)\frac{dy(x)}{dx} + C(x)y(x) = F(x), \quad x \in [a, b],$$

with boundary values $y(a) = y_a$ and $y(b) = y_b$, into a Fredholm equation of the second kind, namely,

$$y(x) = f(x) + \int_a^b K(x, x')y(x')\,dx'.$$

In particular, you should show

$$f(x) = y_a + \frac{(x - a)}{(b - a)}\left\{(y_b - y_a) - \int_a^b (b - x')F(x')\,dx'\right\} + \int_a^b (x - x')F(x')\,dx'$$

and

$$K(x, x') = \begin{cases} \left[\dfrac{x-a}{b-a}\right]\{A(x') - (b-x')[A'(x') - B(x')]\}, & x' > x, \\[4mm] \left[\dfrac{x-a}{b-a} - 1\right] A(x') - [A'(x') - B(x')]\left[\dfrac{(x'-a)(b-x)}{(b-a)}\right], & x' < x. \end{cases}$$

8.7 Verify that particular solutions to Poisson's Eq. (8.82) when $s(\mathbf{r}) = s_o$, a constant, are as follows: (a) $u(x) = -s_o x^2/2$ in 1-D Cartesian geometry, (b) $u(r, \theta) = -s_o r^2/4$ in 2-D polar coordinates, and (c) $u(r, \theta, \phi) = -s_o r^2/6$ in 3-D spherical coordinates.

8.8 Show that for the 1-D Poisson equation with a constant source s_o

$$u(x_o) = \frac{1}{2}[u(x_o - r) + u(x_o + r)] + \frac{s_o r^2}{2}.$$

This result says the value of u at the center of an interval or line is the average of u at the two ends of the interval. Just as the average of u on the perimeter of a circle, given by Eq. (8.112), forms the basis for the 2-D WOC algorithm and the average of u over the surface of a sphere, given by Eq. (8.133), forms the basis for the 3-D WOS algorithm, this equation forms the basis for a continuous "walking on lines" (WOL) algorithm for the 1-D Poisson equation.

8.9 Write a code for the adjoint analysis of the 1-D Poisson equation using continuous RWs based on the result of the previous problem. This code should calculate value $u(x_o)$ for any point x_o in the domain $D : 0 \le x \le L$. In particular, use $L = 20$, $s_o = 10$, with the boundary conditions $u(0) = 20$, $u(L) = 100$. At each step in the random walk, use the maximal-line length, i.e., one end of the line is always at one of the two boundaries so the RW has a 50% chance of ending on each RW step. Note that, unlike the WOC and WOS algorithms, there is no need for a "closeness" parameter ϵ because one end of the maximal line is always at one of the boundaries. Derive an analytic expression for the exact solution and compare your WOL results to the exact value of $u(x_o)$.

8.10 Show that the probability that a diffusing particle born at the center of a diffusing sphere of radius R reaches the surface is

$$P_{esc} = \frac{\beta R}{\sinh(\beta R)},$$

where $\beta^2 = \Sigma_a/D$, Σ_a is the absorption coefficient (the probability of absorption per unit differential distance of travel), and D is the diffusion coefficient.

HINT: Solve the diffusion equation with spherical symmetry

$$\frac{1}{r^2}\frac{d}{dr}\left(r^2\frac{d\phi(r)}{dr}\right) - \beta^2\phi(r) = 0, \quad 0 < r \le R,$$

with a point isotropic source of strength S_o at the origin. First show that the solution of this equation for $\phi(r)$ is

$$\phi(r) = \frac{1}{8\pi D r \sinh \beta R} \left[e^{\beta(R-r)} - e^{-\beta(R-r)} \right].$$

Then the probability the particle is absorbed before it reaches the surface is $P_{abs} = \int \Sigma_a \phi(r) \, dV = \Sigma_a \int_0^R \phi(r) 4\pi r^2 \, dr$. Finally, the probability of reaching the surface is $P_{esc} = 1 - P_{abs}$.

8.11 For diffusion problems, governed by the Helmholtz equation, in which the diffusing particle can interact with the medium and produce new particles, such as occurs for neutrons diffusing in a fissionable medium, the parameter β^2 in Eq. (8.141) may become negative. Using methods analogous to those used in the above problem, show that the probable number of particles reaching the surface per source particle emitted at $r = 0$ is

$$\frac{\beta R}{\sin(\beta R)}, \quad \text{where} \quad \beta = \sqrt{|\beta^2|}.$$

8.12 With the use of L'Hôpital's rule to evaluate indeterminate ratios, show that as $\beta \to 0$, the constant source contribution at each step of the WOS algorithm of Eq. (8.161) reduces to

$$\lim_{\beta \to 0} \frac{S_o}{\beta^2} \left[1 - \frac{\beta d_i}{\sinh(\beta d_i)} \right] = \frac{S_o d_i^2}{6},$$

a result which is consistent with Eq. (8.138).

8.13 For the case that $s(\mathbf{r}) = s_o$, a constant, show that Eq. (8.159) reduces to Eq. (8.161).

8.14 Consider the 3-D diffusion equation for a cube

$$\frac{\partial^2 u}{\partial x^2} + \frac{\partial^2 u}{\partial y^2} + \frac{\partial^2 u}{\partial z^2} - \beta^2 u = -s_o, \quad -\frac{L}{2} \le x, y, z \le \frac{L}{2},$$

where the value on the boundary is $u|_\Gamma = u_o$, a constant. Show that the solution can be expressed as

$$u(x, y, z) = u_o + \sum_{n=1}^{\infty}{}' \sum_{m=1}^{\infty}{}' \frac{S_{nm}}{\gamma_{nm}^2} \left[1 - \frac{\cosh \gamma_{nm} x}{\cosh \gamma_{nm} L/2} \right] \cos\left(\frac{n\pi}{L} y\right) \cos\left(\frac{m\pi}{L} z\right).$$

Here

$$\gamma_{nm}^2 = \left(\frac{n\pi}{L}\right)^2 + \left(\frac{m\pi}{L}\right)^2 + \beta^2$$

and

$$S_{nm} = \frac{16}{nm\pi^2}[s_o - \beta^2 u_o](-1)^{(n+m-2)/2}.$$

Write a Monte Carlo WOS program that evaluates $u(0, 0, 0)$ at the center of the cube for $L = 10$, $u_o = 100$, $\beta^2 = 0.2$, and $s_o = 100$. Construct graphs that show how the error decreases with the number of random walks N and how the converged error varies with ϵ.

8.15 Perform an analysis of the 2-D Helmholtz equation with $\beta^2 < 0$ similar to that of Section 8.3.11.

8.16 For the case that $s(\mathbf{r}) \rightarrow s_o$, i.e., the source in the Helmholtz equations is a constant, show that Eq. (8.159) reduces to Eq. (8.161).

8.17 Derive Eq. (8.165) starting from Eq. (8.164). Here first-order finite difference approximations for the derivatives are to be used so that, for example,

$$\frac{\partial T(\mathbf{r}, t)}{\partial t} \simeq \frac{T(\mathbf{r}, t) - T(\mathbf{r}, t - \Delta t)}{\Delta t}.$$

8.18 Using the method described in Section 8.5.1, estimate the largest eigenvalue of the matrix

$$\mathbf{P} = \begin{pmatrix} 0.7 & 0.1 & 0.2 & 0.1 \\ 0.1 & 0.5 & 0.1 & 0.1 \\ 0.1 & 0.2 & 0.5 & 0.3 \\ 0.2 & 0.2 & 0.2 & 0.5 \end{pmatrix}.$$

Explore the effect of using different values of the maximum number m of steps used per random walk.

8.19 Consider the ODE of Examples 8.3 and 8.4, i.e.,

$$y''(x) + \lambda y(x) = 0,$$

but with boundary conditions $y(a) = y(b) = 0$. (a) Show that it is equivalent to

$$y(x) = \lambda \int_a^b K(x, t) y(t) \, dt$$

with the symmetric kernel

$$K(x, t) = \frac{1}{b - a} \begin{cases} (x - a)(b - t), & a \leq x < t \leq b, \\ (t' - a)(b - x), & a \leq t < x \leq b. \end{cases}$$

(b) Show that the eigenvalues are $\lambda_n = [n\pi/(b - a)]^2$, $n = 1, 2, 3, \ldots$.
(c) Write a program to estimate λ_{min} by the Monte Carlo method.

8.20 Show that the eigenvalue problem

$$y''(x) + y(x) = \lambda y(x) \quad \text{with} \quad y'(0) = 0 \quad \text{and} \quad y'(1) + y(1) = 0$$

is equivalent to

$$y(x) = \lambda \int_{-1}^{1} e^{|t-x|} y(t)\, dt.$$

Show that the smallest eigenvalue is $\lambda_1 = 1.740\,174\ldots$ Write a program to estimate λ_1 by the Monte Carlo method. Investigate the effect on the accuracy of your Monte Carlo estimate on the number k of steps per RW and the number of histories N used.

References

Bahadori, R., Gutierrez, H., Manikonda, S., Meinke, R., 2018. A mesh-free Monte-Carlo method for simulation of three-dimensional transient heat conduction in a composite layered material with temperature dependent thermal properties. Int. J. Heat Mass Transf. 119, 533–541.

Booth, T.E., 1981. Exact Monte Carlo solution of elliptic partial differential equations. J. Comput. Phys. 39, 396–404.

Booth, T.E., 2006. Power iteration method for the several largest eigenvalues and eigenfunctions. Nucl. Sci. Eng. 154, 48–62.

Brown, G.M., 1956. Monte Carlo methods. Chapter 12. In: Brecke, E.F. (Ed.), Modern Mathematics for Engineers. McGraw-Hill, New York, NY.

Chandrasekhar, S., 1943. Stochastic problems in physics and astronomy. Rev. Mod. Phys. 6, 1–89.

Cole, K.D., Beck, J.V., Haji-Sheikh, A., Litouhi, B., 2011. Heat Conduction Using Green's Functions, 2nd ed. CRC Press, Boca Raton, FL.

Curtiss, J.H., 1949. Sampling methods applied to differential and difference equations. In: Proc. Seminar on Scientific Computing. IBM Corp., New York.

Curtiss, J.H., 1954. J. Math. Phys. XXXII, 209–232.

Curtiss, J.H., 1956. A theoretical comparison of the efficiencies of two classical methods and a Monte Carlo method for computing one component of the solution of a set of linear algebraic equations. In: Meyer, H.A. (Ed.), Symp. Monte Carlo Methods. Wiley, New York, pp. 191–233.

Cutkosky, R.E., 1951. A Monte Carlo method for solving a class of integral equations. J. Res. Natl. Bur. Stand. 47 (2), 113–115.

Dimov, I., Alexandrov, V., Kariavanova, A., 1998. Implementation of Monte Carlo algorithms for eigenvalue problem using MPI. In: Alexandrov, V., Dongarra, J. (Eds.), Proc. 5th European PVM/MPI. Springer-Verlag, London, UK, pp. 346–353.

Dimov, I., Philippe, B., Karaivanova, A., Weihrauch, C., 2008. Robustness and applicability of Markov chain Monte Carlo algorithms for eigenvalue problems. Appl. Math. Model. 32, 1511–1529.

Donsker, M.D., Kac, M., 1951. A sampling method for determining the lowest eigenvalue and the principle eigenfunction of Schrödinger's equation. J. Res. Natl. Bur. Stand. 44, 551–557.

Forsythe, G.E., Liebler, R.A., 1950. Matrix inversion by a Monte Carlo method. Math. Tables Other Aids Comput. 4, 127–129.

Fortet, R., 1952. On the estimation of an eigenvalue by an additive functional of a stochastic process, with special reference to the Kac-Donsker process. J. Res. Natl. Bur. Stand. 48, 68–75.

Haji-Sheikh, A., 1965. Application of Monte Carlo Methods to Thermal Conduction Problems. Dissertation. Univ. Minnesota.

Haji-Sheikh, A., Howell, J.R., 2006. Monte Carlo methods. Ch. 8. In: Minkowycz, W.J., Sparrow, E.M., Murthy, J.Y. (Eds.), Handbook of Numerical Heat Transfer, 2nd ed. Wiley, New York, pp. 249–295.

Haji-Sheikh, A., Sparrow, E.M., 1967. The solution of heat conduction problems by probability methods. J. ASME 89, 121–131.

Halton, J.H., 1992. Monte Carlo Techniques for the Solutions of Linear Systems. Report TR-02-033. U. North Carolina at Chapel Hill, Dept. Comp. Sci. 46 pp.

Hammersley, J.M., Handscomb, D.C., 1964. Monte Carlo Methods. Wiley, New York.

Han, J., Lu, J., Zhou, M., 2020. Solving high-dimensional eigenvalue problems using deep neural networks: a diffusion Monte Carlo like approach. arXiv:2002.02600v1 [cs.LG]. 7 Feb 2020.

Hwang, C., Mascagni, M., 2001. Efficient modified 'walk on spheres' algorithm for the linearized Poisson-Boltzmann equation. Appl. Phys. Lett. 78 (6), 787–789.

Ji, H., Mascagni, M., Li, Y., 2013. Convergence analysis of Markov chain Monte Carlo linear solvers using Ulam-von Neumann algorithm. SIAM J. Numer. Anal. 51 (4), 2107–2122.

Lejay, A., Maire, S., 2007. Computing the principal eigenvalue of the Laplace operator by a stochastic method. Math. Comput. Simul. 73 (3), 351–363. https://doi.org/10.1016/j.matcom.2006.06. 011. inria-00092408, Elsevier.

Limy, L.-H., Wearez, J., 2017. Fast randomized iteration: diffusion Monte Carlo through the lens of numerical linear algebra. SIAM Rev. 59 (3), 547–587.

Margenau, H., Murphy, G.M., 1956. The Mathematics of Physics and Chemistry. Van Nostrand, Princeton.

Mikhailov, G.A., 1995. New Monte Carlo Methods with Estimating Derivatives. VSP, Utrecht, The Netherlands.

Muller, M.E., 1956. Some continuous Monte Carlo methods for the Dirichlet problem. Ann. Math. Stat. 27 (3), 569–589.

Nakamura, S., 1977. Computational Methods in Engineering and Science: With Applications to Fluid Dynamics and Nuclear Systems. Wiley, New York.

Özişik, M.N., Orlande, H.R.B., Colaço, M.J., Cotta, R.M., 2017. Finite Difference Methods in Heat Transfer, 2nd ed. CRC Press, Boca Raton, FL.

Patankar, S.V., 1980. Numerical Heat Transfer and Fluid Flow. Hemisphere Publishing, CRC Press, Boca Raton, FL.

Press, W.H., Teukolsky, S.A., Vetterling, W.T., Flannery, B.P., 1996. Numerical Recipes, 2nd ed. Cambridge Univ. Press, Cambridge.

Rubinstein, R.V., 1981. Simulation and the Monte Carlo Method. Wiley, New York.

Sabelfeld, K.K., 1991. Monte Carlo Methods in Boundary Value Problems. Springer-Verlag, Berlin.

Sobol, I.M., 1973. Computational Methods of Monte Carlo. Nauka, Novoosibisk, U.S.S.R. (in Russian).

Troubetzkoy, E.S., Kalos, M.H., Banks, N.E., Steinberg, H.A., Klem, G.T., 1976. Solution of Time-Dependent Heat Conduction Equation in Complex Geometry by the Monte Carlo Method. Report BRL CR 325. BRL, Aberdeen Proving Grounds, MD. https://apps.dtic.mil/dtic/tr/fulltext/ u2/a034513.pdf. (Accessed December 2019).

Vajargah, B.F., Moradi, M., 2007. Monte Carlo algorithms for solving Fredholm integral equations and Fredholm differential integral equations. Appl. Math. Sci. 1 (10), 463–470.

Vladimirov, V.S., 1956. Monte Carlo methods as applied to the calculation of the lowest eigenvalue and the associated eigen-function of a linear integral equation. Theory Probab. Appl. 1 (1), 101–116.

Wachspress, E.L., 1966. Iterative Solution of Elliptic Systems. Prentice-Hall, Englewood Cliffs, NJ.

Wasow, W., 1951. Random walks and the eigenvalues of elliptic difference equations. J. Res. Natl. Bur. Stand. 46, 65–73.

Chapter 9

The Fundamentals of Neutral Particle Transport

Particles that migrate through matter
do so without any clatter.
Now some, very meek,
eventually leak
the rest are absorbed or they scatter.

So far the basis of Monte Carlo has been explored in several contexts, and many of the tricks that make up the toolbox of Monte Carlo have been identified. It is now time to begin the serious applications of Monte Carlo and its tools to important problems that are far more complex than the simple example problems shown to this point. Monte Carlo calculations are used in virtually every discipline involved with quantitative analyses. A book the size of this one cannot consider every application of Monte Carlo. In this and the following chapter, the application of Monte Carlo to the problem of radiation transport is considered, not because it is the most important application, but because we, the authors, are most familiar with this particular problem.

Particle transport refers to the diffusion or transport of small particles, such as protons, neutrons, photons, electrons, ions, or even neutral atoms, within some host medium. What is the host medium? It can be a uniform medium such as air or water, it could be a star, it could be the earth's atmosphere, it could be a lead brick, it could be the human body (or a part of it), or it could be just about anything. A special case is a vacuum, which can be thought of as a medium with no constituents (other than the particles that are streaming through it). The point is that small particles, like neutrons in a reactor or X rays in the body, interact with the host medium (unless it is a vacuum, in which case they simply stream in straight lines away from their point of birth). The problem is to calculate information about the particle field, for example, the number of particles in a certain volume or the energy transferred from the particles to the host medium.

However, before Monte Carlo is applied to radiation transport, it is necessary to introduce some generic aspects of the radiation field and how the radiation particles interact with the host medium. In this generic discussion, the term *radiation* refers to the particles that are migrating though the host medium. This

Exploring Monte Carlo Methods. https://doi.org/10.1016/B978-0-12-819739-4.00017-2

is a slightly loose use of the term (since, for instance, neutral atoms are not typically thought of as radiation), but it is a shorthand notation that serves well as the generic case is considered. Some specific particle types are considered in later subsections. The term *host medium* or simply *medium* refers to the background space through which the particles are *transported*, whether it be a single uniform medium, a heterogeneous medium, or even a medium with one or more regions of vacuum. The discussion that follows is restricted to neutral particles, i.e., neutrons or photons, for simplicity and due to space limitations.

9.1 Description of the Radiation Field

To study particle transport, a few quantities must first be introduced. The "strength" of the radiation field must be quantified, as must also the interactions of the radiation with the host medium. Here are summarized those quantities with which the reader must be aware to understand how to apply Monte Carlo to particle transport problems. The discussion is somewhat abbreviated but should provide the required essentials. For a fuller discussion of terms and concepts, the interested reader is referred to excellent texts such as Duderstadt and Martin (1979), Lewis and Miller (1984), and Bell and Glasstone (1970). Those who are already familiar with this material may skip to later sections.

The first thing to realize is that particle transport attempts to quantify the expected behavior of a large number of particles. The laws that govern particle transport are largely probabilistic in nature. There is always a granularity to the radiation field. Even under steady-state conditions, the number of particles in a small volume fluctuates from one instant of time to another. This stochastic nature of a radiation field arises from both the random emission of radiation by sources and the random manner in which radiation interacts with the medium before reaching the small test volume. Thus, the quantification of a radiation field can be made only in terms of ensemble averages of large numbers of particles. These averages are just expected values and hence are ripe for evaluation by Monte Carlo.

9.1.1 Directions and Solid Angles

The directional properties of radiation fields are almost universally described using spherical polar coordinates as illustrated in Fig. 9.1.

The direction vector $\mathbf{\Omega}$ is a unit vector, given in terms of the orthogonal Cartesian unit vectors $\mathbf{i}$, $\mathbf{j}$, and $\mathbf{k}$ by

$$\mathbf{\Omega} = \mathbf{i}u + \mathbf{j}v + \mathbf{k}w = \mathbf{i}\sin\theta\cos\psi + \mathbf{j}\sin\theta\sin\psi + \mathbf{k}\cos\theta. \qquad (9.1)$$

By increasing θ by $d\theta$ and ψ by $d\psi$, an area $dA = \sin\theta\,d\theta\,d\psi$ is swept out on a unit sphere. The solid angle, contained in the range of directions through some area on a sphere, is defined as the area divided by the square of the sphere's

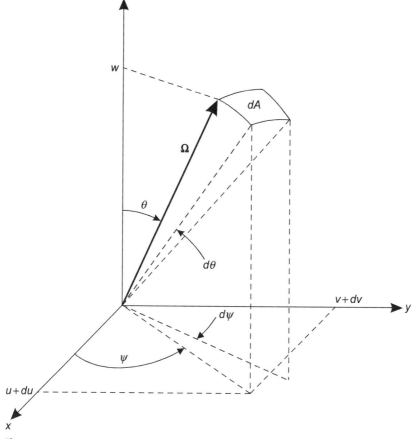

Figure 9.1 Spherical polar coordinate system for specification of the unit direction vector $\mathbf{\Omega}$, polar angle θ, azimuthal angle ψ, and associated direction cosines (u, v, w).

radius. Thus, the differential solid angle associated with differential area dA is

$$d\Omega = \sin\theta \, d\theta \, d\psi. \tag{9.2}$$

From its definition, the solid angle is a dimensionless quantity. However, to avoid confusion when referring to distribution functions of direction, units of *steradians*, with the abbreviation sr, are assigned to the solid angle.

A frequent simplification in notation is realized by using $\omega = \cos\theta$ as an independent variable instead of the angle θ. With this variable, $\sin\theta \, d\theta \equiv -d\omega$. An immediate benefit of this new variable is a more straightforward integration over direction and the avoidance of the common error in which the differential solid angle is incorrectly stated as $d\theta \, d\psi$ instead of $\sin\theta \, d\theta \, d\psi$. The benefit is also evident when the solid angle subtended by "all possible directions" is computed, namely,

$$\Omega = \int_0^{\pi} d\theta \sin\theta \int_0^{2\pi} d\psi = \int_{-1}^{1} d\omega \int_0^{2\pi} d\psi = \int_{4\pi} d\Omega = 4\pi. \qquad (9.3)$$

9.1.2 Particle Density

The expected number of radiation particles per unit volume, i.e., the particle density, can be used to describe a radiation field at some point $\mathbf{r}$ of interest. Generally, the particles in the unit volume have a distribution of energies E and the density will vary in time if the sources of the radiation are time-varying. The *particle density spectrum* $n(\mathbf{r}, E, t)$ is defined such that

$n(\mathbf{r}, E, t)\, dV\, dE =$ the expected number of particles in a differential
volume dV at position $\mathbf{r}$ that have energies in dE
about E at time t.   (9.4)

Unfortunately, there is no exact equation whose solution yields $n(\mathbf{r}, E, t)$. However, if the particle density is generalized somewhat, a quantity can be found that does satisfy an exact equation. To this end, the *angular particle density spectrum* $n(\mathbf{r}, E, \mathbf{\Omega}, t)$ is introduced[1] such that

$n(\mathbf{r}, E, \mathbf{\Omega}, t)\, dV\, dE\, d\Omega =$ the expected number of particles in a differential volume dV at position $\mathbf{r}$ traveling
in a range of directions in $d\Omega$ about $\mathbf{\Omega}$
with energies in dE about E at time t.   (9.5)

If a particle has velocity $\mathbf{v}$, then the unit vector describing its direction of travel is simply $\mathbf{\Omega} = \mathbf{v}/|\mathbf{v}|$. From its definition, $n(\mathbf{r}, E, \mathbf{\Omega}, t)$ has units of $\mathrm{cm}^{-3}\,\mathrm{MeV}^{-1}\,\mathrm{sr}^{-1}$.

Once $n(\mathbf{r}, E, \mathbf{\Omega}, t)$ has been found, it provides a very detailed description of the radiation field (a function of seven independent variables). Usually it is too detailed to be of direct use. Simpler and less detailed descriptions of the radiation field can, however, be readily obtained by integrating the angular density spectrum. For example, the particle *angular density* is

$$n(\mathbf{r}, \mathbf{\Omega}, t) = \int_0^{\infty} n(\mathbf{r}, E, \mathbf{\Omega}, t)\, dE \qquad [\mathrm{cm}^{-3}\,\mathrm{sr}^{-1}]. \qquad (9.6)$$

Similarly, the *density spectrum* is

$$n(\mathbf{r}, E, t) = \int_{4\pi} n(\mathbf{r}, E, \mathbf{\Omega})\, d\Omega \qquad [\mathrm{cm}^{-3}\,\mathrm{MeV}^{-1}], \qquad (9.7)$$

---

[1] The particle density can be further disaggregated to include other properties of the particles besides their direction of travel. For example, the particles' polarizations or spin states, excitation states for composite particles, etc. However, in this chapter such refinements are not needed.

and the *particle density* is

$$n(\mathbf{r}, t) = \int_0^\infty \int_{4\pi} n(\mathbf{r}, E, \mathbf{\Omega}, t) \, d\Omega \, dE \qquad [\text{cm}^{-3}]. \qquad (9.8)$$

9.1.3 Flux Density

Particle density is a straightforward concept. It is a measure of how many particles occupy a unit volume at any time t, just as we might think of the people density as the number of people per unit volume in a high-rise office building on different floors at some time of day. However, as will be seen, a more useful measure of a radiation field is the particle *angular flux density spectrum* defined as

$$\phi(\mathbf{r}, E, \mathbf{\Omega}, t) \equiv v(E) n(\mathbf{r}, E, \mathbf{\Omega}, t) \qquad [\text{cm}^{-2}\,\text{MeV}^{-1}\,\text{sr}^{-1}\,\text{s}^{-1}], \qquad (9.9)$$

where $v(E)$ is the particle speed. For photons $v = c$, the speed of light, while for nonrelativistic neutrons with mass m, $v = \sqrt{2E/m}$. Similarly, one has the *angular flux density*

$$\phi(\mathbf{r}, \mathbf{\Omega}, t) = \int_0^\infty v(E) n(\mathbf{r}, E, \mathbf{\Omega}, t) \, dE \qquad [\text{cm}^{-2}\,\text{sr}^{-1}\,\text{s}^{-1}], \qquad (9.10)$$

the *flux density spectrum*

$$\phi(\mathbf{r}, E, t) = \int_{4\pi} v(E) n(\mathbf{r}, E, \mathbf{\Omega}) \, d\Omega = v(E) n(\mathbf{r}, E, t) \quad [\text{cm}^{-2}\,\text{MeV}^{-1}\,\text{s}^{-1}], \qquad (9.11)$$

and the *scalar flux density* or just the *flux density*

$$\phi(\mathbf{r}, t) = \int_0^\infty \int_{4\pi} v(E) n(\mathbf{r}, E, \mathbf{\Omega}, t) d\Omega \, dE \qquad [\text{cm}^{-2}\,\text{s}^{-1}]. \qquad (9.12)$$

Because $v(E)$ is the distance traveled in a unit time by a particle of energy E, it follows that $\phi(\mathbf{r}, t)$ can be interpreted as the total distance traveled in a unit time by all particles within a unit volume about $\mathbf{r}$ at time t.

To visualize this interpretation, consider the small sphere centered at $\mathbf{r}$ and of volume ΔV (see Fig. 9.2(b)). Let $\sum_i s_i$ be the sum of all the path length segments traversed within the sphere by all particles entering the sphere in a unit time at time t. Then the flux density is

$$\phi(\mathbf{r}, t) = \lim_{\Delta V \to 0} \left[\frac{\sum_i s_i}{\Delta V} \right]. \qquad (9.13)$$

Similar interpretations hold for the angular flux density spectrum, angular flux density, and flux density spectrum. This interpretation of flux density as being

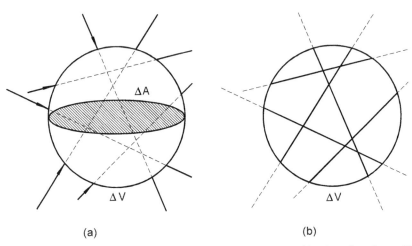

(a)   (b)

Figure 9.2 Two interpretations of the flux density can made by consideration of a sphere with volume ΔV in the form of a sphere with cross-sectional area ΔA. In (a) the number of particles passing through the surface *into* the sphere is of interest. In (b) the paths traveled within the sphere by particles passing through the sphere are of interest.

the particle path length per unit time is important in the next section when the rate of radiation interactions with the medium is discussed.

An alternative interpretation of the concept of flux density is made in terms of the number of particles ΔN_p entering a small sphere of cross-sectional area ΔA in a unit time at time t (see Fig. 9.2(a)). The flux density is then

$$\phi(\mathbf{r}, t) = \lim_{\Delta A \to 0} \left[\frac{\Delta N_p}{\Delta A} \right]. \tag{9.14}$$

In Problem 9.2 at the end of this chapter, the reader is asked to show that this alternative interpretation of the flux density is equivalent to the path length per unit volume concept.

9.1.4 Fluence

Related to flux density is a quantity called *fluence*, which is defined as the integral of the flux density over some time interval $[t_0, t]$, namely,

$$\Phi(\mathbf{r}, t) = \int_{t_0}^{t} \phi(\mathbf{r}, t') dt'. \tag{9.15}$$

Conversely, the flux density is the time rate of change of the fluence, i.e.,

$$\phi(\mathbf{r}, t) = \frac{d\Phi(\mathbf{r}, t)}{dt}. \tag{9.16}$$

9.1.5 Current Vector

Sometimes the flow through a surface is needed in transport calculations. Consider the small surface dA with unit normal $\mathbf{n}$ shown in Fig. 9.3. The surface is located at position $\mathbf{r}$ and the density at time t of particles with energies in dE about E and moving in directions $d\Omega$ about $\mathbf{\Omega}$ is $n(\mathbf{r}, E, \mathbf{\Omega}, t)\, dE\, d\Omega$. The number of these particles that cross dA in the following time interval dt is the number of such particles in the slant cylinder of axial length $v\, dt$ shown in Fig. 9.3. The volume of this cylinder is $(v\, dt)(dA\cos\theta)\, dt = \mathbf{n}\cdot\mathbf{v}\, dA\, dt = v\mathbf{n}\cdot\mathbf{\Omega}\, dA\, dt$. Hence, the number of particles crossing dA in a time of dt is

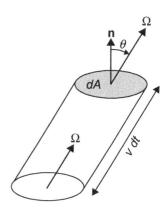

Figure 9.3 Flow through a surface.

$n(\mathbf{r}, E, \mathbf{\Omega}, t)v\, dt\, \mathbf{n}\cdot\mathbf{\Omega}\, dA\, dE = \mathbf{n}\cdot\mathbf{\Omega}\phi(\mathbf{r}, E, \mathbf{\Omega}, t)dA\, dt\, dE$.

The *net* number of particles with energies in dE about E crossing dA per unit time in the direction of positive $\mathbf{n}$ *regardless* of their direction of travel is thus

$$\left[\int_{4\pi} \mathbf{n}\cdot\mathbf{\Omega}\,\phi(\mathbf{r}, E, \mathbf{\Omega}, t)\, d\Omega\right] dE\, dA = \left[\int_{4\pi} \mathbf{n}\cdot\mathbf{J}(\mathbf{r}, E, \mathbf{\Omega}, t)\, d\Omega\right] dE\, dA,$$

where the *current vector spectrum* is defined as

$$\mathbf{j}(\mathbf{r}, E, \mathbf{\Omega}, t) \equiv \mathbf{\Omega}\phi(\mathbf{r}, E, \mathbf{\Omega}, t). \tag{9.17}$$

Then the current vector is

$$\mathbf{j}(\mathbf{r}, E, t) \equiv \int_{4\pi} \mathbf{\Omega}\phi(\mathbf{r}, E, \mathbf{\Omega}, t)\, d\Omega. \tag{9.18}$$

More generally, the net number of particles crossing a unit area of surface with unit normal $\mathbf{n}$ in the positive direction (the direction of $\mathbf{n}$) per unit energy per unit time is

$$j_n(\mathbf{r}, E, t) = \mathbf{n}\cdot\mathbf{J}(\mathbf{r}, E, t) = \int_{4\pi} \mathbf{n}\cdot\mathbf{\Omega}\phi(\mathbf{r}, E, \mathbf{\Omega}, t)\, d\Omega. \tag{9.19}$$

Finally, the *partial currents* are sometimes needed. These quantities are the total flow of particles, per unit energy, through a unit area in the positive direction and in the negative direction, i.e., $J_n(\mathbf{r}, E, t) = J_n^+(\mathbf{r}, E, t) - J_n^-(\mathbf{r}, E, t)$, where the partial currents are given by

$$j_n^+ = \int_{\mathbf{n}\cdot\mathbf{\Omega}>0} \mathbf{n}\cdot\mathbf{\Omega}\phi(\mathbf{r}, E, \mathbf{\Omega}, t)\, d\Omega \quad \text{and} \quad j_n^- = \int_{\mathbf{n}\cdot\mathbf{\Omega}<0} \mathbf{n}\cdot\mathbf{\Omega}\phi(\mathbf{r}, E, \mathbf{\Omega}, t)\, d\Omega. \tag{9.20}$$

The current vector in particle transport is analogous to the heat flux in heat transfer, both being vectors that quantify flow (of particles in the first case and heat energy in the second) per unit area per unit time. Further, the direction of the vector at a point in space and time identifies the direction of the *net* flow there and then.

9.2   Radiation Interactions with the Medium

The quantification of how radiation interacts and is affected by the host medium depends not only on the strength of the radiation field but also on the propensity of the particles to interact with the atoms of the medium. The manner in which radiation interacts with the medium depends on the type of particle (neutral or charged) and the type of atoms composing the medium. For example, electrons interact simultaneously with the electrons of all the nearby ambient atoms and, consequently, their paths through a medium are tortuous, exhibiting countless small angle deflects. Heavy ions plow through the electron clouds of the ambient atoms in straight-line paths leaving in their wake thousands of secondary energetic free electrons and ions. By contrast, neutral particles such as photons and neutrons stream in straight lines through matter and pass countless atoms before undergoing some sort of interaction.

A discussion of how all these types of radiation migrate through an ambient medium is well beyond the scope of this text. To demonstrate, however, how radiation interacts with the host medium, the discussion below is limited to neutral subatomic particles such as neutrons and photons. Unlike charged particles, neutral particles do not interact with the ambient charged electrons and nuclei through the long-range Coulombic force and, hence, they travel in straight-line trajectories punctuated by "point" interactions with a single electron or nucleus.

9.2.1   Interaction Coefficient/Macroscopic Cross Section

The interaction of radiation with the host medium is a statistical (random or stochastic) process. There is no way to predict how a single radiation particle will interact or how far it will travel before interacting. However, the average interaction behavior of a large number of particles can be estimated.

Consider the situation of Fig. 9.4 in which a large number N of identical neutral particles, all with the same energy, are incident on a thin slab of the host medium with thickness Δx. While traversing the slab of thickness Δx, a small number ΔN interacts while most particles simply stream through without interaction. The probability that any one of the incident particles interacts in the slab is $\Delta N/N$. If one were to measure $\Delta N/N$ for different thicknesses Δx and plot the ratio, a plot similar to that of Fig. 9.5 would be obtained. The probability of interaction per unit distance of travel, $(\Delta N/N)/\Delta x$, would appear to vary smoothly for large Δx. But as Δx becomes smaller and smaller, the statistical fluctuations in the number of interactions in the slab would become apparent

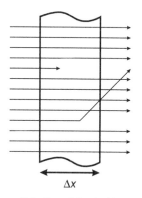

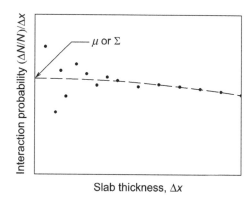

Figure 9.4 N particles incident on a thin slab in which ΔN interacts while traversing the slab.

Figure 9.5 Measured interaction probabilities $\Delta N/N$ divided by the slab thickness.

and the measured interaction probability per unit distance of travel would have larger and larger statistical fluctuations.

If the experiment were repeated a large number of times and the results of each experiment were averaged, the statistical fluctuations at small values of Δx would decrease and approach the dashed line in Fig. 9.5. This dashed line could also be obtained by extrapolating the interaction probability per unit travel distance from those values without large statistical fluctuation obtained at large Δx. The statistically averaged interaction probability per unit travel distance, in the limit of infinitely small Δx, approaches a constant μ, i.e., μ is defined as

$$\mu \equiv \lim_{\Delta x \to 0} \frac{(\Delta N/N)}{\Delta x}. \tag{9.21}$$

The *linear interaction coefficient* μ is a property of the host medium for a specified type of incident particle with a specified energy E. The interaction coefficient can be specialized to specific types of interaction, designated by a subscript on μ, i.e., μ_i, where, for example, $i = s$ for a scatter, $i = a$ for an absorption, and so forth. Thus, in the limit of small path lengths, μ_i is the probability, per unit differential path length, that a particle undergoes an ith type of interaction.

Because μ_i is constant for a given material and for a given type of interaction implies that the probability of interaction, per unit differential path length, is independent of the path length traveled prior to the interaction. In this book, when the interaction coefficient is referred to as the *probability per unit path length* of an interaction, it is understood that this is true only in the limit of very small path lengths.

The constant μ_i is called the *linear interaction coefficient* for reaction i. For each type of reaction, there is a corresponding linear coefficient. For example, μ_a is the *linear absorption coefficient*, μ_s the *linear scattering coefficient*, and so on. Although this nomenclature is widely used to describe photon interactions, μ_i is often referred to as the *macroscopic cross section* for reactions of type i, and is usually given the symbol Σ_i when describing neutron interactions. However, for unity of presentation, the symbol μ_i is used here for all types of radiation.

The probability, per unit path length, that a neutral particle undergoes some sort of reaction, μ_t, is the sum of the probabilities, per unit path length of travel, for each type of possible reaction, i.e.,

$$\mu_t(E) = \sum_i \mu_i(E). \tag{9.22}$$

Often, the subscript t is dropped and the symbol μ is used to represent the total interaction coefficient. Since these coefficients generally depend on the particle's kinetic energy E, this dependence has been shown explicitly. The total interaction probability per unit path length, μ_t, is fundamental in describing how indirectly ionizing radiation interacts with matter and is usually called the *linear attenuation coefficient*. It is perhaps more appropriate to use the words *total linear interaction coefficient* since many interactions do not "attenuate" the particle in the sense of an absorption interaction.

9.2.2 Attenuation of Uncollided Radiation

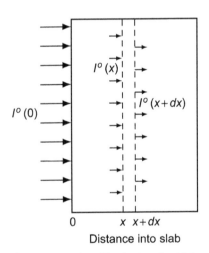

Figure 9.6 Uniform illumination of a slab by radiation.

To see how radiation is attenuated as it travels through matter, consider a plane parallel beam of neutral particles of intensity I^o particles cm^{-2} s^{-1} incident normally on the surface of a slab (see Fig. 9.6). As the particles travel into the slab, some interact with the atoms of the slab, thereby reducing the beam's intensity. In particular, the intensity $I^o(x)$ of *uncollided* particles at depth x into the slab is of interest. At distance x into the slab, some uncollided particles undergo interactions for the first time as they try to traverse the next Δx of distance, thereby reducing the uncollided beam intensity at x, $I^o(x)$, to a lesser value $I^o(x+\Delta x)$ at $x+\Delta x$. The probability an uncollided particle interacts as it traverses Δx is

$$P(\Delta x) = \frac{I^o(x) - I^o(x + \Delta x)}{I^o(x)}.$$

In the limit as $\Delta x \to 0$, we have from Eq. (9.21)

$$\mu_t = \lim_{\Delta x \to 0} \frac{P(\Delta x)}{\Delta x} = \lim_{\Delta x \to 0} \frac{I^o(x) - I^o(x + \Delta x)}{\Delta x} \frac{1}{I^o(x)} \equiv -\frac{dI^o(x)}{dx} \frac{1}{I^o(x)}$$

or

$$\frac{dI^o(x)}{dx} = -\mu_t I^o(x). \tag{9.23}$$

The solution for the uncollided intensity is

$$I^o(x) = I^o(0)e^{-\mu_t x}. \tag{9.24}$$

From this result, it is seen that uncollided neutral-particle radiation is *exponentially* attenuated as it passes through a medium.

Also from this result, the probability $P(x)$ a particle interacts somewhere along a path of length x is

$$P(x) = 1 - \frac{I^o(x)}{I^o(0)} = 1 - e^{-\mu_t x}. \tag{9.25}$$

Conversely, the probability $\overline{P}(x)$ a particle does *not* interact while traveling a distance x is

$$\overline{P}(x) = 1 - P(x) = e^{-\mu_t x}. \tag{9.26}$$

Note that as $x \to dx$, $P(dx) \to \mu_t dx$, a result in agreement with the interpretation that μ_t is the probability, per unit differential path length of travel, the particle interacts.

9.2.3   Average Travel Distance Before an Interaction

Although where a particle interacts cannot be predicted, the probability distribution for how far a neutral particle travels before interacting can be predicted from the above results. Let $p(x)dx$ be the probability that a particle interacts for the first time between x and $x + dx$. Then

$$p(x)dx = \text{Prob\{particle travels a distance } x \text{ without interaction\}} \times$$
$$\text{Prob\{particle interacts in the next } dx\}$$

$$= \{\overline{P}(x)\}\{P(dx)\} = \{e^{-\mu_t x}\}\{\mu_t dx\} = \mu_t e^{-\mu_t x}dx. \tag{9.27}$$

Note that $\int_0^\infty p(x)\,dx = 1$, as it must be for a probability distribution function.

With this probability distribution, the average distance $\overline{x}$ traveled by a neutral particle to the site of its first interaction can be found. The average value of x is simply

$$\overline{x} = \int_0^\infty x\, p(x)\, dx = \mu_t \int_0^\infty x\, e^{-\mu_t x}\, dx = \frac{1}{\mu_t}. \tag{9.28}$$

This average travel distance before an interaction, $1/\mu_t$, is called the *mean free path length*.

9.2.4 Scattering Interaction Coefficients

A radiation field generally has two components, namely, the uncollided particles, which are attenuated exponentially as they pass through the host medium, and secondary particles that generally are a result of scattering interactions.[2] To describe the scattering of radiation, the scattering interaction coefficient must be generalized.

In a scattering interaction, a particle with energy E and direction $\mathbf{\Omega}$ scatters into a new direction $\mathbf{\Omega}'$ and has a new energy E'. The *doubly differential scattering coefficient* is defined such that $\mu_s(E \rightarrow E', \mathbf{\Omega} \rightarrow \mathbf{\Omega}')dE'\,d\mathbf{\Omega}'$ is the probability, per unit differential path length of travel, of a scattering interaction in which the incident particle of energy E and direction $\mathbf{\Omega}$ emerges from the interaction with an energy in dE' about E' and with a direction in $d\mathbf{\Omega}'$ about $\mathbf{\Omega}'$. In an isotropic medium, the probability of interaction at some point is independent of the particle direction and, for the scattering event shown in Fig. 9.7, the scattering probability depends on only the initial and final energies and the scattering angle $\theta_s = \cos^{-1}\mathbf{\Omega}\cdot\mathbf{\Omega}'$, or, equivalently, on $\omega_s \equiv \cos\theta_s$.[3] All changes in the azimuthal angle ψ about the incident direction are equally likely. Thus, the doubly differential scattering coefficient is usually written as $\mu_s(E \rightarrow E', \omega_s)$ or $\mu_s(E \rightarrow E', \mathbf{\Omega}\cdot\mathbf{\Omega}')$. In this form, $\mu_s(E \rightarrow E', \omega_s)$ has units such as $\text{cm}^{-1}\,\text{MeV}^{-1}\,\text{sr}^{-1}$.

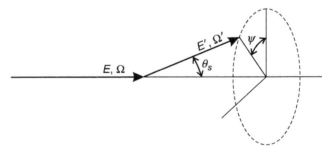

Figure 9.7 A particle moving in direction $\mathbf{\Omega}'$ scatters into a new direction $\mathbf{\Omega}'$. The probability of such a scatter generally depends on the scattering angle $\theta_s = \cos^{-1}\mathbf{\Omega}\cdot\mathbf{\Omega}'$, but in an isotropic medium all azimuthal angles about the incident direction are equally probable.

In any scattering interaction, the constraints of energy and linear momentum conservation require that, if any two of the variables E, E', and ω_s are specified,

---

[2] Other reactions producing secondary particles include stimulated photon emission and neutron-induced fission.

[3] This dependence is true for isotropic media and is assumed to be the case unless otherwise noted specifically. For crystalline and other anisotropic media, μ_s generally depends on the incident radiation direction $\mathbf{\Omega}$ and the exit radiation direction $\mathbf{\Omega}'$.

the third must assume a specific value. Thus, the scattering angle is a function of the initial and final energies, or the final energy is a function of the initial energy and scattering angle. For example, when a neutron scatters elastically from a nucleus with mass number A, the cosine of the scattering angle is given by (Shultis and Faw, 2008)

$$\omega_s = S(E, E') \equiv \frac{1}{2}\left[(A+1)\sqrt{\frac{E}{E'}} - (A-1)\sqrt{\frac{E'}{E}}\right].$$

Other types of photon and neutron scattering have different expressions for $S(E, E')$ as derived by Shultis and Faw (2000).

No matter how a neutron or photon scatters, the differential scattering cross section, thus, has the form

$$\mu_s(E \to E', \omega_s) = \mu_s(E \to E')\delta(\omega_s - S(E, E')). \qquad (9.29)$$

Integration over E' or ω_s eliminates the delta function and produces a *singly differential scattering coefficient*. Often it is of interest to deal only with, say, the energy dependence or the angular distribution of scattered radiation. In this case, one or the other of the following forms of the single differential scattering coefficient may be used:

$$\mu_s(E, E') \equiv \int_{4\pi} d\Omega\, \mu_s(E \to E', \omega_s) \qquad (9.30)$$

or

$$\mu_s(E, \theta_s) \equiv \int_0^\infty dE'\, \mu_s(E \to E', \omega_s). \qquad (9.31)$$

Here $\mu_s(E, E')\, dE'$ is the probability per unit path length for scattering into dE' about E' without regard to scattering angle, and $\mu_s(E, \omega_s)\, d\Omega$ is the like probability for scattering into direction range $d\Omega$ without regard to the energy of the scattered radiation.

Also of interest is

$$\mu_s(E) \equiv \int_0^\infty dE'\, \mu_s(E \to E'), \qquad (9.32)$$

which is just the total linear interaction coefficient for the scattering of incident radiation of energy E, without regard to energy loss or angle of scattering.

9.2.5 Microscopic Cross Sections

The linear coefficient $\mu_i(E)$ depends on the type and energy E of the incident particle, the type of interaction i, and the interacting medium. An important property of the medium that determines μ_i is the density of target atoms or electrons. It is reasonable to expect that μ_i is proportional to the "target" atom density N in the material, i.e.,

$$\mu_i = \sigma_i N = \sigma_i \frac{\rho N_a}{A}, \tag{9.33}$$

where σ_i is a constant of proportionality independent of N. Here ρ is the mass density of the medium, N_a is Avogadro's constant (6.022×10^{23} particles mol^{-1}), and A is the atomic weight of the medium.

The proportionality constant σ_i is called the *microscopic cross section* for reaction i. From Eq. (9.33), it is seen to have dimensions of area. Often σ_i is interpreted as the *effective* cross-sectional area presented by the target atom to the incident particle for a given type of interaction. Indeed, many times σ_i has dimensions comparable to those expected from the physical size of the nucleus or atom. However, this widely used interpretation of the microscopic cross section, while easy to grasp, leads to philosophical difficulties when it is observed that σ_i generally varies with the energy of the incident particle and, for a crystalline material, the particle direction. An alternative interpretation that σ_i is the interaction probability per unit differential path length of travel, normalized to one target atom per unit volume, avoids such conceptual difficulties while emphasizing the statistical nature of the interaction process. Microscopic cross sections are usually expressed in units of *barns*, where the barn equals 10^{-24} cm^2.

Data on cross sections and linear interaction coefficients, especially for photons, are frequently expressed as the ratio of μ_i to the medium mass density ρ. This ratio is called the *mass interaction coefficient* for reaction i. Division of Eq. (9.33) by ρ yields

$$\frac{\mu_i}{\rho} = \frac{\sigma_i N}{\rho} = \frac{N_a}{A} \sigma_i. \tag{9.34}$$

From this result, it is seen that μ_i/ρ is an intrinsic property of the interacting medium—independent of its mass density. For photons μ_i/ρ data are more frequently tabulated, while for neutrons σ_i values are the preferred tabulated quantities.

The linear and mass interaction coefficients in compounds or homogeneous mixtures for interactions of type i are, respectively,

$$\mu_i = \sum_j \mu_i^j = \sum_j N^j \sigma_i^j \tag{9.35}$$

and

$$\frac{\mu_i}{\rho} = \sum_j w_j \left(\frac{\mu_i}{\rho}\right)^j. \tag{9.36}$$

Here the subscript i refers to the type of interaction, the superscript j refers to the jth component of the material, and w_j is the mass fraction of component j.

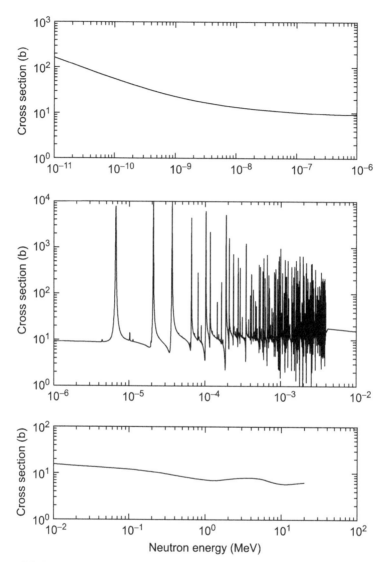

Figure 9.8 Total neutron cross section for uranium computed using NJOY-processed ENDF/B (version V) data. Above 4 keV, the resonances are no longer resolved and only the average cross section behavior is shown.

In Eq. (9.35), the atomic density N^j and the linear interaction coefficient μ_i^j are values for the jth material *after* mixing.

The total mass interaction coefficient μ/ρ for photons in all the elements is shown in Fig. 9.9. A typical neutron microscopic cross section is shown in Fig. 9.8. Unlike photon cross sections, which are generally smooth and slowly varying above 100 keV, neutron cross sections display many resonances. Because photons interact only with atomic electrons, all isotopes of the same element have the same photon cross sections. Moreover, photon cross sections vary slowly as the atomic number of the element changes. By contrast, neutron cross sections vary dramatically from isotope to isotope and from element to element. Consequently, compilations of the empirical cross sections for neutron interactions require huge databases compared to those needed for photon cross sections.

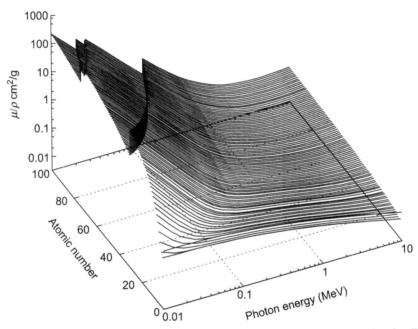

Figure 9.9 Comparison of the total mass interaction coefficients, less coherent scattering, for all the elements. The coefficients generally vary smoothly except for the discontinuities at the *electron shell edges* below which a photon has insufficient energy to remove an electron from the shell in a photoelectric interaction which dominates at low photon energies. At high energies pair production interactions dominate and at intermediate energies photon scattering dominates.

9.2.6 Reaction Rate Density

One of the most fundamental goals of transport theory is to compute the rate at which radiation particles interact with the host medium at some point of interest. For example, the rate of biological damage to some organ of the body is proportional to the rate at which radiation interacts in the organ. Similarly, the effectiveness of a radiation shield depends on the number of interactions that occur in the shield as the radiation migrates through it.

The rate at which radiation interacts with the medium is readily expressed in terms of the flux density and the interaction coefficient. Recall that the flux density spectrum $\phi(\mathbf{r}, E, t)$ has the interpretation that $\phi(\mathbf{r}, E, t)\,dE$ is the total path length traveled, in a unit time at t, by all the particles that are in a unit volume at $\mathbf{r}$ and that have energies in dE about E. Also, recall that the average distance a radiation particle must travel before undergoing an ith-type interaction with the medium is $1/\mu_i(\mathbf{r}, E, t)$, where the interaction coefficient, generally, varies with position, the incident particle energy, and even time if the medium's composition changes. Then the expected number of ith-type interactions occurring in a unit volume at $\mathbf{r}$, in a unit time at time t, caused by those particles with energies in dE about E is

$R_i(\mathbf{r}, E, t)dE = \{$path length traveled by particles in one cm^3 in one second$\}/$

$\{$average travel distance for an ith type interaction$\}$

$= \{\phi(\mathbf{r}, E, t)dE\}/\{1/\mu_i(\mathbf{r}, E, t)\}$

or

$$R_i(\mathbf{r}, E, t) = \mu_i(\mathbf{r}, E, t)\,\phi(\mathbf{r}, E, t). \qquad (9.37)$$

From the interaction rate density, all sorts of useful information can be calculated about the effects of the radiation. For example, the total number of fissions that occur in some volume V inside a reactor core between times t_1 and t_2 by neutrons of all energies is

$$\text{Number of fissions in } (t_1, t_2) = \int_{t_1}^{t_2} dt \iiint_V dV \int_0^{E_{max}} dE\, \mu_f(\mathbf{r}, E, t)\phi(\mathbf{r}, E, t),$$
$$(9.38)$$

where μ_f is the macroscopic fission cross section and E_{max} is the maximum energy a neutron can have.

An expression like that of Eq. (9.38) involving integrals of the reaction rate density is a key result whose evaluation is sought in many radiation transport calculations. That a multidimensional integral has to be evaluated immediately suggests that Monte Carlo may be of some utility. The integral is a "difficult" one because (1) the flux density $\phi(\mathbf{r}, E, t)$ usually has to be obtained by solving numerically the *transport equation* (discussed next), (2) the interaction coefficients $\mu_i(\mathbf{r}, E, t)$ can change abruptly with the particle energy (see Fig. 9.8), and (3) the integral is usually multidimensional. As a consequence, the integrand of Eq. (9.38) is hard to evaluate and, because it generally varies rapidly with energy, it is also hard to integrate. Monte Carlo methods provide relatively simple ways to treat these difficulties.

9.3   Transport Equation

One would ideally like to know the distribution of photons or neutrons everywhere in the medium of interest. For most applications, the spatial and energy distribution of the particle density $n(\mathbf{r}, E, t)$ is all that is needed. Unfortunately, there is no equation for this quantity that holds rigorously in all situations. The simplest equation that accurately describes the particle distribution in a medium is for the differential energy and directional flux density $\phi(\mathbf{r}, E, \mathbf{\Omega}, t)$ (see Section 9.1.3).

To obtain an equation for $\phi(\mathbf{r}, E, \mathbf{\Omega}, t)$, consider the particle balance in an arbitrary volume V with a closed surface S for those particles with energies in dE about E moving in $d\Omega$ about the direction $\mathbf{\Omega}$ at time t. In a radiation field, the rate of change of the number of particles in V that have energies in dE about E and that travel in directions $d\Omega$ about $\mathbf{\Omega}$ at time t is the difference between (1) the creation of particles with these energies and directions (from sources and particles that scatter into these energy and direction intervals) and (2) the loss of these particles (either by leaving or entering the volume V or by changing their energy or direction of travel). Thus, the particles of the radiation field with these energies and directions must satisfy the following balance relation:

$$
\left\{
\begin{array}{l}
\text{(a) the rate of} \\
\text{increase of} \\
\text{particles in } V
\end{array}
\right\}
= -
\left\{
\begin{array}{l}
\text{(b) the net flow} \\
\text{rate out of } V \\
\text{across } S
\end{array}
\right\}
-
\left\{
\begin{array}{l}
\text{(c) the rate at} \\
\text{which particles} \\
\text{interact in } V
\end{array}
\right\}
$$

$$
+
\left\{
\begin{array}{l}
\text{(d) the rate at which} \\
\text{secondary\ \ particles} \\
\text{of energy } E \text{ and di-} \\
\text{rection } \mathbf{\Omega} \text{ are pro-} \\
\text{duced}
\end{array}
\right\}
+
\left\{
\begin{array}{l}
\text{(e) the rate of} \\
\text{production by} \\
\text{sources in } V
\end{array}
\right\}. \quad (9.39)
$$

The five terms have the following interpretations: (a) the rate of increase of the number of particles in V with the specified energies and directions, (b) the net number of such particles flowing out of V across S in a unit time, (c) the total number of these particles suffering collisions in V in a unit time (i.e., those changing their energy and/or direction of travel or simply being absorbed), (d) the number of secondary particles in dE about E moving in $d\Omega$ about $\mathbf{\Omega}$ that are produced by all particle–medium interactions in V in a unit time, and (e) the number of particles in dE about E moving in $d\Omega$ about $\mathbf{\Omega}$ that are introduced in V in a unit time by sources that are independent of the radiation field, e.g., gamma photons from radionuclides. The desired transport equation for $\phi(\mathbf{r}, E, \mathbf{\Omega}, t)$ is then obtained by expressing mathematically each component of this balance relation in terms of the angular flux density $\phi(\mathbf{r}, E, \mathbf{\Omega}, t)$. To obtain the transport equation for $\phi(\mathbf{r}, E, \mathbf{\Omega}, t)$, each of the five terms in the balance relation must be expressed mathematically.

Term (a): The total number of particles in V with energies in dE about E and with directions in $d\Omega$ about Ω are

$$\left[\int_V dV n(\mathbf{r}, E, \Omega, t) \right] dE\, d\Omega = \frac{1}{v} \left[\int_V dV \phi(\mathbf{r}, E, \Omega, t) \right] dE\, d\Omega. \quad (9.40)$$

Thus, the rate of change of this quantity is

$$(a) = \frac{1}{v} \frac{\partial}{\partial t} \left[\int_V dV \phi(\mathbf{r}, E, \Omega, t) \right] dE\, d\Omega = \left[\int_V dV \frac{1}{v} \frac{\partial \phi(\mathbf{r}, E, \Omega, t)}{\partial t} \right] dE\, d\Omega. \quad (9.41)$$

Term (b): As discussed in Section 9.1.5, the quantity $\mathbf{n} \cdot \Omega \phi(\mathbf{r}, E, \Omega)\, dE\, d\Omega$ is the number of particles with energy in dE about E traveling in $d\Omega$ about Ω that cross a unit area of the surface (with unit outward normal $\mathbf{n}$) in a unit time at point $\mathbf{r}$. Thus term (b) of the balance relation may be expressed mathematically as

$$(b) = \left[\int_S dS\, \mathbf{n} \cdot \Omega\, \phi(\mathbf{r}, E, \Omega, t) \right] dE\, d\Omega, \quad (9.42)$$

which can be expressed as a volume integral upon application of the Gauss integral theorem, namely,

$$(b) = \left[\int_V dV\, \nabla \cdot \Omega\, \phi(\mathbf{r}, E, \Omega, t) \right] dE\, d\Omega. \quad (9.43)$$

Because the ∇ operator affects only the spatial variable and not Ω, since $\mathbf{r}$ and Ω are independent variables, this last result may be written as

$$(b) = \left[\int_V dV\, \Omega \cdot \nabla \phi(\mathbf{r}, E, \Omega, t) \right] dE\, d\Omega. \quad (9.44)$$

Term (c): The total interaction rate, term (c) of Eq. (9.39), is

$$(c) = \left[\int_V dV\, \mu_t(\mathbf{r}, E, t) \phi(\mathbf{r}, E, \Omega, t) \right] dE\, d\Omega, \quad (9.45)$$

where μ is the total interaction coefficient or macroscopic cross section. Any interaction in V either changes the energy and direction of the particle or removes it.

Term (d): If it is assumed that the secondary particles appear at the position where the causal interaction occurs, term (d) can be written as

$$(d) = \left[\int_V dV \int_0^\infty dE' \int_{4\pi} d\Omega'\, \mu_s(\mathbf{r}, E' \to E, \Omega' \to \Omega, t) \phi(\mathbf{r}, E', \Omega', t) \right] dE\, d\Omega, \quad (9.46)$$

where $\mu_s(\mathbf{r}, E' \to E, \mathbf{\Omega}' \to \mathbf{\Omega}, t)\, dE\, d\Omega$ is the probable number of secondary particles at point $\mathbf{r}$ with energies in dE about E in direction $d\Omega$ about $\mathbf{\Omega}$ produced by an incident particle of energy E' traveling in direction $\mathbf{\Omega}'$, per unit differential path length of the incident particle. For example, Eq. (9.29) gives this secondary particle distribution for elastic scattering of neutrons.

Term (e): Finally, let $S(\mathbf{r}, E, \mathbf{\Omega}, t)\, dE\, d\Omega$ denote the production rate by nonradiation-induced sources of particles, per unit volume about point $\mathbf{r}$, with energies in dE about E and in directions $d\Omega$ about $\mathbf{\Omega}$. Thus

$$(e) = \left[\int_V dV\, S(\mathbf{r}, E, \mathbf{\Omega}, t) \right] dE\, d\Omega. \qquad (9.47)$$

With these results, the balance relation Eq. (9.39) can be written as

$$\int_V \left[\frac{1}{v} \frac{\partial \phi}{\partial t} - \mathbf{\Omega} \cdot \nabla \phi - \mu_t \phi \right.$$
$$+ \int_0^\infty dE' \int_{4\pi} d\Omega'\, \mu_s(\mathbf{r}, E' \to E, \mathbf{\Omega}' \to \mathbf{\Omega}, t)\phi(\mathbf{r}, E', \mathbf{\Omega}', t)$$
$$\left. + S(\mathbf{r}, E, \mathbf{\Omega}, t) \right] dV = 0. \qquad (9.48)$$

However, the volume V is arbitrary, and hence the integrand of Eq. (9.48) must be identically zero; therefore,

$$\frac{1}{v} \frac{\partial \phi(\mathbf{r}, E, \mathbf{\Omega}, t)}{\partial t} = -\mathbf{\Omega} \cdot \nabla \phi(\mathbf{r}, E, \mathbf{\Omega}, t) - \mu_t(\mathbf{r}, E, t)\phi(\mathbf{r}, E, \mathbf{\Omega}, t)$$
$$+ \int_0^\infty dE' \int_{4\pi} d\Omega'\, \mu_s(\mathbf{r}, E' \to E, \mathbf{\Omega}' \to \mathbf{\Omega}, t)\phi(\mathbf{r}, E, \mathbf{\Omega}', t) + S(\mathbf{r}, E, \mathbf{\Omega}, t). \qquad (9.49)$$

This integro-differential equation for $\phi(\mathbf{r}, E, \mathbf{\Omega})$ is known as the time-dependent *transport equation* or the *linearized Boltzmann equation*, named after Ludwig Boltzmann, who first derived it over a century ago. This equation serves as a precise description of the radiation field in all neutron or photon transport problems. However, in the derivation above, several implicit assumptions have been made. The radiation particles are assumed to be point particles that stream in straight lines between collisions, which occur at distinct points in the system. In other words, the associated wavelength of the particle and the interaction distance between the particle and the atoms of the medium are assumed to be very small compared to the mean free path length of the particle and the size of the system. Also, all internal structure, such as spin or polarization of the diffusing particles, is ignored. More important, the transport equation is linear because all particle–particle interactions are neglected.

The $1/v$ factor in the first term of Eq. (9.49) makes this term generally negligible, unless $\partial\phi/\partial t$ becomes exceptionally large. Such rapid transients occur rarely, such as during the detonation of a nuclear bomb or a supernova. However, in most cases, transients are much slower and the time dependence of the radiation field is that of the primary radiation sources, i.e., any time delays caused by the finite speed of the radiation particles can be ignored in most transport problems. Also, in most transport problems the host medium does not change in time, nor does the primary source of radiation vary with time. Under these conditions in which the first term in Eq. (9.49) can be ignored and the interaction coefficients do not change in time, this equation reduces to the steady-state transport equation

$$\mathbf{\Omega}\cdot\nabla\phi(\mathbf{r}, E, \mathbf{\Omega}) + \mu_t(\mathbf{r}, E)\phi(\mathbf{r}, E, \mathbf{\Omega})$$
$$= \int_0^\infty dE' \int_{4\pi} d\Omega' \, \mu_s(\mathbf{r}, E' \to E, \mathbf{\Omega}' \to \mathbf{\Omega})\phi(\mathbf{r}, E', \mathbf{\Omega}') + S(\mathbf{r}, E, \mathbf{\Omega}).$$

(9.50)

This equation, or a variant thereof, is the basis of many of the large transport codes that are routinely used today (see Appendix E).

The form of the transport equation given by Eq. (9.50) is quite general and applies to any geometry, to either photons or neutrons, and allows for all types of particle–medium interactions, all of whose probabilities are encompassed by the interaction coefficients $\mu(\mathbf{r}, E)$ and $\mu_s(\mathbf{r}, E' \to E, \mathbf{\Omega}' \to \mathbf{\Omega})$. To solve this equation for the differential energy and angular flux density $\phi(\mathbf{r}, E, \mathbf{\Omega})$, one must specify (1) the coordinate system used, so that the streaming term $\mathbf{\Omega}\cdot\nabla\phi$ can be expressed explicitly, (2) the coefficients $\mu_t(\mathbf{r}, E)$ and $\mu_s(\mathbf{r}, E' \to E, \mathbf{\Omega}' \to \mathbf{\Omega})$, which depend on the material composition of the host medium and the interactions of interest, (3) the distribution $S(\mathbf{r}, E, \mathbf{\Omega})$ of the primary radiation sources, which usually arise from the presence of radionuclides in the medium, and (4) the boundary conditions, which determine the incident flux density at the edges of the shielding medium.

Unfortunately, exact analytical solutions of the transport equation are known for only the simplest cases, none of which is directly applicable to realistic transport problems. Consequently, much effort has been directed toward the development of approximations to the transport equation which can then be solved by numerical techniques or, for the simplest approximations, even by analytical methods. These numerical methods for solving the transport equation were important in the first few decades following World War II because they are generally much more computationally efficient than Monte Carlo simulations of the transport process. Today, because of the widespread availability of inexpensive fast computers, Monte Carlo approaches have largely supplanted the numerical approximation methods. Readers interested in numerical approximations of the transport equation are referred to books and review articles devoted to transport theory (e.g., Duderstadt and Martin (1979); Sanchez and McCormick (1982); Lewis and Miller (1984); Adams and Larsen (2002)).

9.3.1   One-Speed Transport Equation in Plane Geometry

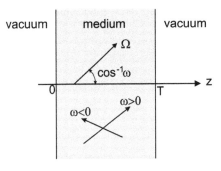

Figure 9.10 1-D plane geometry.

The transport equation of Eq. (9.50) involves six independent variables $(\mathbf{r}, E, \boldsymbol{\Omega}) \equiv (x, y, z, E, \theta, \psi)$, a number that generally makes it difficult to obtain analytical results. Under various assumptions and symmetries, the transport equation can be reduced to one more amenable to analysis. Here the simplest nontrivial transport equation, the so-called one-speed, one-dimensional transport equation with isotropic scattering and azimuthal symmetry, is derived.

First assume the particles do not change energy upon scattering (such as thermal neutrons or low-energy photons). Next assume scattering is isotropic in the laboratory coordinate system so that $\mu_s(\mathbf{r}, E' \to E, \boldsymbol{\Omega}' \to \boldsymbol{\Omega}) \longrightarrow \mu_s/4\pi$. Likewise, assume all sources emit particles isotropically so $S(\mathbf{r}.E, \boldsymbol{\Omega}) \to S(\mathbf{r})/4\pi$. Lastly, assume the medium is homogeneous so that the interaction coefficients μ and μ_s are independent of $\mathbf{r}$. With these assumptions, Eq. (9.50) reduces to

$$\boldsymbol{\Omega}\cdot\nabla\phi(\mathbf{r}, \boldsymbol{\Omega}) + \mu_t\phi(\mathbf{r}, \boldsymbol{\Omega}) = \frac{\mu_s}{4\pi}\int_{4\pi}\phi(\mathbf{r}, \boldsymbol{\Omega}')\,d\Omega' + \frac{S(\mathbf{r})}{4\pi}. \qquad (9.51)$$

In plane geometry, the transport medium is an infinite slab of thickness T with one spatial coordinate, say z, perpendicular to the slab surfaces. See Fig. 9.10. If the source and boundary conditions do not vary with the other two coordinate directions x and y, then $\phi(\mathbf{r}, \boldsymbol{\Omega}) \to \phi(x, \theta, \psi)$, so $\partial\phi/\partial x = \partial\phi/\partial y = 0$. The streaming term becomes

$$\boldsymbol{\Omega}\cdot\nabla\phi = (\cos\psi\sqrt{1-\omega^2}, \sin\psi\sqrt{1-\omega^2}, \omega)\cdot(0, 0, \partial\phi/\partial z) = \omega\partial\phi/\partial z,$$

where $\omega = \cos\theta$. Hence in plane geometry, Eq. (9.51) becomes

$$\omega\frac{\partial\phi(z, \omega, \psi)}{\partial z} + \mu_t\phi(z, \omega, \psi) = \frac{\mu_s}{4\pi}\int_{-1}^{1}\int_{0}^{2\pi}\phi(z, \omega', \psi')\,d\psi'\,d\omega' + \frac{S(z)}{4\pi}.$$

Finally, integrate this result over all azimuthal angles $0 < \psi \leq 2\pi$, divide through by μ_t, and let $\hat{z} = \mu_t z$ (distance measured in mean free path lengths) to obtain the much simpler transport equation

$$\omega \frac{\partial \phi(\hat{z}, \omega)}{\partial \hat{z}} + \phi(\hat{z}, \omega) = \frac{c}{2} \int_{-1}^{1} \phi(\hat{z}, \omega') d\omega' + \frac{S(\hat{z})}{2\mu_t},$$

(9.52)

where $c = \mu_s / \mu_t$. To specify a unique solution to this transport equation values of $\phi(0, \omega)$, $\omega > 0$, and $\phi(\widehat{T}, \omega)$, $\omega < 0$, must be specified. For *vacuum boundary conditions*, these values are set to zero, i.e., no particles are incident upon the slab.

9.4 Integral Forms of the Transport Equation

The integro-differential form of the transport equation, Eq. (9.50), can be converted into a pure integral equation. Because this alternative integral form of the transport equation is equivalent to the differential-integral form, exact solutions to it are also not available, except for the simplest of cases. The utility of the integral form is that analytical and numerical techniques can be developed from it for obtaining approximate solutions for realistic problems. In particular, approximations to the integral transport equation are capable of more accurately treating highly anisotropic angular flux densities and scattering cross sections than are approximations based on the integro-differential transport equation. Also the integral form of the transport equation can be used directly by Monte Carlo methods using the techniques discussed in the next chapter.

9.4.1 Integral Equation for the Angular Flux Density

To derive an integral equation for the angular flux density spectrum $\phi(\mathbf{r}, E, \boldsymbol{\Omega})$, begin by writing Eq. (9.50) in the form

$$\boldsymbol{\Omega} \cdot \nabla \phi(\mathbf{r}, E, \boldsymbol{\Omega}) + \mu_t(\mathbf{r}, E)\phi(\mathbf{r}, E, \boldsymbol{\Omega}) = \chi(\mathbf{r}, E, \boldsymbol{\Omega}), \quad \mathbf{r} \in V, \quad (9.53)$$

where

$$\chi(\mathbf{r}, E, \boldsymbol{\Omega}) \equiv \int_0^\infty dE' \int_{4\pi} d\Omega' \, \mu_s(\mathbf{r}, E' \to E, \boldsymbol{\Omega}' \to \boldsymbol{\Omega})\phi(\mathbf{r}, E', \boldsymbol{\Omega}') + S(\mathbf{r}, E, \boldsymbol{\Omega}).$$

(9.54)

The quantity $\chi(\mathbf{r}, E, \boldsymbol{\Omega})$ is the *emission rate density*, i.e., the rate density of particles leaving a source or a collision with coordinates $(\mathbf{r}, E, \boldsymbol{\Omega})$ in *phase* space.

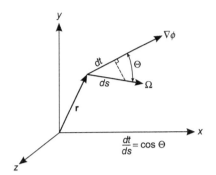

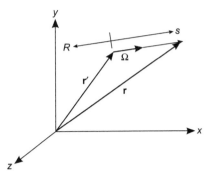

Figure 9.11 Orientation of the unit vector $\mathbf{\Omega}$ and the gradient of $\phi(\mathbf{r}, E, \mathbf{\Omega})$.

Figure 9.12 Coordinate system for integrating the transport equation.

It is assumed that V is a convex surface and that no particles are incident on it. Thus, the boundary condition is[4]

$$\phi(\mathbf{r}_s, E, \mathbf{\Omega}) = 0, \quad \text{for } \mathbf{n} \cdot \mathbf{\Omega} < 0, \tag{9.55}$$

where $\mathbf{n}$ is the outward normal on the surface at $\mathbf{r}_s$.

Suppose that distances s and t are measured along the directions of $\mathbf{\Omega}$ and $\nabla\phi$, as in Fig. 9.11, and that Θ is the angle between these two vectors. Then, because $\nabla\phi = d\phi/dt$, $dt/ds = \cos\Theta$, and $\mathbf{\Omega}$ is a unit vector, it follows that

$$\mathbf{\Omega} \cdot \nabla\phi = \cos\Theta\frac{d\phi}{dt} = \cos\Theta\frac{ds}{dt}\frac{d\phi}{ds} = \frac{d\phi}{ds}. \tag{9.56}$$

Thus, the streaming term $\mathbf{\Omega} \cdot \nabla\phi$ is seen to be the rate of change of ϕ along the direction of particle travel $\mathbf{\Omega}$, namely, $d\phi/ds$, where s is measured along the direction of $\mathbf{\Omega}$. Now define $\mathbf{r} \equiv \mathbf{r}' - R\mathbf{\Omega}$, where $R = -s$ is measured along the direction opposite to $\mathbf{\Omega}$ (see Fig. 9.12). Then $d/ds = -d/dR$ and Eq. (9.53) can be rewritten as

$$-\frac{d}{dR}\phi(\mathbf{r}' - R\mathbf{\Omega}, E, \mathbf{\Omega}) + \mu_t(\mathbf{r}' - R\mathbf{\Omega}, E)\phi(\mathbf{r}' - R\mathbf{\Omega}, E, \mathbf{\Omega}) = \chi(\mathbf{r}' - R\mathbf{\Omega}, E, \mathbf{\Omega}).$$
$$\tag{9.57}$$

Now multiply this result by the integrating factor $\exp[-\int_0^R \mu_t(\mathbf{r}' - R'\mathbf{\Omega}, E)\,dR']$, which has the property

$$\frac{d}{dR}\exp\left[-\int_0^R \mu_t(\mathbf{r}' - R'\mathbf{\Omega}, E)\,dR'\right]$$

---

[4] Concavities can always be filled with void regions to produce a volume V that is convex. Moreover, if radiation is incident from outside of V, the source S can be modified to include the first collision distribution of such incident radiation and the homogeneous boundary condition then applies.

$$= -\mu_t(\mathbf{r'} - R\mathbf{\Omega}, E) \exp\left[-\int_0^R \mu_t(\mathbf{r'} - R'\mathbf{\Omega}, E)\,dR'\right],$$

(9.58)

to obtain

$$-\frac{d}{dR}\left\{\phi(\mathbf{r'} - R\mathbf{\Omega}, E, \mathbf{\Omega})\exp\left[-\int_0^R \mu_t(\mathbf{r'} - R'\mathbf{\Omega}, E)\,dR'\right]\right\}$$
$$= \chi(\mathbf{r'} - R\mathbf{\Omega}, E, \mathbf{\Omega})\exp\left[-\int_0^R \mu_t(\mathbf{r'} - R'\mathbf{\Omega}, E)\,dR'\right].$$

(9.59)

Next replace the dummy variable $\mathbf{r'}$ by $\mathbf{r}$ and integrate over R from zero to the distance to the boundary R_s (or to infinity if the medium is infinite) to obtain the following integral equation for the angular flux density:

$$\phi(\mathbf{r}, E, \mathbf{\Omega}) = \int_0^{R_s} dR'\,\chi(\mathbf{r} - R'\mathbf{\Omega}, E, \mathbf{\Omega})\exp\left[-\int_0^{R'} \mu_t(\mathbf{r} - R''\mathbf{\Omega}, E)\,dR''\right].$$

(9.60)

To obtain this result, the boundary condition of Eq. (9.55) has been used, i.e., $\phi(\mathbf{r} - R_s\mathbf{\Omega}, E, \mathbf{\Omega}) \equiv \phi(\mathbf{r}_s, E, \mathbf{\Omega}) = 0$, so that the term in the curly brackets in Eq. (9.59) vanishes when $R = R_s$.

To simplify notation, define the *optical thickness* as

$$\tau(E, \mathbf{r}, \mathbf{r'}) \equiv \int_0^{|\mathbf{r} - \mathbf{r'}|} \mu_t(\mathbf{r} - R''\mathbf{\Omega}, E)\,dR'',$$

(9.61)

which is the number of mean free path lengths of material through which a particle must pass in traveling from $\mathbf{r'}$ to $\mathbf{r}$ along the direction $\mathbf{\Omega}$. For a homogeneous medium $\tau(E, \mathbf{r}, \mathbf{r'}) = \mu_t(E)|\mathbf{r} - \mathbf{r'}|$. Then Eq. (9.60) can be written as

$$\phi(\mathbf{r}, E, \mathbf{\Omega}) = \int_0^{R_s} dR'\,\chi(\mathbf{r} - R'\mathbf{\Omega}, E, \mathbf{\Omega})e^{-\tau(E, \mathbf{r}, \mathbf{r'})}.$$

(9.62)

This result says the flux at some point $\mathbf{r}$ with a given direction and energy equals the production of particles from all positions backwards along the direction $-\mathbf{\Omega}$ to the surface times the probability these particles reach $\mathbf{r}$ without collision (the exponential term).

Equation (9.62) can be expressed in a slightly different form by defining $\mathbf{r'} \equiv \mathbf{r} - R'\mathbf{\Omega}$, so that $R' = |\mathbf{r} - \mathbf{r'}|$ and $\mathbf{\Omega} = (\mathbf{r} - \mathbf{r'})/|\mathbf{r} - \mathbf{r'}|$. The integrals over R' can be converted into volume integrals over the volume V of the system by using the two-dimensional delta function.[5] Thus from Eq. (9.62)

$$\phi(\mathbf{r}, E, \mathbf{\Omega}) = \int_0^{R_s} dR' \, \chi(\mathbf{r}', E, \mathbf{\Omega}) e^{-\tau(E,\mathbf{r},\mathbf{r}')}$$

$$= \int_{4\pi} d\Omega' \int_0^{R_s} dR' \, \chi(\mathbf{r}', E, \mathbf{\Omega}') e^{-\tau(E,\mathbf{r},\mathbf{r}')} \delta(\mathbf{\Omega}' - \mathbf{\Omega})$$

$$= \int_{4\pi} d\Omega' \int_0^{R_s} dR' \, \chi(\mathbf{r}', E, \mathbf{\Omega}') e^{-\tau(E,\mathbf{r},\mathbf{r}')} \delta\left(\mathbf{\Omega}' - \frac{(\mathbf{r} - \mathbf{r}')}{|\mathbf{r} - \mathbf{r}'|}\right).$$
$$\tag{9.63}$$

Then multiply the integrand in the above result by $R'^2/|\mathbf{r} - \mathbf{r}'|^2 = 1$ and recognize that $\int_{4\pi} \int_0^{R_s} R'^2 dR' d\Omega' = \int_V dV'$, so that the above results become

$$\phi(\mathbf{r}, E, \mathbf{\Omega}) = \int_V dV' \frac{e^{-\tau(E,\mathbf{r},\mathbf{r}')}}{|\mathbf{r} - \mathbf{r}'|^2} \delta\left(\mathbf{\Omega}' - \frac{(\mathbf{r} - \mathbf{r}')}{|\mathbf{r} - \mathbf{r}'|}\right) \chi(\mathbf{r}', E, \mathbf{\Omega}'). \tag{9.64}$$

Substitution for $\chi(\mathbf{r}', E, \mathbf{\Omega})$ from Eq. (9.54) then gives an explicit integral form of the transport equation, namely,

$$\phi(\mathbf{r}, E, \mathbf{\Omega}) = \int_V dV' \int_0^\infty dE' \int_{4\pi} d\Omega' \frac{\mu_t(\mathbf{r}', E' \to E, \mathbf{\Omega}' \to \mathbf{\Omega})}{|\mathbf{r} - \mathbf{r}'|^2} \exp[-\tau(E, \mathbf{r}, \mathbf{r}')] \times$$

$$\delta\left(\mathbf{\Omega}' - \frac{(\mathbf{r} - \mathbf{r}')}{|\mathbf{r} - \mathbf{r}'|}\right) \phi(\mathbf{r}', E', \mathbf{\Omega}')$$

$$+ \int_V dV' \exp[-\tau(E, \mathbf{r}, \mathbf{r}')] \delta\left(\mathbf{\Omega}' - \frac{(\mathbf{r} - \mathbf{r}')}{|\mathbf{r} - \mathbf{r}'|}\right) \frac{S(\mathbf{r}', E, \mathbf{\Omega}')}{|\mathbf{r} - \mathbf{r}'|^2},$$
$$\tag{9.65}$$

where $\mathbf{\Omega} = (\mathbf{r} - \mathbf{r}')/|\mathbf{r} - \mathbf{r}'|$ in the first term.

This integral transport equation can be written much more compactly by using P to denote all the phase space variables $(\mathbf{r}, E, \mathbf{\Omega})$. Equation (9.64) can be written as

$$\boxed{\phi(P) = \int dP' K(P' \to P)\phi(P') + \tilde{S}(P),} \tag{9.66}$$

---

[5] The two-dimensional Dirac delta function $\delta(\mathbf{\Omega} - \mathbf{\Omega}_o)$ is defined such that
$$\int_{4\pi} d\Omega' \, f(\mathbf{\Omega}')\delta(\mathbf{\Omega}' - \mathbf{\Omega}_o) = f(\mathbf{\Omega}_o).$$

where the kernel is given by

$$K(P' \to P) \equiv \frac{\mu_t(\mathbf{r}', E' \to E, \mathbf{\Omega}' \to \mathbf{\Omega})}{|\mathbf{r} - \mathbf{r}'|^2} \exp[-\tau(E, \mathbf{r}, \mathbf{r}')] \delta\left(\mathbf{\Omega} - \frac{(\mathbf{r} - \mathbf{r}')}{|\mathbf{r} - \mathbf{r}'|}\right),$$

(9.67)

and the contribution to ϕ from the source is

$$\widetilde{S}(P) \equiv \int_V dV' \exp[-\tau(E, \mathbf{r}, \mathbf{r}')] \delta\left(\mathbf{\Omega} - \frac{(\mathbf{r} - \mathbf{r}')}{|\mathbf{r} - \mathbf{r}'|}\right) \frac{S(\mathbf{r}', E, \mathbf{\Omega})}{|\mathbf{r} - \mathbf{r}'|^2}.$$

(9.68)

9.4.2 Integral Equations for Integrals of $\phi(\mathbf{r}, E, \mathbf{\Omega})$

With the integral equation of Eq. (9.66), various averages or integrated values of $\phi(\mathbf{r}, E, \mathbf{\Omega})$ can be estimated formally. Consider the integral transport equation in the general form of Eq. (9.66), where P and P' represent six-dimensional phase space, without time. If the integration of Eq. (9.66) was made over some subset of P, such as solid angle, an equation for an expected flux density, such as the flux density spectrum, is obtained, namely,

$$\phi(\mathbf{r}, E) = \int_{4\pi} \left[\int_V dV' \int_0^\infty dE' \int_{4\pi} d\mathbf{\Omega}' K(\mathbf{r} \to \mathbf{r}', E' \to E, \mathbf{\Omega}' \to \mathbf{\Omega}) \phi(\mathbf{r}', E', \mathbf{\Omega}') \right.$$

$$\left. + \widetilde{S}(\mathbf{r}, E, \mathbf{\Omega}) \right] d\mathbf{\Omega}.$$

(9.69)

This can be decomposed into the following:

$$\phi(\mathbf{r}, E) = \Phi(\mathbf{r}, E) + Q(\mathbf{r}, E),$$

(9.70)

where

$$\Phi(\mathbf{r}, E) = \int_{4\pi} d\mathbf{\Omega} \int_V dV' \int_0^\infty dE' \int_{4\pi} d\mathbf{\Omega}' K(\mathbf{r} \to \mathbf{r}', E' \to E, \mathbf{\Omega}' \to \mathbf{\Omega}) \phi(\mathbf{r}', E', \mathbf{\Omega}')$$

(9.71)

and

$$Q(\mathbf{r}, E) = \int_{4\pi} \widetilde{S}(\mathbf{r}, E, \mathbf{\Omega}) d\mathbf{\Omega}.$$

(9.72)

Note that Eq. (9.71) is a slightly generalized form of an eightfold expectation, with Φ in some sense taking the place of $\langle z \rangle$ and K taking the place of a probability density function, within a normalization factor. Further, Eq. (9.72) is just a definite integral, which Monte Carlo can readily evaluate. Hence, the integral form of the transport equation makes it clear that Monte Carlo is a natural method to solve radiation transport problems. We could alternatively integrate over an arbitrary volume or over an energy interval and similarly obtain equations that are in the natural form of expectations.

It, of course, is not necessary to explicitly write equations such as Eq. (9.69) in order to solve radiation transport problems, but this approach does provide a tie to the mathematical formulation provided in Chapter 2. In fact, most radiation transport simulation proceeds on physical grounds, in which a radiation particle is tracked through its history from birth to death and the results of many histories are averaged to estimate some quantity of interest. In the next chapter, it is shown how a formal solution of the integral equation can be interpreted as a simulation of particle transport. Also, the quantity of interest often is not the flux density but rather a quantity that depends on the flux density such as the response of a detector, the energy deposited in a region, or a certain type of reaction rate density.

9.4.3 Explicit Form for the Scalar Flux Density

As an example of the above integration of $\phi(\mathbf{r}, E, \boldsymbol{\Omega})$ over some subset of P, consider the integral equation for the flux density spectrum $\phi(\mathbf{r}, E)$. Upon integration of Eq. (9.65) over all directions, the delta functions immediately integrate to unity to give

$$\phi(\mathbf{r}, E) = \int_V dV' \int_0^\infty dE' \int_{4\pi} d\Omega' \, \frac{\mu_t(\mathbf{r}', E' \to E, \boldsymbol{\Omega}' \to \boldsymbol{\Omega})}{|\mathbf{r} - \mathbf{r}'|^2} e^{-\tau(E, \mathbf{r}, \mathbf{r}')} \phi(\mathbf{r}', E', \boldsymbol{\Omega}')$$

$$+ \int_V dV' \frac{e^{-\tau(E, \mathbf{r}, \mathbf{r}')}}{|\mathbf{r} - \mathbf{r}'|^2} S(\mathbf{r}', E, \boldsymbol{\Omega}),$$
(9.73)

where $\boldsymbol{\Omega} = (\mathbf{r} - \mathbf{r}')/|\mathbf{r} - \mathbf{r}'|$.

Note that this integral equation for the scalar flux density still depends on the angular flux density because $\chi(\mathbf{r}, E, \boldsymbol{\Omega})$ generally depends on $\phi(\mathbf{r}, E, \boldsymbol{\Omega})$ as shown in Eq. (9.54). For the special case of isotropic sources (i.e., $S(\mathbf{r}, E, \boldsymbol{\Omega}) = S_o(\mathbf{r}, E)/4\pi$) and for isotropic scattering (i.e., $\mu_s(\mathbf{r}, E' \to E, \boldsymbol{\Omega}' \to \boldsymbol{\Omega}) = \mu_{so}(\mathbf{r}, E' \to E)/4\pi$), this integral equation reduces to one involving only the scalar flux density. Under the isotropy assumptions, Eq. (9.73) reduces to

$$\phi(\mathbf{r}, E) = \int_V dV' \frac{S_o(\mathbf{r}', E) + \int_0^\infty dE' \, \mu_{so}(\mathbf{r}', E' \to E)\phi(\mathbf{r}', E)}{4\pi |\mathbf{r} - \mathbf{r}'|^2} e^{-\tau(E, \mathbf{r}, \mathbf{r}')}.$$
(9.74)

9.4.4 Alternate Forms of the Integral Transport Equation

Historically, different forms of the integral transport equation have been used by introducing different dependent variables. Here two such variations are introduced. First, the *collision rate density*, defined as the number of interactions, per unit time, experienced by particles, in a unit volume about $\mathbf{r}$, that have energies in a unit energy about E and directions in a unit solid angle about $\boldsymbol{\Omega}$ is given by

$$F(\mathbf{r}, E, \mathbf{\Omega}) = \mu(\mathbf{r}, E) \, \phi(\mathbf{r}, E, \mathbf{\Omega}). \qquad (9.75)$$

Multiplication of Eq. (9.66) by $\mu_t(\mathbf{r}, E)$ gives

$$F(\boldsymbol{P}) = \int d\boldsymbol{P}' \overline{K}(\boldsymbol{P}' \to \boldsymbol{P}) F(\boldsymbol{P}') + \overline{S}(\boldsymbol{P}), \qquad (9.76)$$

where

$$\overline{K}(\boldsymbol{P}' \to \boldsymbol{P}) = \frac{\mu_t(\mathbf{r}, E)}{\mu_t(\mathbf{r}', E')} K(\boldsymbol{P}', \boldsymbol{P}) \qquad \text{and} \qquad \overline{S}(\boldsymbol{P}) = \mu_t(\mathbf{r}, E) \widetilde{S}(\boldsymbol{P}). \qquad (9.77)$$

The integral transport equation is sometimes written in terms of the *emissions rate density* $\chi(\mathbf{r}, E, \mathbf{\Omega})$ defined by Eq. (9.54). Substitution of Eq. (9.64) into Eq. (9.54) immediately gives an integral equation for $\chi(\mathbf{r}, E, \mathbf{\Omega})$, namely,

$$\chi(\boldsymbol{P}) = \int d\boldsymbol{P}' K(\boldsymbol{P}' \to \boldsymbol{P}) \chi(\boldsymbol{P}') + S(\boldsymbol{P}), \qquad (9.78)$$

where the kernel $K(\boldsymbol{P}' \to \boldsymbol{P})$ is given by Eq. (9.67).

9.5 Adjoint Transport Equation

In the next chapter, which discusses the use of Monte Carlo techniques for particle transport problems, the concept of the *adjoint* flux density is very important when importance sampling is used for variance reduction. Readers unfamiliar with adjoint operators are referred to Appendix D which is a general tutorial on linear In this section, the adjoint transport equation is derived and an important property of the adjoint flux density is introduced.

Operators

First some mathematical preliminaries. An *operator* $\mathcal{O}$ maps a function $f(\boldsymbol{P})$ into another function $g(\boldsymbol{P})$, i.e.,

$$\mathcal{O}f(\boldsymbol{P}) = g(\boldsymbol{P}), \qquad \boldsymbol{P} \in R,$$

where $\boldsymbol{P}$ are all the independent variables of phase space and R is some region of phase space. For example, in this section $\boldsymbol{P} = (\mathbf{r}, E, \mathbf{\Omega})$, the independent variables of the transport equation. An operator is *linear* if $\mathcal{O}[f_1(\boldsymbol{P}) + f_2(\boldsymbol{P})] = \mathcal{O}f_1(\boldsymbol{P}) + \mathcal{O}f_2(\boldsymbol{P})$.

Inner Product

An *inner product* assigns a number to a pair of functions,[6] namely,

$$(f, g) \equiv \int_R f(\boldsymbol{P}) g(\boldsymbol{P}) \, d\boldsymbol{P} = \int_V dV \int_0^\infty dE \int_{4\pi} d\Omega f(\mathbf{r}, E, \boldsymbol{\Omega}) g(\mathbf{r}, E, \boldsymbol{\Omega}),$$

where it is assumed that such integrals exist.

Adjoint Operator

Let $\mathcal{O}$ be a real continuous linear operator that operates on square integrable functions $f(\boldsymbol{P})$; $\mathcal{O}^\dagger$ is the *adjoint* of $\mathcal{O}$ if

$$(g^\dagger, \mathcal{O}f) = (\mathcal{O}^\dagger g^\dagger, f) \tag{9.79}$$

or, equivalently,

$$\int_R g^\dagger(\boldsymbol{P})[\mathcal{O}f(\boldsymbol{P})] \, d\boldsymbol{P} = \int_R [\mathcal{O}^\dagger g^\dagger(\boldsymbol{P})] f(\boldsymbol{P}) \, d\boldsymbol{P}. \tag{9.80}$$

Here $g^\dagger$ are square integrable functions of $\boldsymbol{P}$ often with different properties (although not necessarily) than those of the functions $f(\boldsymbol{P})$. For example, if $\mathcal{O}$ is a differential operator, $f(\boldsymbol{P})$ and $g^\dagger(\boldsymbol{P})$ may have different behavior at the boundary of V. Many examples of linear adjoint operators and demonstrations of why they are so useful are presented in Appendix D.

9.5.1 Derivation of the Adjoint Transport Equation

With the above definitions, the adjoint of the various operators in the transport equation Eq. (9.50) can be found. First consider the streaming operator $\boldsymbol{\Omega} \cdot \nabla$. Because $\boldsymbol{\Omega} \cdot \nabla = \nabla \cdot \boldsymbol{\Omega}$, the vector identity

$$\nabla \cdot (\boldsymbol{\Omega} \phi^\dagger \phi) = \phi^\dagger \boldsymbol{\Omega} \cdot \nabla \phi + \phi \boldsymbol{\Omega} \cdot \nabla \phi^\dagger$$

can be integrated to give

$$\int_S dS \, \mathbf{n} \cdot \boldsymbol{\Omega} \phi^\dagger \phi = \int_V dV \, \phi^\dagger \boldsymbol{\Omega} \cdot \nabla \phi + \int_V dV \, \phi \boldsymbol{\Omega} \cdot \nabla \phi^\dagger, \tag{9.81}$$

where the divergence theorem has been used to convert the volume integral on the left to a surface integral and $\mathbf{n}$ is a unit outward normal on the surface of the medium. Note that the boundary condition of Eq. (9.55) makes the integrand of this surface integral vanish for $\mathbf{n} \cdot \boldsymbol{\Omega} < 0$. This integrand can be forced to vanish entirely if the adjoint flux density is required to satisfy the boundary condition

$$\phi^\dagger(\mathbf{r}_s, E, \boldsymbol{\Omega}) = 0, \qquad \mathbf{r}_s \in S, \qquad \mathbf{n} \cdot \boldsymbol{\Omega} \geq 0. \tag{9.82}$$

---

[6] Throughout this section, it is assumed that all operators and functions are real.

Finally, integrating Eq. (9.81) over all E and $\mathbf{\Omega}$ gives

$$\int_R \phi^\dagger(P)[\mathbf{\Omega}\cdot\nabla\phi(P)]\,dP = \int_R [-\mathbf{\Omega}\cdot\nabla\phi^\dagger(P)]\phi(P)\,dP. \qquad (9.83)$$

Comparison of this result to the definition of the adjoint given by Eq. (9.80) immediately reveals that $(\mathbf{\Omega}\cdot\nabla)^\dagger = -(\mathbf{\Omega}\cdot\nabla)$.

The operator for the collision term $\mu_t\phi$ in Eq. (9.54) is simply $\mathcal{O} = \mu_t(\mathbf{r}, E)$, a multiplicative operator. This is a self-adjoint operator because $\phi^\dagger \mu_t \phi = [\mu_t\phi^\dagger]\phi$.

Finally, consider a general integral operator $\mathcal{O}[\cdot] \equiv \int H(P' \to P)[\cdot]\,dP'$. By definition

$$(\phi^\dagger, \mathcal{O}\phi) = \int_R \phi^\dagger(P)\left[\int_R H(P' \to P)\phi(P')\,dP'\right]dP.$$

Interchange the order of integration and interchange P and P' to obtain

$$(\phi^\dagger, \mathcal{O}\phi) = \int_R\left[\int_R H(P \to P')\phi^\dagger(P')\,dP'\right]\phi(P)\,dP.$$

This must equal $(\mathcal{O}^\dagger\phi^\dagger, \phi)$, so it is seen that

$$\mathcal{O}^\dagger[\cdot] = \int H(P \to P')[\cdot]\,dP'. \qquad (9.84)$$

Thus, the effect of taking the adjoint of an integral operator is to interchange the primed and unprimed variables in the kernel H.

From these results, the integro-differential adjoint equation is found from Eq. (9.50) to be

$$-\mathbf{\Omega}\cdot\nabla\phi^\dagger(\mathbf{r}, E, \mathbf{\Omega}) + \mu_t(\mathbf{r}, E)\phi^\dagger(\mathbf{r}, E, \mathbf{\Omega})$$

$$= \int_0^\infty dE'\int_{4\pi} d\Omega'\,\mu_s(\mathbf{r}, E \to E', \mathbf{\Omega} \to \mathbf{\Omega}')\phi^\dagger(\mathbf{r}, E', \mathbf{\Omega}') + S^\dagger(\mathbf{r}, E, \mathbf{\Omega}),$$
$$(9.85)$$

where $S^\dagger(\mathbf{r}, E, \mathbf{\Omega})$ is the distributed adjoint source and the boundary condition is

$$\phi^\dagger(\mathbf{r}_s, E, \mathbf{\Omega}) = 0, \qquad \mathbf{n}\cdot\mathbf{\Omega} \geq 0. \qquad (9.86)$$

Similarly, the adjoint integral transport equation for the adjoint flux density is obtained from Eq. (9.66) as

$$\phi^\dagger(P) = \int dP'\,K(P \to P')\phi^\dagger(P') + \widetilde{S}^\dagger(P). \qquad (9.87)$$

Later, the adjoint equation for the collision density is needed, which from Eq. (9.76) is

$$F^\dagger(P) = \int dP' \overline{K}(P \to P') F^\dagger(P') + \overline{S}^\dagger(P). \qquad (9.88)$$

9.5.2 Utility of the Adjoint Solution

Suppose it is desired to evaluate some functional of the flux density

$$I = \int_R f(P)\phi(P)\, dP, \qquad (9.89)$$

where $f(P)$ is a specified *weight function*. For example, if $f(P) = 1/V$, then I would give the average scalar flux density in the system volume V. Likewise, if $f(P) = \delta(P - P_o)$, then $I = \phi(P_o)$, the flux density at a particular position, energy, and direction. Almost any property of the particle field can be obtained with the proper choice of the weight function.

To evaluate I, one could first find ϕ by solving the transport equation

$$\mathcal{O}\phi(P) = S(P), \qquad (9.90)$$

where the transport operator $\mathcal{O}$ is defined by Eq. (9.85) or (9.87). Then once ϕ is known, Eq. (9.89) could be evaluated numerically. It so happens that the same value of I can be found from the corresponding adjoint transport problem

$$\mathcal{O}^\dagger \phi^\dagger = f(P), \qquad (9.91)$$

where the adjoint source is the weight function in I. To see this alternate method for evaluating I, multiply Eq. (9.90) by $\phi^\dagger$ and Eq. (9.91) by ϕ and integrate the results over all P. The results are

$$(\phi^\dagger, \mathcal{O}\phi) = \int_R S(P)\phi^\dagger(P)\, dP \quad \text{and} \quad (\mathcal{O}^\dagger \phi^\dagger, \phi) = \int_R f(P)\phi(P)\, dP.$$

From the definition of the adjoint, the two inner products on the left-hand sides are equal. Thus, it is seen that

$$\boxed{I = \int_R f(P)\phi(P)\, dP = \int_R S(P)\phi^\dagger(P)\, dP.} \qquad (9.92)$$

This important result has practical applications and provides an insight into the significance of the adjoint flux.

To understand the significance of the adjoint flux density, suppose that $f(P)$ is some detector response function $\mu_d(\mathbf{r}, E)$ defined for some subregion of V

and is zero outside that subregion. The units of μ_d are the probability, per unit path length, that a particle interacts with the detector material and produces a "count," i.e., cm^{-1}. Hence, the units of I, from the first integral in Eq. (9.92), are "counts per unit time." Further suppose that one particle is born per unit time at $\mathbf{r}_o$ with energy E_o and in direction $\mathbf{\Omega}_o$, i.e., $S(P) = \delta(\mathbf{r} - \mathbf{r}_o)\delta(E - E_o)\delta(\mathbf{\Omega} - \mathbf{\Omega}_o) \{s^{-1}\}$. Then, from the second integral in Eq. (9.92),

$$I = \phi^\dagger(\mathbf{r}_o, E_o, \mathbf{\Omega}_o) = \text{detector count rate.} \qquad (9.93)$$

From Eq. (9.85) with $S^\dagger = \mu_d$, it is seen that $\phi^\dagger$ is dimensionless and, from the above result, it is seen that the adjoint flux density gives the *importance* of a particle born at $(\mathbf{r}_o, E_o, \mathbf{\Omega}_o)$ to the detector response. This interpretation is consistent with the adjoint boundary condition that $\phi^\dagger = 0$ for particles reaching the surface traveling in the outward direction since such particles can never return and contribute to the detector response. That the adjoint flux density describes how important different regions of phase space are to the tally or score in a Monte Carlo simulation gives a powerful tool for how to pick the importance functions (Coveyou et al., 1967). Indeed it will be shown that if $\phi^\dagger$ is used as the importance function, the tally has zero variance!

Another very practical use of the adjoint transport equation is to transform a hard problem into an easier one. For example, if the average flux in some small region of phase space is sought and the source is widely distributed, then few histories in a MC simulation will even reach the small region and contribute to the tally. In other words, the first integral in Eq. (9.92) is difficult to evaluate. By contrast, the adjoint problem has the source in the small volume and the source region becomes the tally region, making the second integral in Eq. (9.92) much easier to evaluate.

9.6 Summary

In this chapter, the fundamentals are presented of (1) how the strength of a radiation field of neutral particles, such as neutrons and photons, is characterized by the flux density $\phi(\mathbf{r}, E, \mathbf{\Omega})$, (2) how the flow across a surface is characterized by the current vector $\mathbf{J}(\mathbf{r}, E, \mathbf{\Omega})$, (3) how the interaction coefficient determines the probability a radiation particle interacts with the host medium, and (4) how the reaction rate per unit volume of the radiation with the medium is given by $R_i(\mathbf{r}, E) = \mu_i(\mathbf{r}, E)\phi(\mathbf{r}, E)$. It was then shown that uncollided radiation is attenuated exponentially as it traverses the host medium, a key result that is used in the next chapter on MC transport simulation. The interaction coefficient and associated microscopic cross sections for both photons and neutrons were illustrated.

Two forms of the linearized Boltzmann equation are derived that describe exactly how radiation migrates through a host medium. The integro-differential form, which is widely used in deterministic transport calculations, was then converted into an integral form, the form first used in Monte Carlo analysis of radiation transport and that is also used in the following chapter. Both the direct and adjoint integral forms are introduced. As shown in the next chapter, the adjoint solution plays an important role in maximizing variance reduction of Monte Carlo evaluation of the direct solution.

Problems

9.1 Show that the solid angle subtended by the disk shown to the right at the point P which is a distance z from the disk on a line perpendicular to the center of the disk is

$$\Omega = 2\pi(1 - \cos\vartheta).$$

9.2 Equations (9.13) and (9.14) are two definitions of the fluence. Consider the sphere illustrated to the right, with a uniform parallel beam of particles incident on its surface, so that ΔN_p particles enter the sphere. Verify that the mean chord length is $\langle L \rangle = 4R/3$ and use this result to demonstrate the equivalence of the two definitions of the two definitions.

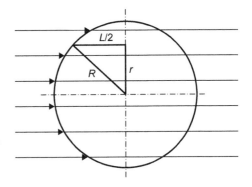

9.3 Show that if the angular flux density $\phi(\mathbf{r}, \mathbf{\Omega})$ is isotropic, the current density $\mathbf{j}(\mathbf{r})$ is equal to zero.

9.4 A broad beam of neutrons is normally incident on a 6-cm thick homogeneous slab. The intensity of neutrons transmitted through the slab without interactions is found to be 30% of the incident intensity. (a) What is the total interaction coefficient μ_t for the slab material? (b) What is the average distance a neutron travels in this material before undergoing an interaction?

9.5 A small homogeneous sample of mass m (g) with atomic mass A is irradiated uniformly by a constant flux density ϕ (cm^{-2}s^{-1}). If the total atomic cross section for the sample material with the irradiating particles is denoted by σ_t (cm^2), derive an expression for the fraction of the atoms in the sample that interact during a 1-h irradiation. State any assumptions made.

9.6 The transport equation derived in Section 9.3 is based on the angular flux density $\phi(\mathbf{r}, E, \mathbf{\Omega})$ in terms of the particle energy E and direction of travel $\mathbf{\Omega}$. However, an angular flux density $\phi(\mathbf{r}, \mathbf{v})$ based on the particle velocity $\mathbf{v}$ could be used instead. Derive the relation between these two flux densities.

9.7 Show that $\nabla \cdot \mathbf{\Omega} f(\mathbf{r}, \mathbf{\Omega}) = \mathbf{\Omega} \cdot \nabla f(\mathbf{r}, \mathbf{\Omega})$, where f is a continuous function in $\mathbf{r}$.

9.8 In many radiation shielding analyses, the uncollided flux density is of interest. The uncollided angular flux density can be obtained from the transport equation by treating the medium as a purely absorbing one (i.e., any interaction removes the particle from its uncollided status and no secondary uncollided particles are created). Show that for a homogeneous medium, the uncollided angular flux density is given by

$$\phi(\mathbf{r}, E, \mathbf{\Omega}) = \int_0^\infty dR \ S(\mathbf{r} - R\mathbf{\Omega}, E, \mathbf{\Omega}) e^{-\mu_t(E)R}.$$

9.9 Consider a half-space of purely absorbing material in which there is a uniformly distributed, isotropic, volumetric source which emits S_v photons of energy E_o per unit time. From Eq. (9.52), determine the angular flux density and current of photons at the surface of the half-space.

9.10 Generalize the previous problem with the half-space replaced by a slab of thickness T (mfp). Determine the angular fluxes and currents leaving both surfaces. Show that your results reduce to those for a half-space as $T \to \infty$.

9.11 Just as the transport equation of Eq. (9.50) can be reduced to the simpler version of Eq. (9.52) with many assumptions, so can the integral equation Eq. (9.62) be reduced to a much simpler form. Show that Eq. (9.52), with vacuum boundary conditions and with $\hat{z}$ replaced by just z for convenience, can be integrated over z to give the following integral equation for $\phi(z, \omega)$:

$$\phi(z, \omega) = \begin{cases} \dfrac{1}{\omega} \displaystyle\int_0^z e^{(z'-z)/\omega} \widehat{S}(z') \, dz', & \omega > 0, \\[4mm] -\dfrac{1}{\omega} \displaystyle\int_z^T e^{(z'-z)/\omega} \widehat{S}(z') \, dz', & \omega < 0, \end{cases}$$

where

$$\widehat{S}(z) = \frac{c}{2} \int_{-1}^1 \phi(z, \omega') \, d\omega' + \frac{1}{2} \frac{S(z)}{\mu_t}.$$

9.12 Use the integral equation derived in Problem 9.11 to solve Problem 9.10.

References

Adams, M.L., Larsen, E.L., 2002. Fast iterative methods for discrete-ordinates particle transport calculations. Prog. Nucl. Energy 40 (1), 3–159.

Bell, G.I., Glasstone, S., 1970. Nuclear Reactor Theory. Van Nostrand Reinhold, New York.

Coveyou, R.R., Cain, V.R., Yost, K.J., 1967. Adjoint and importance in Monte Carlo applications. Nucl. Sci. Eng. 27, 219–234.

Duderstadt, J.J., Martin, W.R., 1979. Transport Theory. Wiley, New York.

Lewis, E.E., Miller, W.F., 1984. Computational Methods of Neutron Transport Theory. Wiley, New York.

Sanchez, R., McCormick, N.J., 1982. A review of neutron transport approximations. Nucl. Sci. Eng. 80, 481–535.

Shultis, J.K., Faw, R.E., 2000. Radiation Shielding. American Nuclear Society, LaGrange Park, IL.

Shultis, J.K., Faw, R.E., 2008. Fundamentals of Nuclear Science and Engineering, 2nd ed. CRC Press, Boca Raton, FL.

Chapter 10

Monte Carlo Simulation of Neutral Particle Transport

Particle transport calculations
do not always meet expectations.
The trouble, of course,
is using brute force
rather than variance reductions.

Numerical solutions of the transport equation, called *deterministic* calculations, usually require a discretization of the time, energy, angular, and spatial variables. Each such approximation introduces a *systematic* error. Moreover, the geometry for deterministic calculations must be rather simple, for example, planar, cylindrical, or spherical surfaces, and a full three-dimensional calculation requires tremendous computer memories to hold all the discretized values of the energy and angular flux density.

Monte Carlo calculations, by contrast, are capable of treating very complex three-dimensional geometries. The continuous treatment of time, energy, direction, and space avoids all systematic discretization errors. However, the stochastic simulation of the transport process introduces *statistical* errors in the results, errors which are not present in deterministic calculations. Historically, Monte Carlo analyses could be performed only on large mainframe computers because of the intensive computational requirements. But today's inexpensive personal computers are capable of running very complex Monte Carlo simulations.

Monte Carlo calculations are not always better than deterministic approaches. Monte Carlo excels at estimating some quantity averaged over, at most, a few regions of phase space $P = (\mathbf{r}, E, \mathbf{\Omega}, t)$. But, if, for example, detailed three-dimensional spatial profiles of the flux in a reactor core are needed, deterministic methods often are superior. The reason for this is that, to obtain with Monte Carlo the value of the flux density at a large number of points, the spatial domain would have to be broken into a vast number of small contiguous volumes ΔV and the flux density estimated in each of these volumes. The number of histories that contribute to the flux estimates in the volumes decrease rapidly as ΔV is made smaller to increase the spatial resolution, thereby causing the statistical uncertainty in the estimated flux densities to increase, sometimes

Exploring Monte Carlo Methods. https://doi.org/10.1016/B978-0-12-819739-4.00018-4

to unacceptable values. However, Monte Carlo techniques are, or can be made, efficient when the number of results or tallies are small.

10.1 Basic Approach for Monte Carlo Transport Simulations

A Monte Carlo radiation transport calculation simulates a finite number N of particle histories by sampling from the appropriate probability density functions (PDFs) governing the various events that may happen to a particle from its birth by a source to its eventual demise by being absorbed or escaping through the problem boundary. A history begins by randomly selecting a particle's initial position, energy, and direction from the PDFs that describe the sources for the problem. As the particle travels on various legs of its trajectory, sampling is performed to determine random values for distance to the next interaction, type of interaction, scattering angle, energy change, and so on. Between interactions, the neutrons or photons travel in straight lines. On some interactions, secondary particles are created, such as neutrons from fission reactions or X-rays from photon photoelectric interactions. These secondary particles are "banked" and their initial energies, interaction positions, and travel directions are recorded. After the original particle finishes its history, these secondary banked particles are processed sequentially to create a history for each.

After each history is terminated (or after its interactions or surface crossings are determined), its contribution z_i is added to the tally being used to estimate the quantity of interest, for example, average flux in a cell, probability of escape, energy deposited in a cell, current through a surface, or many other quantities related to the radiation field. After N histories, the estimate of $\langle z \rangle$ is

$$\langle z \rangle \simeq \overline{z} = \frac{1}{N} \sum_{i=1}^{N} z_i.$$

Along with such an estimate, various measures of its statistical reliability must be made (see Section 4.2).

Such an analog simulation mimics the stochastic events that befall an actual particle if the problem was converted to the equivalent experiment. In many such calculations, extremely few of the histories contribute anything to the tally of interest. For example, if the probability a neutron incident on a thick shield eventually reaches the other side is 1 in 10^9, then in an analog simulation only one out of 10^9 histories, on average, terminates on the back side of the shield. For such problems, it is vital to use variance reduction techniques (see Chapter 5) and perform *nonanalog* Monte Carlo calculations.

10.2 Geometry

Almost all Monte Carlo codes use a Cartesian coordinate system, even if the problem possesses cylindrical or spherical symmetry. The reason for this is that, as a particle travels along the direction $\boldsymbol{\Omega} = \mathbf{i}u + \mathbf{j}v + \mathbf{k}w$, the direction cosines $u = \mathbf{i} \cdot \boldsymbol{\Omega}$, $v = \mathbf{j} \cdot \boldsymbol{\Omega}$, and $w = \mathbf{k} \cdot \boldsymbol{\Omega}$ with respect to the coordinate axes do not change. If a particle moves a distance s from $\mathbf{r}_0$ in direction $\boldsymbol{\Omega}$, its new position $\mathbf{r}$ is

$$\mathbf{r} = \mathbf{r}_0 + s\boldsymbol{\Omega} = (x_0 + su, y_0 + sv, z_0 + sw). \tag{10.1}$$

This simplicity in describing a particle's position as it moves in a straight line is not possible in other coordinate systems.

The three-dimensional space in which the simulated histories are constructed is usually assumed to be composed of contiguous homogeneous volumes or *cells*, each cell being bounded by one or more surfaces (or portions of surfaces). A cell can be a void or composed of any homogeneous material. All of three-dimensional space must belong to some cell—there can be no "holes" where a particle would become "lost." Although not necessary, the geometry of the problem is often surrounded by a *problem boundary*, usually far from tally regions. The cell containing all space beyond the problem boundary is the "graveyard," whose purpose is to "kill" any particle that enters it, thereby saving time by not tracking particles beyond the boundary and that have negligible chance of contributing to the tally. For example, in calculating the doses to a patient from an X-ray therapy unit, the walls of the treatment room should be modeled, but nearby buildings and airplanes passing overhead have little chance of affecting the dose to the patient. Similarly, objects far from tally regions often can be modeled crudely compared to objects near the important regions. For example, door knobs in the treatment room need not be modeled unless the problem is to determine how much radiation leaks through the key hole. Constructing an effective but efficient geometry model for a particular problem is a matter of skill and experience on the modeler's part.

10.2.1 Combinatorial Geometry

Surfaces are usually defined by first-degree functions (planes) or second-degree functions (cylinders, cones, ellipsoids, etc.) as $f(x, y, z) = 0$. Sometimes fourth-degree functions (toroids) are used, although such surfaces present greater numerical difficulties for evaluating the intersection point with a particle trajectory. Commonly used surfaces are given in Table 10.1. Surface A, defined by $f_A(x, y, z) = 0$, divides all space into two subdomains, namely, those points $\mathbf{r} = (x, y, z)$ for which $f_A(x, y, z) > 0$ (denoted by A^+) and those points for which $f_A(x, y, z) < 0$ (denoted by A^-). For example, consider the spherical surface $S : f_S(x, y, z) = x^2 + y^2 + z^2 - R^2 = 0$. All points inside the sphere (S^-) are to the negative side of the surface, while all points outside (S^+) are to the positive side of the surface.

By using Boolean unions ($\cup$ "and") and intersections ($\cap$ "or") of the two subdomains for each surface, volumes of quite complex shapes can be approximated. As a simple example, consider the construction of a finite cylinder of

Table 10.1 Some first-, second-, and fourth-degree surfaces. The location and orientation of the surfaces are determined by the values of the surface parameters denoted by capital and Greek letters. Extracted from MCNP (2003).

Type	Description	Surface $f(x, y, z) = 0$
plane	general	$Ax + By + Cz - D = 0$
	normal to x-axis	$x - D = 0$
	normal to y-axis	$y - D = 0$
	normal to z-axis	$z - D = 0$
sphere	centered at origin	$x^2 + y^2 + z^2 - R^2 = 0$
	general	$(x-A)^2 + (y-B)^2 + (z-C)^2 - R^2 = 0$
	centered on x-axis	$(x-A)^2 + y^2 + z^2 - R^2 = 0$
	centered on y-axis	$x^2 + (y-B)^2 + z^2 - R^2 = 0$
	centered on z-axis	$x^2 + y^2 + (z-C)^2 - R^2 = 0$
cylinder	parallel to x-axis	$(y-B)^2 + (z-C)^2 - R^2 = 0$
	parallel to y-axis	$(x-A)^2 + (z-C)^2 - R^2 = 0$
	parallel to z-axis	$(x-A)^2 + (y-B)^2 - R^2 = 0$
	on x-axis	$y^2 + z^2 - R^2 = 0$
	on y-axis	$x^2 + z^2 - R^2 = 0$
	on z-axis	$x^2 + y^2 - R^2 = 0$
cone	parallel to x-axis	$(y-B)^2 + (z-C)^2 - D(x-A)^2 = 0$
	parallel to y-axis	$(x-A)^2 + (z-C)^2 - D(y-B)^2 = 0$
	parallel to z-axis	$(x-A)^2 + (y-B)^2 - D(z-C)^2 = 0$
	on x-axis	$y^2 + z^2 - D(x-A)^2 = 0$
	on y-axis	$x^2 + z^2 - D(y-B)^2 = 0$
	on z-axis	$x^2 + y^2 - D(z-C)^2 = 0$
ellipsoid, hyperboloid, paraboloid	axis parallel to x-, y-, or z-axis	$A(x-\alpha)^2 + B(y-\beta)^2 + C(z-\gamma)^2$ $+2D(x-\alpha) + 2E(y-\beta)$ $+2F(z-\gamma) + G = 0$
cylinder, cone, ellipsoid, paraboloid, hyperboloid	axis not parallel to x-, y-, or z-axis	$Ax^2 + By^2 + Cz^2 + Dxy + Eyz$ $+Fzx + Gz + Hy + Jz + K = 0$
elliptical or circular torus. Axis is parallel to x-, y-, or z-axis		$(x-\alpha)^2/B^2 + (\sqrt{(y-\beta)^2 + (z-\gamma)^2} - A)^2/C^2 - 1 = 0$ $(y-\beta)^2/B^2 + (\sqrt{(x-\alpha)^2 + (z-\gamma)^2} - A)^2/C^2 - 1 = 0$ $(z-\gamma)^2/B^2 + (\sqrt{(x-\alpha)^2 + (y-\beta)^2} - A)^2/C^2 - 1 = 0$

radius R and height $2H$, centered at the origin, and with its axis on the x-axis. Such a cylindrical volume is bounded by two parallel planes perpendicular to the x-axis, $f_{P_1}(x, y, z) \equiv x - H = 0$ and $f_{P_2}(x, y, z) \equiv x + H = 0$, and the infinite cylinder $f_C(x, y, z) \equiv y^2 + z^2 - R^2 = 0$. The space inside this cylinder is specified as

$$P_1^- \cap P_2^+ \cap C^-,$$

while all of the space outside this finite cylinder is described as

$$P_1^+ \cup P_2^- \cup C^+.$$

To permit rapid particle tracking through the geometry, each region (or cell) and each surface are uniquely numbered. Then, for each cell, the numbers of the surfaces bounding it are tabulated as are the specifications of the regions on the other side of the bounding surfaces. Although somewhat redundant, this bookkeeping makes tracking particles very efficient because when a particle leaves a region, the adjacent regions are immediately known as are the new surfaces toward which the particle is traveling.

10.3 Sources

Each particle history is begun by sampling from the spatial, energy, and angular distributions of the source to determine the starting position, energy, and direction of each particle history. The source may be specified explicitly (a *fixed source* problem) or calculated by the simulation itself (an *eigenvalue* problem). In the latter case, initial simulations are used to estimate the spatial distribution of sources such as a fission source and the resulting equilibrium distribution in a critical system. However, in this chapter only the fixed source problem is discussed.

Sources in a transport calculation can be distributed over many regions, be localized to a few cells, or be singular sources such as point, line, and plane sources. The energy of particles emitted by the sources can consist of a discrete set of energies (typical of gamma rays produced by radioactive decay) or have a continuous distribution of energies (such as neutrons produced by spontaneous fission). While many sources emit radiation isotropically, some sources can be anisotropic (such as a beam of radiation from a reactor beam port). To start a particle history, one must sample from the specified spatial, energy, and angular distributions. To complicate matters, these three distributions are often interrelated. For example, different source regions may emit particles with different energy and angular distributions. Modeling the source distributions can often be quite a complex task.

10.3.1 Isotropic Sources

Many physical sources emit radiation isotropically. The probability $p(\mathbf{\Omega})d\Omega$ a source particle is emitted in $d\Omega$ about a direction $\mathbf{\Omega}$ is thus

$$p(\mathbf{\Omega})d\Omega = \frac{d\Omega}{4\pi} = \frac{d\psi}{2\pi}\frac{d\omega}{2}, \tag{10.2}$$

where $\omega = \cos\theta$. Thus, the PDF of the azimuthal angle is seen to be $p(\psi) = 1/(2\pi)$, $0 \le \psi < 2\pi$, and the PDF for the cosine of the polar angle θ is $p(\omega) = 1/2$, $-1 \le \omega \le 1$, both uniform distributions. Sampling from these distributions gives

$$\psi_i = 2\pi\rho_j \quad \text{and} \quad \omega_i = -1 + 2\rho_{j+1}, \tag{10.3}$$

where ρ_j and ρ_{j+1} are two random numbers uniformly distributed between $(0, 1)$. The directed cosines of the initial direction $\mathbf{\Omega}_i$ of a source particle are then calculated as

$$u_i = \mathbf{i}\cdot\mathbf{\Omega} = \cos\psi_i\sqrt{1 - \omega_i^2}, \quad v_i = \mathbf{j}\cdot\mathbf{\Omega} = \sin\psi_i\sqrt{1 - \omega_i^2}, \quad w_i = \mathbf{k}\cdot\mathbf{\Omega} = \omega_i. \tag{10.4}$$

10.4 Path Length Estimation

A key component for calculating a particle history is a geometry module that can identify for a particle at $\mathbf{r}_0$ (1) what surface (if any) it will first hit if it travels a distance s in direction $\mathbf{\Omega}$, (2) what is the distance to the intersection point if it hits the surface, and (3) what region or cell is on the other side.

10.4.1 Travel Distance in Each Cell

The distance a particle travels before interacting is a random variable. The path length s of each leg of a particle's history is estimated from its PDF (see Eq. (9.27)) $f(s) = \mu e^{-\mu s}$, where μ is the total interaction coefficient. Because μ generally varies from region to region and because a track segment may span multiple regions, the PDF for travel distance is expressed in terms of mean free path lengths $\lambda \equiv \mu s$. Thus, the distance to the next collision, in mean free path lengths, has a PDF $f(\lambda) = e^{-\lambda}$ and is independent of the medium. The length of each track segment is then obtained by sampling from an exponential distribution (see Example 4.2) as $\lambda = -\ln\rho$, where ρ is from $\mathcal{U}(0, 1)$.

Let lengths $s_1, s_2, \ldots, s_n$ be the segments of the flight path in each region along the particle's direction of travel (see Fig. 10.1). Then, if a particle starts at $\mathbf{r}_0$ and if

$$\sum_{i=1}^{n-1} \mu_i s_i \le \lambda < \sum_{i=1}^{n} \mu_i s_i, \tag{10.5}$$

the next collision occurs in region n at a distance

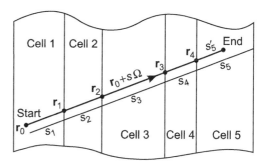

Figure 10.1 Path lengths in each cell along a straight-line segment of a particle trajectory.

$$s_n' = \frac{1}{\mu_n} \left(\lambda - \sum_{i=1}^{n-1} \mu_i s_i \right) \tag{10.6}$$

beyond the entrance point to the nth region. Here μ_i is the interaction coefficient for the ith region.

The distances s_i are readily calculated from the intersection of the straight-line trajectory of Eq. (10.1) with the surfaces $f_i(x, y, z)$ through which it passes. The straight line, along which the particle moves, can alternatively be written as

$$\mathbf{r} = \mathbf{r}_{i-1} + \mathbf{\Omega} s, \qquad i = 1, 2, \ldots, n, \tag{10.7}$$

where $\mathbf{r}_{i-1}$ is the position where the particle begins its flight in region i. Then the distance to the point of exit s_i from this region is found by (1) first solving

$$f_m^i(x_{i-1} + u s_{i_m}, y_{i-1} + v s_{i_m}, z_{i-1} + w s_{i_m}) = 0, \tag{10.8}$$

where f_m^i, $m = 1, 2, \ldots$, are all the surfaces bounding region i, and (2) then by picking s_i as the smallest positive real solution s_{i_m}, $m = 1, 2, \ldots$. If $f_m^i(x, y, z)$ are quadratic surfaces,[1] then Eq. (10.8) is a quadratic equation for s_{i_m}, namely,

$$a s_{i_m}^2 + 2b s_{i_m} + c = 0, \tag{10.9}$$

where a, b, and c depend on the parameters of f_m^i and on $(x_{i-1}, y_{i-1}, z_{i-1})$. Of course, in performing this calculation for all the surfaces bounding region i, most of the double roots s_{i_m} will be imaginary (meaning no intersection) or negative (an intersection in the backward direction). The smallest positive real root gives the path distance s_i in region i.

---

[1] The most general form is $f(x, y, z) = Ax^2 + By^2 + Cz^2 + Dxy + Exz + Fyz + Gx + Hy + Iz + J$. The 10 parameters A–J define 10 possible surfaces: ellipsoids, cones, cylinders, hyperboloids of one sheet, hyperboloids of two sheets, elliptic paraboloids, hyperbolic cylinders, hyperbolic paraboloids, parabolic cylinders, and planes (Olmsted, 1947).

Figure 10.2 A concave cell whose concavities are single surfaces.

Figure 10.3 A concave cell whose concavities are from multiple surfaces.

10.4.2   Convex Versus Concave Cells

So far, it has been tacitly assumed that the various regions are all convex, i.e., a straight line between any two points lies wholly within the region. Actually, concavities in a cell due to single surfaces (see Fig. 10.2) present no difficulties with the previous procedure for calculating path lengths in a cell. However, if the concavities are produced by multiple surfaces (the dashed lines in Fig. 10.3), additional tests are required to eliminate irrelevant surface intersections by determining if the same cell is on the other side of these spurious intersections.

10.4.3   Effect of Computer Precision

Although the previous description of how a particle is traced along its trajectory of straight-line segments, punctuated by point interactions, is conceptually simple, its implementation in computer code is not so easy because of the finite precision with which numbers can be represented. The position $\mathbf{r}_i$ of the intersection of a trajectory with surface f_i, bounding the cells i and $i + 1$, may be rounded by the finite precision of the computer to (1) still remain in region i, (2) be placed in cell $i + 1$, or (3) actually be on the surface (see Fig. 10.4). In all cases, the computer "thinks" the particle is entering cell $i + 1$.

The calculation of the distance s_{i+1} of the path length in the next cell $i + 1$ is done by finding the smallest positive distance to the surfaces bounding cell $i + 1$ (here surfaces f_i and f_{i+1}). All three cases give essentially the correct distance to surface f_{i+1}; however, the distance to f_i is a small positive value for case 1 (undershoot), a small negative value for case 2 (overshoot), and zero for case 3 (exact).[2] Only in cases 2 and 3 is the correct distance s_{i+1} to the exit point in cell $i + 1$ obtained.

It is tempting to add a small positive distance to s_i to ensure there are never any undershoots. But such a fix prevents the code from being "scale-invariant," i.e., the size of the correction depends on the size of the geometry. Also use of higher-precision arithmetic does not fix this problem, but makes the under/overshoots only smaller. Fortunately there is a simple remedy to avoid these problems.

---

[2] Round-off also affects these small or zero distances to f_i, and any one of them could be rounded into a small positive distance and, thus, gives the incorrect distance to the exit point in cell $i + 1$.

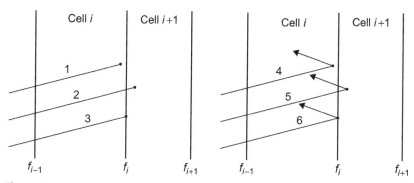

Figure 10.4 Three possibilities when a particle is transported to surface f_i.

Figure 10.5 Three possibilities when a particle is transported to and reflected from surface f_i.

Rule 1. *Do not calculate distances to boundaries from which the particle is receding based on its known direction Ω and cell the particle thinks it is in. This resolves all undershoot problems and small or zero distances to surfaces.*

A further complicating factor occurs if the particle changes direction on reaching the boundary. This change in direction occurs if the boundary is a "reflecting" boundary (used, for example, to take advantage of problem symmetries) or if (very rarely) the next interaction is on a boundary and is a scattering event. The ambiguities on where near the boundary the particle is and a possible change in its direction can cause serious errors in the calculation of the next leg of the trajectory. In Fig. 10.5, the particle has been transported to surface f_i, its direction reversed, but the computer still thinks it is entering region $i + 1$. In the next step, only the distance to surface f_i is made because the particle is receding from surface $i + 1$. Cases 4, 5, and 6 return a small negative, a small positive, and a zero value, respectively. Such problems are resolved by using another tracking rule.

Rule 2. *For a flight distance that is negative or zero, use a zero step size and make a region change. This resolves all undershoot or exact intersections when the particle is backscattered at the surface.*

Finally, it should be mentioned that, with finite-precision arithmetic, it is possible in complex geometries for a particle to become "lost." For example, if several cells have a common vertex, a particle reaching the vertex (within machine precision) may not be able to resolve which region it is entering on the other side of the vertex. Such problems, however, are extremely rare (the probability any trajectory exactly passes through any specified point is zero). For problems using millions of histories, one or two "lost" particles have negligible effect on the results.

10.5 Purely Absorbing Media

A degenerate form of transport theory is to calculate the probability that a neutron born in a region escapes without an interaction. In essence, each collision is treated as an absorption, and the problem is to estimate the probability of escape from the region without undergoing any collision. In other words, what is the fraction of the particles born in the region that have a path length to the first collision site that lies outside the region? For this special case, exact analytic solutions for simple geometries are known (Case et al., 1953). More important, such escape-probability calculations are readily evaluated with the Monte Carlo method using only track length simulation. Such a calculation is given in Example 10.1 and compared to the analytical result.

Example 10.1 Escape Probability From an Infinite Slab

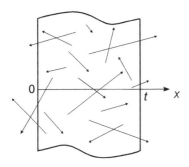

A simple application of sampling path lengths of neutral particles is the determination of the escape probability P_{esc} of particles emitted uniformly in some purely absorbing body, for which $\mu = \mu_a$. For simple bodies, such as infinite slabs, spheres, and infinite cylinders, analytical solutions of the transport equation Eq. (9.50) are known (Case et al., 1953). As an example, consider the case in which particles are emitted uniformly and isotropically throughout an infinite absorbing slab of thickness t as shown in the figure. Because the slab is infinite in the x- and y-directions, only the x-position where the particle is born affects its chance of escape. Further, for isotropic emission, the cosine of the emission angle θ with respect to the positive polar x-axis is uniformly distributed in $(-1, 1)$. The azimuthal angle of emission does not affect the escape probability, so only the polar angle θ need be simulated.

The following Monte Carlo game is played N times to determine the number of simulated particles that escape, N_{esc}.

1. Select an emission point uniformly in $(0, t)$ as $x_i = t\rho_i$.
2. Select an emission direction $\omega \equiv \cos\theta$ as $\omega_i = 2\rho_i - 1$.
3. Calculate the distance d to the surface. If $\omega_i > 0$, then $d_i = (t - x_i)/\omega_i$; otherwise $d_i = x_i/\omega_i$.
4. Generate a random path length s for the particle's first collision (absorption) in the material from the distribution $f(s) = \mu e^{-\mu s}$ as (see Example 4.2) $s_i = [-\ln\rho_i]/\mu$.
5. If $s_i < d_i$, the particle is absorbed and $z_i = 0$; otherwise the particle escapes and $z_i = 1$.

After performing this simulation N times the escape probability is estimated from $\bar{z} = (1/N) \sum_{i=1}^{N} z_i$ (which also equals $\overline{z^2} = (1/N) \sum_{i=1}^{N} z_i^2$) as

$$P_{esc} \simeq \bar{z} \pm \frac{1}{\sqrt{N}} \left[\overline{z^2} - \bar{z}^2 \right].$$

The analytic solution for this problem is (Case et al., 1953)

$$P_{esc} = \mu t \left[\frac{1}{2} - E_3(\mu t) \right],$$

where $E_3(x) \equiv x^2 \int_x^\infty u^{-3} e^{-u} \, du$ is the exponential integral function of order 3. Results using $N = 1000$ particles are compared to the analytical result for several slab widths in Fig. 10.6.

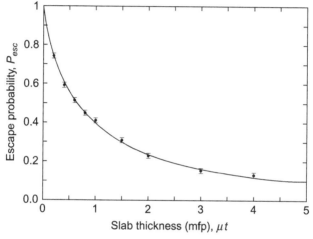

Figure 10.6 The escape probability P_{esc} for a purely absorbing slab with a uniform isotropic source of particles. The line is the analytical result, and the data are Monte Carlo simulation results using 1000 histories for each point. Error bars are the estimated standard deviations.

10.6 Type of Collision

At the end of a straight-line path segment, the particle undergoes an interaction with the material in the current cell.[3] The probability that the particle with energy E interacts with species j (an isotope for neutrons or an element for photons) in a reaction of type i is

$$p_i^j = \frac{\mu_i^j(E)}{\sum_i \sum_j \mu_i^j(E)}, \tag{10.10}$$

---

[3] Of course, if the collision site is outside the problem boundary, the history is terminated, tallies are updated, and a new source particle history is started.

where the summations are over all species in the cell and over all possible reaction types.

If the reaction is one in which no particle of the type being tracked emerges from the interaction, the particle is absorbed and the history ends. If more than one particle of the type being considered results from the interaction (e.g., fission for neutrons or fluorescence and X-rays for photons), the energy and direction of all but one are "banked" for later processing and one particle continues the presented history.

10.6.1 Scattering Interactions

Scattering is a frequent type of interaction in which the incident particle is deflected into a new direction $\mathbf{\Omega}'$ from its old direction $\mathbf{\Omega}$ through a scattering angle $\theta_s = \cos^{-1} \mathbf{\Omega}{\cdot}\mathbf{\Omega}$. In a scattering interaction with species j, the constraints of conservation of energy and momentum force the doubly differential scattering interaction coefficient (see Section 9.2.4) $\mu_s^j(E \to E', \mathbf{\Omega}{\cdot}\mathbf{\Omega}')$ to be proportional to $\delta(\mathbf{\Omega}{\cdot}\mathbf{\Omega}' - S(E, E'))$, i.e., if any two of the variables E, E', and $\cos\theta_s = \mathbf{\Omega}{\cdot}\mathbf{\Omega}'$ are specified, the third is forced to have a value that makes the argument of the delta function zero. Thus, the scattering angle $\cos\theta_s$ is sampled (usually with the rejection technique) from the PDF formed from the empirical[4] singly differential scattering interaction coefficient

$$f(\mathbf{\Omega}{\cdot}\mathbf{\Omega}') = \frac{\mu_s^j(E, \mathbf{\Omega}{\cdot}\mathbf{\Omega}')}{\int_{-1}^{1} \mu_s^j(E, \mathbf{\Omega}{\cdot}\mathbf{\Omega}')\, d(\mathbf{\Omega}{\cdot}\mathbf{\Omega}')}. \tag{10.11}$$

Once $\cos\theta_s$ is determined, the new energy of the particle is obtained by solving

$$\mathbf{\Omega}{\cdot}\mathbf{\Omega}' = \cos\theta_s = S(E, E') \tag{10.12}$$

for E'.

Direction Cosines of the Scattered Particle

Once the scattering angle and energy of the scattered particle are determined, the direction cosines (u', v', w') of $\mathbf{\Omega}'$ need to be calculated. The geometry is illustrated in Fig. 10.7. The change in azimuth $\Delta\psi_s = \psi' - \psi$ is uniformly distributed in $(0, 2\pi)$ radians. Let θ and ψ be the polar and azimuthal angles before scatter, with direction cosines $u = \sin\theta\cos\psi$, $v = \sin\theta\sin\psi$, and $w = \cos\theta$.

To calculate the direction cosines (u', v', w') of $\mathbf{\Omega}'$ it is first necessary to introduce rotation matrices $\mathbf{R}$. Consider first a rotation about the z-axis through an angle ψ, as shown in Fig. 10.8. The coordinates of the point P in original

---

[4] Only for the case of Compton scattering of photons is there an analytical expression for the differential scattering interaction coefficient.

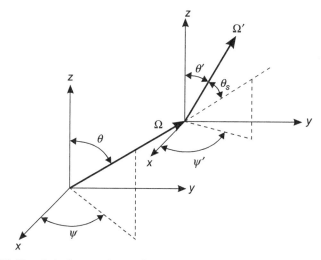

Figure 10.7 The relation between the coordinates of the particle in direction $\boldsymbol{\Omega}$ before a scatter and the direction $\boldsymbol{\Omega}'$ after the scatter. The angle between $\boldsymbol{\Omega}$ and $\boldsymbol{\Omega}'$ is the scattering angle θ_s, namely, $\cos\theta_s = \boldsymbol{\Omega} \cdot \boldsymbol{\Omega}'$.

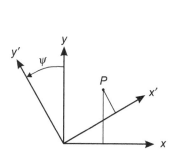

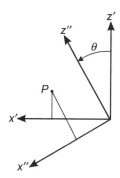

Figure 10.8 A rotation in the positive sense about the z-axis.

Figure 10.9 A rotation about the y-axis in the positive sense.

system (x, y, z) and in the rotated systems (x', y', z') are readily shown to be related by

$$
\begin{pmatrix} x \\ y \\ z \end{pmatrix} = \begin{pmatrix} \cos\psi & -\sin\psi & 0 \\ \sin\psi & \cos\psi & 0 \\ 0 & 0 & 1 \end{pmatrix} \begin{pmatrix} x' \\ y' \\ z' \end{pmatrix} \equiv \mathbf{R}_z(\psi) \begin{pmatrix} x' \\ y' \\ z' \end{pmatrix}. \quad (10.13)
$$

If this rotated coordinate system is then itself rotated about the y'-axis by an angle θ, as shown in Fig. 10.9, the relation between the coordinates of the point P

in both systems is given by

$$
\begin{pmatrix} x' \\ y' \\ z' \end{pmatrix} = \begin{pmatrix} \cos\theta & 0 & \sin\theta \\ 0 & 1 & 0 \\ -\sin\theta & 0 & \cos\theta \end{pmatrix} \begin{pmatrix} x'' \\ y'' \\ z'' \end{pmatrix} \equiv \mathbf{R}_y(\theta) \begin{pmatrix} x'' \\ y'' \\ z'' \end{pmatrix}. \tag{10.14}
$$

Combining these two rotations then gives

$$
\begin{pmatrix} x \\ y \\ z \end{pmatrix} = \begin{pmatrix} \cos\theta\cos\psi & -\sin\psi & \sin\theta\cos\psi \\ \cos\theta\sin\psi & \cos\psi & \sin\theta\sin\psi \\ -\sin\theta & 0 & \cos\theta \end{pmatrix} \begin{pmatrix} x'' \\ y'' \\ z'' \end{pmatrix}. \tag{10.15}
$$

The resulting rotation matrix is defined as

$$
\mathbf{R}_{zy}(\theta,\psi) = \mathbf{R}_z(\psi)\mathbf{R}_y(\theta) = \begin{pmatrix} \cos\theta\cos\psi & -\sin\psi & \sin\theta\cos\psi \\ \cos\theta\sin\psi & \cos\psi & \sin\theta\sin\psi \\ -\sin\theta & 0 & \cos\theta \end{pmatrix}.
$$
$$\tag{10.16}$$

Rotation matrices have several interesting properties such as the fact that their determinant is always ± 1. More useful is the fact that their inverse $\mathbf{R}^{-1} = \mathbf{R}^{\mathrm{T}}$, i.e., the inverse is the transpose of the matrix.[5]

Now consider a particle moving in direction $\mathbf{\Omega}(u, v, w) = \mathbf{i}\sin\theta\cos\psi + \mathbf{j}\sin\theta\cos\psi + \mathbf{k}\cos\theta \equiv (\sin\theta\cos\psi, \sin\theta\cos\psi, \cos\theta)$ which is scattered through an angle θ_s with a change in azimuth of $\Delta\psi_s$. How does one calculate $\mathbf{\Omega}'(u', v', w')$? First, rotate $\mathbf{\Omega}$ in the laboratory system to a local coordinate system in which the vector lies along the z-axis. It is readily verified that $\mathbf{\Omega}(\theta,\psi) = \mathbf{R}_{zy}(\theta,\psi)(0,0,1)$, so it follows that $\mathbf{R}_{zy}^{-1}(\theta,\psi)\mathbf{\Omega} = (0,0,1)$.

Next, the transformation $\mathbf{R}_{zy}(\theta_s, \Delta\psi_s)(0,0,1)$ produces a vector, in the local coordinate system, with the direction cosines given by the scattering angles, namely, $(\sin\theta_s\cos\Delta\psi_s, \sin\theta_s\sin\Delta\psi_s, \cos\theta_s)$. Finally, this scattered vector is rotated back to the laboratory system with the rotation matrix $\mathbf{R}_{zy}(\theta,\psi)$. Combining these three steps, the direction of the scattered particle is given by

$$
\mathbf{\Omega}'(u', v', w') = \mathbf{R}_{zy}(\theta,\psi)\mathbf{R}_{zy}(\theta_s, \Delta\psi_s)\mathbf{R}_{zy}^{-1}(\theta,\psi)\mathbf{\Omega}. \tag{10.17}
$$

To relate the new direction cosines (u', v', w') to the before-scattered direction cosines (u, v, w), the rotation matrix of Eq. (10.16) can be written as

$$
\mathbf{R}_{zy}(\theta,\psi) = \begin{pmatrix} w\cos\psi & -\sin\psi & u \\ w\sin\psi & \cos\psi & v \\ -\sin\theta & 0 & w \end{pmatrix}. \tag{10.18}
$$

---

[5] Such matrices are often called *orthogonal* matrices.

With this result and the relation $\mathbf{R}_{zy}^{-1}(\theta, \psi)\mathbf{\Omega} = (0, 0, 1)$, Eq. (10.17) gives the following explicit expressions:

$$u' = u \cos\theta_s + \sin\theta_s (w \cos\psi \cos\Delta\psi_s - \sin\psi \sin\Delta\psi_s), \qquad (10.19)$$

$$v' = v \cos\theta_s + \sin\theta_s (w \sin\psi \cos\Delta\psi_s + \cos\psi \sin\Delta\psi_s), \qquad (10.20)$$

$$w' = w \cos\theta_s - \sin\theta \sin\theta_s \cos\Delta\psi_s. \qquad (10.21)$$

Because $\sin\theta = \sqrt{1 - w^2}$, $u = \sin\theta \cos\psi$, and $v = \sin\theta \sin\psi$, these results can be written in an alternative form that minimizes trigonometric function calls, namely,

$$u' = u\mu_s + \sqrt{1 - \mu_s^2}(wu \cos\Delta\psi_s - v \sin\Delta\psi_s)/\sqrt{1 - w^2}, \qquad (10.22)$$

$$v' = v\mu_s + \sqrt{1 - \mu_s^2}(wv \cos\Delta\psi_s + u \sin\Delta\psi_s)/\sqrt{1 - w^2}, \qquad (10.23)$$

$$w' = w\mu_s - \sqrt{1 - w^2}\sqrt{1 - \mu_s^2}\cos\Delta\psi_s, \qquad (10.24)$$

where $\mu_s \equiv \cos\theta_s$.

From Eq. (10.21), explicit expressions for the polar and azimuthal angles after scatter can be obtained, namely,

$$\cos\theta' = \cos\theta_s \cos\theta - \sin\theta_s \sin\theta \cos\Delta\psi_s \qquad (10.25)$$

and

$$\psi' = \psi + \cos^{-1}\left[\frac{\cos\theta \cos\theta_s - \cos\theta'}{\sin\theta \sin\theta_s}\right]. \qquad (10.26)$$

The above results for calculating (u', v', w') are sufficient for most cases. However, if $\mathbf{\Omega}(u, v, w)$ is almost parallel to the z-axis, then $w \simeq 1$ and the $1/\sqrt{1 - w^2}$ term becomes very large and leads to numerical errors. For the case that $w \gtrsim 0.9$, say, an alternative approach is to use a rotation matrix that first rotates about the x-axis followed by a rotation about the y-axis to create a local coordinate system in which $\mathbf{\Omega}$ is pointed along the y-axis. In this case, the rotation matrix is

$$\mathbf{R}_{yx}(\theta, \psi) = \begin{pmatrix} \cos\psi & \sin\theta \sin\psi & \cos\theta \sin\psi \\ 0 & \cos\theta & -\sin\theta \\ -\sin\psi & \sin\theta \cos\psi & \cos\theta \cos\psi \end{pmatrix}. \qquad (10.27)$$

Now $\mathbf{R}_{yx}^{-1}(\theta, \psi)\mathbf{\Omega} = (0, 1, 0)$, and the direction vector, as above, is given by

$$\mathbf{\Omega}' = \mathbf{R}_{yx}(\theta, \psi)\mathbf{R}_{yx}(\theta_s, \Delta\psi_s)\mathbf{R}_{yx}^{-1}(\theta, \psi)\mathbf{\Omega}. \qquad (10.28)$$

Explicitly, it is found that, with $\mu_s = \cos\theta_s$,

$$u' = u\mu_s + \sqrt{1 - \mu_s^2}(vu\cos\Delta\psi - w\sin\Delta\psi_s)/\sqrt{1 - v^2}, \qquad (10.29)$$

$$v' = v\mu_s - \sqrt{1 - v^2}\sqrt{1 - \mu_s^2}\cos\Delta\psi_s, \qquad (10.30)$$

$$w' = w\mu_s + \sqrt{1 - \mu_s^2}(wv\cos\Delta\psi - u\sin\Delta\psi_s)/\sqrt{1 - v^2}. \qquad (10.31)$$

In the subsections below, specific approaches for simulating neutron and photon scattering are detailed. For sampling from other reactions that produce secondary particles (e.g., fission, $(n, 2n)$, fluorescence, etc.), the reader is referred to the MCNP manual (MCNP, 2003).

10.6.2   Photon Scattering From a Free Electron

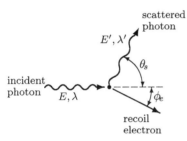

Figure 10.10 A photon with energy E and wavelength λ is scattered by a free electron. After scattering, the photon has a reduced energy E' and larger wavelength λ'.

In Compton scattering, a photon of energy E and dimensionless wavelength $\lambda = m_e c^2/E$ scatters from a free electron as shown in Fig. 10.10. Here $m_e c^2$ is the rest mass energy equivalent of the electron, namely, 0.511 MeV. The differential scattering cross section is given by the Klein–Nishina formula (Shultis and Faw, 2000). The PDF to sample from this cross section formula can be written in terms of the ratio $r \equiv \lambda'/\lambda = E/E'$ as

$$f(r|\lambda) = \frac{K(\lambda)}{r^2}\left[\frac{1}{r} + r - 1 + (1 + \lambda - r\lambda)^2\right], \qquad (10.32)$$

where $K(\lambda)$ is the normalization constant to make $f(r|\lambda)$ a PDF (see Problem 10.2). The cosine of the scattering angle is given by

$$\cos\theta_s = 1 + \lambda - r\lambda. \qquad (10.33)$$

The range of the scattering angle $0 \le \theta_s \le \pi$ thus gives the constraint $1 \le r \le 1 + (2/\lambda)$. This PDF can be separated into a sum of two PDFs amenable to sampling by the composition-rejection method (see Section 4.1.8) as follows (Kahn, 1954):

$$f(r|\lambda) = K\left\{\left(\frac{1}{r} - \frac{1}{r^2}\right) + \frac{1}{r^2}\left(\frac{1}{r} + [1 + \lambda - r\lambda]^2\right)\right\}$$

$$= A_1 g_1(r)f_1(r) + A_2 g_2(r)f_2(r). \qquad (10.34)$$

The constants $A_1 \equiv K/(2\lambda)$ and $A_2 \equiv 4K/(\lambda + 2)$ have the properties

$$\frac{A_1}{A_1 + A_2} = \frac{\lambda + 2}{2 + 9\lambda} \equiv A_1' \quad \text{and} \quad \frac{A_2}{A_1 + A_2} = \frac{8\lambda}{2 + 9\lambda} \equiv A_2', \tag{10.35}$$

where $A_1' + A_2' = 1$. The functions f_1 and f_2 are defined as

$$f_1(r) \equiv 4\left(\frac{1}{r} - \frac{1}{r^2}\right) \quad \text{and} \quad f_2(r) \equiv \frac{1}{2}\left[\frac{1}{r} + (1 + \lambda - r\lambda)^2\right], \tag{10.36}$$

and have the property that $f_i(r) \leq 1$ for $1 \leq r \leq 1 + (2/\lambda)$. Finally, the PDFs $g_1(r) \equiv \lambda/2$ and $g_2(r) \equiv (\lambda + 2)/(2r^2)$ have the easily invertible CDFs

$$G_1(r) = \frac{\lambda}{2}(r - 1) \quad \text{and} \quad G_2(r) = \frac{\lambda + 2}{2}\left(1 - \frac{1}{r}\right). \tag{10.37}$$

Thus Eq. (10.32) can be written as

$$f(r|\lambda) = K'(\lambda)[A_1' g_1(r) f_1(r) + A_2' g_2(r) f_2(r)], \tag{10.38}$$

where the PDF normalization constant $K'(r) = K(r)(9\lambda + 2)/[2\lambda(\lambda + 2)]$.

The algorithm for sampling from Eq. (10.34) is as described in Section 4.1.8, namely,

1. If $\rho_i < (\lambda + 2)/(2 + 9\lambda)$, calculate $r = 1 + (2\rho_{i+1} + 1)/\lambda$ and accept this value if $\rho_{i+2} < (4/r)[1 - (1/r)]$.
2. Else, if $\rho_i \geq (\lambda + 2)/(2 + 9\lambda)$, calculate $r = (\lambda + 2)/(2\rho_{i+1} + \lambda)$ and accept this value if $\rho_{i+2} < [(1/r) + (1 + \lambda - r\lambda)^2]/2$.
3. If no acceptable value for r is found, return to step 1 and try again.

Once a value of r is accepted, the wavelength of the scattered photon is simply $\lambda' = r\lambda$ corresponding to an energy of $E' = E/r$. Lastly, $\cos\theta_s$ is found from Eq. (10.33) and $\sin\theta_s = \sqrt{1 - \cos^2\theta_s}$. A comparison of this and other sampling methods for the Klein–Nishina cross section is given by Blomquist and Gelbard (1983).

10.6.3 Neutron Scattering

Neutron scattering data are almost always tabulated in the center-of-mass coordinate system (Shultis and Faw, 2000) in which there is zero net linear momentum before and after the scatter. The microscopic differential cross section for either elastic or inelastic scattering is most often approximated as

$$\sigma_s(E, \omega_c) \simeq \frac{\sigma_s(E)}{4\pi} \sum_{n=0}^{N} (2n + 1) f_n(E) P_n(\omega_c), \tag{10.39}$$

in which $P_n(\omega_c)$ is the Legendre polynomial of order n and ω_c is the cosine of the scattering angle in the center-of-mass system. Cross section data consist of tabulating the Legendre coefficients $f_n(E)$ (with f_0 always equal to 1). The reason for this representation is that, in the center-of-mass system, scattering is usually more isotropic than in the laboratory system (target at rest system) so that only a low-order expansion is needed. Indeed, for isotropic scattering $N = 0$, and most data libraries seldom exceed $N = 8$ even for highly anisotropic scattering.[6]

The first step in simulating a neutron scattering event is to construct a PDF with the same shape as Eq. (10.39) and to obtain a sample of ω_c from it. The corresponding cosine of the scattering angle in the laboratory system is (Shultis and Faw, 2000)

$$\mathbf{\Omega \cdot \Omega'} \equiv \cos\theta_s = \frac{\gamma + \omega_c}{\sqrt{1 + 2\gamma\omega_c + \gamma^2}}, \tag{10.40}$$

where

$$\gamma = \left[A^2 + \frac{A(A+1)Q}{E}\right], \tag{10.41}$$

with Q being the Q-value (Shultis and Faw, 2007) of the scatter ($= 0$ for elastic scatter and < 0 for inelastic scatter) and A being the ratio of the atom's mass to that of the neutron ($\simeq$ atomic mass number). From the conservation of energy and momentum, the energy of the scattered neutrons is given by (Shultis and Faw, 2000)

$$E' = \frac{1}{2}E(1+\alpha) + \frac{1}{2}(1-\alpha)E\omega_c\sqrt{1+\Delta} + \frac{QA}{A+1}, \tag{10.42}$$

in which $\alpha \equiv (A-1)^2/(A+1)^2$ and $\Delta \equiv Q(A+1)/(AE)$.

10.7 Time Dependence

Although most transport problems are independent of time, Monte Carlo simulations are easily extended to time-dependent cases when needed, for example, to estimate when a particle reaches some surface. For neutrons, the energy determines the neutron's speed from which the time for various segments of its history are easily calculated. Tallies are then binned by the time the contribution to the tally was made. Photons are even easier because they all travel at the speed of light. Even time-varying sources can be treated by sampling from an appropriate time distribution to obtain the starting time for each history and tracking the times of the various events that occur during the history. In the following discussion, however, it is assumed that the tallies and flux density are independent of time.

---

[6] For the case of isotropic scattering (often a good approximation for heavier atoms since the center-of-mass system becomes the target-at-rest system for infinitely heavy scattering centers), the directed cosines of the scattered particle can be calculated in the same manner as used for the directed cosines of an isotropically emitted source particle (see Section 10.3.1).

10.8 Particle Weights

The efficiency of a Monte Carlo transport calculation depends on both the speed with which the calculation can be made and the variance of the result. The speed is, to a large extent, controlled by the hardware and, to a lesser extent, by the computer programmer and the operating system. The variance is controlled primarily by the type of tally used and the number of particles that contribute to the tally. In a strict analog simulation, the only way to reduce the variance of a tally is to run more histories. But this brute-force approach only reduces the variance as $1/N$. As explained in Chapter 5, a better way to increase the precision of the tally is to use nonanalog techniques to force more particles to the regions of phase space where particles are more likely to score without increasing (and perhaps decreasing) the sampling in less important regions of phase space.

In a nonanalog simulation, as explained in Section 4.2.1, samplings for the various physical events used to create a particle history are biased so that events more likely to contribute to the tally are produced more frequently than in an analog simulation. This bias introduced into particle histories must, of course, be corrected when scoring the histories so as to produce an unbiased estimator. For example, if a particular bias makes a history α times more likely to score, then its contribution to the tally must be multiplied by a factor of $1/\alpha$. This removal of biases is readily accomplished by assigning a *weight* W to each particle, and, at every event in its history that introduces a bias, its weight is adjusted so that its score *times* its weight produces an unbiased contribution to the score. For example, if, in a biased sampling for some event, the particle's chance of scoring is changed by a factor of α, its current weight W is modified to W/α. If there is no bias in starting a particle history, the history begins with a weight of unity. Note that weights are not constrained to integer values and may be greater or less than unity.

In short, in a nonanalog Monte Carlo simulation, the scoring of the particles reaching the tally region or detector must be modified from just the sum of the particles' scores to the sum of the particles' weighted scores. In this way, unbiased results can be achieved with less computing time, if proper variance reduction methods are used, compared to an analog calculation producing the same results and precision.

10.9 Scoring and Tallies

After a particle history ends or, more usually, after a particle leaves a cell, its contribution to the score (or tally) is added to the running score of interest. Almost anything of interest that depends on the radiation field can be estimated from the particle histories used in a Monte Carlo simulation. Here, some of the more common tallies are reviewed. For more thorough treatments and other estimators, the reader is referred to Kalos et al. (1968), Carter and Cashwell (1975), and MCNP (2003).

10.9.1 Fluence Averaged Over a Surface

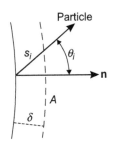

Figure 10.11 Particle crossing a surface at an angle θ_i from the outward normal **n**.

Often the fluence, averaged over some surface (or portion of a surface), is sought. Imagine a parallel surface, a very small distance δ from the surface of interest as shown in Fig. 10.11. From Eq. (9.13), the fluence (flux density integrated over all time or all simulated histories) is just the weighted sum of the path lengths of all the particles passing through this incremental volume $\Delta V = A\delta$ divided by the volume of the extended region, i.e., $\Phi = \sum_i W_i s_i / \Delta V$, where W_i is the weight of particle i. The path length for the particle shown is $s_i = \delta / |\cos \theta_i|$, where θ_i is the angle between the particle's exit direction and the outward normal **n**. Thus, the fluence contribution of the ith particle crossing the surface is $\Phi_i = \lim_{\delta \to 0} (W_i \delta / |\cos \theta_i|)/A\delta = W_i/(A|\cos \theta_i|) = W_i/(A|\mathbf{\Omega} \cdot \mathbf{n}|)$. Every time a particle crosses the surface, the value of its $W_i / \cos \theta_i$ is added to the tally. Of course, many histories may not cross the surface in question, so no score is contributed to the tally. Finally, after N histories the average fluence, *per source particle*, is estimated as

$$\overline{\Phi}_S \equiv \frac{1}{A} \int_A dA \int_0^\infty dE \int_{4\pi} d\mathbf{\Omega}\, \Phi(\mathbf{r}, E, \mathbf{\Omega}) \simeq \frac{1}{NA} \sum_{i=1}^{N} \sum_{j=1}^{n_i} \frac{W_i^j}{|\cos \theta_i^j|}, \qquad (10.43)$$

where n_i is the number of surface crossing made by the ith particle,[7] W_i^j is the ith particle's weight on its jth crossing, and θ_i^j is the angle with respect to the outward normal at the jth crossing. Here $\Phi(\mathbf{r}, E, \mathbf{\Omega})$ is the energy and angular fluence normalized to one source particle.

It should be noted that the variance of this fluence estimator is infinite and, hence, the central limit theorem cannot be used to estimate confidence intervals. If many particles cross the surface in nearly tangential directions, i.e., $\cos \theta$ is very small, the tally contributions become very large. To avoid such large scoring events, many Monte Carlo transport codes use an ad hoc fixup procedure to avoid large variances. If $|\cos \theta| < \epsilon$, where ϵ is some small value, say 0.01, then replace the $|1/\cos \theta_i^j|$ in Eq. (10.43) by $2/\epsilon$, the expected value of $1/|\cos \theta|$ when $|\cos \theta| < \epsilon$ and the fluence is nearly isotropic. Although this procedure is not rigorous, in most cases reliable estimates of the surface-averaged fluence are obtained.

---

[7] Typically n_i is 0 or 1. But a particle may make several crossing, especially if the surface is reentrant.

If the angular distribution and energy spectra of the particles crossing the surface are of interest, then the tally procedure requires the use of a multi-dimensional array for scoring. Polar angle cosines, $\cos\theta$, can be divided into some number of intervals from -1 to 1. Likewise, the azimuthal angle about the outward normal to the surface can be divided into some number of intervals from 0 to 2π radians. Finally, the full range of possible particle energies may also be divided into some number of intervals. When a particle is found to cross the surface, a score is made in the *bin* associated with the energy and direction of the particle. Binning by energy also allows each bin to be multiplied by an energy-dependent response function $\mathcal{R}(E)$, so that, for example, summing over all energy and angular bins, the average dose at the surface can be estimated. In a similar fashion, multiplication by a reaction interaction coefficient $\mu_j(E)$ and summation of the bin results yields the average reaction density $\int \mu_j(E)\overline{\Phi}(E)\,dE$ at the surface.

10.9.2 Fluence in a Volume: Path Length Estimator

The idea behind Eq. (9.13) can also be extended to determine the average fluence in the volume of a cell. In this type of tally, the sum of path lengths s_i each particle makes in the volume V of interest is accumulated (see Fig. 10.12). Again, if a particle history never reaches the cell in question, then $s_i = 0$. The average fluence, per source particle, is then estimated as

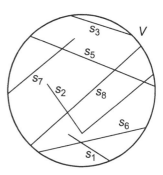

$$\overline{\Phi}_V \simeq \frac{1}{NV} \sum_{i=1}^{N} \sum_{j=1}^{n_i} W_i^j s_i^j, \qquad (10.44)$$

Figure 10.12 Particle paths in a cell of volume V.

where n_i is the number of times the ith particle enters V, s_i^j is the jth track length in V, and W_i^j is the particle's weight when entering V for the jth time. By breaking the energy range into a contiguous set of bins and adding each weighted path length to the tally in the appropriate energy bin, the energy spectrum of the volume-averaged fluence can be estimated as

$$\overline{\Phi}_V(E) \equiv \frac{1}{V} \int_V dV \int_{4\pi} d\Omega \, \Phi(\mathbf{r}, E, \Omega) \simeq \frac{S_i}{NV\Delta E}, \qquad (10.45)$$

where ΔE is the width of the energy bin containing E and S is the weighted sum of path lengths tallied in that energy bin. As before the tally for each particle can be multiplied by some function $\mathcal{R}(E)$ so that when the results in each energy bin are summed, the average value of $\int dE\, \mathcal{R}(E)\overline{\Phi}_V(E)$ is estimated.

If the volume V is large so that many histories enter it, the path length estimator is generally a good tally. If the region is a thin curved shell, most track lengths have similar lengths and the estimator has a small variance. By contrast, for a region bounded by two closely spaced parallel planes, there may be a wide variation in the track lengths through the region, and the resulting path length estimator may have a large variance. The path length estimator is also computationally quite efficient, since particle tracking already computed the track lengths in the regions, so little extra effort is needed for this tally.

10.9.3 Fluence in a Volume: Reaction Density Estimator

An alternative estimator for the average fluence in a region is obtained by tallying the number of interactions that occur in the region. From Eq. (9.37), the average density of reactions in a homogeneous volume V is found to be

$$\overline{R}(E) \equiv \frac{1}{V}\int_V \mu_t(\mathbf{r}, E)\Phi(\mathbf{r}, E)dV = \frac{\mu_t(E)}{V}\int_V \Phi(\mathbf{r}, E)dV \equiv \mu_t(E)\overline{\Phi}_V(E). \tag{10.46}$$

Thus, the contribution to the average fluence in V made by a particle whose energy E lies in ΔE and that collides in V is made by adding $1/\mu_t(E)$ to the tally energy bin. Thus, after N histories, the estimate of the average fluence is

$$\overline{\Phi}_V(E) \equiv \frac{1}{V}\int_V dV \int_{4\pi} d\mathbf{\Omega}\, \Phi(\mathbf{r}, E, \mathbf{\Omega}) \simeq \frac{1}{NV\Delta E}\sum_{i=1}^{N}\sum_{j=1}^{n_i}\frac{W_i^j}{\mu_t(E)}, \tag{10.47}$$

where n_i is the number of collisions in V made by particle i when it had an energy in ΔE and the superscript j refers to the jth such interaction in V made by the ith particle.

The probability a particle interacts while crossing a cell is

$$p = 1 - \exp[-\mu_t(E)s_{\max}],$$

where $s_{\max}$ is the maximum distance the particle could travel in the cell along its flight path. For thin or small regions, this probability is small and the above estimator will produce poor results. A more efficient way to estimate the fluence is to tally the interaction contribution times the probability the particle interacts, i.e.,

$$\overline{\Phi}_V(E) \simeq \frac{1}{NV\Delta E} \sum_{i=1}^{N} \sum_{j=1}^{n_i} W_i^j \frac{1 - e^{-\mu_t(E)s_{\max,j}}}{\mu_t(E)}, \qquad (10.48)$$

where n_i now is the number of cell crossings made by particle i when it had energy E in ΔE.

For the reaction density estimator of Eq. (10.47) to work properly, particles must cross the cell of interest. If the cell is far from the source, even if a particle heads toward the cell, it is unlikely to reach it and contribute to the tally. But all is not lost. The contribution to the fluence in the cell, however, can be calculated for particles heading toward the cell but that interact at some point before reaching it. Denote the distance from the interaction point to the nearest cell surface by d along the original trajectory. One simply includes in the estimator of Eq. (10.47) a term that gives the probability p_{hit} that the particle *does* reach the cell,

$$p_{hit} = \exp\left[-\int_0^d \mu_t(\mathbf{r}, E)ds\right],$$

where the integration is along the trajectory. Thus, the argument of the exponential is just the number of mean free path lengths the particle would have to travel from its interaction point to reach the cell. The average fluence in the cell is then estimated as

$$\overline{\Phi}_V(E) \simeq \frac{1}{NV\Delta E} \sum_{i=1}^{N} \sum_{j=1}^{n_i} W_i^j \frac{1 - e^{-\mu_t(E)s_{\max,j}}}{\mu_t(E)} \exp\left[-\int_0^{d_j} \mu_t(\mathbf{r}, E)ds\right].$$

$$(10.49)$$

Here n_i is the number of collision, both inside and outside the cell, made by particle i. If a collision is inside the cell, then, of course, $d_j = 0$.

For a large thin region, subtending a large solid angle, Eq. (10.49) may be a good estimator; but for small distant regions few particles will even interact while heading toward it. Even this can be overcome by including the probability that, if the collision was a scatter, the particle would have an energy in ΔE and head toward the cell. This idea of forcing a scatter toward a cell of interest is developed further in Section 10.9.5.

10.9.4 Average Current Through a Surface

Sometimes the average total flow or current through a surface is sought (see Section 9.1.5). For particles with energy E in ΔE and direction $\mathbf{\Omega}$ in $\Delta\Omega$, the

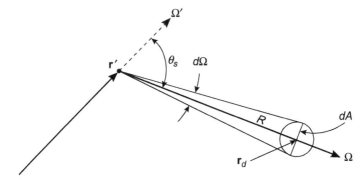

Figure 10.13 A particle moving in direction $\mathbf{\Omega}'$ makes a collision at $\mathbf{r}'$ and scatters toward a small spherical detector at $\mathbf{r}_d$ of cross-sectional area dA and a distance $R = |\mathbf{r}' - \mathbf{r}_d|$ from the scattering point.

estimate of the average current, per source particle, is[8]

$$\overline{J}(E, \mathbf{\Omega}) \equiv \frac{1}{A} \int_A dA \int_{\Delta\Omega} d\mathbf{\Omega} |\mathbf{\Omega}\cdot\mathbf{n}| \Phi(\mathbf{r}, E, \mathbf{\Omega}) \simeq \frac{1}{N A \, \Delta E \Delta\Omega} \sum_{i=1}^{N} n_i W_i, \quad (10.50)$$

where n_i is number of crossings of the surface made by particle i when it had energy in ΔE and direction in $\Delta\Omega$. Typically, n_i is 0 or 1, but a particle may cross a re-entrant surface several times.

If the net flows across a surface are wanted (see Eq. (9.20)), then use the above procedure with only the two cosine ($\mu \equiv \mathbf{n}\cdot\mathbf{\Omega}$) angular bins $-1 \leq \mu < 0$ and $0 < \mu \leq 1$ to estimate $J^-(E_i)$ and $J^+(E_i)$, respectively. The net flow is $J_{net} = J^+ - J^-$.

10.9.5   Fluence at a Point: Next-Event Estimator

A very powerful technique for scoring is to combine deterministic tally contributions and the stochastic collisions that occur during a history. One such estimator is the *next-event estimator*, also known as the no-further-collision estimator. Consider the small spherical detector, with cross-sectional area dA, shown in Fig. 10.13. A particle traveling in direction $\mathbf{\Omega}'$ has a collision at $\mathbf{r}'$. The collision may not be a scatter, and even if it was a scatter, it is very unlikely to scatter in the direction of dV or even reach it to contribute to the fluence tally for dV. However, one can analytically calculate the probability the particle will scatter at $\mathbf{r}'$ and reach the detector without further interaction, and thus provide a contribution to the fluence tally for the detector, thereby short-cutting the Monte Carlo process.

---

[8] Here, J is the time-integrated current vector discussed in Section 9.1.4 just as the fluence Φ is the time-integrated flux ϕ (see Eq. (9.15)).

The probability the particle with energy E scatters at $\mathbf{r}'$ through a scattering angle θ_s and has a new direction of travel in $d\Omega$ about $\mathbf{\Omega}$ with energy E' is

$$
\begin{aligned}
p_1 &= \frac{\mu_s(\mathbf{r}', E, \omega_s)\, d\Omega}{\int_0^{2\pi} d\psi \int_{-1}^{1} d\omega_s \mu_s(\mathbf{r}', E, \omega_s)} = \frac{1}{2\pi} \left[\frac{\mu_s(\mathbf{r}', E, \omega_s)}{\int_{-1}^{1} d\omega_s \mu_s(\mathbf{r}', E, \omega_s)} \right] d\Omega \\
&= \frac{p(\omega_s)}{2\pi} d\Omega,
\end{aligned}
$$

where $p(\omega_s)$ is the PDF (defined by the terms in the square brackets) for scattering through an angle $\theta_s = \cos^{-1} \mathbf{\Omega} \cdot \mathbf{\Omega}'$. The distance between $\mathbf{r}'$ and $\mathbf{r}_d$ is $R = |\mathbf{r}' - \mathbf{r}_d|$ so that $d\Omega = dA/R^2$. The probability that such a scattered particle actually reaches the detector without further collision is

$$
p_2 = \exp\left[-\int_0^R \mu_t(\mathbf{r}, E')\, ds \right],
$$

where the argument of the exponential is the total number of mean free path lengths between $\mathbf{r}'$ and $\mathbf{r}_d$.

As stated by Eq. (9.14), the flux density can be interpreted as the number of particles entering a spherical detector of cross-sectional area dA divided by dA. Thus, the contribution to the fluence made by the particle with weight W that interacts at $\mathbf{r}'$ is

$$
\delta\Phi = W \frac{\mu_s(\mathbf{r}', E)}{\mu_s(\mathbf{r}', E')} \frac{p_1 p_2}{dA} = W \frac{\mu_s(\mathbf{r}', E)}{\mu_s(\mathbf{r}', E')} \frac{p(\omega_s)}{2\pi R^2} \exp\left[-\int_0^R \mu_t(\mathbf{r}, E)\, ds \right].
\tag{10.51}
$$

This result is independent of dA and, thus, is used as a fluence tally for a point detector at $\mathbf{r}_d$. After each interaction in a particle's history, an estimate is made deterministically for the expected contribution of that interaction to the fluence at point $\mathbf{r}_d$.

A difficulty with this point detector tally occurs when the tally point lies within a scattering medium. Because of the $1/R^2$ term in Eq. (10.51), an enormous contribution is made to the tally if an interaction occurs very near the tally point. In fact, in a scattering region the variance of this estimator is infinite! An infinite variance, however, does not mean the tally cannot be used; indeed the tally will converge, but the asymptotic behavior is reduced to $1/\sqrt[3]{N}$, slower than the $1/\sqrt{N}$ convergence for a tally with a finite variance (Kalos et al., 1968). Of course, if the tally point is in a vacuum or nonscattering medium no such problem occurs.

The way to avoid an infinite variance in a scattering medium is to surround the tally point by a small sphere of radius R_o, and do no estimation if an interaction occurs at $R < R_o$. Alternatively, if an interaction occurs with $R < R_o$, tally the average fluence uniformly distributed in the volume, i.e.,

$$\delta\Phi(R < R_o) = \frac{\int \Phi\, dV}{\int dV} = W\frac{\mu_s(\mathbf{r}', E)}{\mu_s(\mathbf{r}', E')}\frac{p(\omega_s)}{2\pi}\frac{\int_0^{R_o}\frac{1}{r^2}e^{-\mu_t(E')r}(4\pi r^2)\,dr}{\frac{4}{3}\pi R_0^3}$$

$$= W\frac{\mu_s(\mathbf{r}', E)}{\mu_s(\mathbf{r}', E')}p(\omega_s)\frac{3(1 - e^{-\mu_t(E')R_o})}{2\pi R_o^3}. \tag{10.52}$$

Although the above next-event estimator can produce good prediction of the fluence at a point, its convergence may be rather slow. This estimator needs many statistical tests to ensure that sound results are achieved. Lux and Koblinger (1991) and MCNP (2003) discuss variance issues associated with the use of next-event estimators to determine point values of the fluence and related quantities.

10.9.6 Flow Through a Surface: Leakage Estimator

Another partially deterministic method is the *leakage estimator* used to find the probability P_{esc} that a particle entering a region leaves through some surface of the region. The idea is very similar to the next-event estimator. At the point of entry of a particle into the region (either entering from outside or born in the region), and after each scattering interaction in the region, a determination is made of the probability that, without further interaction, the photon can escape from the slab. That probability is then scored as a contribution to the flow across the surfaces.

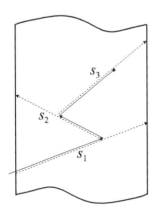

Figure 10.14 A three-segment particle path ending in an absorption.

Thus, for the three-segment path ending in an absorption shown in Fig. 10.14, the leakage tally for this particle history is

$$e^{-\mu s_1} + e^{-\mu s_2} + e^{-\mu s_3}.$$

Also note that if the particle passes through the surface on its last segment in the region, no score is tallied. This method can also be applied to scoring the energy spectrum and angular distribution of the flow rates through each slab face. For deep penetration problems, score contributions may vary considerably among histories, so that large variances result. An example of an analog leakage calculation and one using the leakage estimator is given in Example 10.2.

Example 10.2 Escape Probability From an Infinite Slab

Consider an infinite homogeneous slab surrounded by a vacuum. One-speed particles are born uniformly throughout the slab. The particles scatter isotropically with $c = \mu_s/\mu$, the scattering probability per interaction. The problem is to calculate the probability P_{esc} that a particle born in the slab eventually escapes of leaks from it.

In a straight analog simulation of N histories in which n_{esc} particles leak from the slab,

$$p_{esc} = \frac{n_{esc}}{N}.$$

However, for thick slabs with small values of c only particles born near the surface have much chance of escape and of contributing to the tally. With a leakage estimator, *every* particle contributes something to the tally, although many contributions are quite small. In Fig. 10.15 results are shown for the case $c = 0.1$ using analog scoring (left-hand figure) and the leakage estimator (right figure). It is seen that, although both methods give comparable results for thin slabs, the leakage estimator is better for the thick slabs.

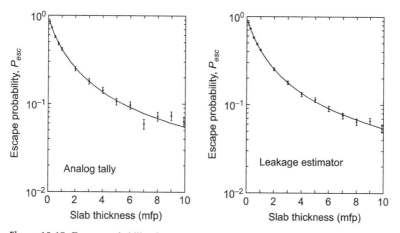

Figure 10.15 Escape probability for particles born uniformly in an infinite homogeneous slab with $c = 0.1$. Monte Carlo results with $\pm\sigma$ error bars are for 1000 particles. Results on the left are for an analog tally, i.e., $P_{esc} = $ (no. escaping)/(no. born). Results on the right were obtained using the leakage estimator. The solid lines on these plots are cubic spline fits to results obtained with 10^7 histories and that contain negligible error.

10.10 An Example of One-Speed Particle Transport

Before venturing into more detailed aspects of Monte Carlo transport theory, it might be useful to consider a simple example that uses many of the concepts introduced up to now. In Example 10.3, a homogeneous half-space is normally and uniformly illuminated by incident particles. The problem is to find the distribution of reflected particles.

Example 10.3 Particle Reflection From a Half-Space

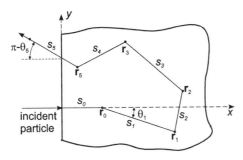

The challenge for solving the transport equation occurs when particles can scatter as well as be absorbed. Analytic solutions are known only for highly simplified forms of the transport equation Eq. (9.50), and then only with substantial mathematical effort. To allow a comparison between a Monte Carlo calculation and an exact analytical solution, the following simplifications are applied to Eq. (9.50): (1) particles are assumed to have only a single energy, (2) scattering is isotropic, (3) the medium is a source-free homogeneous half-space $x \geq 0$, and (4) particles are incident uniformly and normally on the surface of the half-space. Under these conditions, Eq. (9.50) can be written as

$$\omega \frac{\partial \phi(x,\omega)}{\partial x} + \phi(x,\omega) = \frac{c}{2} \int_{-1}^{1} \phi(x,\omega')\,d\omega', \qquad 0 \leq x < \infty, \quad -1 \leq \omega \leq 1, \quad (10.53)$$

where x is measured in mean free path lengths, $\omega = \mathbf{i} \cdot \mathbf{\Omega}$, and $c = \mu_s/\mu \leq 1$. The boundary conditions of this *albedo problem* are (1) $\phi(0,\omega) = \delta(\omega - 1)$ and (2) $\lim_{x\to\infty} \phi(x,\omega) < \infty$.[9] In this formulation, the units of $\phi(x,\omega)$ are no longer $cm^{-2}\,s^{-1}\,sr^{-1}$ but $cm^{-2}\,s^{-1}$ per unit ω, since, to obtain Eq. (10.53), integration over all azimuthal angles has been done.

Several different (and complex) analytical methods can be used to obtain equivalent closed-form analytical expressions for the angular flux density $\phi(x,\omega)$. In particular, the reflected flux density is (Chandrasekhar, 1960)

$$\phi(0,\omega) = \frac{c}{2} \frac{1}{1+|\omega|} H(c,|\omega|) H(c,1), \qquad -1 \leq \omega \leq 0, \qquad (10.54)$$

where $H(c,\cos\theta)$ is Chandrasekhar's H-function, which can be evaluated by an iterative solution of the nonlinear integral equation (Shultis and Faw, 2000)

$$\frac{1}{H(c,\cos\theta)} = \sqrt{1-c} + \frac{c}{2} \int_{0}^{1} \frac{u H(c,u)}{u+\cos\theta}\,du. \qquad (10.55)$$

---

[9] This albedo problem is used, for example, to describe light reflected from a planetary atmosphere or thermal neutrons reflected from the surface of a shield.

Not only is the analysis to obtain Eq. (10.54) difficult, so is its numerical evaluation, which requires a numeral solution of Eq. (10.55). By contrast, the reflected flux density is easily calculated by a Monte Carlo simulation if the problem is first transformed to an easier equivalent one, as discussed in Section 5.1. Rather than having particles incident all over the infinite surface of the half-space and scoring only those that flow back through a given unit area on the surface after scattering in the half-space, it is far better to have all incident particles impinge on the same point on the surface (say, $y = z = 0$) and to score particles that emerge anywhere on the surface. Because the half-space is infinite in the y- and z-directions, these coordinates have no effect on whether a particles escape, and, hence, only the x-coordinate of a particle's interaction site needs to be calculated.

The Monte Carlo simulation proceeds as follows. An incident particle penetrates a distance s_0, randomly sampled from $f(s) = \mu e^{-s}$, where it is scattered, with probability c, or absorbed, with probability $1 - c$. If a scatter occurs a new direction θ_0 is selected, where $\cos\theta_0$ is sampled from $\mathcal{U}(-1, 1)$, and another path length is randomly chosen to the next interaction point $\mathbf{r}_1$. This procedure is continued, as shown in the figure on the previous page, until the particle either is absorbed or escapes through the surface. The escaping particle is then binned according to its angle of escape.

One possible algorithm for this albedo simulation is given below. Here each ρ refers to a new random number from $\mathcal{U}(0, 1)$.

1. Set up and initial to zero, an M-component vector $\mathbf{g}$ in which g_m accumulates reflected angular flux estimates of $\phi(0, -|\omega|)$ with $0 < \omega_m < |\omega| \leq \omega_{m+1} \leq 1$. Note $\omega_1 = 0$ and $\omega_{M+1} = 1$, so bin g_k accumulates flux tallies for reflected particles whose $|\omega|$ is in the range $(\omega_k, \omega_{k+1}]$.
2. Generate the first collision site $x_0 = s_0 = -\ln\rho_0$. Set $i = 0$.
3. If $\rho > c$ the particle is absorbed, so go to step 1; else:
   (a) Pick a new cosine direction $\omega_{i+1} = 2\rho - 1$.
   (b) Pick the distance to the next interaction site as $s_{i+1} = -\ln\rho$.
   (c) Calculate the x-coordinate of the interaction site as $x_{i+1} = x_i + \omega_{i+1}s_{i+1}$.
   (d) If $x_{i+1} \leq 0$, the particle escapes through the illuminated surface, i.e., it is reflected. Note ω_{i+1} must be negative to escape the half-space. Bin the flux density estimator $1/|\omega_{i+1}|$ in the reflection tally vector $g_k = g_k + 1/|\omega_{i+1}|$ if $\omega_k < \omega_{i+1} \leq \omega_{k+1}$. Time for another history, so go to step 2.
   (e) Else, because $0 < x_{i+1} < \infty$, the particle is still in the half-space. If $\rho < c$, the particle scatters so set $i = i + 1$ and go to step 3a; otherwise the particle is absorbed so go to step 2.

After a large number of histories N, the estimate of the reflected angular flux density $\phi(0, \omega)$, $\omega \leq 0$, is simply $g(\Delta\omega_i)/N$. In Fig. 10.16, a comparison is shown between the analytical result of Eq. (10.54) and the above Monte Carlo procedure. To obtain such good agreement, 10^7 histories were used.

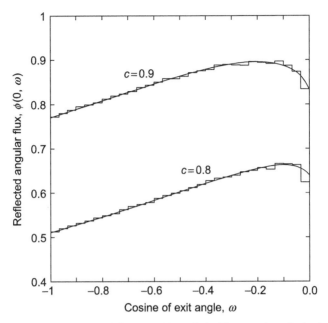

Figure 10.16 Analytical (smooth lines) and Monte Carlo (histograms) results for the albedo problem. The Monte Carlo results used 10^7 histories and 30 cosine bins.

10.11 Monte Carlo Based on the Integral Transport Equation

As discussed in earlier chapters, Monte Carlo is ideally suited for evaluating multidimensional integrals (see Section 2.6) and for solving integral equations (see Section 8.2.4). Thus, it is not surprising that early transport calculations used Monte Carlo techniques on the integral form of the transport equation (Section 9.4).

10.11.1 The Integral Transport Equation

One widely used way to develop Monte Carlo calculational procedures for particle transport is to begin with some form of the integral transport equation (see Section 9.4). It does not really matter which form of the integral equation is used, but here the procedure of Carter and Cashwell (1975) is used, which is based on the integral equation for the collision rate density $F(\mathbf{r}, E, \mathbf{\Omega})$ that is given by Eq. (9.76), namely,

$$F(\boldsymbol{P}) = \int_R d\boldsymbol{P}' \overline{K}(\boldsymbol{P}' \to \boldsymbol{P}) F(\boldsymbol{P}') + \overline{S}(\boldsymbol{P}). \tag{10.56}$$

Here $\boldsymbol{P}$ are the particle phase space coordinates $(\mathbf{r}, E, \mathbf{\Omega})$ and region R of phase space contains the volume V of the ambient medium, the energy range of

E, and the 4π sr of directions; $\overline{S}(P)$ is the first-flight collision rate density of particles streaming from the source and is given by

$$\overline{S}(P) \equiv \mu(\mathbf{r}, E) \int_V dV' \exp[-\tau(E, \mathbf{r}, \mathbf{r}')] \delta\left(\mathbf{\Omega} - \frac{\mathbf{r} - \mathbf{r}'}{|\mathbf{r} - \mathbf{r}'|}\right) \frac{S(\mathbf{r}', E, \mathbf{\Omega})}{|\mathbf{r} - \mathbf{r}'|^2}$$

and $\overline{K}(P \to P)$ is the next-flight collision rate density at P due to a collision at P', given by

$$\overline{K}(P' \to P) \equiv \frac{\mu(\mathbf{r}, E)}{\mu(\mathbf{r}', E')} \frac{\mu(\mathbf{r}', E' \to E, \mathbf{\Omega}' \to \mathbf{\Omega})}{|\mathbf{r} - \mathbf{r}'|^2} \exp[-\tau(E, \mathbf{r}, \mathbf{r}')] \delta\left(\mathbf{\Omega} - \frac{\mathbf{r} - \mathbf{r}'}{|\mathbf{r} - \mathbf{r}'|}\right).$$

Further, suppose the purpose of the Monte Carlo calculation is to estimate the value of some weighted average of $F(P)$, namely,

$$G = \int dP\, g(P) F(P), \tag{10.57}$$

where $g(P)$ is a prescribed weight function, such as a detector response function, that gives the contribution to G from a collision at P.

Formal Solution of the Integral Transport Equation

The integral equation Eq. (10.56) can be solved using the Neumann iteration scheme of Eq. (8.53), namely,

$$F_n(P) = \int dP' \overline{K}(P' \to P) F_{n-1}(P') + \overline{S}(P), \quad n = 1, 2, 3, \ldots, \tag{10.58}$$

where $F_0(P) = 0$. Thus

$$F_1(P) = \overline{S}(P),$$

$$F_2(P) = \overline{S}(P) + \int dP_1 \overline{K}(P_1 \to P) \overline{S}(P_1),$$

$$F_3(P) = \overline{S}(P) + \int dP_2 \overline{K}(P_2 \to P) F_2(P_2)$$

$$= \overline{S}(P) + \int dP_1 \overline{K}(P_1 \to P) \overline{S}(P_1)$$

$$+ \int dP_1 \int dP_2 \overline{K}(P_1 \to P_2) \overline{K}(P_2 \to P) \overline{S}(P_1),$$

or, in general,

$$F_n(P) = \overline{S}(P) + \sum_{m=1}^{n-1} \int dP_1 \ldots \int dP_m \overline{K}(P_1 \to P_2) \ldots \overline{K}(P_m \to P) \overline{S}(P_1).$$

If the series converges, the solution of Eq. (10.56) is $F(P) = \lim_{n \to \infty} F_n(P)$, i.e.,

$$F(P) = \overline{S}(P) + \sum_{m=1}^{\infty} \int dP_1 \ldots \int dP_m \overline{K}(P_1 \to P_2) \ldots \overline{K}(P_m \to P)\overline{S}(P_1).$$

(10.59)

Formal Evaluation of G

The value of G can now be formally obtained by substitution of the collision rate density given by Eq. (10.59) into Eq. (10.57). The result is

$$G = \int dP_1 \overline{S}(P) + \sum_{m=1}^{\infty} \int dP_1 \ldots \int dP_{m+1} \times$$
$$. \ \overline{S}(P_1)\overline{K}(P_1 \to P_2) \ldots \overline{K}(P_m \to P_{m+1})\overline{S}(P_1)g(P_{m+1}),$$

which can be written more succinctly as

$$G = \sum_{m=0}^{\infty} \int dP_1 \ldots \int dP_{m+1} \overline{S}(P_1)\overline{K}(P_1 \to P_2) \ldots \overline{K}(P_m \to P_{m+1})\overline{S}(P_1)g(P_{m+1}).$$

(10.60)

Evaluation of G by Monte Carlo

To evaluate Eq. (10.60) by Monte Carlo methods, the kernel $\overline{K}(P' \to P)$ is first decomposed into a product of a nonabsorption probability, a normalization factor to account for multiplying interactions such as fission or $(n, 2n)$ reactions, and a normalized kernel, namely,

$$\overline{K}(P' \to P) = [1 - \alpha(P')]\eta(P')\kappa(P' \to P).$$ (10.61)

Here $\alpha(P') = \mu_a(P')/\mu(P')$ is the absorption (capture) probability at P'[10] and $\eta(P')$ is the normalization factor

$$\eta(P') = \frac{\int dP \, \overline{K}(P' \to P)}{1 - \alpha(P')}.$$ (10.62)

The normalized collision kernel $\kappa(P' \to P)$ is

$$\kappa(P' \to P) = \frac{\overline{K}(P' \to P)}{\int dP'' \overline{K}(P' \to P'')},$$ (10.63)

---

[10] If the spatial coordinates of P' lie outside V, the particle has leaked from the system and $\alpha(P')$ is set to unity, the same as if the particle had been absorbed.

which has the property that $\int d\mathbf{P}\,\kappa(\mathbf{P}'\to\mathbf{P})=1$ so that $\kappa(\mathbf{P}'\to\mathbf{P})$ is a conditional PDF from which the next collision point $\mathbf{P}$ can be sampled given the previous nonabsorption collision was at $\mathbf{P}'$.

Now apply the factorization of Eq. (10.61) to Eq. (10.60) to obtain

$$G = \sum_{m=0}^{\infty} \int d\mathbf{P}_1 \int d\mathbf{P}_2 \dots \int d\mathbf{P}_{m+1} \overline{S}(\mathbf{P}_1)[1-\alpha(\mathbf{P}_0)]\kappa(\mathbf{P}_1\to\mathbf{P}_2) \times$$

$$[1-\alpha(\mathbf{P}_2)]\kappa(\mathbf{P}_2\to\mathbf{P}_3)\dots[1-\alpha(\mathbf{P}_m)]\kappa(\mathbf{P}_m\to\mathbf{P}_{m+1}) \times$$

$$\alpha(\mathbf{P}_{m+1})W(\mathbf{P}_0,\mathbf{P}_1,\dots,\mathbf{P}_{m+1}), \tag{10.64}$$

where W is defined as

$$W(\mathbf{P}_1,\mathbf{P}_2,\dots,\mathbf{P}_{m+1}) \equiv \frac{g(\mathbf{P}_{m+1})}{\alpha(\mathbf{P}_{m+1})} \prod_{j=1}^{m} \eta(\mathbf{P}_j). \tag{10.65}$$

Equation (10.64) suggests a Monte Carlo algorithm for evaluating G. Each particle history is created in the following manner. (1) The coordinates of the first collision, $\mathbf{P}_1$, are obtained by sampling from the first-flight collision density rate $\overline{S}(\mathbf{P}_1)$ (after renormalization of $\overline{S}(\mathbf{P}_0)$ to make it a proper PDF). (2) Then for $i = 1, 2, \dots, m+1$, sample $\alpha(\mathbf{P}_i)$ to see if the history ends in an absorption at the ith collision. (3) Sample $\kappa(\mathbf{P}_{i-1}\to\mathbf{P}_i)$ to find the next collision coordinates (assuming the particles survived the previous collision and the chain is to be continued). With this procedure, the quantity

$$\{\overline{S}(\mathbf{P}_1)[1-\alpha(\mathbf{P}_1)]\kappa(\mathbf{P}_1\to\mathbf{P}_2)\dots$$

$$\dots[1-\alpha(\mathbf{P}_m)]\kappa(\mathbf{P}_m\to\mathbf{P}_{m+1})\alpha(\mathbf{P}_{m+1})\}d\mathbf{P}_1,d\mathbf{P}_2,\dots,d\mathbf{P}_{m+1} \tag{10.66}$$

can be identified as the probability that the initial coordinates are in $d\mathbf{P}_1$ about $\mathbf{P}_1$, the second in $d\mathbf{P}_2$ about $\mathbf{P}_2$, and so on until the chain is terminated by an absorption at collision $m+1$.

The quantity $W(\mathbf{P}_1,\mathbf{P}_2,\dots,\mathbf{P}_{m+1})$ is the contribution of the history to the tally for G. Moreover, from the interpretation of the expression inside the brackets of Eq. (10.66) as the joint PDF for $(\mathbf{P}_1,\dots,\mathbf{P}_{m+1})$, it is seen that the right side of Eq. (10.64) is just the expected value of W, i.e., $\langle W\rangle = G$. Thus, sampling from the PDFs $\overline{S}$, α, and κ and scoring the resulting value of W defined by Eq. (10.65) yield an unbiased estimator for G. The Monte Carlo estimate of G is then obtained by sampling N histories, calculating W for each history, and averaging the scores, i.e.,

$$G \simeq \frac{1}{N}\sum_{i=1}^{N} W(\mathbf{P}_1^i,\dots,\mathbf{P}_{n_i}^i),$$

where n_i is the collision number when the ith particle was absorbed or escaped.

10.11.2 The Integral Equation Method as Simulation

It may not be readily apparent how the above sampling scheme, based on the integral transport equation, is related to the Monte Carlo simulation method, outlined in Section 10.1. The simulation approach had no need for a transport equation. However, the above scheme is really an equivalent approach for transport simulations. Here it is shown that the scheme based on the integral transport equation for estimating the functional G can be cast into a Monte Carlo simulation whose histories are constructed much more physically.

For simplicity, it is assumed here that the medium is nonmultiplying, i.e., the medium can only absorb or scatter particles. For such a medium $\eta(P) = 1$. A history begins by sampling the source function $S(\mathbf{r}, E, \mathbf{\Omega})$ (properly normalized) to select starting values $(\mathbf{r}_0, E_0, \mathbf{\Omega}_0)$. Then, the flight distance R to the first collision point $\mathbf{r}_1$ is found by sampling R from

$$\mu(\mathbf{r}_0, E_0) \exp\left\{ -\int_0^R \mu(\mathbf{r}_0 - R'\mathbf{\Omega}_0, E_0)\, dR' \right\}$$

(again properly normalized) to obtain the first collision point $\mathbf{r}_1 = \mathbf{r}_0 + R\mathbf{\Omega}_0$.

The next step is to determine if the first collision is an absorption so that the history should be terminated. This is done by sampling from the absorption probability $\alpha(\mathbf{r}_1, E_0) = \mu_a(\mathbf{r}_1, E_0)/\mu(\mathbf{r}_1, E_0)$. However, as discussed later in Section 10.12.4, it is a bad idea to terminate a history when an absorption event occurs. Rather, it is more efficient to continue the particle history after reducing the weight of the particle by multiplying its weight by the nonabsorption probability $[1 - \alpha(\mathbf{r}_0, E_0)] = \mu_s(\mathbf{r}_0, E_0)/\mu(\mathbf{r}_0, E_0)$. In this manner, every interaction is deemed to be a scatter.

After this (perhaps forced) scatter, a new particle energy and direction are determined from the appropriate scattering distributions. The new energy is obtained by sampling from the PDF

$$\frac{\mu_s(\mathbf{r}_1, E_0 \to E)}{\int dE\, \mu_s(\mathbf{r}_1, E_0 \to E)},$$

and the new direction is found by sampling from the PDF

$$\frac{\mu_s(\mathbf{r}_0, E_0 \to E_1, \mathbf{\Omega}_0 \cdot \mathbf{\Omega})}{\int_{4\pi} \mu_s(\mathbf{r}_0, E_0 \to E_1, \mathbf{\Omega}_0 \cdot \mathbf{\Omega})}.$$

The above sampling procedure is continued to successive collision events until the particle weight falls below some predetermined minimum value (at which point, Russian roulette—see Section 10.12.3—is used to, perhaps, end the history) or until the particle leaves the phase space of interest in the problem.

When the particle history is ended, its contribution is added to the tally for G, and a new particle history is begun.

10.12 Variance Reduction and Nonanalog Methods

To define the efficiency of a Monte Carlo calculation, one must take into account both speed and variance. The speed is, to a large extent, determined by the hardware. Of course, the skill of the computer programmer and the operating system of the computer also determine speed. The emphasis in this section, however, is the variance of the result of a calculation and ways to minimize it.

As discussed in Chapter 5, the power of Monte Carlo depends on using many nonanalog techniques to reduce the variance of the estimator. In this section, some of the more useful variance reduction techniques for particle transport are briefly reviewed.

10.12.1 Importance Sampling

The use of importance sampling can greatly increase the flexibility of scoring and sampling procedures. For any single step in the Monte Carlo process involving the probability of survival of a particle or of its reaching a certain point, the result of the sampling process, insofar as its effect on the final answer is concerned, may be taken as the product of the particle weight and the probability of occurrence of the event. Thus, if a PDF $f(x)$ describes some random process, one is allowed to use some alternative, perhaps simpler, PDF $\tilde{f}(x)$; however, the ultimate expectation of the process must be left unchanged, and therefore the particle's weight before entering the process must be adjusted by the factor $w(x)\tilde{f}(x)$, where the weight function $w(x) \equiv f(x)/\tilde{f}(x)$. The function $\tilde{f}(x)$ is somewhat arbitrarily chosen, although it should not deviate too markedly from $f(x)$ and should have the nature of a proper PDF. In an analog simulation, the use of an alternative PDF instead of $f(x)$ is compensated by the subsequent testing for rejection or acceptance. In the importance sampling technique, the value of x selected is accepted without further ado, but the particle weight after the process is obtained by multiplying its previous weight by $w(x)$, where the argument is the value of x selected.

The use of importance sampling, which is a very effective variance reduction technique for any Monte Carlo problem, can also be used with great effect for particle transport. Consider the problem of Section 10.11.1 in which

$$G = \int d\boldsymbol{P}\, g(\boldsymbol{P})F(\boldsymbol{P}) \tag{10.67}$$

is to be estimated. The collision density $F(\boldsymbol{P})$ is given by the integral transport equation of Eq. (9.76)

$$F(\boldsymbol{P}) = \int d\boldsymbol{P}'\overline{K}(\boldsymbol{P}' \to \boldsymbol{P})F(\boldsymbol{P}') + \overline{S}(\boldsymbol{P}). \tag{10.68}$$

The function $g(P)$ can be viewed as the contribution to G from a collision at P. In many problems, $g(P)$ is very small over large portions of phase space, and, to reduce the number of collisions in such regions, it is desirable to bias the samplings that generate the simulated particle histories so that more collisions occur in regions where $g(P)$ is large. Ideally, if one could find an importance function $I(P)$ that is proportional to the expected contribution to G from a particle at P, this function could be used to bias the density functions used to generate the sequence of points in phase space that define a particle history.

The importance function can be used to define the following two functions:

$$\widetilde{S}(P) \equiv \frac{\overline{S}(P)I(P)}{\int \overline{S}(P')I(P')\,dP'}, \tag{10.69}$$

$$\widetilde{K}(P' \to P) \equiv \frac{\overline{K}(P' \to P)I(P)}{I(P')}. \tag{10.70}$$

As in Section 10.11.1, these functions can be used to construct biased histories with the weight of each particle adjusted at each step to compensate for the bias. Thus a source particle has its weight modified by a multiplicative factor of $\overline{S}(P)/\widetilde{S}(P)$, and each particle that enters a collision from a previous collision has its weight multiplied by $\overline{K}(P' \to P)/\widetilde{K}(P' \to P)$.

Now what function to pick for $I(P)$? It can be shown (Carter and Cashwell, 1975) that the optimum importance function I_{opt} is the solution of the equation

$$I_{opt}(P) = \int dP'\overline{K}(P \to P')I_{opt}(P') + g(P). \tag{10.71}$$

This is just the integral adjoint transport equation of Eq. (9.88) with the adjoint source specified as $\overline{S}^{\dagger}(P) = g(P)$. With the optimum importance function, the Monte Carlo calculation gives a score of G for *every* history and, thus, is an estimator with zero variance! However, solving Eq. (10.71) is as much work as solving the original transport equation Eq. (10.68). Indeed, if $I_{opt}(P)$ were known, G could be estimated with simple numerical quadrature by using Eq. (9.92), namely,

$$G = \int g(P)F(P)\,dP = \int \overline{S}(P)I_{opt}(P)\,dP. \tag{10.72}$$

Although this optimum importance function is not known beforehand, its existence tempts one to try to develop various approximations for it without unduly increasing the computation time. These approximations are often based on deterministic methods such as discrete ordinate calculations using a simplified geometry (Tang et al., 1976). Alternatively, a Monte Carlo calculation of the adjoint flux can be performed using simplified physics models such as energy multigroup cross sections (MCNP, 2003).

10.12.2   Truncation Methods

During the course of its random walk, a particle may reach a location, direction, energy, or weight for which the likelihood of scoring is negligible. The energy may be so low or the position so remote that terminating the tracking of the particle is justified. Truncation is a particularly valuable tool in the tracking of charged particles, which have reasonably well-defined ranges, so that a truncation decision is straightforward. A special type of truncation usually associated with small particle weights is called *Russian roulette*. The Russian roulette method of truncation is addressed along with *splitting*, its companion procedure used for variance reduction.

10.12.3   Splitting and Russian Roulette

Suppose that by virtue of its position, direction, or some other feature, a particle is relatively likely to make an appreciable contribution to a score. The variance may be reduced by *splitting* the particle into two or more, say m, particles each having the same characteristics as the original particle, except each has a weight given by the fraction $1/m$ of the weight of the original particle. Each particle is then tracked independently. On the other hand, a particle may be relatively unlikely to make an appreciable contribution to a score. Then, a random number ρ may be selected, and if less than some fraction, say $1/m$, the particle may be *killed* by the Russian roulette scheme. Otherwise, tracking the particle would continue, but the particle's weight would be multiplied by the factor m.

Splitting and Russian roulette are normally used together. Different spatial or energy regions, for example, may be assigned *importances I*. As a particle moves from region j to region k, if $n \equiv I_k/I_j$ exceeds unity, the particle is split into n particles[11] with weights adjusted by the factor $1/n$. If n is less than unity, then Russian roulette may be played, with the particle surviving with probability $1 - n$ and its weight multiplied by the factor $1/n$. In a shielding problem with particles penetrating a thick-slab shield, Carter and Cashwell (1975) suggest that, for geometric splitting and Russian roulette, the slab be divided into layers of one mean free path thickness. As particles pass from region to region away from the entry face, they should be split 2 for 1. Russian roulette with 1/2 survival probability should be applied as particles pass from region to region toward the entrance face.

10.12.4   Implicit Absorption

An example of the usefulness of weighting is in the avoidance of the absorption process for a particle. Since particles absorbed in a medium give zero contribution to the final answer in many transport calculations, one intuitively has

---

[11] If n is not integer, the particle may be split into one of the bounding-integer number of particles and the weights adjusted as discussed in Section 5.8.

the feeling that a history terminating in this way is in a sense a waste of computer time. Thus, if a collision is deemed an absorption, the particle weight is reduced by multiplying its incident weight by the probability of survival, i.e., $[1 - (\mu_a/\mu)]$, and the type of collision is redetermined by sampling from all the nonabsorption interaction probabilities. This numerical avoidance of absorption is sometimes called *implicit absorption* or *survival biasing*, as distinct from *analog absorption* in which the particle history is actually terminated.

However, in such a scheme the only way for a history to terminate is for the particle to leak from the system, and this can result in the buildup of a large number of particles with very low weights and that contribute little to the tally. This implicit capture is almost always combined with Russian roulette (see Section 10.12.3) to artificially reduce the population of low-weight particles when a particle's weight falls below some specified value.

10.12.5 Interaction Forcing

Consider, for example, the passage of photons or neutrons through important regions, but regions with dimensions of only one or so mean free paths, so that a particle is likely to pass through the region without interaction. The effort wasted by tracking a particle through such a region can be eliminated by *interaction forcing*. Suppose that the distance through the region along the particle's path is S; collisions may be forced if path lengths are selected by sampling the following alternative PDF:

$$\tilde{f}(s) = \frac{\mu e^{-\mu s}}{1 - e^{-\mu S}}, \quad 0 \le s \le S. \tag{10.73}$$

The particle weight must then be adjusted by the factor $(1 - e^{-\mu S})$.

10.12.6 Exponential Transformation

For Monte Carlo calculations involving penetration or reflection from a slab shield, there is motivation to artificially increase or decrease, respectively, the distance between collisions. To do this, the normal PDF for a sampling path length, $f(s) = \mu \exp(-\mu s)$, is modified by replacing μ by $(1 - \beta\omega)\mu$, in which ω is the cosine of the angle between the preferred direction (toward the scoring region) and the direction of flight of the particle. The factor β is a biasing parameter, normally restricted to values less than unity. For deep penetration problems, this *exponential transformation* is often applied only to particles traveling in the generally favored direction (i.e., with $\omega > 0$). In other words, path lengths may be selected by sampling the following alternative PDF:

$$\tilde{f}(s) = (1 - p\omega)\mu e^{-(1-\beta\omega)\mu s}. \tag{10.74}$$

For $\omega > 0$, the effective interaction coefficient $(1 - p\omega)\mu$ is less than μ; thus, the particle path is longer than would be the case in a strictly analog calculation.

The corresponding weight factor is

$$w(s) = \frac{e^{-\beta\omega\mu s}}{1 - p\omega}.$$ (10.75)

By contrast, for $\omega < 0$, as may be the case for particles moving away from the scoring region, the effective interaction coefficient is greater than μ and the particle path is foreshortened. The parameter p must not be so near unity that excessively large weight factors result. In a particular application, it may be necessary to adjust p by trial and error to optimize calculations.

10.13 Summary

In a Monte Carlo neutral-particle transport simulation, the geometry of the system is first specified, typically by combinatorial geometry, although other approaches can be used. Then by sampling from many distributions, a complete simulation of a particle's track as it migrates through phase space can be obtained. Source sampling is required to pick the initial starting location, direction, and energy of a particle. Then sampling is used to pick a flight distance before a collision, followed by sampling to determine the type of collision. If the particle is not absorbed, more sampling is done to determine the type, direction, and energy of secondary particles. Each subsequent leg of the simulated track continues as for the first leg. The particle is tracked until it is absorbed or leaves the problem boundary. As a particle moves along its trajectory, various tallies are updated so that, after many histories, some desired properties of the radiation field can be estimated.

To reduce computational effort, transport simulations often change the physical sampling distributions to bias a particle's track to increase the chance of it scoring. To remove the bias from the score, a weight is assigned to each particle, and as the particle track evolves, the weight is adjusted to remove the bias each time a biased sampling is used. Such biasing can greatly reduce the variance of the tally and, consequently, most Monte Carlo simulations employ one or more variance reduction techniques.

With the advent of inexpensive and powerful personal computers or computer clusters and with the availability of powerful general-purpose Monte Carlo codes, Monte Carlo has become the method of choice for the vast majority of neutron and photon transport analyses.

Problems

10.1 Verify the functions $f_1(r)$ and $f_2(r)$ defined by Eq. (10.36) are ≤ 1 for $1 \leq r \leq 1 + 2/\lambda$.

10.2 Evaluate the constant K required to make Eq. (10.32) a properly normalized PDF. Then derive an expression for the efficiency of the sampling method described in Section 4.1.8. Plot the efficiency as a function of the incident photon energy.

10.3 In Monte Carlo particle transport, the distance s_{int} a particle must travel from a point $\mathbf{r}_o$ in direction $\boldsymbol{\Omega}$ to reach the nearest surface (in the direction of $\boldsymbol{\Omega}$) is of paramount importance if a particle is to be tracked as it moves through the problem geometry. Once s_{int} is known, the intersection point on the surface is readily found as $\mathbf{r}_{int} = \mathbf{r}_o + s_{int}\boldsymbol{\Omega}$. Derive formulas for s_{int} if $\mathbf{r}_o = (x_o, y_o, z_o)$ is inside (a) an infinite slab of thickness $2T$ centered at the origin, perpendicular to the y-axis, and parallel to the z-axis, (b) an infinite cylinder of radius R whose axis is the z-axis, and (c) a sphere of radius R centered on the origin.

10.4 Perform a Monte Carlo analysis, similar to that of Example 10.1, to determine the escape probability of a particle born uniformly in a purely absorbing sphere of radius R. Compare your results to the analytical result (Case et al., 1953)

$$P_{esc} = \frac{3}{8(\mu R)^3}\left[2(\mu R)^2 - 1 + (1 + 2\mu R)e^{-2\mu R}\right].$$

10.5 Perform a Monte Carlo analysis, similar to that of Example 10.1, to determine the escape probability of a particle born uniformly in a purely absorbing infinite cylinder of radius R. Compare your results to the analytical result (Case et al., 1953)

$$P_{esc} = \frac{2\mu R}{3}\,\{2[\mu R\{K_1(\mu R)I_1(\mu R) + K_0(\mu R)I_0(\mu R)\} - 1]$$
$$+ K_1(\mu R)I_1(\mu R)/(\mu R) - K_0(\mu R)I_1(\mu R) + K_1(\mu R)I_0(\mu R)\},$$

where K_n and I_n are the nth-order modified Bessel functions of the first and second kind, respectively.

10.6 Verify Eqs. (10.19) to (10.21) by evaluating Eq. (10.17).

10.7 Derive the rotation matrix $\mathbf{R}_{xy}(\theta, \psi)$ for first a rotation about the x-axis followed by a rotation about the z-axis. Verify that the inverse of this matrix transforms the vector $\boldsymbol{\Omega}(\theta, \psi)$ in the vector $(0, 1, 0)$.

10.8 Consider a purely absorbing slab of thickness z mean free path lengths that is uniformly and normally illuminated by particles. The average flux density in the slab can be estimated analytically by (a) the path length estimator and (b) the reaction density estimator. Show that the relative error (ratio of the standard deviation to the mean) of these two estimators is as follows:

$$\text{collision estimator} = \sqrt{\frac{e^{-z}}{(1 - e^{-z})}},$$

$$\text{path length estimator} = \frac{\sqrt{1 - 2ze^{-z} - e^{-2z}}}{(1 - e^{-z})}.$$

(a) Plot these results as a function of z. (b) Also determine the limiting values as $z \to 0$ and as $z \to \infty$. (c) Discuss which estimator is best for optically thick and thick media and explain why these results are expected.

10.9 Write a code to produce the results of Example 10.2. Add to Fig. 10.15 results for $c = 0.5$ and $c = 0.9$.

10.10 Write a code to produce escape probabilities similar to those of Example 10.2 but for a sphere of radius R. (a) Verify your code by comparing the output for a very small c to the analytic expression given in Problem 10.4 for $c = 0$. (b) Besides using an analog tally $P_{esc} = N_{esc}/N_{total}$, i.e., the number of escaping particles to the total number of histories, add to your code a leakage estimator tally. For both tallies, the relative error of the estimated escape probabilities should be given.

10.11 This problem requires development of a computer code for calculation of transmission and reflection of photons normally incident (parallel to the x-axis) on a homogeneous shielding slab surrounded by a vacuum. It is a generalization of the problem of Example 10.3. For simplicity, assume one-speed particles and isotropic scattering.

(a) Calculate the reflected and transmitted angular flux densities. In particular, for $c = 0.9$, plot the angular distribution of the escaping angular flux density $\phi(0, \omega)$, $-1 < \omega < 0$, for $c = 0.9$ for different slab thicknesses.

(b) Estimate the reflected and transmitted currents $j_x(0, \omega)$ and $j_x(T, \omega)$. Plot the total reflected and transmitted leakage currents $j_x^-(0)$ and $j_x^+(T)$ (see Eq. (9.20)) as a function of slab thickness for various values of $c = \mu_s/\mu$.

References

Blomquist, R.N., Gelbard, E.M., 1983. An assessment of existing Klein-Nishina Monte Carlo sampling methods. Nucl. Sci. Eng. 83, 380–384.

Carter, L.L., Cashwell, E.D., 1975. Particle-Transport Simulation with the Monte Carlo Method. TID-26607. National Technical Information Service, U.S. Department of Commerce, Springfield, VA.

Case, K.M., de Hoffman, F., Placzek, G., 1953. Introduction to the Theory of Neutron Diffusion, vol. I. Los Alamos Scientific Laboratory, NM.

Chandrasekhar, S., 1960. Radiative Transfer. Dover, New York.

Kahn, H., 1954. Applications of Monte Carlo. Report AECU-3259. The Rand Corp., Santa Monica, CA.

Kalos, M.H., Nakache, F.R., Celnik, J., 1968. Monte Carlo methods in reactor computations. In: Greenspan, H., Kelber, C.N., Okrent, D. (Eds.), Computing Methods in Reactor Physics. Gordon and Breach, New York.

Lux, I., Koblinger, L., 1991. Monte Carlo Particle Transport Methods: Neutron and Photon Calculations. CRC Press, Boca Raton, FL.

MCNP, 2003. A General Monte Carlo N-Particle Transport Code, Version 5, vol. I: Overview and Theory. LA-UR-03-1987. Los Alamos National Lab.

Olmsted, J.M.H., 1947. Solid Analytic Geometry. Appelton-Centruy-Crafts Inc., New York.

Shultis, J.K., Faw, R.E., 2000. Radiation Shielding. American Nuclear Society, LaGrange Park, IL.
Shultis, J.K., Faw, R.E., 2007. Fundamentals of Nuclear Science and Engineering, 2nd ed. CRC Press, Boca Raton, FL.
Tang, J.S., Stevens, P.N., Hoffman, T.J., 1976. Methods of Monte Carlo Biasing Using Two-Dimensional Discrete Ordinates Adjoint Flux. ORNL-TM-5454. Oak Ridge National Laboratory, TN.

Appendix A

Some Common Probability Distributions

In this appendix, several frequently encountered probability distributions are presented. For each distribution, explicit expressions are given for both the probability density function (PDF) and the cumulative distribution function (CDF), and the mean value μ and variance σ^2 are given or derived. More detail on these and on other distributions can be obtained from references such as Kendall and Stuart (1977), Snedecor and Cochran (1971), and the truly definitive set of books by Johnson and Kotz (1987) and Johnson et al., (1992, 1994, 1995, 1997, 2004).

Also presented here are some methods for generating random values from these distributions; however, the emphasis is on simplicity and, consequently, the suggested sampling schemes may not be the fastest or most efficient. Indeed, the literature is filled with literally hundreds of research papers on "better" random variate generators, many duplicative because of the widespread use of Monte Carlo methods in very different disciplines. Also many are of incremental improvement, a result of a minor tweak in some line of computer code. The reader interested in random number generators faster (and more complex) than those given here are referred to several excellent works on the subject, notably Fishman (2003), Gentle (1998), Devroye (1986), McGrath and Irving (1975), and Rubinstein (1981).

A.1 Discrete Distributions

Seven common univariate discrete distributions are first presented: the Bernoulli, the binomial, the geometric, the negative binomial, the hypergeometric, the negative hypergeometric, and the Poisson. More thorough coverage of these and many other discrete distributions is provided by Johnson et al. (1992). For many discrete distributions, the inverse CDF method is a simple, although not always the most efficient, method for sampling from these distributions. In this method (see Section 4.1.2), one needs the first density function $f(0)$ in the sequence $f(x), x = 0, 1, 2, \ldots$, and the ratio between successive values, i.e., $R(x) = f(x+1)/f(x)$. In Table A.1, the results of the inverse CDF method are summarized for all but the simplest (Bernoulli) discrete distributions considered in this appendix.

Exploring Monte Carlo Methods. https://doi.org/10.1016/B978-0-12-819739-4.00019-6

Table A.1 Summary of properties of six discrete distributions whose mass points are $x = 0, 1, 2, \ldots$. The quantities $f(0)$ and $R(x)$ are used to sample the PDFs using the inverse CDF method described in Section 4.1.2.

Distribution in $x=0,1,\ldots,x_{\max}$	PDF $f(x)$	$f(0)$	$R(x) = \dfrac{f(x+1)}{f(x)}$	mean μ	variance σ^2
binomial: $\mathcal{B}(x\|n,p)$ $x_{\max}=n;\ 0<p<1$	$\dbinom{n}{x} p^x (1-p)^{n-x}$	$(1-p)^n$	$\dfrac{(n-x)p}{(x+1)(1-p)}$	np	$np(1-p)$
geometric: $\mathcal{G}_e(x\|p)$ $x_{\max}=\infty;\ 0<p<1$	$p(1-p)^x$	p	$1-p$	$\dfrac{1}{p}$	$\dfrac{(1-p)}{p^2}$
negative binomial: $\mathcal{NB}(x\|s,p)$ $x_{\max}=\infty;\ 0<p<1$	$\dbinom{x+s-1}{x} p^x (1-p)^{n-x}$	p^s	$\dfrac{s+x}{x+1}(1-p)$	$\dfrac{sp}{1-p}$	$\dfrac{sp}{(1-p)^2}$
hypergeometric: $\mathcal{H}(x\|N,n,K)$. Let $\beta \equiv N-K$. $\max(0,n-\beta) \le x \le \min(K,n)$ $N,n=0,1,\ldots;\ K=0,1,\ldots,N$.	$\dbinom{K}{x}\dbinom{\beta}{n-x} \Big/ \dbinom{N}{n}$	$\dfrac{(\beta)!(N-n)!}{N!(\beta-n)!}$	$\dfrac{(K-x)(n-x)}{(x+1)(\beta-n+x+1)}$	$\dfrac{nK}{N}$	$\mu\,\dfrac{\beta}{N}\dfrac{N-n}{N-1}$
negative hypergeometric: $\mathcal{NH}(x\|N,r,K)$. Let $\beta \equiv N-K$. $N=0,1,\ldots;\ K=0,1,\ldots,N$ $r=0,1,2,\ldots,N-K$.	$\dbinom{x+r-1}{x}\dbinom{N-r-x}{K-x} \Big/ \dbinom{N}{K}$	$\dfrac{(N-r)!}{N!(\beta-r)!}$	$\dfrac{(x+r)(K-x)}{(x+1)(\beta-r)}$	$\dfrac{rK}{\beta+1}$	$\dfrac{\mu(\beta-r+1)}{(\beta+1)(\beta-2)}$
Poisson: $\mathcal{P}(x\|\mu)$ $x_{\max}=\infty;\ \mu>0$	$\dfrac{\mu^x}{x!}e^{-\mu}$	$e^{-\mu}$	$\dfrac{\mu}{x+1}$	μ	μ

A.1.1 Bernoulli Distribution

The Bernoulli (Fig. A.1) distribution applies to a variable that can assume either of two mutually exclusive values. The value $x = 1$ is assigned to one outcome, which is called a "success," and the value $x = 0$ is given to the other outcome, which is called a "failure." Let $0 < p < 1$ be the probability of success and $q = 1 - p$ the probability of failure. An obvious example is flipping a (fair) coin, for which $p = q = 0.5$ and success can be assigned to either "heads" or "tails." Another example might be a baseball player getting a hit, given that his or her average is 300 (or $p = 0.30$). In any given at-bat, a hit is success, $x = 1$, and an out is a failure, $x = 0$ (a walk does not count).

Figure A.1 Jacob Bernoulli (1655–1705).

The Bernoulli PDF $f(x)$ is defined only for $x = 0$ and $x = 1$ and is given by

$$
\begin{aligned}
f_1 &= 1 - p, & \text{associated with } x = 0, \\
f_2 &= p & \text{associated with } x = 1.
\end{aligned}
\tag{A.1}
$$

Another way to express this is

$$
f(x) = (1 - p)\delta(x) + p\delta(x - 1).
\tag{A.2}
$$

The CDF is simply

$$
F(x) = \begin{cases}
0, & \text{if } x < 0, \\
1 - p, & \text{if } 0 \leq x < 1, \\
1, & \text{if } x \geq 1.
\end{cases}
\tag{A.3}
$$

The mean value is

$$
\langle x \rangle \equiv \mu \equiv \sum_{i=1}^{2} x_i f_i = 0(1 - p) + 1(p) = p,
\tag{A.4}
$$

and the variance is

$$
\sigma^2(x) \equiv \sum_{i=1}^{2} (x_i - \mu)^2 f_i = p^2(1 - p) + (1 - p)^2 p = p(1 - p).
\tag{A.5}
$$

The PDF and CDF for a Bernoulli distribution are shown in Fig. A.2.

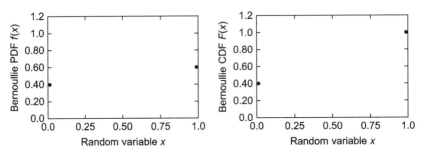

Figure A.2 The PDF and the CDF for the Bernoulli distribution with parameter $p = 0.6$. There are only two outcomes, $x = 0$ (with probability $1 - p$) and $x = 1$ (with probability p).

The Bernoulli distribution is important because it gives rise to the concept of "Bernoulli trials," in which independent experiments are run, each presuming a constant probability p of a successful outcome. The baseball example is interesting because the value of p is not known until after the season (or even career). However, one would expect a baseball hitter with a cumulative average of 243 to get a hit about once every four times at-bat.[1] By contrast, a 330-hitter gets a hit about once every three times at-bat. Of course, the result of any given at-bat cannot be predicted with certainty.

A.1.2 Binomial Distribution

Another basic discrete probability distribution is the binomial distribution. This PDF gives the probability of obtaining x successes in n *independent* random trials or draws of "objects," a fraction p of which have a specified "success" feature. Most important, p is a constant, unaffected by the outcome of any draw. For a finite number of objects, the constancy of p requires the drawn object, after recording if it has the desired "success" property, to be returned to the pool from which objects are drawn.[2]

Consider an experiment in which n independent trials are made and for which each trial has a probability p of success (such as rolling a "3" with a pair of dice and, to ensure subsequent rolls have the same probability, the "1" and "2" spots are left on the dice). Then the PDF that x successes are obtained in n trials is given by the binomial distribution

$$f(x) \equiv f(x|n, p) = \frac{n!}{x!(n-x)!} p^x (1-p)^{n-x}, \qquad x = 0, 1, 2, \ldots, n, \quad \text{(A.6)}$$

---

[1] For readers unfamiliar with baseball, the *batting average* sports statistic means this batter, from past performance, has a probability of 0.234 of hitting the ball and safely reaching a base.

[2] The case of x successes in n trials *without* replacement is described by the hypergeometric distribution, discussed later in Section A.1.5.

which is denoted by $\mathcal{B}(n, p)$. The fraction in this distribution is called the *binomial coefficient* and is often given the short-hand notation

$$\binom{n}{x} \equiv \frac{n!}{x!(n-x)!}. \tag{A.7}$$

Note that the binomial distribution is not defined for noninteger or negative values of x.

The binomial distribution is the governing distribution whenever the following experimental conditions are fulfilled:

1. The experiment consists of a fixed number of n trials, where n is specified before the experiment begins.
2. Each trial is identical in condition, and each trial can result in one of two possible outcomes. Generally, the outcome is denoted as a success S or a failure F.
3. The trials are independent, and the outcome of each trial does not affect the outcomes of other trials.
4. The probability of a success p is the same from trial to trial.

The binomial CDF is given by

$$F(x) = \sum_{m=0}^{x} \frac{n!}{m!(n-x)!} p^m (1-p)^{n-m}$$

$$= (n-x)\binom{n}{k} \int_0^{1-p} t^{n-k-1}(1-t)^k \, dt = I_{1-p}(n-k, k+1), \tag{A.8}$$

where I_{1-p} is the *regularized incomplete beta function*.

Mean and Variance

The mean or average value is defined as

$$\langle x \rangle = \mu = \sum_{x=0}^{n} x f(x|n, p) = \sum_{x=0}^{n} \frac{xn!}{x!(n-x)!} p^x (1-p)^{n-x}. \tag{A.9}$$

Let $y = x - 1$ and $m = n - 1$; then $x = y + 1$, $n = m + 1$, and it is easy to show that $n - x = m - y$. Thus,

$$\mu = \sum_{y=0}^{m} \frac{(y+1)(m+1)!}{(y+1)!(m-y)!} p^{y+1}(1-p)^{m-y} = \sum_{y=0}^{m} \frac{(m+1)!}{y!(m-y)!} pp^y (1-p)^{m-y}$$

$$= (m+1)p \sum_{y=0}^{m} \frac{m!}{y!(m-y)!} p^y (1-p)^{m-y} = np \sum_{y=0}^{m} f(y|m, p). \tag{A.10}$$

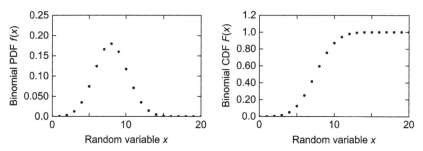

Figure A.3 The PDF and the CDF for the binomial distribution with parameters $n = 20$ and $p = 0.4$.

Note that the summations on y begin at $y = 0$ because the binomial PDF is not defined for negative values of the argument. Because the summation over all outcomes of a discrete PDF is unity, the mean is given simply as

$$\mu = np. \tag{A.11}$$

In a similar manner, the variance of the binomial distribution is found to be

$$\sigma^2 = np(1 - p) = npq. \tag{A.12}$$

An example of the binomial distribution is shown in Fig. A.3.

Radioactive Decay and the Binomial Distribution

Radionuclides are governed by binomial statistics because a radionuclide either decays in a time interval t or it does not. Consider a radioactive sample initially containing N_o identical radionuclides with a decay constant of λ. The probability a radionuclide does not decay in time t is $(1 - p) = e^{-\lambda t}$, so the probability it does decay in time t is $p = (1 - e^{-\lambda t})$.[3] From Eq. (A.6), the probability $f(x)$ that x atoms decay in time T is

$$f(x) = \frac{N_o!}{(N_o - x)!x!}(1 - e^{-\lambda T})^x (e^{-\lambda T})^{N_o - x}, \qquad x = 0, 1, 2, \ldots, N_o. \tag{A.13}$$

From Eq. (A.11), the average number decaying in time T is

$$\mu(x) = N_o(1 - e^{-\lambda T}) \tag{A.14}$$

and the variance, from Eq. (A.12), is

$$\sigma^2(x) = N_o(1 - e^{-\lambda T})e^{-\lambda T} = \mu(x)e^{-\lambda T}. \tag{A.15}$$

---

[3] The expected number $\langle N(t) \rangle$ of radionuclides at time t is given by the relation that the sample decay rate equals $-\lambda \langle N(t) \rangle$, or $d \langle N(t) \rangle / dt = -\lambda \langle N(t) \rangle$, whose solution is $\langle N(t) \rangle = \langle N(0) \rangle e^{-\lambda t} = N_o e^{-\lambda t}$. Thus, the probability one radionuclide decays in time t is $p = \langle N(t) \rangle / N_o = e^{-\lambda t}$.

If $\lambda T \ll 1$, i.e., the measurement time is very much less than the half-life of the radionuclide, $\sigma(x) = \sqrt{\mu(x)}$.

Let c be the probability that a decay produces a count in a radiation detection system. Then the probability a count is observed in time T is $p_c = c(1 - e^{-\lambda T}) \simeq c\lambda T$ if $\lambda T \ll 1$. Thus, the probability of obtaining x counts in time T is

$$f(x) \simeq \frac{N_o!}{(N_o - x)!x!}(c\lambda T)^x (1 - c\lambda T)^{N_o-x}, \qquad x = 0, 1, 2, \ldots, N_o. \quad \text{(A.16)}$$

Although this approximation for $f(x)$ and the exact result of Eq. (A.13) describe the statistics of radioactive decay, these descriptions have little utility. Because N_o is typically very large (10^5–10^{15}), the terms involving N_o typically are incalculably huge or ridiculously small. Another difficulty with using the binomial distribution to describe radioactive decays occurs if a radioactive sample has multiple radioactive species. In this case, the probability of obtaining x decays in time T is no longer described by a binomial distribution. However, both of these problems are avoided by using the Poisson distribution discussed later in Section A.1.7.

On a more philosophical level, the use of binomial statistics to describe radioactive decay is also open to additional criticism. Inherent in the binomial distribution is the requirement that the population N_o remain constant trial after trial, i.e., the loss of an atom through radioactive decay should, in theory, be replaced. This occurs only if the production rate (say by neutron absorption that produces new radionuclides) exactly equals the decay rate of the radionuclides. In most radionuclide samples, N_o is so large (10^{10}–10^{15}) that the loss of a few hundred or thousand radionuclides leaves N_o essentially unchanged. But in the case in which nuclear accelerators are used to produce only a few (say five) exotic short-lived radionuclides, the hypergeometric distribution may be more appropriate.

Sampling From a Binomial Distribution

The simplest way to generate a random sample from a binomial distribution is to conduct n independent Bernoulli trials, each with probability of success p, and let x be the number of successes. By definition, the x is a random sample from the binomial distribution. This method works well provided n is not very large, since the time to generate one sample is clearly proportional to n.

Alternatively, one could use the inverse CDF method (see Section 4.1.2). For the binomial $f(0) = (1 - p)^n$ and the ratio of successive values of the PDF is

$$R(x) \equiv \frac{f(x + 1)}{f(x)} = \frac{n - x}{x + 1}\frac{p}{1 - p}. \quad \text{(A.17)}$$

Because the inverse CDF method is essentially a table search, for large n, starting the search at the mode (maximum) of the binomial reduces the time by more than half compared to a sequential search starting from $x = 0$ (Kemp, 1986).

For values of $n(1 - p) \gtrsim 500$, other methods are available that are not so dependent on n (Stadlober, 1991; Kachitvichyanukul and Schmeiser, 1988).

A.1.3 Geometric Distribution

Consider a sequence of Bernoulli trials in which p is the probability of success. Then the probability of obtaining x failures before getting the first success on trial $x + 1$ is

$$f(x) = p(1 - p)^x, \qquad 0 < p < 1, \quad x = 0, 1, 2, \ldots. \tag{A.18}$$

This is known as the *geometric distribution* and is denoted by $\mathcal{G}_e(p)$. The corresponding CDF is

$$F(x) = p \sum_{y=0}^{x} p(1 - p)^x = 1 - (1 - p)^x, \tag{A.19}$$

and the mean and variance are $\mu = 1/p$ and $\sigma^2 = (1 - p)/p^2$, respectively.

Sampling From a Geometric Distribution

To generate samples from $\mathcal{G}_e(p)$, one of the best ways is to use the property that if x is a random variate from the exponential distribution $\mathcal{E}(\lambda)$ (see Section A.2.2) with $\lambda \equiv -1/\ln p$, then x is distributed as $\mathcal{G}_e(p)$. Thus, generate ρ_i from $\mathcal{U}(0, 1)$ and use as the exponential variate[4]

$$x_i = \left\lfloor \frac{\ln \rho_i}{\ln p} \right\rfloor. \tag{A.20}$$

Alternatively, one could use the inverse CDF method (see Section 4.1.2) in which the starting density value is $f(0) = p$ and the ratio of successive values is $R(x) = f(x + 1)/f(x) = (1 - p)$. Although this is simple to implement, it can be very inefficient if p is very small.

A.1.4 Negative Binomial Distribution

Closely related to the geometric distribution is the *negative binomial distribution*, which gives the probability of x failures prior to obtaining s successes in a sequence of Bernoulli trials. Suppose that p is the probability of success. Then the probability of obtaining x failures before obtaining the sth success on trial $x + s$ is

$$f(x) \equiv f(x|s, p)) = \frac{(s + x - 1)!}{x!(s - 1)!} p^s (1 - p)^x, \quad x = 0, 1, 2, \ldots. \tag{A.21}$$

---

[4] Here the *floor function* $\lfloor x \rfloor$ is used to indicate the largest integer $\leq x$. Thus, $\lfloor 5.2 \rfloor = 5$. By contrast, the *ceiling function* $\lceil x \rceil$ gives the smallest integer $\geq x$.

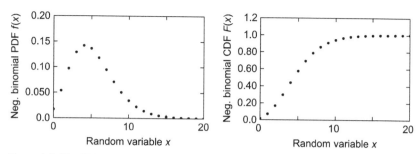

Figure A.4 The PDF and CDF for the negative binomial distribution with parameters $p = 0.6$ and $s = 8$.

This distribution is denoted by $\mathcal{NB}(p)$. This PDF can be written in an alternate form from which it gets its name. The binomial coefficient in Eq. (A.21) can be written as follows:

$$\binom{s+x-1}{x} = \frac{(s+x-1)!}{(s-1)!x!} = \frac{(s+x-1)(s+x-2)(s+x-3)\ldots(s)}{x!}$$

$$= \frac{(-1)^x}{x!}[(-s)(-s-1)(-s-3)\ldots(-s-x-1)]$$

$$= (-1)^x \binom{-s}{x}. \tag{A.22}$$

Hence Eq. (A.21) can be written in terms of *negative* binomial coefficients (whence comes the name of the PDF) as

$$f(x|s, p) = (-1)^x \binom{-s}{x} p^s (1-p)^x. \tag{A.23}$$

The corresponding CDF is

$$F(x) = \sum_{y=0}^{x} \frac{s+y-1}{y!(s-1)!} p^s (1-p)^y, \quad x = 0, 1, 2, \ldots$$

$$= 1 - I_p(x+1, s) = I_{1-p}(s, x+1), \tag{A.24}$$

where $I_p(a, b)$ is the *regularized incomplete beta function* defined as

$$I_p(a, b) = \frac{\Gamma(a+b)}{\Gamma(a)\Gamma(b)} \int_0^P t^{a-1}(1-t)^{b-1} dt. \tag{A.25}$$

To show this CDF has the correct asymptotic behavior, i.e., $F(\infty) = 1$, begin with the *binomial series* which can be written as

$$(1+z)^\alpha = \sum_{x=0}^{\infty} \binom{\alpha}{x} z^x. \tag{A.26}$$

This series converges for $|z| < 1$, when α is any complex number, and x is a nonnegative integer. Let $z = p - 1$ and $\alpha = -s$, so Eq. (A.26) gives

$$p^{-s} = \sum_{x=0}^{\infty} \binom{-s}{x} (p-1)^x \tag{A.27}$$

or division by p^{-s} and use of Eq. (A.23) produce

$$1 = \sum_{x=0}^{\infty} (-1)^x \binom{-s}{x} (1-p)^x p^s = \sum x = 0^{\infty} f(x|s, p) = F(\infty). \tag{A.28}$$

An example of the negative binomial distribution is shown in Fig. A.4.

Some Properties

The negative binomial PDF has the following properties:

mean: $\mu = \dfrac{sp}{(1-p)}$,

mode: $\begin{cases} \left\lfloor \dfrac{p(s-1)}{1-p} \right\rfloor, & s > 1, \\ 0, & s \leq 1, \end{cases}$

variance: $\sigma^2 = \dfrac{sp}{(1-p)^2}$.

Other Forms and Related Distributions

Note that when $s = 1$, the negative binomial is just the geometric distribution. However, the parameter s in Eq. (A.21) need not be a positive integer, and, in general, can be any real positive number. When the draw stopping parameter s is an integer, then Eq. (A.21) is also known as the *Pascal distribution*. When s, however, is a real continuous number, the negative binomial PDF is often called the Pólya distribution.

The negative binomial distribution is closely related to the Poisson distribution, discussed later in Section A.1.7. First consider the asymptotic behavior as the stopping parameter $s \to \infty$. Suppose at the same time $p \to 0$ in such a way that mean $sp/(1-p)$ remains constant at a value λ or $p = \lambda/(s+\lambda)$. With this parameterization, the negative binomial PDF can be written as

$$F(x|s, p) = \frac{\Gamma(s+x)}{x!\,\Gamma(s)} p^x (1-p)^s = \frac{\lambda^x}{x!} \cdot \frac{\Gamma(s+x)}{\Gamma(s)(s+\lambda)^x} \cdot \frac{1}{\left(1+\frac{\lambda}{s}\right)^s}. \tag{A.29}$$

Now as $s \to \infty$, the middle factor converges to unity, so that

$$\lim_{s \to \infty} f(x|s, p) = \frac{\lambda^x}{x!} \cdot 1 \cdot \frac{1}{e^\lambda} = \frac{\lambda^x}{x!} e^{-\lambda}, \tag{A.30}$$

which is the PDF for the Poisson distribution.

Hence, the reparameterized negative binomial PDF converges for large s to a Poisson PDF. Indeed, the negative binomial is a robust alternative to the Poisson PDF which approaches the Poisson distribution for large s but has a larger variance for small s.

Gamma-Poisson Mixture

The negative binomial PDF also appears in a model with a continuous mixture of Poisson distributions whose variation of the Poisson rates λ is treated as a random variable and is described by a gamma distribution (see Section A.2.3) in which the shape parameter is λ and the rate parameter $\beta = (1 - p)/p$. Formally, the PDF of such a gamma-Poisson mixture model is

$$f(x|s, p) = \int_0^\infty f_{\mathcal{G}(s,(1-p)/p)}(\lambda) \cdot f_{\mathcal{P}(\lambda)}(s) \, d\lambda$$

$$= \int_0^\infty \frac{\lambda^{s-1} \exp[-\lambda(1 - p)/p]}{[p/(1 - p)]^s \, \Gamma(s)} \frac{\lambda^x}{x!} e^{-\lambda} d\lambda$$

$$= \frac{(1 - p)^s p^{-s}}{x! \Gamma(s)} \int_0^\infty \lambda^{s+x-1} e^{-\lambda/p} d\lambda$$

$$= \frac{(1 - p)^s p^{-s}}{x! \Gamma(s)} p^{s+x} \Gamma(s + x) = \frac{\Gamma(s + x)}{x! \Gamma(s)} p^x (1p)^s, \qquad \text{(A.31)}$$

which is a negative binomial distribution.

Sampling From a Negative Binomial

The inverse CDF method of Section 4.1.2 with $f(0) = p^s$ and the ratio

$$R(x) \equiv f(x + 1)/f(x) = (s + x)(1 - p)/(x + 1).$$

However, this approach becomes very inefficient as p becomes very small. Alternatively, if generators for the Poisson distribution $\mathcal{P}(\mu)$ (see next section) and for the gamma distribution $\mathcal{G}(\alpha, \beta)$ (see Section A.2.3) are available proceed as follows. First generate x_i from $\mathcal{G}(s, 1)$ and then from $\mathcal{P}(x_i(1 - p)/p)$ generate y_i, which is distributed according to the negative binomial $\mathcal{NB}(p)$.

A.1.5 Hypergeometric Distribution

This discrete PDF gives the probability of obtaining x successes (random draws for which the drawn object has some specified characteristic deemed "success") in n draws, *without* replacement, from a pool initially containing N objects of which $K \leq N$ objects have the success characteristic. This is the same problem addressed by the binomial distribution, except in the latter case, the drawn object is placed back into the pool after each draw.

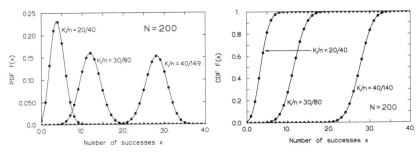

Figure A.5 The PDF (left) and the CDF (right) for the hypergeometric distribution with parameters for sampling from a finite population with replacement. Parameters are population size $N = 200$, number K of success states in the population, and the number of draws n. The smooth lines connecting the discrete distribution points are for visual convenience of identifying the various cases.

For example, one may want to survey a sample of n sophomores at a particular university with N sophomores. In particular, opinions of the K sophomores with GPAs over 3.5 (the "success" characteristic) are sought. Further, each sophomore is to be surveyed only once (the "without replacement" condition). The probability that x "successful" students are included in the survey is given by the hypergeometric PDF. This PDF, denoted by $\mathcal{H}(N, n, K)$ or $\mathcal{H}(x|N, n, K)$, is

$$f(x) = \binom{K}{x}\binom{N-K}{n-x} \bigg/ \binom{N}{n}, \quad \max(0, K+n-N) \le x \le \min(K, n).$$
$$(A.32)$$

The parameters are $N \in \{0, 1, 2, \ldots\}$, $K \in \{0, 1, 2, \ldots, N\}$, and $n \in \{0, 1, 2, \ldots, N\}$. The hypergeometric PDF has the following properties:

mean: $\mu = nK/N$,

mode: $\left\lceil \dfrac{(n+1)(K+1)}{N+2} \right\rceil - 1, \left\lfloor \dfrac{(n+1)(K+1)}{N+2} \right\rfloor$,

variance: $\sigma^2 = n\dfrac{K}{N}\dfrac{(N-K)}{N}\dfrac{N-n}{N-1}$.

The cumulative distribution

$$F(x) = \sum_{m=0}^{x} \binom{K}{m}\binom{N-X}{n-m} \bigg/ \binom{N}{n} \qquad (A.33)$$

can be written, in closed form, in terms of the *generalized hypergeometric function* $_pF_q$ (Wikipedia, 2020a) but its numerical evaluation offers no advantage over performing numerically the summation in Eq. (A.33). Examples of the PDF and CDF hypergeometric distribution are shown in Fig. A.5.

Related Distributions

Distributions related to the hypergeometric distribution include the following:

1. If $n = 1$, then x has a Bernoulli distribution with $p = K/N$.
2. Let y have a binomial distribution with parameters n and $p = K/N$ and consider a similar problem addressed by the hypergeometric PDF, but with *replacement* after each trial. If N and K are large compared to n and if $p = K/N$ is not close to 0 or 1, then x and y are similar in that $\mathrm{Prob}(x \leq k) \simeq \mathrm{Prob}(y \leq k)$.
3. If N and K are large compared to n and $p = K/N$ is not close to 0 or 1, then, for $p = K/N$,

$$\mathrm{Prob}(x \leq k) = \Phi\left(\frac{k - np}{\sqrt{np(1-p)}}\right),$$

   where $\Phi(x) = e^{(-x^2/2)}/(\sqrt{2\pi})$ is the standard normal distribution function, namely, $\mathcal{N}(0, 1)$.
4. The beta-binomial distribution is the conjugate prior for the hypergeometric distribution.

Sampling the Hypergeometric PDF

To sample $\mathcal{H}(N, n, K)$ one simulates an experiment, without replacement, which is just a way of generating Bernoulli trials from $\mathcal{B}(N, p)$. But instead of N and p being constant for all trials, they vary in accordance with the number of objects and the number of objects with the "success" characteristic that have previously been drawn.

The initial value of $N = N_0$ is reduced trial after trial as

$$N_i = N_{i-1} - 1, \qquad i = 1, \ldots, n,$$

whenever an object is removed from the pool. Similarly, the initial value of $p = p_0 = K/N_0$ is altered after each draw as

$$p_i = \frac{N_{i-1}p_{i-1} - \delta}{N_{i-1} - 1}, \qquad i = 1, \ldots, n,$$

where $\delta = 1$ if the drawn object belongs to the "success" class and $\delta = 0$ otherwise.

A.1.6   Negative Hypergeometric Distribution

The negative hypergeometric distribution describes almost the same problem that the negative binomial PDF does. The negative hypergeometric PDF describes the probabilities in sampling, *without* replacement, from a finite pool or population of, initially, N objects, each of which has one of two mutually exclusive properties which here are called abstractly "success" and "failure" but

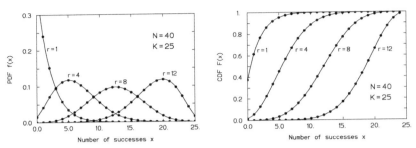

Figure A.6 The PDF (left) and the CDF (right) for the negative hypergeometric distribution with parameters for sampling from a finite population without replacement. Parameters are the initial population size $N = 200$, initial number K of success states in the population, and the number of failures r. The smooth lines connecting the discrete distribution points are for visual convenience of identifying the various cases.

could just as well be male/female, red/green, lepton/hadron, and so on. In the negative hypergeometric problem, objects are randomly drawn from the pool (Bernoulli trials), without replacement, until exactly r failures have been drawn. The negative hypergeometric PDF gives the probability that such a sample also contains x successes. In other words, the negative geometric distribution gives the likelihood that a sample with exactly r failures contains x successes.

In the initial pool of N objects or elements, K are termed "successes" and the rest are "failures." The construction of the sample requires x successes in $(x + r - 1)$ draws and the $(x + r)$th draw must be a failure. The probability of the former is given by the hypergeometric PDF $\mathcal{H}(x|N, x + r - 1, K)$ of Eq. (A.32) and the probability of the latter is the number of failures remaining in the pool, namely, $N - x - (r - 1)$, divided by the number of objects in the pool, namely, $N - (x + r - 1)$. The product of these two probabilities then gives the probability $f(x)$ that the sample contains x successes when the rth failure was drawn (and ended the sampling) as

$$f(x) = \left[\frac{\binom{K}{x}\binom{N-K}{x+r-1-x}}{\binom{N}{x+r-1}}\right]\left[\frac{N-K-(r-1)}{N-(x+r-1)}\right] = \frac{\binom{x+r-1}{x}\binom{N-r-x}{K-x}}{\binom{N}{K}}.$$

$$(A.34)$$

Here the distribution parameters have integer values so that $N \in 0, 1, 2, \ldots$ and $K \in 0, 1, 2, \ldots, N$. The number of failures r realized when the sampling ceases are limited to $r \in 0, 1, 2, \ldots, N - K$.

The mean of this distribution is

$$\mu(x) \equiv \langle x \rangle = \sum_{x=0}^{K} x f(x) = \frac{rK}{N - K + 1}, \qquad (A.35)$$

and the variance is

$$\sigma^2(x) = \langle x^2 \rangle - \langle x \rangle^2 = \frac{rK(N+1)(N-K-r+1)}{(N-K+1)^2(N-K+2)}. \tag{A.36}$$

The derivation of these results is given by Wikipedia (2020b). Examples of the negative hypergeometric distribution are shown in Fig. A.6.

A.1.7 Poisson Distribution

The Poisson distribution was first published in 1837 by the French mathematician Siméon Denis Poisson (1781–1840) (Fig. A.7). This distribution gives the probability of observing a given number x of events in some specified time interval if (1) the events occur at a constant mean rate λ and (2) the occurrence of an event is independent of when a previous event occurred. Of course, the mean rate can have units of events or numbers per unit anything that can have events in it, such as radioactive decays per gram of some material, supernovas per galaxy, DNA mutations per chromosome, and so forth.

The Poisson PDF, often denoted by $\mathcal{P}(\lambda)$ or $\mathcal{P}(x|\lambda)$, is

Figure A.7 Siméon Denis Poisson (1781–1840).

$$f(x) = \frac{\lambda^x e^{-\lambda}}{x!}, \qquad \lambda > 0, \quad x = 0, 1, 2, \ldots. \tag{A.37}$$

The associated CDF is

$$F(x) = \sum_{k=0}^{x} \frac{\lambda^k}{k!} e^{-\lambda} = \frac{\Gamma(\lfloor k+1 \rfloor, \lambda)}{\lfloor x \rfloor!}, \tag{A.38}$$

where $\Gamma(\alpha, \beta)$ is the *upper incomplete gamma function* defined as

$$\Gamma(s, x) = \int_{\beta}^{\infty} t^{\alpha-1} e^{-t} \, dt, \qquad \alpha, \beta > 0. \tag{A.39}$$

The limiting value of the CDF is

$$F(\infty) = \sum_{x=0}^{\infty} f(x) = e^{-\lambda} \sum_{x=0}^{\infty} \frac{\lambda^x}{x!} = e^{-\lambda} e^{\lambda} = 1, \tag{A.40}$$

as expected. Examples of the Poisson distribution are shown in Fig. A.8.

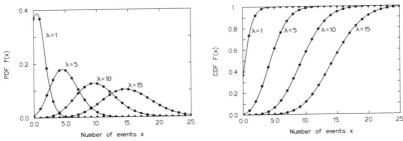

Figure A.8 The PDF (left) and the CDF (right) for the Poisson distribution for four values of the rate parameter λ. The smooth line between the discrete distribution points is for the visual convenience of connecting the discrete points with the same λ.

Approximating a Binomial PDF by a Poisson PDF

When n is large, the evaluation of the binomial distribution of Eq. (A.6) becomes computationally difficult because the various factorials in the binomial coefficient become enormous. However, when p is very small and n very large, the binomial distribution can be well approximated by the Poison distribution which is computationally much simpler. To see this simplification, consider Eq. (A.6) for a fixed x and let $n \to \infty$ and $p \to 0$ in such a manner that np remains constant, say $np = \lambda$. Then with $p = \lambda/n$, the limit of Eq. (A.6) yields

$$
\lim_{n \to \infty} f_B(x) = \lim_{n \to \infty} \frac{n!}{x!(n-x)!} \left(\frac{\lambda}{n}\right)^x \left(1 - \frac{\lambda}{n}\right)^{n-x}
$$

$$
= \lim_{n \to \infty} \frac{n(n-1)\cdots(n-x+1)}{n \quad n \quad \cdots \quad n} \frac{\lambda^x}{x!} \left(1 - \frac{\lambda}{n}\right)^n \left(1 - \frac{\lambda}{n}\right)^{-x}
$$

$$
= \lim_{n \to \infty} \frac{\left(1 - \frac{1}{n}\right)\cdots\left(1 - \frac{x-1}{n}\right)}{\left(1 - \frac{\lambda}{n}\right)^x} \frac{\lambda^x}{x!} \left(1 - \frac{\lambda}{n}\right)^n . \tag{A.41}
$$

As $n \to \infty$, the terms $(1 - 1/n)$ and $(1 - \lambda/n)$ approach unity. Also, from the properties of the base of the natural logarithms e, it is known that

$$
\lim_{n \to \infty} \left(1 - \frac{\lambda}{n}\right)^n = e^{-\lambda}. \tag{A.42}
$$

Hence Eq. (A.41), in the limit, yields

$$
\lim_{n \to \infty} f_B(x) = \frac{\lambda^x}{x!} e^{-\lambda}, \qquad x = 0, 1, 2, \dots . \tag{A.43}
$$

This is the PDF $f(x)$ of the Poisson distribution, which has only a single parameter λ. In Fig. A.9, a comparison of the binomial PDF and its approximating

Poisson PDF is shown. It is seen that the two distributions are very nearly the same, even for this rather large value of p. For many applications, the Poisson approximation is adequate with n as small as 20 provided $\lambda \equiv np < 5$.

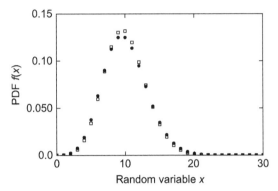

Figure A.9 Comparison of PDFs for the Poisson distribution (circles) with $\lambda = 10$ and the binomial PDF (open squares) with $n = 100$ and $p = 0.1$ (and hence $\lambda = np = 10$).

Mean and Variance of the Poisson PDF

For a large number of trials, p represents the proportion of time an event (success) occurs. Then for a large number of trials n, the *average* or *expected* number of successes would be $np \equiv \lambda$. Thus the expected value $\langle x \rangle = \lambda$, which is the single parameter of the Poisson distribution. This result can be derived more formally as shown below:

$$\langle x \rangle = \sum_{x=0}^{\infty} x \frac{\lambda^x}{x!} e^{-\lambda} = \sum_{x=1}^{\infty} \frac{\lambda^x}{(x-1)!} e^{-\lambda} = \lambda \sum_{y=0}^{\infty} \frac{\lambda^y}{y!} e^{-\lambda} = \lambda F(\infty) = \lambda. \quad \text{(A.44)}$$

In a similar fashion, the variance of the Poisson distribution can be found. From Eq. (2.15), the variance is

$$\sigma^2 = \langle x^2 \rangle - \langle x \rangle^2 = \langle x(x-1) + x \rangle - \langle x \rangle^2 = \langle x(x-1) \rangle + \langle x \rangle - \langle x \rangle^2$$
$$= \langle x(x-1) \rangle + \lambda - \lambda^2.$$

Here

$$\langle x(x-1) \rangle = \sum_{x=0}^{\infty} x(x-1) \frac{e^{-\lambda}\lambda^x}{x!} = \sum_{x=2}^{\infty} \frac{e^{-\lambda}\lambda^x}{(x-2)!} = \lambda^2 \sum_{x'=0}^{\infty} \frac{e^{-\lambda}\lambda^{x'}}{x'!}$$
$$= \lambda^2 F(\infty) = \lambda^2.$$

Combining these last two results gives

$$\sigma^2(x) = \langle x(x-1) \rangle + \lambda - \lambda^2 = \lambda^2 + \lambda - \lambda^2 = \lambda. \quad \text{(A.45)}$$

This is a very important property of Poisson statistics, namely, that the variance is equal to its mean, i.e., $\sigma^2 = \mu$.

Moreover, because the Poisson distribution is the asymptotic distribution for the binomial distribution, this result must agree with the asymptotic value of the binomial variance for large n, p small, and $np = \lambda$. From Eq. (A.12), one has

$$\sigma_{\mathcal{B}}^2 = np(1 - p) = \lambda(1 - p) = \lim_{p \to 0} \lambda(1 - p) = \sigma_{\mathcal{P}}^2. \qquad (A.46)$$

Finally, the mode of the PDF occurs in the interval $[\lceil \lambda \rceil - 1, \lfloor \lambda \rfloor]$.

The Poisson distribution has several interesting properties (Fishman, 2003):

1. If x_i are independent samples from $\mathcal{E}(1)$ and y is the smallest integer such that $\sum_1^{y+1} > \lambda$, then y is distributed as $\mathcal{P}(\lambda)$.
2. If ρ_i are independent samples from $\mathcal{U}(0, 1)$ and y is the smallest integer such that $\prod_{i=1}^{y+1} \rho_i < e^{-\lambda}$, then y is distributed according to the Poisson distribution $\mathcal{P}(\lambda)$.
3. If x is from $\mathcal{P}(\lambda)$, then $(x - \lambda)/\sqrt{\lambda}$ has a distribution that converges to $\mathcal{N}(0, 1)$ as $\lambda \to \infty$.
4. As discussed in Section 6.4.4 about Bayesian inference for the failure rate analyses of normally operating systems, the conjugate prior distribution for a Poisson likelihood function is the gamma distribution. Thus the posterior PDF is also a gamma function.

Sampling From the Poisson Distribution

Often item 2 of the above properties of the Poisson distribution has been used to generate Poisson deviates. However, the inverse CDF method has comparable speed because of the large number of samples from $\mathcal{U}(0, 1)$ required by property 2 (Schmeiser and Kachitvichyanukul, 1981). In the inverse CDF method (see Section 4.1.2), the initial value $f(0) = e^{-\lambda}$ and the ratio of successive PDF values is $R(x) \equiv f(x + 1)/f(x) = \lambda/(x + 1)$. The inverse CDF method has a computational time proportional to λ. For large values of λ, there are generating schemes that are relatively insensitive to λ, all of which are rather complex (Atkinson, 1979; Devroye, 1981; Schmeiser and Kachitvichyanukul, 1981; and Ahrens and Dieter, 1991).

A.2 Continuous Distributions

In this section, 12 continuous univariate probability distributions are presented, namely, the uniform, the exponential, the gamma, the beta, the Weibull, the normal, the lognormal, the Cauchy or Lorentzian, the chi-squared, the Student's t, and the Pareto. There are many other continuous distributions; the interested reader is referred to Johnson et al., (1994, 1995), Kendall and Stuart (1977), McGrath and Irving (1975), or many other books on probability and statistics.

Sampling and Linear Transformations

In many of the sampling schemes for continuous distributions, the sampling is first performed from a simpler *standardized* form of the distribution with some parameters set to zero or unity. Then a linear transformation is used to obtain a sample from the more general form of the distribution. For example the shifted gamma distribution,

$$f(x) = \frac{1}{\beta \Gamma(\alpha)} \left(\frac{x - a}{\beta} \right)^{\alpha - 1} e^{-(x-a)/\beta}, \qquad a \leq x \leq \infty,$$

has a shift parameter a, a shape parameter α, and a scale parameter β. The linear transformation $(x - a)/\beta = y$ transforms the above PDF to the much simpler standardized PDF

$$g(y) = \frac{1}{\Gamma(\alpha)} y^{\alpha - 1} e^{-y}, \qquad 0 \leq y \leq \infty.$$

It is much easier to obtain a sample y from this single-parameter PDF and then transform it back to obtain $x = \beta y + a$ as a sample from the three-parameter form of the distribution. This trick is useful for many of the distributions discussed below.

Inverse CDF Method

The inverse CDF method (see Section 4.1.1) is an excellent way to generate samples if the inverse CDF function is easily calculated or if an accurate approximation is known. Thus, one simply has to evaluate $F^{-1}(\rho)$ to obtain a sample from the distribution. In Table A.2 the inverse functions for several continuous distributions are summarized.

A.2.1 Uniform Distribution

The uniform or rectangular distribution can be defined over any range $[a, b]$, where a and b are finite and $b > a$. In this distribution, denoted by $\mathcal{U}(a, b)$, all values within the range are equally likely. Thus, the PDF is

$$f(x) = \frac{1}{b - a}, \qquad a \leq x \leq b, \tag{A.47}$$

and the CDF is

$$F(x) = \int_a^b \frac{1}{b - a} dx = \frac{x - a}{b - a}, \qquad a \leq x \leq b. \tag{A.48}$$

The moments are

$$\langle x^n \rangle = \int_a^b x^n f(x) dx = \frac{b^{n+1} - a^{n+1}}{(n+1)(b - a)} = \frac{1}{n+1} \sum_{k=0}^n a^k b^{n-k}, \tag{A.49}$$

Table A.2 Some continuous distributions for which the inverse CDF method can be used to obtain random variates $x = F^{-1}(\rho)$, where ρ is from $\mathcal{U}(0, 1)$. After Fishman (2003).

Distribution	PDF $f(x)$	Inverse $F^{-1}(\rho)$	Efficient form
uniform: $\mathcal{U}(x\|a, b)$ $a \leq x \leq b$	$\dfrac{1}{b-a}$	$a + (b-a)\rho$	—
exponential: $\mathcal{E}(x\|\lambda)$ $\lambda > 0;\ x \geq 0$	$\lambda e^{-\lambda x}$	$-\dfrac{1}{\lambda}\ln(1-\rho)$	$-\dfrac{1}{\lambda}\ln(\rho)$
beta: $\mathcal{B}_e(x\|\alpha, 1)$ $\alpha > 0;\ 0 \leq x \leq 1$	$\alpha x^{\alpha-1}$	$\rho^{1/\alpha}$	—
beta: $\mathcal{B}_e(x\|1, \beta)$ $\beta > 0;\ 0 \leq x \leq 1$	$\beta(1-x)^{\beta-1}$	$1 - (1-\rho)^{1/\beta}$	$1 - \rho^{1/\beta}$
Logistic: $\mathcal{L}(x\|\alpha, \beta)$ $\beta>0;\ -\infty < x, \alpha < \infty$	$\dfrac{e^{-(x-\alpha)/\beta}}{\beta[1 + e^{-(x-\alpha)/\beta}]^2}$	$\alpha + \beta\ln[\rho/(1-\rho)]$	—
Weibull: $\mathcal{W}(x\|\alpha, \beta)$ $\alpha, \beta > 0;\ x \geq 0$	$\dfrac{\alpha}{\beta}\left(\dfrac{x}{\beta}\right)^{\alpha-1} e^{-(x/\beta)^{\alpha}}$	$\beta[-\ln(1-\rho)]^{1/\alpha}$	$\beta[-\ln\rho]^{1/\alpha}$
normal:[a] $\mathcal{N}(x\|\mu, \sigma)$ $\sigma^2 > 0;\ -\infty < x, \mu < \infty$	$\dfrac{e^{(\mu-x)^2/(2\sigma^2)}}{\sqrt{2\pi\sigma^2}}$	$\mu + \text{sign}(\rho - 1/2)\sigma \times$ $\left(t - \dfrac{c_0 + c_1 t + c_2 t^2}{1 + d_1 t + d_2 t^2 + d_3 t^3}\right)$ $t \equiv \sqrt{-\ln[\min(\rho, 1-\rho)]^2}$	—
Cauchy: $\mathcal{C}(x\|\alpha, \beta)$ $\beta>0;\ -\infty < x, \alpha < \infty$	$\dfrac{1}{\pi}\dfrac{\beta}{(x-\alpha)^2 + \beta^2}$	$\alpha + \beta\tan\pi[\rho - (1/2)]$	$\alpha + \beta\tan\pi\rho$
Pareto: $\mathcal{P}_a(x\|\alpha, \beta)$ $\alpha>0;\ x \geq \beta > 0$	$\dfrac{\alpha\beta^{\alpha}}{x^{\alpha+1}}$	$\dfrac{\beta}{(1-\rho)^{1/\alpha}}$	$\dfrac{\beta}{\rho^{1/\alpha}}$

[a] $c_0 = .515517, c_1 = .802853, c_2 = .010328, d_1 = 1.432788, d_2 = .189269, d_3 = .001308$; absolute error $< 0.45 \times 10^{-3}$; this approximation is from Hastings (1955).

from which the mean value is

$$\mu = \frac{b+a}{2} = a + \frac{b-a}{2}, \qquad (A.50)$$

and the variance is

$$\sigma^2 = \langle x^2 \rangle - \mu^2 = \frac{b^3 - a^3}{3(b-a)} - \left(\frac{b+a}{2}\right)^2 = \frac{(b-a)^2}{12}. \qquad (A.51)$$

An example uniform distribution over the interval $(-6, 5)$ is shown in Fig. A.10.

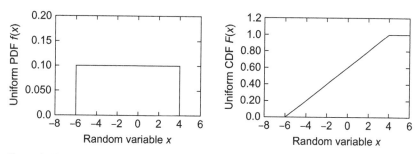

Figure A.10 The PDF and the CDF for the uniform distribution with $a = -6$ and $b = 4$.

Sampling From the Rectangular Distribution

Of obvious importance is the unit rectangular distribution $\mathcal{U}(0, 1)$, which forms the basis of almost all sampling schemes for other probability distributions. Random variates from the unit rectangular distribution are denoted in this book by ρ_i to distinguish them from other random variates. The generation of sequences of ρ_i is the subject of Chapter 3. To obtain a random sample x_i from $\mathcal{U}(a, b)$, one first takes a sample ρ_i from $\mathcal{U}(0, 1)$ and then applies a linear transformation to obtain

$$x_i = a + \rho_i(b - a). \tag{A.52}$$

This result is equivalent to the inverse CDF method (see Example 4.1).

A.2.2   Exponential Distribution

The PDF of the exponential distribution is given by

$$f(x) = \lambda e^{-\lambda x}, \qquad x \geq 0, \tag{A.53}$$

and is denoted by $\mathcal{E}(\lambda)$ or, more explicitly, by $\mathcal{E}(x|\lambda)$. This PDF is a special case of the gamma distribution, which is discussed in the next section. The *scale* parameter $\lambda > 0$ determines the rate of descent of the exponential. The CDF is

$$F(x) = \int_0^x \lambda e^{-\lambda u} \, du = 1 - e^{-\lambda x}. \tag{A.54}$$

The moments of the PDF are

$$\langle x^n \rangle = \int_0^\infty u^n \lambda e^{-\lambda u} \, du = \frac{n!}{\lambda^n}. \tag{A.55}$$

It then follows that the mean is $\mu \equiv \langle x \rangle = 1/\lambda$ and the variance is $\sigma^2 = \langle x^2 \rangle - \mu^2 = 1/\lambda^2$. Figure A.11 shows the exponential distribution for three values of λ. Clearly, λ serves as a scaling parameter.

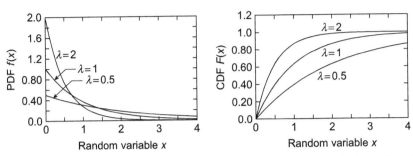

Figure A.11 The PDF and the CDF for the unshifted exponential distribution for three values of the scaling parameter λ.

The exponential distribution is very important because many processes behave exponentially. In Chapter 2, radioactive decay was considered. In the decay process, the number of radioactive atoms at time t is given by

$$N(t) = N(0)e^{-\lambda t}, \tag{A.56}$$

where $N(0)$ is the number of radioactive atoms at time 0 and λ is the decay constant of the particular radioactive species. Also, in Chapter 9 we found that neutral particles from a source are attenuated exponentially according to

$$\phi(r) = \phi^o(0)e^{-\mu r}, \tag{A.57}$$

where $\phi^o(0)$ is the particle flux at $r = 0$, μ is the total linear attenuation coefficient of the medium, and r is the distance from the source. More generally, the exponential distribution applies to any process for which the rate of decrease of a quantity f with respect to a variable x is proportional to the quantity, i.e.,

$$-\frac{df}{dx} = \lambda f, \tag{A.58}$$

where λ is the constant of proportionality.

Sampling From an Exponential Distribution

Generation of variates from the exponential distribution is readily done using the inverse CDF method (see Example 4.2), namely,

$$x_i = -\ln(\rho_i)/\lambda, \tag{A.59}$$

where the ρ_i are sampled from $\mathcal{U}(0, 1)$.

Shifted Exponential Distribution

A slightly more generalized version of the exponential distribution is defined on the half-open interval $[\theta, \infty)$. The PDF in this case is just shifted (or translated) along the x-axis by an amount θ, where θ is called the *shift* parameter, and the PDF takes the form

$$f(x) = \lambda e^{-\lambda(x-\theta)}, \qquad x \geq \theta, \;\; \lambda > 0. \tag{A.60}$$

This distribution is denoted by $\mathcal{E}(\lambda, \theta)$. The CDF is seen to be

$$F(x) = 1 - e^{-\lambda(x-\theta)}. \tag{A.61}$$

Note that the PDF integrates to unity regardless of the value of θ and the CDF reduces to Eq. (A.54) for $\theta = 0$. The mean value is

$$\mu = \int_\theta^\infty \lambda e^{-\lambda(u-\theta)} \, du = \frac{1}{\lambda} + \theta, \tag{A.62}$$

and the variance is

$$\sigma^2 = \frac{2e^{\lambda\theta} - (\lambda\theta + 1)^2}{\lambda^2}. \tag{A.63}$$

In particular, the (central) moments of the shifted exponential PDF are found from the recursion relation

$$\omega_n' \equiv \langle x^n \rangle = \frac{n!}{\lambda^n} - \sum_{i=0}^n (1)^i \binom{n}{i} \theta^i \omega_{n-1}', \qquad n = 1.2.\ldots, \tag{A.64}$$

with $\omega_0' = 1$.

Sampling From a Shifted Exponential Distribution

To generate variates from the shifted exponential distribution, first sample from the unshifted distribution (see Eq. (A.59)) and add the shift, i.e.,

$$x_i = -\ln(\rho_i)/\lambda + \theta, \tag{A.65}$$

where the ρ_i are uniformly distributed in $(0, 1)$.

A.2.3 Gamma Distribution

A generalization of the exponential distribution is the gamma distribution $\mathcal{G}(\alpha, \beta, \theta)$, whose PDF can be written as

$$f(x) = \frac{1}{\beta\Gamma(\alpha)} \left(\frac{x-\theta}{\beta} \right)^{\alpha-1} e^{-(x-\theta)/\beta}, \qquad \alpha, \beta > 0, \;\; \theta \geq 0, \;\; x \geq \theta, \tag{A.66}$$

where $\Gamma(\alpha)$ is the gamma function defined by

$$\Gamma(u) = \int_0^\infty x^{u-1} e^{-x} dx. \tag{A.67}$$

Here α, β, and θ are the *shape*, *scale*, and *shift* parameters, respectively. This PDF is often reparameterized with $\beta' = 1/\beta$, so Eq. (A.66) assumes the form

$$f(x) = \frac{(x - \theta)^{\alpha - 1}\beta'^{\alpha}}{\Gamma(\alpha)} e^{-\beta'(x-\theta)}, \tag{A.68}$$

where β' is now referred to as the *rate* parameter since it has units of "per unit x." The gamma distribution for $\alpha = 1$ is in fact the exponential distribution of Eq. (A.60) with $\beta = 1/\beta' = 1/\lambda$. This close relationship is, however, somewhat muddied by the fact that no explicit closed-form expression for the CDF is available for general values of α. Thus, the best one can say is, for $x \geq \theta$,

$$\begin{aligned} F(x) &= \frac{1}{\beta\Gamma(\alpha)} \int_{\theta}^{x} \left(\frac{x - \theta}{\beta}\right)^{\alpha - 1} e^{-(x-\theta)/\beta} \, dx \\ &= \frac{1}{\Gamma(\alpha)} \int_{0}^{(x-\theta)/\beta} t^{\alpha - 1} e^{-t} \, dt = P(\alpha, (x - \theta)/\beta), \end{aligned} \tag{A.69}$$

where $P(\alpha, y)$ is the *incomplete gamma function* defined as

$$P(\alpha, y) \equiv \frac{1}{\Gamma(\alpha)} \int_{0}^{y} e^{-u} u^{\alpha - 1} \, du. \tag{A.70}$$

Special Cases

If $\theta = 0$, Eq. (A.66) becomes the unshifted gamma function $\mathcal{G}(\alpha, \beta)$, and if, in addition, $\beta = 1$, the *standardized* gamma function $\mathcal{G}(\alpha, 1)$ is obtained. This standardized gamma function, when α is a positive integer, gives the distribution of waiting times x to the αth Poisson event. Thus, for $\alpha = 1$, $\mathcal{G}(1, 1) \rightarrow e^{-x}$, which is the distribution of waiting times for the first event, such as a radioactive decay in a radionuclide sample.

Another special case for the gamma distribution occurs when the scale parameter $\beta = 2$ and the shift parameter $\theta = 0$. The gamma distribution then becomes the chi-squared distribution with 2α equal to the ν degrees of freedom, i.e., $\mathcal{G}(\alpha, 2) \rightarrow \chi^2(\nu = 2\alpha)$ (see Section A.2.10).

Mean, Variance, and Moments

The (central) moments of the gamma PDF are given recursively by

$$\omega_n' \equiv \langle x^n \rangle = \beta^n \prod_{i=0}^{n-1}(\alpha + i) - \sum_{i=0}^{n}(1)^i \binom{n}{i}\theta^i \omega_{n-1}', \quad n = 1.2. \ldots, \tag{A.71}$$

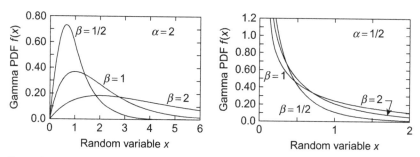

Figure A.12 The PDF for the unshifted gamma distribution for $\alpha > 1$ (left) and for $\alpha < 1$ (right).

with $\omega_0' = 1$. For the unshifted gamma function, these moments reduce to

$$\langle x^n \rangle \equiv \int_0^\infty x^n f(x)\, dx = \beta^n \prod_{i=0}^{n-1} (\alpha + i). \tag{A.72}$$

From this it follows that, for the unshifted gamma function, the mean value is $\mu = \langle x \rangle = \alpha\beta$ and the variance is $\sigma^2 = \langle x^2 \rangle - \mu^2 = \alpha\beta^2$, which, of course, reduce to the appropriate results for an exponential distribution when $\alpha = 1$. Examples of the gamma distribution are plotted in Fig. A.12. Note that when the shape parameter $\alpha > 1$, $f(x|\alpha, \beta)$ is unimodal with a maximum at $x_{max} = (1 - \alpha^{-1})/\beta$. For $\alpha < 1$, $f(x|\alpha, \beta)$ is a monotonic decreasing function that is unbounded at $x = 0$.

Sampling From a Gamma Distribution

Generating samples from the gamma distribution is not straightforward because the sampling method depends on whether α is greater or less than unity. Special and very efficient techniques can be used if α is a positive integer (Press et al., 1996). For the general case, methods for sampling from the gamma distribution have been described by McGrath and Irving (1975), Rubinstein (1981), and Fishman (2003). Gentle (1998) reports the following algorithms for sampling from the unshifted standardized gamma distribution.

Case $\alpha > 1$

The Cheng–Feast (1979) algorithm:

1. Generate ρ_1 and ρ_2 from a uniform distribution on $(0, 1)$ and set

$$\xi = \left[(\alpha - (1/6\alpha))\,\rho_1\right] \Big/ \left[(\alpha - 1)\rho_2\right].$$

2. If $\frac{2(\rho_2 - 1)}{\alpha - 1} + \xi + \frac{1}{\xi} \le 2$, then return $x = (\alpha - 1)\xi$;
   otherwise, if $\frac{2\ln\rho_2}{\alpha - 1} - \ln\xi + \xi \le 1$, then return $x = (\alpha - 1)\xi$.
3. If no value is returned, go to step 1 and try again.

Case $\alpha < 1$

The Ahrens–Dieter (1974) algorithm, modified by Best (1983):

1. Set $t = 0.07 + 0.75\sqrt{1-\alpha}$ and $b = \alpha[1 + (e^t/t)]$.
2. Generate ρ_1 and ρ_2 from a uniform distribution on $(0, 1)$ and set $\xi = b\rho_1$.
3. If $\xi \leq 1$, then set $x = t\xi^{1/\alpha}$; then, if $\rho_2 \leq (2-x)/(2+x)$, return x;
   otherwise, if $\rho_2 \leq e^{-x}$, return x;
   otherwise, set $x = \ln[t(b-\xi)/\alpha]$ and $y = x/t$ and, if $\rho_2[\alpha + y(1-\alpha)] \leq 1$, return x.
4. If no value is returned, go to step 2 and try again.

To generate samples z_i from the nonstandardized gamma distribution ($\beta \neq 1$), first generate samples x_i from the standardized distribution using the above algorithms and set $z_i = x_i/\beta$. Finally, to obtain samples from a shifted gamma distribution, first obtain a sample from the unshifted distribution, as described above, and then add the shift to the unshifted sample.

A.2.4 Beta Distribution

This is an important PDF in Bayesian analysis because it is a conjugate prior to a binomial likelihood distribution. The beta distribution $\mathcal{B}_e(\alpha, \beta)$ has the PDF

$$f(x) = \frac{\Gamma(\alpha + \beta)}{\Gamma(\alpha)\Gamma(\beta)} x^{\alpha-1}(1-x)^{\beta-1}, \quad \alpha > 0, \quad \beta > 0, \quad 0 \leq x \leq 1, \quad \text{(A.73)}$$

in which α and β are shape parameters. Like the gamma distribution, numerical integration must be used to evaluate the CDF. The CDF is

$$F(x) = I_x(\alpha, \beta), \quad \text{(A.74)}$$

where $I_x(\alpha, \beta)$ is the *incomplete beta function*

$$I_x(\alpha, \beta) \equiv \frac{\Gamma(\alpha + \beta)}{\Gamma(\alpha)\Gamma(\beta)} \int_0^x u^{\alpha-1}(1-u)^{\beta-1} \, du. \quad \text{(A.75)}$$

The moments of the beta distribution are

$$\langle x^n \rangle \equiv \int_0^1 x^n f(x) \, dx = \prod_{i=0}^{n-1} \frac{\alpha + i}{\alpha + \beta + i}, \quad \text{(A.76)}$$

from which the mean value is

$$\mu \equiv \langle x \rangle = \frac{\alpha}{\alpha + \beta}, \quad \text{(A.77)}$$

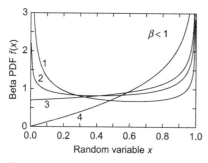

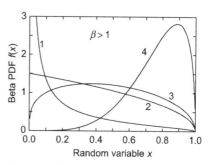

Figure A.13 The PDF for the beta distribution for $\beta < 1$. Here $\beta_1 = \beta_2 = \beta_3 = \beta_4 < 1$ and $\alpha_1 < \beta_i < \alpha_2 < \alpha_3 = 1 < \alpha_4$.

Figure A.14 The PDF for the beta distribution for $\beta > 1$. Here $\beta_1 = \beta_2 = \beta_3 = \beta_4 > 1$ and $\alpha_1 < \alpha_2 = 1 < \alpha_3 < \beta_i < \alpha_4$.

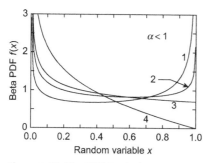

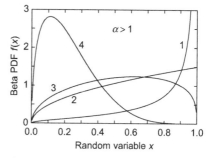

Figure A.15 The PDF for the beta distribution for $\alpha < 1$. Here $\alpha_1 = \alpha_2 = \alpha_3 = \alpha_4 < 1$ and $\beta_1 < \alpha_i < \beta_2 < \beta_3 = 1 < \beta_4$.

Figure A.16 The PDF for the beta distribution for $\alpha > 1$. Here $\alpha_1 = \alpha_2 = \alpha_3 = \alpha_4 > 1$ and $\beta_1 < \beta_2 = 1 < \beta_3 < \alpha_i < \beta_4$.

and the variance is

$$\sigma^2 \equiv \langle x^2 \rangle - \mu^2 = \frac{\alpha\beta}{(\alpha + \beta)^2(\alpha + \beta + 1)}. \tag{A.78}$$

Examples of the beta function are shown in Figs. A.13–A.16. It is seen from these figures that the beta PDF can assume a wide variety of shapes over the interval [0, 1]. For this reason the beta function is often used to represent, for example, the distributions of component failure probabilities for components in a complex mechanical system. The parameter α controls the behavior of the function near $x = 0$ while β determines the behavior near $x = 1$. If $\alpha < 1$, $f(x)$ is unbounded at $x = 0$ and if $\beta < 1$, $f(x)$ is unbounded at $x = 1$. If $\alpha, \beta > 1$, there is a maximum (or mode) at $x = (\alpha - 1)/(\alpha + \beta - 2)$. If $\beta > \alpha$ or $\alpha > \beta$, the distribution is skewed toward $x = 0$ or toward $x = 1$, respectively. If $\alpha = \beta$, the distribution is symmetric about $x = 0.5$.

Sampling From a Beta Distribution

Generating samples from the beta distribution is not straightforward, except for the simple cases of $\alpha = 1$, $\beta = 1$, or $\alpha = \beta = 1$ (see Table A.2). For $\max(\alpha, \beta) < 1$, Jöhnk (1964) gives the following simple algorithm as reported by Gentle (1998).

Case $\max(\alpha, \beta) < 1$

1. Generate ρ_1 and ρ_2 from a uniform distribution on $(0, 1)$ and set $v_1 = \rho_1^{1/\alpha}$ and $v_2 = \rho_2^{1/\beta}$.
2. Set $w = v_1 + v_2$.
3. If $w < 1$, return $x = v_1/w$; otherwise go to step 1 and try again.

For this case of $\max(\alpha, \beta) < 1$, Atkinson and Whittaker (1976) also give a very fast algorithm. For other cases in which either or both of the shape parameters are greater than unity, fast algorithms have been devised by Cheng (1978), Schmeiser and Babu (1980), and Zechner and Stadlober (1993). Fishman (2003) gives explicit algorithms for these cases.

Restricted Range Beta Distribution

With the beta PDF above, the range of the random variable x is the interval $[0, 1]$. Sometimes, very small values near $x = 0$ or very large values near $x = 1$ are so improbable that they can be excluded from consideration and a restricted range $0 \leq x_{min} = a \leq x \leq b = x_{max} \leq 1$ is desired. For such situations, the four-parameter beta function $\mathcal{B}_e(\alpha, \beta, a, b)$ can be used. The PDF is

$$f(x) = \frac{\Gamma(\alpha + \beta)}{\Gamma(\alpha)\Gamma(\beta)} (b - a)^{1-\alpha-\beta} (x - a)^{\alpha-1} (b - x)^{\beta-1}, \qquad (A.79)$$

where the shape parameters $\alpha, \beta > 0$ and the scale parameters $0 \leq a < b \leq 1$. The shape of the density distributions are the same as those in Figs. A.13–A.16, except the plots are squeezed into the finite subinterval $[a, b]$.

The corresponding CDF is

$$F(x) = \int_a^x f(x)\,dx = I_{x*}(\alpha, \beta), \qquad (A.80)$$

where $x* = (x - a)/(b - a)$ and I_x is the incomplete beta function defined by Eq. (A.75). The central moments are

$$\omega'_n \equiv \langle x^n \rangle = (b - a)^n \prod_{i=0}^{n-1} \frac{\alpha + i}{\alpha + \beta + 1}. \qquad (A.81)$$

To generate samples x'_i, first generate a sample x_i for the interval $[0, 1]$, as described above, and then transform it to the restricted interval $[a, b]$ by $x'_i = a + (b - a)x_i$.

A.2.5 Weibull Distribution

This distribution is named after Swedish engineer and mathematician Waloddi Weibull (Fig. A.17), who described it in detail in 1951, although it was first identified by Fréchet in 1927 and used by Rosin and Rammler in 1933 to describe a particle size distribution. The PDF of the Weibull distribution $\mathcal{W}(\alpha, \beta)$, or more explicitly $\mathcal{W}(x|\alpha, \beta)$, can be written as

Figure A.17 Waloddi Weibull (1887–1979).

$$f(x) = \frac{\alpha}{\beta}\left(\frac{x-\theta}{\beta}\right)^{\alpha-1} \exp[-\{(x-\theta)/\beta\}^{\alpha}], \ x \geq 0, \tag{A.82}$$

with a scale parameter $\beta > 0$, a shape parameter $\alpha > 0$, and a shift parameter $\theta > 0$. Often x is the time to some event, so the Weibull distribution is widely used, for example, to describe the distribution of lifetimes such as those encountered in fatigue analyses. It is also used to describe the probability that large acute radiation doses x lead to deterministic biological effects, such as cataract formation (Faw and Shultis, 1999). The Weibull CDF is readily found to be

$$F(x) = 1 - \exp[-\{(x-\theta)/\beta\}^{\alpha}]. \tag{A.83}$$

Examples of the Weibull PDF are shown in Figs. A.18 and A.19. For $\alpha < 1$, the Weibull PDF is a monotonic decreasing function of x that is unbounded at $x = 1$, while for $\alpha > 1$ the Weibull distribution is unimodal with a maximum at $x = \alpha[1 - (1/\beta)]^{1/\beta}$.

Special Cases

For the special case that $\alpha = 2$, $\beta = \sqrt{2}\sigma$, and $\theta = 0$, the Weibull distribution becomes the *Rayleigh distribution* with a PDF of

$$f(x) = \frac{x}{\sigma^2} \exp\left[-\frac{x^2}{2\sigma^2}\right]. \tag{A.84}$$

This distribution naturally arises when the wind velocity is described in two dimensions. If each wind speed component is uncorrelated and normally distributed with equal variance and zero mean, then the overall wind speed (vector magnitude) is characterized by the Rayleigh distribution.

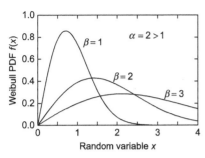
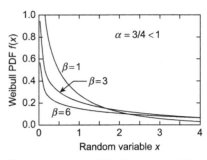

Figure A.18 The PDF for the unshifted Weibull distribution for $\alpha > 1$.

Figure A.19 The PDF for the unshifted Weibull distribution for $\alpha < 1$.

Moments

The central moments of the Weibull distribution are given recursively by

$$\omega_n' \equiv \langle x^n \rangle \equiv \int_0^\infty x^n f(x)\, dx \beta^n \Gamma(1 + \frac{1}{\alpha}) - \sum_{i=1}^n (-1)^i \theta^i \binom{n}{i} \omega_{n-1}', \quad i = 1, 2, \dots,$$
(A.85)

where $\omega_0' = 1$. For an unshifted PDF, this result reduces to

$$\langle x^n \rangle = \alpha^n \Gamma\left(1 + \frac{n}{\beta}\right).$$
(A.86)

The mean value is thus given by

$$\mu \equiv \langle x \rangle = \beta\, \Gamma\left(1 + \frac{1}{\alpha}\right),$$
(A.87)

and the variance can be written as

$$\sigma^2 \equiv \langle x^2 \rangle - \mu^2 = \beta^2 \left[\Gamma\left(1 + \frac{2}{\alpha}\right) - \Gamma^2\left(1 + \frac{1}{\alpha}\right) \right].$$
(A.88)

Sampling From a Weibull Distribution

To generate samples from a Weibull PDF, the inverse CDF method is quite efficient, namely,

$$x_i = -\beta[\ln \rho_i]^{1/\alpha},$$
(A.89)

where the ρ_i are random samples from $\mathcal{U}[0, 1]$.

A.2.6 Normal Distribution

The *normal distribution*, also often called the *Gaussian distribution* after J. Gauss (Fig. A.20), is denoted as $\mathcal{N}(\mu, \sigma)$ and has the PDF

$$f(x) = \frac{1}{\sqrt{2\pi}\sigma} e^{-(\mu-x)^2/(2\sigma^2)}, \quad -\infty < x < \infty. \tag{A.90}$$

The mean μ and variance σ^2 act, respectively, as the shift and shape parameters of this distribution. A special case is the *unit normal* distribution $\mathcal{N}(0, 1)$. Numerical integration of the PDF must be used to obtain the CDF. Thus,

Figure A.20 Johann Carl Friedrich Gauß (1777–1855).

$$F(x) = \frac{1}{\sqrt{2\pi}\sigma} \int_{-\infty}^{x} e^{-(\mu-u)^2/(2\sigma^2)} du$$

$$= \frac{1}{2}\left[1 - \text{sign}(x - \mu)\text{erf}\left(\left|\frac{x - \mu}{\sqrt{2}\sigma}\right|\right)\right], \tag{A.91}$$

where $\text{erf}(x)$ is the *error function*

$$\text{erf}(x) \equiv \frac{2}{\sqrt{\pi}} \int_{0}^{x} e^{-t^2} dt, \qquad x > 0. \tag{A.92}$$

The error function is related to the incomplete gamma function of Eq. (A.70) by $\text{erf}(x) = P(1/2, x^2)$, $x \geq 0$. The central moments of the normal distribution are

$$\langle (x - \mu)^n \rangle = \frac{1}{\sqrt{2\pi}\sigma} \int_{-\infty}^{\infty} u^n e^{-u^2/(2\sigma^2)} dx$$

$$= \begin{cases} 0, & n \text{ odd,} \\ 1 \cdot 3 \cdot 5 \cdots (n - 1)\sigma^n, & n \text{ even.} \end{cases}$$

A normal distribution in which $\sigma = \sqrt{\mu}$ is a very good approximation to the Poisson distribution with mean μ, especially for large μ.[5] Hence, we often treat processes (like radioactive decay) that behave according to Poisson statistics with the continuous normal distribution *for the special case that* $\sigma = \sqrt{\mu}$.

Sampling From a Normal Distribution

To generate variates from a normal distribution, first sample y_i from the unit normal $\mathcal{N}(0, 1)$ and then take $x_i = \sigma y_i + \mu$ as a sample from $\mathcal{N}(\mu, \sigma)$. The

---

[5] For most purposes, this approximation is quite good if $\mu > 25$.

Box–Muller (1958) scheme can be used to sample from the unit normal as follows. Select two random numbers ρ_1 and ρ_2 distributed uniformly on $[0, 1]$. Then *two* unit normal deviates are

$$
\begin{aligned}
y_1 &= \sqrt{-2\ln\rho_1}\,\cos 2\pi\rho_2, \\
y_2 &= \sqrt{-2\ln\rho_1}\,\sin 2\pi\rho_2.
\end{aligned}
\tag{A.93}
$$

To avoid the computer expense of evaluating trigonometric functions, this scheme can be modified. In the so-called *rejection polar method*, two uniform variates ξ_1 and ξ_2 are sampled from the unit disk using the rejection technique of Example 3.4. Then set $r^2 = \xi_1^2 + \xi_2^2$, which is a uniform deviate. The unit normal samples are then given by

$$
\begin{aligned}
y_1 &= \xi_1\sqrt{-2\ln(r^2)/r^2}, \\
y_2 &= \xi_2\sqrt{-2\ln(r^2)/r^2}.
\end{aligned}
\tag{A.94}
$$

This polar rejection method is the same as the Box–Muller method since r^2 can be used as x_1 and the quantities $\xi_1/\sqrt{r^2}$ and $\xi_2/\sqrt{r^2}$ are the cosine and sine in Eq. (A.93). Note that this Box–Muller algorithm returns two samples from the unit normal.

Finally to obtain a sample from $\mathcal{N}(\mu, \sigma)$, simply transform one of the unit normal samples y_i to obtain $x_i = \sigma y_i + \mu$ as the sample. Although faster sampling algorithms are available (see, for example, Ahrens and Dieter (1988)), that of Eq. (A.94) is sufficient for most purposes.

A.2.7   Lognormal Distribution

In many physical systems, the logarithm of some quantity of interest is found to be distributed normally, i.e., $\ln x \sim \mathcal{N}(\alpha, \beta)$. For example, the radon concentration in US homes is observed to have a lognormal distribution (Faw and Shultis, 1999). The lognormal distribution has the PDF

$$
f(x) = \frac{1}{\sqrt{2\pi}\,\sigma x}\,e^{-(\ln x - \alpha)^2/(2\beta^2)}, \qquad 0 \le x < \infty, \tag{A.95}
$$

and is denoted by $\mathcal{LN}(\alpha, \beta)$. Unlike the normal distribution, where the mean μ acts as a shift parameter, here α is a scale parameter because it stretches and compresses the PDF. By contrast, β is a shape parameter that determines how skewed and heavy-tailed the lognormal distribution is. The CDF can be expressed as

$$F(x) = \frac{1}{2}\left[1 + \text{sign}(\ln x - \alpha)\text{erf}\left(\frac{\ln x - \alpha}{\sqrt{2}}\beta\right)\right]. \tag{A.96}$$

Examples of the lognormal distribution are shown in Fig. A.21.

Moments

The moments of the lognormal distribution about zero are

$$\langle x^n \rangle = e^{n\alpha + n^2\beta^2/2}, \tag{A.97}$$

from which the mean and variance are found to be

$$\mu = e^{\alpha + \beta^2/2} \tag{A.98}$$

and

$$\sigma^2 = \langle x^2 \rangle - \mu^2 = (e^{\beta^2} - 1)e^{2\alpha + \beta^2}, \tag{A.99}$$

respectively.

Sampling From the Lognormal Distribution

To generate samples from a lognormal distribution $\mathcal{LN}(\alpha, \beta)$, first generate x_i from the unit normal $\mathcal{N}(0, 1)$ and set $y_i = \alpha x_i + \beta$. Finally, use $z_i = e^{y_i}$ as the desired sample.

Shifted Lognormal Distribution

Sometimes data fit a lognormal distribution better if the PDF is shifted along the x-axis in a small amount, i.e., $\ln(x - \theta) \sim \mathcal{N}(\alpha, \beta)$. The PDF of this shifted lognormal distribution, denoted by $\mathcal{LN}(\alpha, \beta, \theta)$, is

$$f(x) = \frac{1}{\sqrt{2\pi}\beta(x - \theta)}e^{[(\ln(x-\theta)-\alpha)^2/(2\beta^2)]}, \quad \alpha, \beta > 0 \quad x \geq \theta. \tag{A.100}$$

This PDF has a mode at $x_{mode} = \theta + \exp[-\beta^2 + \alpha]$. Increasing β or decreasing α shifts $f(x)$ to smaller x-values. The CDF is given by

$$F(x) = \frac{1}{2}\left[1 - \text{sgn}(x^*)\text{erf}(|x^*|/\sqrt{2})\right], \tag{A.101}$$

where $x^* = [-\ln(x - \theta + \alpha]/\beta$. The central moments are given recursively by

$$\omega_n' \equiv \langle x^n \rangle = \exp[n\alpha + n^2\beta^2/2] - \sum_{i=1}^{n}(-1)^i\theta^i\binom{n}{i}\omega_{n-i}', \quad \text{with } \omega_0' = 1. \tag{A.102}$$

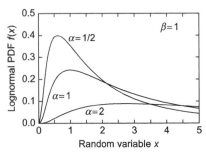

Figure A.21 The PDF for the lognormal distribution.

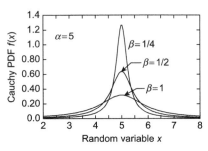

Figure A.22 The PDF for the Cauchy distribution.

A.2.8 Cauchy Distribution

Figure A.23 Augustin-Louis Cauchy (1789–1857).

The *Cauchy distribution* $\mathcal{C}(\alpha, \beta)$ is named after the French mathematician A. Cauchy (see Fig. A.23) who made contributions in many areas, most notably in his pioneering work on complex analysis. This distribution (also sometimes called the *Lorentzian distribution* or *Breit–Wigner distribution*) is often used to describe the behavior of resonance phenomena. The Lorentzian PDF can be written as

$$f(x) = \frac{1}{\pi} \frac{\beta}{(x - \alpha)^2 + \beta^2}, \qquad (A.103)$$

where $-\infty < \alpha < \infty$, $\beta > 0$, and $-\infty < x < \infty$. The corresponding CDF is

$$F(x) = \frac{1}{2} + \frac{1}{\pi} \tan^{-1}\left(\frac{x - \alpha}{\beta}\right). \qquad (A.104)$$

Examples of the Cauchy distribution are shown in Fig. A.22. The α parameter determines the location of the resonance peak and the width parameter β is twice the peak's full-width-at-half-maximum (FWHM).

Mean and Variance

Unlike most continuous distributions, the mean and variance are undefined, since the moments

$$\langle x^n \rangle = \frac{\beta}{\pi} \int_{-\infty}^{\infty} \frac{x^n}{(x - \alpha)^2 + \beta^2} \, dx \qquad (A.105)$$

diverge for all $n \geq 1$. Because the mean and variance do not exist, the central limit theorem does not apply to samples taken from the Cauchy distribution!

Sampling From the Cauchy Distribution

The inverse CDF method can be used to generate samples from the Cauchy distribution, namely,

$$x_i = \alpha + \beta \tan \pi (\rho_i - 1/2), \qquad (A.106)$$

where ρ_i is from $\mathcal{U}(0, 1)$. To avoid the expense of the tangent evaluation, take two samples y_1 and y_2 from the unit normal $\mathcal{N}(0, 1)$. The ratio $z = y_1/y_2$ is then from the unit Cauchy distribution $\mathcal{C}(0, 1)$, so that $x = \alpha + \beta z$ is from $\mathcal{C}(\alpha, \beta)$. Alternatively, sample y_1 and y_2 uniformly from a circle with diameter one (see Example 4.4) and the ratio y_1/y_2 is from $\mathcal{C}(0, 1)$. Another fast sampling scheme is offered by Kronmal and Peterson (1981).

A.2.9   Logbeta Distribution

A distribution which can assume a rich variety of shapes over a *finite* range is the one in which $\ln x \sim \mathcal{B}_e(\alpha, \beta, a, b)$. The PDF of this function, denoted by $\mathcal{LB}_e(\alpha, \beta, a, b)$, is

$$f(x) = \frac{\Gamma(\alpha + \beta)}{\Gamma(\alpha)\Gamma(\beta)} (b - a)^{1-\alpha-\beta} (\ln x - a)^{\alpha-1} (b - \ln x)^{\beta 1}. \qquad (A.107)$$

The range of x is $0 \leq x_{min} \leq x \leq x_{max} < \infty$, where $x_{min} = e^a$ and $x_{max} = e^b$. The shape parameters $\alpha, \beta > 0$ and the range parameters are $-\infty < a < b < \infty$.

As in the case of the beta function, the logbeta function has a wide variety of shapes that depend on the values of the two shape parameters α and β. This shape changing ability makes this PDF very useful in modeling data distributions. In particular, several general features should be noted (see Figs. A.24–A.29):

1. If $\alpha < 1$, $f(x)$ is unbounded at $x_{min} = e^a$.
2. If $\beta < 1$, $f(x)$ is unbounded at $x_{max} = e^b$.
3. If $\alpha > 1$ and $\beta > 1$, $f(x)$ has a mode between x_{min} and x_{max}.
4. If $\alpha > 1$ and $\beta < 1$, $f(x)$ either has both a minimum and a maximum or increases monotonically.
5. If $\alpha < 1$ and $\beta > 1$, $f(x)$ decreases monotonically.

The CDF is given in terms of the incomplete beta function of Eq. (A.75) as

$$F(x) = I_{x^*}(\alpha, \beta), \qquad (A.108)$$

where $x^* = (\ln x - a)/(b - a)$.

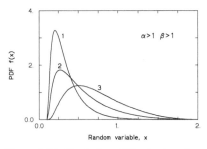

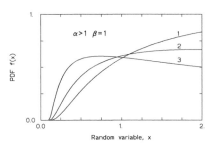

Figure A.24 The logbeta PDF for $(\alpha_1, \beta_1 = 3, 5)$, $(\alpha_2, \beta_2 = 3, 3)$, and $(\alpha_3, \beta_3 = 5, 3)$. Range restricted to $e^a = 0.1 \le x \le 2 = e^b$.

Figure A.25 The logbeta PDF for $(\alpha_1, \beta_1 = 5, 1)$, $(\alpha_2, \beta_2 = 1 + b - a, 1)$, and $(\alpha_3, \beta_3 = 3, 1)$. Range restricted to $e^a = 0.1 \le x \le 2 = e^b$.

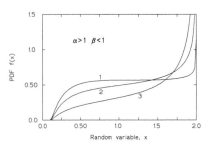

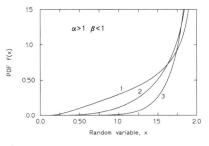

Figure A.26 The logbeta PDF for $(\alpha_1, \beta_1 = 5, 0.5)$, $(\alpha_2, \beta_2 = 10, 0.5)$, and $(\alpha_3, \beta_3 = 20, 0.5)$. Range restricted to $e^a = 0.1 \le x \le 2 = e^b$.

Figure A.27 The logbeta PDF for $(\alpha_1, \beta_1 = 3, 5)$, $(\alpha_2, \beta_2 = 3, 5)$, and $(\alpha_3, \beta_3 = 3, 5)$. Range restricted to $e^a = 0.1 \le x \le 2 = e^b$.

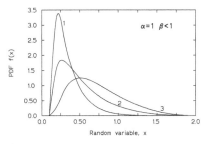

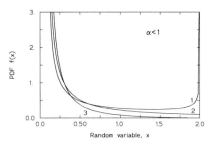

Figure A.28 The logbeta PDF for $(\alpha_1, \beta_1 = 1, 0.7)$, $(\alpha_2, \beta_2 = 1, 0.5)$, and $(\alpha_3, \beta_3 = 1, 0.3)$. Range restricted to $e^a = 0.1 \le x \le 2 = e^b$.

Figure A.29 The logbeta PDF for $(\alpha_1, \beta_1 = 0.6, 0.6)$, $(\alpha_2, \beta_2 = 0.6, 1)$, and $(\alpha_3, \beta_3 = 0.6, 2)$. Range restricted to $e^a = 0.1 \le x \le 2 = e^b$.

The logbeta distribution has been proposed as an alternative to the lognormal distribution. It has the advantage of being able to model both left and right skewness (the lognormal can only model right skewness). It may also be more appropriate when the data have an upper bound.

Moments

No closed-form expression is known to exist that gives the central moments of $f(x)$. Numerical techniques must be used to evaluate

$$\omega'_n \equiv \langle x^n \rangle = \int_{x_{\min}}^{x_{\max}} x^n f(x)\, dx$$

$$= \frac{\Gamma(\alpha + \beta)}{\Gamma(\alpha)\Gamma(\beta)} \left[\int_0^1 y^{\alpha - 1} \left\{ (1 - y)^{\beta - 1} \left[\widehat{f}(y) - \widehat{f}(1) y^{1 - \alpha} \right] - \widehat{f}(0) \right\} dy \right.$$

$$\left. + \left\{ \widehat{f}(1)/\beta + \widehat{f}(0)/\alpha \right\} \right], \quad \text{(A.109)}$$

where $\widehat{f}(y) \equiv \exp[n(a + (b - a)y)]$.

A.2.10 Chi-Squared Distribution

The chi-squared statistic can be defined as

$$\chi^2 = \sum_{i=1}^{N} \frac{(x_i - \mu)^2}{\sigma^2}, \quad \text{(A.110)}$$

where x_i is the ith sample of x from the distribution that has mean μ and variance σ^2 and N is the number of samples. Use of this statistic in data analysis is wide-spread because the "reduced" chi-square, defined as

$$\chi^2_\nu = \frac{\chi^2}{\nu}, \quad \text{(A.111)}$$

where ν is the number of degrees of freedom, approaches unity if the "data" are drawn from the correct distribution with mean μ and variance σ^2. For instance, if data d_i are fit to a model $m(x_i)$ and the square of the deviations of the data from the model value (the numerator $[d_i - m(x)]^2$) is about the same as the variance in the data (the denominator $\sigma^2(d_i)$), normalized by the number of degrees of freedom (the factor $1/\nu$), then χ^2_ν will be approximately unity and one can conclude that the model is a good fit to the data. If the model is not a good fit, the numerator will generally exceed the denominator and χ^2_ν will be greater than unity. This concept has been used widely in data analysis as, for instance, in the rolling-window template-matching procedure of Dunn (2004).

The number of degrees of freedom ν is simply a measure of the number of samples, N, less the number of constraints, c, or

$$\nu = N - c. \quad \text{(A.112)}$$

There are no constraints when only the data themselves are used to determine a statistic, but there is one constraint when the value of a single quantity is estimated from the data and is then used to compute a statistic. This accounts for why the sample mean is given by

$$\bar{x} = \frac{1}{N} \sum_{i=1}^{N} x_i, \qquad (A.113)$$

since only the N data points are used, whereas the sample variance is given by

$$s^2 = \frac{1}{N-1} \sum_{i=1}^{N} [x_i - \bar{x}]^2. \qquad (A.114)$$

Both of these quantities can be written as sums divided by v, where in the former case $v = N$ ($c = 0$) and in the latter case $v = N - 1$ ($c = 1$). See Section 2.1.1 for a more complete discussion of degrees of freedom.

For the chi-squared statistic defined by Eq. (A.110), the *chi-squared* PDF is

$$f(\chi^2|v) = \left[2^{v/2} \Gamma\left(\frac{v}{2}\right) \right]^{-1} \left(\chi^2\right)^{v/2-1} e^{-\chi^2/2}, \qquad v > 0 \quad x \geq 0, \quad (A.115)$$

and is denoted by $\chi^2(v)$. This is simply a special case of the gamma distribution $\mathcal{G}(\alpha, \beta)$, namely,

$$f(\chi^2|v) = \mathcal{G}(v/2, 2). \qquad (A.116)$$

Hence the results obtained for the gamma distribution in Section A.2.3 can be used to obtain the moments and CDF for $\chi^2(v)$. Also sampling methods for the gamma distribution can be used to obtain samples from the chi-squared distribution.

The mean value of the chi-squared distribution is $\mu = v$, and the variance is $\sigma^2 = 2v$. For large v (> 30), the chi-squared distribution is well approximated by a normal distribution with mean v and variance $2v$. See Fig. A.30 for examples of $\chi^2(v)$ and the approximating normal distribution for $v = 40$.

A.2.11   Student's *t* Distribution

The central limit theorem states that, for very large samples, the distribution of the deviations between the sample and population means is described by the unit normal distribution. However, the behavior of smaller samples is only "semi-normal." William Gosset (Fig. A.32), who worked for the Guiness brewery, developed an interesting alternative to the normal distribution. His employer did not want him publishing his findings as an identified employee, so he published several seminal statistical works under the pseudonym "Student." As a result, a distribution he developed is known today as Student's *t* distribution.

Gosset recognized that for small populations, the normal distribution overestimates the achieved results near the mean and underestimates them far from the

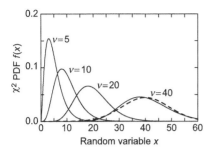

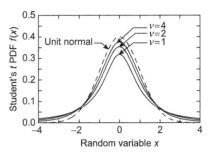

Figure A.30 The PDF for the chi-squared distribution. Shown by the heavy dashed line is the approximating normal distribution.

Figure A.31 The PDF for the Student's t distribution. The heavy dashed line is the limiting unit normal distribution.

mean. He thus devised a distribution that has the desirable properties of (1) more closely fitting the results of finite samples and (2) containing an adjustable parameter.

Student's t distribution, denoted by $\mathcal{S}_t(v)$, has a PDF of the form

$$f(x) = \frac{\Gamma\left(\frac{v+1}{2}\right)}{\Gamma\left(\frac{v}{2}\right)\sqrt{\pi v}}\left(1 + \frac{x^2}{v}\right)^{-(v+1)/2},$$

(A.117)

where the parameter $v > 0$ is known as the number of degrees of freedom. This function is symmetric about $x = 0$ and unimodal and extends to infinity in both directions. The CDF can be written in terms of the incomplete beta function defined by Eq. (A.75) as (Kendall and Stuart, 1977)

Figure A.32 William Sealy Gosset (aka Student) (1876–1937).

$$F(x) = 1 - \frac{1}{2}I_{x^*}\left(\frac{v}{2}, \frac{1}{2}\right),$$

(A.118)

where $x^* = [1 + (x/v)]^{-1}$.

Moments

Moments $\langle x^n \rangle$ about zero exist only for $n < v$. Because of the symmetry of $f(x)$, all odd-order moments vanish. The even moments are

$$\langle x^{2n} \rangle = v^n \frac{\Gamma(n + \frac{1}{2})\Gamma(\frac{v}{2} - n)}{\Gamma(\frac{1}{2})\Gamma(\frac{v}{2})}.$$

(A.119)

Special Cases

When $\nu = 1$, the Student's t distribution is the same as the unit Cauchy distribution $\mathcal{C}(0, 1)$. Also, as $\nu \to \infty$, the Student's t distribution becomes the unit normal $\mathcal{N}(0, 1)$. In Fig. A.31 examples of the Student's t distribution are shown. For $\nu \gtrsim 10$, the unit normal is a good approximation for $\mathcal{S}_t(\nu)$.

Sampling From the Student's t Distribution

If generators are available for the unit normal and the gamma distributions, a convenient way to generate samples from the Student's t distribution is as follows (Fishman, 2003): sample y_1 and y_2 from $\mathcal{N}(0, 1)$ and $\mathcal{G}(\nu/2, 1/2)$, respectively; then $x = y_1/\sqrt{y_2}$ is a sample from $\mathcal{S}_t(\nu)$. Other faster, but more complex, generators are available (see Kinderman and Monahan (1980), Marsaglia (1984), and Stadlober (1982)).

A.2.12   Pareto Distribution

Figure A.33 Vilfredo Federico Damaso Pareto (1848–1923).

The Pareto distribution is a skewed, heavy-tailed distribution, which, like the lognormal distribution, is used to describe rare but important events or conditions such as sizes of meteorites striking earth (many small, few large), radon concentrations in houses (many low, few high), areas burned in forest fires (many small, few large), etc. Vilfredo Pareto (1848–1923) (Fig. A.33) was an economist who used this PDF to describe the distribution of wealth in Italy (many poor, few wealthy) and from which he devised the "80-20" rule that says 20% of the population control 80% of the society's wealth.

There are at least six slightly different PDFs in the Pareto family. The simplest (Type I) is

$$f(x) = \frac{\alpha\beta^\alpha}{x^{\alpha+1}}, \quad \alpha > 0, \quad x \geq \beta > 0, \tag{A.120}$$

and the CDF has the form

$$F(x) = 1 - \left[\frac{\beta}{x}\right]^a. \tag{A.121}$$

This distribution is a power-law distribution, which has been applied in many fields of the physical and social sciences. Sometimes it is referred to as the "Bradford" distribution. The PDF and CDF of the Pareto distribution for $\beta = 1$ are shown in Fig. A.34 for four values of α. On a linear-linear scale, the PDF has the usual sideways "J" shape approaching the two orthogonal axes asymptotically. On a log-log plot, a straight line with slope $-(1 + \alpha)$ results.

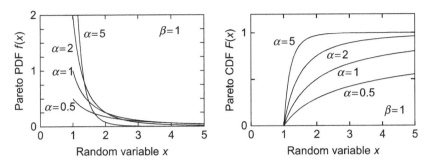

Figure A.34 The PDF and CDF of the Pareto distribution for $\beta = 1$.

Moments and Median

The moments of the Pareto distribution for $\alpha > n$ can be expressed as

$$\langle x^n \rangle = \frac{\alpha \beta^n}{\alpha - n}.$$ (A.122)

For $\alpha \leq n$, the nth moment is infinite or nonexistent.

From this result for the moments, it is easy to show that the mean value is given by

$$\mu = \frac{\alpha \beta}{\alpha - 1}, \quad \alpha > 1,$$ (A.123)

and the mean does not exist for $\alpha \leq 1$. The variance has the form

$$\sigma^2(x) = \frac{\alpha \beta^2}{(\alpha - 1)^2 (\alpha - 2)}, \quad \alpha > 2,$$ (A.124)

and does not exit for $\alpha \leq 2$.

The median of this Pareta distribution is $x_{median} = \beta 2^{1/\alpha}$, and the mode is at $x = \beta$.

Sampling From the Pareto Distribution

The inverse CDF sampling method can be used to obtain samples from the Pareto distribution. From Table A.2, it follows that

$$x_i = \frac{\beta}{(1 - \rho_i)^{1/\alpha}},$$ (A.125)

where ρ_i is a pseudorandom number selected uniformly from the unit interval.

A.3 Joint Distributions

Finally three examples of joint or multivariate distributions are presented: the multivariate normal distribution, the multinomial distribution, and the Dirichlet distribution. A vast number of other joint distributions are covered comprehensively in two definitive books (Johnson and Kotz, 1987; Johnson et al., 2004).

Sampling from multivariate distributions is often done by writing the joint distribution as a product of the marginal distribution of one of the variates and conditional distributions for the other variates. Each univariate distribution is then sampled sequentially to create a sample from the joint distribution. This is the approach discussed earlier in Section 4.1.10 and is especially popular for analyses based on Markov chain Monte Carlo methods. An alternative sampling approach is to build a sample by a direct transformation of a vector of independent identically distributed scalar variates, as discussed in the following section.

A.3.1 Multivariate Normal Distribution

The k-dimensional multivariate normal distribution with mean vector $\boldsymbol{\mu}$ and a nonsingular symmetric covariance matrix $\boldsymbol{\Sigma}$ has the PDF

$$f(\mathbf{x}) = \frac{1}{(2\pi)^{n/2}\sqrt{|\boldsymbol{\Sigma}|}} \exp\left(-[(\mathbf{x} - \boldsymbol{\mu})^T \boldsymbol{\Sigma}^{-1}(\mathbf{x} - \boldsymbol{\mu})]/2\right), \qquad (A.126)$$

where $\mathbf{x}$ is a real k-dimensional column vector with components x_i, $-\infty < x_i < \infty$, and $|\boldsymbol{\Sigma}|$ is the determinant of the $k \times k$ covariance matrix. Equation (A.126) reduces to that of the univariate normal distribution if $\boldsymbol{\Sigma}$ is a 1×1 matrix (i.e., a single real number).

Sampling From the Multivariate Normal Distribution

A direct way of generating random vectors $\mathbf{x}$ from this PDF is to generate a vector $\mathbf{y} = (y_1, y_2, \ldots, y_n)$, where each y_i is a sample from the unit normal distribution $N(0, 1)$. A sample of $\mathbf{x}$ is then formed by the transformation

$$\mathbf{x} = \mathbf{T}^T\mathbf{y} + \boldsymbol{\Sigma}, \qquad (A.127)$$

where $\mathbf{T}$ is an $n \times n$ matrix such that $\mathbf{T}^T\mathbf{T} = \boldsymbol{\Sigma}$ and can be calculated by a Cholesky decomposition of $\boldsymbol{\Sigma}$ (Press et al., 1996).

Another approach for generating samples from a multinomial distribution is to generate x_1 from $N(\mu_1, \sigma_1)$, generate x_2 conditionally on x_1, generate x_2 conditionally on x_1 and x_2, and so on.

The Bivariate Normal Distribution

For the special case of $k = 2$, the bivariate normal distribution is obtained. The covariance matrix for this case is

$$\begin{pmatrix} \sigma_1^2 & \rho\sigma_1\sigma_2 \\ \rho\sigma_1\sigma_2 & \sigma_2^2 \end{pmatrix}, \qquad (A.128)$$

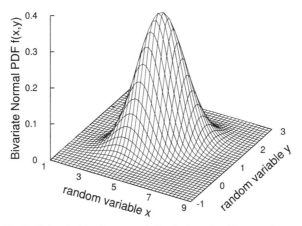

Figure A.35 The PDF for the bivariate normal distribution. For this example $\mu_1 = 5$, $\mu_2 = 1$, $\sigma_1 = 1$, $\sigma_2 = 0.5$, and $\rho = 0.5$.

where ρ is the correlation coefficient defined as $\mathrm{covar}(x_1, x_2)/\sigma_1\sigma_2$. Then $\mathbf{\Sigma}^{-1}$ is given by

$$\mathbf{\Sigma}^{-1} = \begin{pmatrix} \dfrac{1}{\sigma_1^2(1-\rho^2)} & \dfrac{-\rho}{\sigma_1\sigma_2(1-\rho^2)} \\ \dfrac{-\rho}{\sigma_1\sigma_2(1-\rho^2)} & \dfrac{1}{\sigma_2^2(1-\rho^2)} \end{pmatrix}. \tag{A.129}$$

With this result, Eq. (A.126) reduces to

$$f(x_1, x_2) = \frac{1}{2\pi\sigma_1\sigma_2\sqrt{1-\rho^2}} \exp\left\{-\frac{\xi}{2(1-\rho^2)}\right\}, \tag{A.130}$$

where

$$\xi = \frac{(x_1-\mu_1)^2}{\sigma_1^2} - \frac{2\rho(x_1-\mu_1)(x_2-\mu_2)}{\sigma_1\sigma_2} + \frac{(x_2-\mu_2)^2}{\sigma_2^2}. \tag{A.131}$$

An example of the bivariate normal is shown in Fig. A.35.

A.3.2 Multinomial Distribution

The multinomial discrete distribution is a generalization of the binomial distribution in which a trial outcome resulted in one of two outcomes termed "success" or "failure." Now assume there are outcomes such as extracting a colored ball from an urn that has k distinct colors. After each draw the extracted ball is replaced; so the probability of extracting a ball of a particular does not change. Then the multinomial distribution gives the probabilities of extracting n_i balls of color i in a specified number n of trials.

To be more specific, consider a situation in which k mutually exclusive events $E_1, E_2, \ldots, E_k$ can occur with $p_i > 0$ representing the probability event E_i occurs on any given trial. Because an event must occur on every trial, $\sum_{i=1}^{k} p_i = 1$. Define the vector $\mathbf{p} = (p_1, p_2, \ldots, p_k)$. The joint distribution of the random variables $\mathbf{x} = (x_1, x_2, \ldots, x_k)$, where x_i is the number of occurrences of event E_i in n independent trials with $\sum_{i=1}^{k} x_i = n$, is given by the multinomial density distribution

$$f(\mathbf{x}) \equiv f(\mathbf{x}|n, \mathbf{p}) = n! \prod_{i=1}^{k} \frac{p_i^{x_i}}{(x_i)!}, \tag{A.132}$$

which is denoted as $\mathcal{MN}(\mathbf{x}|\mathbf{p})$. This PDF can be written in terms of the gamma function as

$$f(\mathbf{x}|n, \mathbf{p}) = \frac{\Gamma(n+1)}{\prod_i \Gamma(x_i + 1)} \prod_{i=1}^{k} p_i^{x_i}. \tag{A.133}$$

This form shows the resemblance to the Dirichlet distribution, which is its conjugate prior distribution. The Dirichlet distribution is discussed later in Section A.3.3. Because this distribution is a generalization of the binomial distribution (see Section A.1.2), the above reduces to Eq. (A.6) when $k = 2$.

Properties
The expected number of times event i is observed in n trials is

$$\langle x_i \rangle = np_i. \tag{A.134}$$

The covariance matrix has diagonal elements that are the variances of binomially distributed random variables, i.e.,

$$\mathrm{var}(x_i) = np_i(1 - p_i). \tag{A.135}$$

The off-diagonal components are the covariances, which for distinct i and j are

$$\mathrm{cov}(x_i, x_j) = -np_i p_j. \tag{A.136}$$

Note that all the covariances are negative because, for fixed n, a decrease in one component of the multicomponent vector $\mathbf{x}$ requires an increase in another component. The components of the associated correlation matrix are

$$\rho(x_i, x_i) = 1 \tag{A.137}$$

and

$$\rho(x_i, x_j) = \frac{\mathrm{cov}(x_i, x_j)}{\sqrt{\mathrm{var}(x_i)\,\mathrm{var}(x_j)}} = \frac{-p_i p_j}{\sqrt{p_i(1 - p_i)p_j(1 - p_j)}}$$

$$= -\sqrt{\frac{p_i p_j}{(1 - p_i)(1 - p_j)}}.$$ (A.138)

Note that the correlation matrix is independent of the sample size n.

Sampling From the Multinomial Distribution

An important property of the multinomial distribution is that the marginal and conditional distributions of any x_i are binomial distributions. Thus, to generate a multinomial sample, one can simply sample, sequentially, from the binomial marginal and conditional distributions. For the best efficiency, the first marginal considered should be that for the x_i with the largest p_i.

A.3.3 Dirichlet Distribution

The continuous Dirichlet PDF of order k has k parameters $\boldsymbol{\alpha} = (\alpha_1, \ldots, \alpha_k)$, with $\alpha_i > 0$. It is denoted by $\mathcal{D}_i(\boldsymbol{\alpha})$ and is given by

$$f(\mathbf{x}) \equiv f(\mathbf{x}|\boldsymbol{\alpha}) = \frac{1}{B(\boldsymbol{\alpha})} \prod_{i=1}^{k} x_i^{\alpha_i - 1},$$ (A.139)

where, for $i = 1, 2, \ldots, k$, $x_i \geq 0$ and $\sum_{i=1}^{k} x_i = 1$. Here the normalizing constant $B(\boldsymbol{\alpha})$ is the multivariate *beta function*, which can be written in terms of the gamma function as

$$B(\boldsymbol{\alpha}) = \frac{\prod_{i=1}^{k} \Gamma(\alpha_i)}{\Gamma\left(\sum_{i=1}^{k} \alpha_i\right)}.$$ (A.140)

The *support* or domain of the Dirichlet distribution is the set of k-dimensional vectors $\mathbf{x}$ whose components are real numbers in the interval $[0, 1]$ such that $\sum_{i=1}^{k} x_i = 1$.

Difference Between Dirichlet and Multinomial Distributions

A cursory comparison of Eqs. (A.133) and (A.139) would lead one to think both PDFs are nearly the same. Both are distributions over vectors whose components x_i sum to a constant. In both cases, the normalization is the beta function. So what is the difference between these two distributions?

The primary difference between the two distributions is that Dirichlet random variables x_i are *continuous* and real-valued in the interval $[0, 1]$. By contrast, the multinomial random variables x_i are integer-valued (number of events). In other words, the Dirichlet PDF is a continuous PDF whereas the multinomial distribution is a discrete PDF.

Another salient difference is their parameterizations. The multinomial distribution is parameterized using a number of trials n (a positive integer) and a vector of probabilities $p_i \in [0, 1]$ that must sum to one. The Dirichlet distribution, however, is parameterized by a vector $\boldsymbol{\alpha}$ whose components are positive real-valued numbers.

Nevertheless, the two distributions are connected because Dirichlet random variables are often used to parameterize the probabilities in the multinomial distribution. In other words, the vector of probabilities $p_i \in [0, 1]$ for a multinomial random variable is itself a Dirichlet random variable. This is done because it is a pretty natural assumption to make and because it makes the mathematics of inference much simpler.

Properties

If $\mathbf{x} = (x_1, \ldots, x_k)$ are distributed according to $\mathcal{D}_i(\mathbf{x}|\boldsymbol{\alpha})$, the first $k - 1$ components have the density function of Eq. (A.139) and $x_k = 1 - \sum_{i=1}^{k-1} x_i$. Let $\alpha_0 = \sum_{i=1}^{k} \alpha_i$. Then the Dirichlet PDF has the following properties (Wikipedia, 2020c).

mean: $\quad \langle x_i \rangle = \alpha_i / \alpha_0$.

mode: The mode of Eq. (A.139) is the vector $\mathbf{x}$ with

$$x_i = \frac{\alpha_i - 1}{\alpha_0 - k}, \quad i = 1, \ldots, k.$$

variance: $\quad \text{var}(x_i) = \dfrac{\alpha_i(\alpha_0 - \alpha_i)}{\alpha_0^2(\alpha_0 + 1)}.$

covariance: $\quad \text{covar}(x_i, x_j) = \dfrac{-\alpha_i \alpha_j}{\alpha_0^2(\alpha_0 + 1)}, \quad i \neq j.$

The covariance matrix formed from these expressions is singular, because x_k is linearly dependent on the other x_i.

marginals: The marginal distributions $f_{x_i}(x_i)$ are beta distributions, i.e.,

$$f_{x_i}(x_i) = \mathcal{B}_e(x_i|\alpha_i, \alpha_0 - \alpha_i)$$
$$= \frac{\Gamma(\alpha_0)}{\Gamma(\alpha_i)\Gamma(\alpha_0 - \alpha_i)} x_i^{\alpha_i - 1}(a - x_i)^{\alpha_0 - \alpha_i - 1}.$$

Use as a Conjugate Prior Distribution

In Bayesian inference applications, the Dirichlet PDF is the conjugate prior PDF to both the categorical[6] and multinomial distributions. This means if observa-

---

[6] The categorical distribution is the generalization of the Bernoulli distribution for a discrete variable with $k > 2$ possible outcomes, such as the roll of a die. The distribution parameters are the probabilities $0 \leq p_i \leq 1$ of each outcome constrained by $\sum_{i=1}^{k} p_i = 1$. The categorical distribution is a special case of the multinomial distribution, in that it gives the probabilities of potential outcomes of a single drawing rather than multiple drawings.

tions (or data) are governed by either of these two distributions, the posterior distribution is also a Dirichlet PDF. Then starting with what one knows about the prior parameters (usually very little), one can refine values of the parameters as data are obtained because the posterior, which is also a Dirichlet PDF, can be used as a new prior, all without having to alter the form of the equation for the posterior PDF. This feature means knowledge about parameter values can be periodically updated, with almost no mathematical effort, as more observations are accumulated.

Sampling From a Dirichlet Distribution

It is easy to generate samples $\mathbf{x} = (x_1, \ldots, x_k)$ from a k-dimensional Dirichlet PDF with parameters $(\alpha_1, \ldots, \alpha_k)$ (Wikipedia, 2020c). First generate k independent samples $y_1, \ldots, y_k$ from gamma distributions each with a density

$$\mathcal{G}(y_i|\alpha_i, 1) = \frac{y_i^{\alpha_i - 1} e^{-y_i}}{\Gamma(\alpha_i)}. \tag{A.141}$$

Then take

$$x_i = y_i \bigg/ \sum_{j=1}^{k} y_j. \tag{A.142}$$

Note that this procedure is independent of how the gamma distribution is parameterized (shape/scale versus shape/rate) since $\beta' = \beta = 1$.

References

Ahrens, J.H., Dieter, U., 1974. Computer methods for sampling from gamma, beta, poisson, and binomial distributions. Computing 25, 193–208.

Ahrens, J.H., Dieter, U., 1988. Efficient table-free sampling methods for the exponential, Cauchy and normal distributions. Commun. ACM 31, 1330–1337.

Ahrens, J.H., Dieter, U., 1991. A convenient sampling method with bounded computing times for Poisson distributions. In: Nelson, P.R., Dudewicz, E.J., Öztürk, A., van der Meulen, E.C. (Eds.), The Frontiers of Statistical Computation, Simulation and Modeling. American Science Press, Columbus, OH, pp. 137–149.

Atkinson, A.C., Whittaker, J., 1976. A switching algorithm for the computer generation of beta random variables with at least one parameter less than one. J. R. Stat. Soc. A 139, 462–467.

Atkinson, A.C., 1979. The computer generation of Poisson random variables. Appl. Stat. 28, 29–35.

Best, D.J., 1983. A note on gamma variate generators with shape parameter less than unity. Computing 30, 152–157.

Box, G.E.P., Muller, M.E., 1958. A note on the generation of normal deviates. Am. Math. Stat. 29, 610–611.

Cheng, R.C.H., 1978. Generating beta variates with nonintegral shape parameters. Commun. ACM 21, 290–295.

Cheng, R.C.H., Feast, G.M., 1979. Some simple gamma variate generators. Appl. Stat. 28, 290–295.

Devroye, Luc, 1981. The computer generation of Poisson random variables. Computing 26, 197–205.

Devroye, Luc, 1986. Nonuniform Random Variate Generation. Springer-Verlag, New York.

Dunn, W.L., 2004. Flaw detection using the rolling-window template-matching procedure. Appl. Radiat. Isot., 1–10.

Faw, R.E., Shultis, J.K., 1999. Radiological Assessment: Sources and Doses. Am. Nuclear Society, LaGrange Park, IL.

Fishman, G.S., 2003. Monte Carlo Concepts, Algorithms, and Applications. Corrected Printing. Springer, New York.

Gentle, J.E., 1998. Random Number Generation and Monte Carlo Methods. Springer-Verlag, New York.

Hastings, C., 1955. Approximations for Digital Computers. Princeton Univ. Press, Princeston, NJ, p. 192.

Jöhnk, M.D., 1964. Erzeugung von Betaverteilter und Gammaverteilter Zufallszahlen. Metrika 8, 5–15.

Johnson, N.L., Kotz, S., 1987. Distributions in Statistics, Continuous Multivariate Distributions. John Wiley & Sons, New York.

Johnson, N.L., Kotz, S., Balakrishnan, N., 1992. Univariate Discrete Distributions, 2nd ed. Wiley, New York.

Johnson, N.L., Kotz, S., Balakrishnan, N., 1994. Continuous Univariate Distributions, vol. 1, 2nd ed. Wiley, New York.

Johnson, N.L., Kotz, S., Balakrishnan, N., 1995. Continuous Univariate Distributions, vol. 2, 2nd ed. Wiley, New York.

Johnson, N.L., Kotz, S., Balakrishnan, N., 1997. Discrete Multivariate Distributions. Wiley, New York.

Johnson, N.L., Kotz, S., Balakrishnan, N., 2004. Discrete Multivariate Distributions. Wiley, New York.

Kachitvichyanukul, V., Schmeiser, B.C., 1988. Binomial random variate generation. Commun. ACM 31, 216–223.

Kemp, C.D., 1986. A modal method for generating binomial variables. Commun. Stat. 15, 805–813.

Kendall, M., Stuart, A., 1977. The Advanced Theory of Statistics, vol. 1, 3rd ed. MacMillan Publ., New York.

Kinderman, A.J., Monahan, J.F., 1980. New methods for generating Student's t and gamma variables. Computing 25, 367–377.

Kronmal, J.C., Peterson, A.V., 1981. A variant of the acceptance/rejection method for generating random variables. J. Am. Stat. Assoc. 76, 446–451.

Marsaglia, G., 1984. The exact-approximation method for generating random variables in a computer. J. Am. Stat. Assoc. 79, 218–221.

McGrath, E.J., Irving, D.C., 1975. Techniques for Efficient Monte Carlo Simulations, vol. II, Random Number Generation for Selected Probability Distributions. Report RSIC-38. Radiation Shielding Information Center, Oak Ridge National Lab.

Press, W.H., Teukolsky, S.A., Vetterling, W.T., Flannery, B.P., 1996. Numerical Recipes, 2nd ed. Cambridge Univ. Press, Cambridge.

Rubinstein, R.Y., 1981. Simulation and the Monte Carlo Method. Wiley, New York.

Schmeiser, B.W., Babu, A.J.G., 1980. Beta variate generation via exponential majorizing functions. Oper. Res. 28, 719–926. Corrected Oper. Res. 31, 802.

Schmeiser, B.W., Kachitvichyanukul, V., 1981. Poisson Random Variate Generation. Res. Memo. School of Industrial Engineering, Purdue Univ., Lafayette, IN, pp. 81–84.

Snedecor, G.W., Cochran, W.G., 1971. Statistical Methods, 6th ed. The Iowa State University Press, Ames, IA.

Stadlober, E., 1982. Generating Student's t variates by a modified rejection method. In: Grossmann, W., Plug, G., Wertz, W. (Eds.), Probability and Statistical Inference. Reidel, Dordrecht, Netherlands, pp. 349–360.

Stadlober, E., 1991. Binomial variate generation: a method based on ratios of uniforms. In: Nelson, P.R., Dudewicz, E.J., Öztürk, A., van der Meulen, E.C. (Eds.), The Frontiers of Statistical Computation, Simulation and Modeling. American Science Press, Columbus, OH, pp. 93–112.

Wikipedia Contributors, 2020a. Generalized hypergeometric function. In: Wikipedia, The Free Encyclopedia. https://en.wikipedia.org/wiki/Generalized_hypergeometric_function. (Accessed 25 June 2020).

Wikipedia Contributors, 2020b. Negative hypergeometric distribution. In: Wikipedia, The Free Encyclopedia. https://en.wikipedia.org/wiki/Negative_hypergeometric_distribution. (Accessed 27 June 2020).

Wikipedia Contributors, 2020c. Dirichlet distribution. In: Wikipedia, The Free Encyclopedia. https://en.wikipedia.org/wiki/Negative_hypergeometric_distribution. (Accessed 4 July 2020).

Zechner, H., Stadlober, E., 1993. Generating beta variates via patchwork rejection. Computing 50, 1–18.

Appendix B

The Weak and Strong Laws of Large Numbers

In Chapter 1, it is stated that the history of Monte Carlo dates, in some sense, to 1689 when Jacob Bernoulli first enunciated the law of large numbers (Bernoulli, 1713). Bernoulli spent almost 20 years devising a rigorous proof of what he called his "golden theorem" but which was generally referred to as "Bernoulli's theorem." Simèon-Denis Poisson referred to this theorem, in 1835, as "*la loi des grands nombres*" and this label "the law of large numbers" is commonly used today.

Two forms of this law, weak and strong, are now distinguished. The differences between these forms are unimportant for Monte Carlo applications, as both state that the sample mean $\overline{x}_N$ approaches the population mean μ in the limit of large N, i.e., $\lim_{N\to\infty}\overline{x}_N = \mu$. It almost seems intuitive that, as long as the process being simulated has a finite population mean, the sample mean should approach the true mean as the number of samples increase. The distinction between the two forms of the law has to do with the manner in which this convergence takes place. For the interested reader, the distinction between the two is briefly discussed here.

B.1 The Weak Law of Large Numbers

Consider a sequence of N random samples, denoted by $\{x_i\}$, $i = 1, 2, ..., N$, that are uncorrelated so that $\langle x_k x_j \rangle = \langle x_k \rangle \langle x_j \rangle$ and $\sigma^2(x_k + x_j) = \sigma^2(x_k) + \sigma^2(x_j)$, i.e., the covariance between any two samples x_k and x_j is zero. Another way to state this restriction is that all members of the random sequence are independent and identically distributed. The population mean is

$$\mu = \langle x_i \rangle \equiv E(x_i), \tag{B.1}$$

while the sample mean is given by

$$\overline{x}_N = \frac{1}{N} \sum_{i=1}^{N} x_i. \tag{B.2}$$

The weak law states that, for all positive values of ϵ, no matter how small,

$$\lim_{N\to\infty} \text{Prob}\{|\overline{x}_N - \mu| \geq \epsilon\} = 0. \tag{B.3}$$

Exploring Monte Carlo Methods. https://doi.org/10.1016/B978-0-12-819739-4.00020-2

Proof

To prove the weak law, consider a PDF $f(x)$ with finite mean μ and finite variance σ^2. If $g(x)$ is any nonnegative function and k is any positive constant, then

$$\langle g(x) \rangle = \int_{-\infty}^{\infty} g(x) f(x)\, dx = \int_{\{x:g(x) \geq k\}} g(x) f(x)\, dx \quad + \quad \int_{\{x:g(x) < k\}} g(x) f(x)\, dx$$

$$\geq \int_{\{x:g(x) \geq k\}} g(x) f(x)\, dx \geq \int_{\{x:g(x) \geq k\}} k f(x)\, dx = k\, \mathrm{Prob}\{g(x) \geq k\}$$

or

$$\mathrm{Prob}\{g(x) \geq k\} \leq \langle g(x) \rangle / k. \tag{B.4}$$

In particular, let $g(x) = (x - \mu)^2$ and $k = \lambda^2 \sigma^2$, so that Eq. (B.4) gives

$$\mathrm{Prob}\{|x - \mu| \geq \lambda \sigma(x)\} = \mathrm{Prob}\{(x - \mu)^2 \geq \lambda^2 \sigma^2(x)\} \leq \frac{\langle (x - \mu)^2 \rangle}{\lambda^2 \sigma^2(x)} = \frac{1}{\lambda^2} \tag{B.5}$$

for all $\lambda > 0$. This result is known as the Bienaymé–Chebyshev inequality and leads directly to the weak law. The quantity $\overline{x}_N$ is itself a random variable with mean μ and variance $\sigma_{\overline{x}}^2 = \sigma^2(x)/N$. Set $\lambda \sigma_{\overline{x}} = \epsilon$ and use Eq. (B.5) to obtain

$$\mathrm{Prob}\{|\overline{x}_N - \mu| \geq \epsilon\} \leq \frac{\sigma^2(x)}{N\epsilon^2}. \tag{B.6}$$

Because it is implicitly assumed that $\sigma^2(x)$ is bounded, the right side approaches zero as N approaches infinity, thus proving the weak law. This law can be written as

$$\boxed{\lim_{N \to \infty} \mathrm{Prob}\{|\overline{x}_N - \mu| \geq \epsilon\} = 0} \tag{B.7}$$

or, equivalently,

$$\boxed{\lim_{N \to \infty} \mathrm{Prob}\{|\overline{x}_N - \mu| < \epsilon\} = 1.} \tag{B.8}$$

In words, the weak law states that for any nonzero tolerance ϵ, no matter how small, with a sufficiently large sample there is a high probability that the sample average is close to the expected value, i.e., within the tolerance margin. This type of convergence for $\overline{x}_N$ is technically called *convergence in probability* or *weak convergence* of a random variable.

Example B.1 Use of the Weak Law

What should the sample size N be to ensure with 99% probability that $\bar{x}_N$ is within 0.2σ of the population mean μ?

From Eq. (B.6), one immediately obtains

$$\text{Prob}\{|\bar{x}_N - \mu| < \epsilon\} \geq 1 - \frac{\sigma^2}{N\epsilon^2(x)} \equiv 1 - \delta$$

for $\delta > \sigma^2/(N\epsilon^2)$ or $N > \sigma^2(x)/(\epsilon^2\delta)$. Here one has $\epsilon = 0.2\sigma$ and $\delta = 0.01$, so

$$N > \frac{\sigma^2}{\delta\epsilon^2} = \frac{\sigma^2}{(0.01)(0.2\sigma)^2} = 2500.$$

B.2 The Strong Law of Large Numbers

The weak law is a statement about the *limiting* behavior of sums of random variables. It is possible, however, to make a stronger statement and say something about the behavior of the sum $\sum_{i=1}^{N} x_i$ on the way to the limit. The derivation of the strong law is much more complex than that for the weak law and is simply stated here as follows (Kendall and Stuart, 1977). For any small $\epsilon > 0$ and $0 < \delta < 1$, a value of n exists such that, for any specified $m > 0$,

$$\text{Prob}\{|\bar{x}_N - \mu| \geq \epsilon\} \leq \delta, \tag{B.9}$$

where $N = n, n + 1, ..., n + m$. Equivalently, the strong law requires

$$\boxed{\text{Prob}\{\lim_{N \to \infty} \bar{x}_N = \mu\} = 1.} \tag{B.10}$$

This theorem says more than does the weak law. In particular, it asserts that, for any specified $\epsilon > 0$ and $0 < \delta < 1$, one can identify a number of histories n such that Eq. (B.9) holds for a sequence of values of $N > n$ of arbitrary finite length m. Occasional deviations $|\bar{x}_N - \mu|$ may exceed ϵ but these are few. The strong law, thus, requires that the sample mean approaches the population mean in a more demanding way. Technically, the sample average is said to *converge almost surely* or to *converge strongly*. A proof is given by Etemadi (1981).

B.2.1 Difference Between the Weak and Strong Laws

The weak form of the law of large numbers states that, for a specified large N, the sample mean $\bar{x}_N$ is likely to be near μ. But it leaves open the possibility that cases when $|\bar{x}_N - \mu| > \epsilon$, i.e., when the deviation of $\bar{x}_N$ from μ is outside some small tolerance ϵ, can occur an arbitrary (even infinite) number of times as N increases; however, such occurrences happen at infrequent intervals. The strong law shows that such an infinite number of occurrences cannot happen, i.e., $|\overline{x}_N - \mu|$ will lie outside ϵ at most only a finite number of times.

In other words, the weak law states that $|\overline{x}_N - \mu|$ eventually becomes small but that not every value on the way is small. It may be that, for some N, the difference is relatively large. The strong law says that the probability of such a large event, however, is extremely small. Either form of the law is sufficient to form a basis for the Monte Carlo method, because both indicate that $\lim_{N \to \infty} \overline{x}_N = \mu$.

B.2.2 Other Subtleties

It should be noted that the law of large numbers also holds even for an unbounded variance σ^2 (Cremér, 1947). However, for an infinite variance, the proof of the law is considerably more complex that than given above for the weak law. A very large or infinite variance simply slows the convergence of $\overline{x}_N$ to μ.

Thus, both the strong and weak laws are valid for independent random variables that are identically distributed under the *sole* requirement that the mean exists. If the mean does not exist, then the sample mean does not have a distribution that becomes narrower as the sample size N increases. An example of a distribution without a mean is the Cauchy distribution. The distribution of the sample mean is always the same for all N, namely, the Cauchy distribution itself.

Often the law of large numbers, also known as the "law of averages," is misused, as exemplified by the gambler's fallacy. Gamblers often mistakenly believe that, in a fair game of chance, for example, a long series of losses must be followed by a series in which wins are more frequent in order to balance out the even win/loss ratio required by the law of averages for a large number of games. As another example, consider the tossing of a coin whose probability of a head is p. The law of large numbers does *not* imply that the number of heads in N tosses necessarily deviate little from the expected number of heads Np; rather it states that the relative frequency of heads to flips is close to p. Nor does this law imply that, if the ratio of heads to flips is $> p$ after N flips, the probability of tails on subsequent flips becomes larger to compensate for the surplus of heads. Nature "averages out" by swamping (large N), rather than by fluctuating (keeping $\overline{x}_N$ close to μ within every subset of samples)!

References

Bernoulli, J., 1713. Ars Conjectandi. Impensis Thurnisiorum, Fratrun, Basilea. This posthumus Latin work was subsequently published in German as Wahrscheinlichkeitsrechnung. Englemann, Leipzig, 1899, two volumes.

Cremér, H., 1947. Mathematical Methods for Statistics. Princeton University Press, Princeton, New Jersey.

Etemadi, N., 1981. An elementary proof of the strong law of large numbers. Z. Wahrscheinlichkeitstheor. Verw. Geb. 55, 119–122.

Kendall, M., Stuart, A., 1977. The Advanced Theory of Statistics, vol. 1: Distribution Theory, fourth ed. MacMillan Publishing Co., Inc., New York, NY.

Appendix C

Central Limit Theorem

In this appendix, the validity of the central limit theorem (CLT) is demonstrated. Although the proof introduces some statistical concepts not previously discussed, the extreme importance that this theorem plays in Monte Carlo calculations merits examining the underpinnings of this amazing theorem. One, of course, may just accept the validity of the central limit theorem, but, for those interested, the pertinent details are presented in this appendix.

C.1 Moment-Generating Functions

The *moment-generating function* (MGF) of the random variable x with a probability density function (PDF) $f(x)$ is defined by

$$M_x(t) \equiv \langle e^{tx} \rangle = \begin{cases} \sum_j e^{tx_j} f(x_j), & \text{for a discrete distribution,} \\ \int e^{tx} f(x)\, dx, & \text{for a continuous distribution.} \end{cases} \tag{C.1}$$

This function plays a central role in probability theory and its applications because it has many useful properties.

One of the more important properties of the MGF is that it readily yields all the moments of the PDF. From its definition

$$M_x(t) = \langle e^{tx} \rangle = \left\langle 1 + tx + \frac{(tx)^2}{2!} + \frac{(tx)^3}{3!} + \cdots \right\rangle$$

$$= 1 + \langle tx \rangle + \langle (tx)^2/2! \rangle + \langle (tx)^3/3! \rangle + \cdots$$

$$= 1 + t\langle x \rangle + \frac{t^2}{2!}\langle x^2 \rangle + \frac{t^3}{3!}\langle x^3 \rangle + \cdots$$

$$= 1 + \mu_1 t + \mu_2 \frac{t^2}{2!} + \mu_3 \frac{t^3}{3!} + \cdots, \tag{C.2}$$

where $\mu_n = \langle x^n \rangle$ is the nth moment of the PDF $f(x)$. Now differentiate this last expression n times and set $t = 0$ in the result to obtain

$$\left. \frac{d^n M_x(t)}{dt^n} \right|_{t=0} = \mu_n. \tag{C.3}$$

Exploring Monte Carlo Methods. https://doi.org/10.1016/B978-0-12-819739-4.00021-4

Thus, if the characteristic function is known, it is a simple matter to compute all the moments of the PDF.

If all the moments are known, then $M_x(t)$ is also known because a Maclaurin series gives

$$M_x(t) = M_x(0) + M_x'(0)t + M_x''(0)\frac{t^2}{2!} + M_x'''(0)\frac{t^3}{3!} + \cdots$$

$$= 1 + \mu_1 t + \mu_2 \frac{t^2}{2!} + \mu_3 \frac{t^3}{3!} + \cdots . \tag{C.4}$$

Thus, knowing all the moments of a PDF is equivalent to knowing the PDF itself.

C.1.1 Central Moments

In some applications, the *central* moment μ_n' is defined by $\langle (x - \mu)^n \rangle$, where $\mu = \mu_1$ is the first moment or mean of the distribution. These central moments can be related to the moments μ_n. For example, the first central moment is always zero ($\langle x - \mu \rangle = \langle x \rangle - \mu = 0$) and the second central moment is the variance given by

$$\text{var}(x) \equiv \sigma^2 \equiv \mu_2' \equiv \langle (x - \mu)^2 \rangle = \langle x^2 - 2\mu x + \mu^2 \rangle$$

$$= \langle x^2 \rangle - 2\mu \langle x \rangle + \mu^2 = \langle x^2 \rangle - 2\mu^2 + \mu^2 = \mu_2 - \mu^2. \tag{C.5}$$

In general, the central moments μ_n' can be related to the moments μ_n and vice versa. It can be shown (Kendall and Stuart, 1977) that

$$\mu_n = \sum_{j=0}^{n} \binom{n}{m} \mu_{n-m}' \mu^m, \tag{C.6}$$

$$\mu_n' = \sum_{m=0}^{n} \binom{n}{m} \mu_{n-m} (-\mu)^m. \tag{C.7}$$

C.1.2 Some Properties of the Moment-Generating Function

The MGF has many other useful properties. Here two simple properties are derived. The MGF of cx, where c is a constant, is

$$M_{cx}(t) = \langle e^{cxt} \rangle = \langle e^{x(ct)} \rangle = M_x(ct). \tag{C.8}$$

Similarly, the MGF for $x + c$ is

$$M_{x+c}(t) = \langle e^{(x+c)t} \rangle = e^{ct} \langle xt \rangle = e^{ct} M_x(t). \tag{C.9}$$

Another property, needed for the proof of the central limit theorem, is the MGF for a sum of independent random variables. Let $S = \sum_{i=1}^{N} x_i$, where x_i are independent random variables. Then

$$M_S(t) = \langle e^{tS} \rangle = \langle e^{t(x_1 + x_2 + \ldots + x_N)} \rangle = \left\langle \prod_{i=1}^{N} e^{tx_i} \right\rangle.$$

Because x_i are independent,

$$M_S(t) = \prod_{i=1}^{N} \langle e^{tx_i} \rangle = \prod_{i=1}^{N} M_{x_i}(t). \tag{C.10}$$

In words, the MGF of a sum of independent random variables is the product of their individual MGFs. From this result and Eq. (C.9), the MGF of $S = c_1 x_1 + c_2 x_2 + \ldots + c_N x_N$, where c_i are constants, is

$$M_S(t) = \prod_{i=1}^{N} M_{x_i}(c_i t). \tag{C.11}$$

Example C.1 Moment-Generating Function for a Normal Distribution

The PDF for a normal distribution with mean μ and variance σ^2 is

$$f(x) = \frac{1}{\sqrt{2\pi}\sigma} \exp[-(x - \mu)^2/\sigma^2].$$

The random variable $y = (x - \mu)/\sigma$ is distributed according the unit normal

$$g(y) = \frac{1}{\sqrt{2\pi}} e^{-y^2/2},$$

which has the MGF

$$M_y(t) = \langle e^{yt} \rangle = \frac{1}{\sqrt{2\pi}} \int_{-\infty}^{\infty} e^{yt} e^{-y^2/2} \, dy = e^{t^2/2} \frac{1}{\sqrt{2\pi}} \int_{-\infty}^{\infty} e^{-(y-t)^2/2} \, d(y - t)$$

$$= e^{t^2/2} \frac{1}{\sqrt{2\pi}} \int_{-\infty}^{\infty} e^{-z^2/2} \, dz = e^{t^2/2}. \tag{C.12}$$

Finally, use Eqs. (C.8) and (C.9) with $x = \sigma y + \mu$ to obtain

$$M_x(t) = e^{\mu t} M_{\sigma y}(t) = e^{\mu t} e^{\sigma^2 t^2/2} = \exp\left[\frac{\sigma^2 t^2}{2} + \mu t\right]. \tag{C.13}$$

C.1.3   Uniqueness of the Moment-Generating Function

Does the MGF or the moments μ_n completely characterize the PDF $f(x)$? Could there be two PDFs that have the same MGF and the same moments? The answer is no. If the MGF for a random variable x_1 is the same as that for another random variable x_2, then the PDFs for the two random variables are identical. Although this is intuitively reasonable, a rigorous proof is rather complicated (see, for example, Cramér (1946)). Here, we quote two related theorems about the uniqueness of the MGF and a distribution's moments (Cramér, 1946).

- **Theorem 1:** If an MGF $M_x(t)$ exists in an interval that includes the origin $t = 0$, then there is only one PDF having that MFG.
- **Theorem 2:** If the moments $\mu_1, \mu_2, \ldots$ exist such that the right-hand side of Eq. (C.4) converges in an interval containing the origin $t = 0$, there is only one distribution having those moments.

Thus, if two distributions have the same MGFs or the same moments, they must be the same. This is the key fact needed to prove the central limit theorem.

C.2   The Central Limit Theorem

Pierre Simon de Laplace (1749–1827) is largely responsible for discovering and demonstrating the central role the normal distribution plays in the mathematical theory of probability. He also proved what is today called the central limit theorem (Laplace, 1810). This theorem was a generalization of the special case of Moivre's (1733) earlier result. A complete statement of the central limit theorem is as follows (Riley et al., 2002).

Central Limit Theorem. *Let* x_i, $i = 1, 2, \ldots, N$, *be* independent *random variables, each of which is described by a PDF* $f_i(x)$ *(these may be all different) with a mean* μ_i *and variance* σ_i^2. *The random variable* $z = \sum_i x_i/N$, *i.e., the average of the* x_i, *has the following properties:*

1. *its expected value is given by* $\langle z \rangle = (\sum_i \mu_i)/N$;
2. *its variance is given by* $\mathrm{var}(z) = (\sum_i \sigma_i^2)/N^2$;
3. *as* $N \to \infty$, *the PDF of* z *tends to a normal distribution with the same mean and variance.*

It should be noted that for this theorem to hold, the PDFs $f_i(x)$ must have well-defined means and variances. If, for example, the $f_i(x)$ were Cauchy distributions, which have no mean, then the theorem would not apply.

The first two properties are easily proved. The first is proved as follows:

$$\langle z \rangle = \frac{1}{N}(\langle x_1 \rangle + \langle x_2 \rangle + \cdots + \langle x_N \rangle) = \frac{1}{N}(\mu_1 + \mu_2 + \cdots + \mu_N) = \frac{\sum_i \mu_i}{N}.$$
(C.14)

Note that this result does *not* require the x_i to be *independent*. Indeed if $\mu_i = \mu$ for all i, then the above result reduces to $\langle z \rangle = N\mu/N = \mu$. The second property, however, does require the x_i to be independent so that all covariances are zero. Then

$$\text{var}(z) = \text{var}\left(\frac{1}{N}[x_1 + x_2 + \ldots + x_N]\right)$$

$$= \frac{1}{N^2}[\text{var}(x_1) + \text{var}(x_2) + \ldots + \text{var}(x_N)] = \frac{\sum_i \sigma_i^2}{N^2}. \qquad \text{(C.15)}$$

If x_i are all from the same distribution, then $\text{var}(z) = \sigma^2/N$.

Now examine property 3, a property that accounts for the ubiquity of the normal distribution throughout statistics. Begin by considering the MGF of z. From Eq. (C.11), this MGF is given by

$$M_z(t) = \prod_{i=1}^N M_{x_i}\left(\frac{t}{N}\right), \qquad \text{(C.16)}$$

where $M_{x_i}(t)$ is the MGF of $f_i(x)$. Following the expansion in Eq. (C.4), expand $M_{x_i}(t/N)$ as

$$M_{x_i}\left(\frac{t}{N}\right) = 1 + \frac{t}{N}\langle x_i \rangle + \frac{t^2}{2!N^2}\langle x_i^2 \rangle + \ldots = 1 + \mu_{1i}\frac{t}{N} + \mu_{2i}\frac{t^2}{2!N^2} + \ldots. \qquad \text{(C.17)}$$

Here $\mu_{1i}, \mu_{2i}, \ldots$ are the moments of $f_i(x)$. With Eq. (C.5), this result can be written in terms of the central moments as

$$M_{x_i}\left(\frac{t}{N}\right) = 1 + \mu_i\frac{t}{N} + \frac{1}{2}(\sigma_i^2 + \mu_i^2)\frac{t^2}{N^2} + \ldots. \qquad \text{(C.18)}$$

As N becomes large,

$$M_{x_i}\left(\frac{t}{N}\right) \simeq \exp\left[\frac{\mu_i t}{N} + \frac{1}{2}\sigma_i^2\frac{t^2}{N^2}\right],$$

which can be verified by expanding the exponential up to terms of order $(t/N)^2$ to obtain Eq. (C.18). Then from Eq. (C.16) one obtains, for large N,

$$M_z(t) \simeq \prod_{i=1}^N \exp\left(\frac{\mu_i t}{N} + \frac{1}{2}\sigma_i^2\frac{t^2}{2}\right) = \exp\left(\frac{\sum_i \mu_i}{N}t + \frac{1}{2}\frac{\sum_i \sigma_i^2}{N^2}t^2\right). \qquad \text{(C.19)}$$

Upon comparison of this MGF to that of the normal distribution (see Eq. (C.13)), it is observed that they are the same. Thus, the PDF of the random variable z, as N becomes large, approaches a normal distribution with mean $\sum_i \mu_i/N$ and variance $\sum_i \sigma_i^2/N^2$.

In particular, if z is the sum of N independent measurements of the same random variable, i.e., from the same PDF $f(x)$, then, as $N \to \infty$, z has a normal distribution with mean μ and variance σ^2.

References

Cramér, H., 1946. Mathematical Methods of Statistics. Princeton University Press, Princeton, New Jersey.

Kendall, M., Stuart, A., 1977. The Advanced Theory of Statistics, vol. 1: Distribution Theory, fourth ed. MacMillan Publishing Co., Inc., New York, NY.

Laplace, P.S., 1810. 1898 Mémoire sur Les Approximations des Formules qui Sont Fonctions de Trés Grand Nombres et sur Leur Application aux Probabilités, vol. 12. Gauthier-Villars, Paris, pp. 301–345.

Moivre, Abraham De, 1733. Approximatio ad Summam Terminorum Binomii $(a + b)^n$ in Seriem Expansi;

Republished: Smith, D.E. A Method of Approximating the Sum of Terms of the Binomial $(a + b)^n$ Expanded into a Series, from Whence Are Deduced Some Practical Rules to Estimate the Degree of Assent Which Is to Be Given to Experiments. A Source Book in Mathematics, vol. 2. Dover, New York, 1959, pp. 566–575.

Riley, K.F., Hobson, M.P., Bence, S.J., 2002. Mathematical Methods for Physics and Engineering, 2nd ed. Cambridge University Press, Cambridge.

Appendix D

Linear Operators

D.1 Linear Operators

An operator $\mathcal{O}$ maps a function $f(\boldsymbol{\xi})$ into another function $g(\boldsymbol{\xi})$, i.e.,

$$\mathcal{O}f(\boldsymbol{\xi}) = g(\boldsymbol{\xi}), \qquad \boldsymbol{\xi} \in \mathcal{R}, \tag{D.1}$$

where $\boldsymbol{\xi}$ are the independent variables. For example, $\boldsymbol{\xi} \equiv x, y, z, t$ for a function that depends on position (x, y, z) and time t. The operator is *linear* if $\mathcal{O}[f_1(\boldsymbol{\xi}) + f_2(\boldsymbol{\xi})] = \mathcal{O}[f_1(\boldsymbol{\xi})] + \mathcal{O}[f_2(\boldsymbol{\xi})]$. Some simple operators are

$$\mathcal{O} = \frac{d}{dx}, \qquad \mathcal{O} = \frac{d^n}{dx^n}, \qquad \mathcal{O} = \nabla^2 + \frac{\partial}{\partial t}, \qquad \mathcal{O}[\cdot] = \int_a^b K(x, x')[\cdot]dx'.$$

As a special case, matrix multiplication by an $n \times n$ matrix $\mathbf{A}$ is also a linear operator, whereby an n-component vector $\mathbf{x}$ is mapped into another n-component vector $\mathbf{y}$, i.e.,

$$\mathbf{Ax} = \mathbf{y}.$$

D.2 Inner Product

An *inner product* assigns a number to a pair of functions, namely,

$$(f, g) \equiv \int_{\mathcal{R}} f^*(\boldsymbol{\xi})g(\boldsymbol{\xi})d\boldsymbol{\xi}, \tag{D.2}$$

where f^* denotes the complex conjugate of f and $\mathcal{R}$ is the range of the independent variables. In this treatise, it is assumed that all functions are *square integrable* or, in math jargon, $f, g \in L_2$. For a function to be square integrable

$$\int_{\mathcal{R}} f^*(\boldsymbol{\xi})f(\boldsymbol{\xi})d\boldsymbol{\xi} < \infty.$$

Many common functions like e^x are not square integrable if the range $\mathcal{R} = [a, b]$ becomes infinite, i.e., if $[a, b] \to [-\infty, \infty]$.

The inner product has the following three properties:

$$(1)(\alpha g, f) = \alpha^*(g, f),$$

Exploring Monte Carlo Methods. https://doi.org/10.1016/B978-0-12-819739-4.00022-6

$$(2)(g, \alpha f) = \alpha(g, f),$$

$$(3)(g, \alpha_1 f_1 + \alpha_2 f_2) = \alpha_1(g, f_1) + \alpha_2(g, f_2),$$

where α is any constant.

D.3 Adjoint of a Linear Operator

Let $\mathcal{O}$ be a continuous linear operator that operates on square integrable functions $f(\boldsymbol{\xi})$; $\mathcal{O}^\dagger$ is the *adjoint* of $\mathcal{O}$ if

$$(g^\dagger, \mathcal{O}f) = (\mathcal{O}^\dagger g^\dagger, f), \tag{D.3}$$

or, equivalently,

$$\int_{\mathcal{R}} g^{\dagger *}(\boldsymbol{\xi})[\mathcal{O}f(\boldsymbol{\xi})] d\boldsymbol{\xi} = \int_{\mathcal{R}} [\mathcal{O}^\dagger g^\dagger(\boldsymbol{\xi})]^* f(\boldsymbol{\xi}) d\boldsymbol{\xi}. \tag{D.4}$$

Here $g^\dagger$ are L_2 integrable functions of $\boldsymbol{\xi}$, often with different properties (although not necessarily) than those of the functions $f(\boldsymbol{\xi})$. For example, if $\mathcal{O}$ is a differential operator, $f(\boldsymbol{\xi})$ and $g^\dagger(\boldsymbol{\xi})$ may have different behavior at the boundary of $\mathcal{R}$.

Example D.1 Adjoint of a Differential Operator

Let $\boldsymbol{\xi} \to x$ and $\mathcal{O} = d/dx$, $x \in [a, b]$, with $f(x)$ and $g^\dagger(x)$ being *real* functions that both vanish at the endpoints. Then by definition

$$(g^\dagger, \mathcal{O}f) \equiv \int_a^b g^\dagger(x) \frac{df(x)}{dx} dx$$

$$\underset{\text{by parts}}{=} g^\dagger(x) f(x) \Big|_a^b - \int_a^b \frac{dg^\dagger(x)}{dx} f(x) dx$$

$$= -\int_a^b \frac{dg^\dagger(x)}{dx} f(x) dx, \tag{D.5}$$

where the boundary values of f and $g^\dagger$ make the surface term vanish. This result for $(g^\dagger, \mathcal{O}f)$ must equal, from the definition of the adjoint,

$$(\mathcal{O}^\dagger g^\dagger, f) = \int_a^b [\mathcal{O}^\dagger g^\dagger(x)] f(x) dx. \tag{D.6}$$

Comparison of Eq. (D.6) to Eq. (D.5) immediately identifies $\mathcal{O}^\dagger$ as

$$\mathcal{O}^\dagger = -\frac{d}{dx} = -\mathcal{O}.$$

Example D.2 A Self-Adjoint Differential Operator

Let $\boldsymbol{\xi} \to x$ and $\mathcal{O} = d^2/dx^2$, $x \in [a, b]$, with $f(x)$ and $g^\dagger(x)$ again being *real* functions that vanish at the endpoints. As before, use the definition of the inner product, integrate by parts twice, and use the boundary conditions to make surface terms vanish. Thus

$$(g^\dagger, \mathcal{O}f) \equiv \int_a^b g^\dagger(x) \frac{d^2 f(x)}{dx^2} \, dx$$

$$= g^\dagger(x) \frac{df(x)}{dx} \bigg|_a^b - \int_a^b \frac{dg^\dagger(x)}{dx} \frac{df(x)}{dx} \, dx = -\int_a^b \frac{dg^\dagger(x)}{dx} \frac{df(x)}{dx} \, dx$$

$$= -\frac{dg^\dagger(x)}{dx} f(x) \bigg|_a^b + \int_a^b \frac{d^2 g^\dagger(x)}{dx^2} f(x) \, dx = \int_a^b \frac{d^2 g^\dagger(x)}{dx^2} f(x) \, dx. \quad (D.7)$$

Again, by comparing this result to Eq. (D.6), it is seen that the adjoint operator is

$$\mathcal{O}^\dagger = \frac{d^2}{dx^2} = \mathcal{O}, \qquad \text{i.e., self-adjoint or } \textit{Hermitian.}$$

This result can be generalized when $d^2/dx^2 \to \nabla^2$, which can be shown to have the property

$$\nabla^2 \equiv \mathcal{O} = \mathcal{O}^\dagger$$

provided the functions $f(\boldsymbol{\xi})$ and $g^\dagger(\boldsymbol{\xi})$ vanish at the boundaries.

Example D.3 Importance of Boundary Conditions

If f and $g^\dagger$ are complex functions of $x \in [a, b]$, what boundary conditions are required for $\mathcal{O} \equiv d^2/dx^2$ to be self-adjoint? Integration by parts twice gives

$$(g^\dagger, \mathcal{O}f) \equiv \int_a^b [g^\dagger(x)]^* \frac{d^2 f(x)}{dx^2} \, dx$$

$$= [g^\dagger(x)]^* \frac{df(x)}{dx} \bigg|_a^b - \int_a^b \left[\frac{dg^\dagger(x)}{dx} \right]^* \frac{df(x)}{dx} \, dx$$

$$= [g^\dagger(x)]^* \frac{df(x)}{dx} \bigg|_a^b - \left[\frac{dg^\dagger(x)}{dx} \right]^* f(x) \bigg|_a^b + \int_a^b \left[\frac{d^2 g^\dagger(x)}{dx^2} \right]^* f(x) \, dx.$$

$$(D.8)$$

By the definition of the adjoint operator, this must equal

$$(\mathcal{O}^\dagger g^\dagger, f) = \int_a^b (\mathcal{O}^\dagger g(x)^\dagger)^* f(x) \, dx. \quad (D.9)$$

Comparison of Eq. (D.8) to Eq. (D.9) shows that $\mathcal{O}^\dagger = \mathcal{O}$ provided

$$[g^\dagger(x)]^* \frac{df(x)}{dx}\bigg|_a^b = \left[\frac{dg^\dagger(x)}{dx}\right]^* f(x)\bigg|_a^b. \qquad (D.10)$$

This is not very restrictive. Many boundary conditions for functions f and $g^\dagger$ can satisfy this condition. For example, $f = g^\dagger = 0$ or $df/dx = dg^\dagger/dx = 0$ at $x = a$ and $x = b$ makes $\mathcal{O} = d^2/dx^2$ self-adjoint.

Example D.4 Adjoint of an Integral Operator

Consider the integral operator $\mathcal{O}[\cdot] = \int_a^b K(x, x')[\cdot]\,dx'$. By definition

$$(g, \mathcal{O}f) = \int_a^b g^*(x)\left[\int_a^b K(x, x')f(x')\,dx'\right]dx.$$

Interchanging the order of integration yields

$$(g, \mathcal{O}f) = \int_a^b \left[\int_a^b K(x, x')g^*(x)\,dx\right]f(x')\,dx',$$

which, upon interchanging x and x', becomes

$$(g, \mathcal{O}f) = \int_a^b \left[\int_a^b K(x', x)g^*(x')\,dx'\right]f(x)\,dx$$

$$= \int_a^b \left[\int_a^b K^*(x', x)g(x')\,dx'\right]^* f(x)\,dx. \qquad (D.11)$$

This must equal

$$(\mathcal{O}^\dagger g, f) \equiv \int_a^b [\mathcal{O}^\dagger g(x)]^* f(x)\,dx. \qquad (D.12)$$

Comparing Eq. (D.12) with Eq. (D.11), it is seen that

$$\mathcal{O}^\dagger[\cdot] = \int_a^b K^*(x', x)[\cdot]\,dx'.$$

For the special case that $K(x, x') = K(x', x)$ and K is a real kernel, this integral operator is self-adjoint. Note that, unlike differential operators, no boundary conditions need be specified.

Example D.5 Adjoint of a Matrix Operator

Let $\mathcal{O}$ be an $n \times n$ matrix $\mathbf{A}$. For a matrix operator, the mapping functions f and g become n-component vectors $\mathbf{f}$ and $\mathbf{g}$. The inner product of two vectors is defined as

$$(\mathbf{g}, \mathbf{f}) \equiv \mathbf{g}^{T*} \cdot \mathbf{f} = \sum_{i=1}^{n} g_i^* f_i. \tag{D.13}$$

To find the adjoint operator $\mathbf{A}^{\dagger}$, first evaluate

$$(\mathbf{g}, \mathbf{Af}) \equiv \sum_{i=1}^{n} g_i^* \sum_{k=1}^{n} A_{ik} f_k. \tag{D.14}$$

This must equal, by definition,

$$(\mathbf{A}^{\dagger}\mathbf{g}, \mathbf{f}) = \sum_{i=1}^{n} \left(\sum_{k=1}^{n} A_{ik}^{\dagger} g_k \right)^* f_i = \sum_{i=1}^{n} \sum_{k=1}^{n} A_{ik}^{\dagger *} g_k^* f_i$$

$$= \sum_{k=1}^{n} \sum_{i=1}^{n} A_{ki}^{\dagger *} g_i^* f_k = \sum_{i=1}^{n} g_i^* \sum_{k=1}^{n} A_{ki}^{\dagger *} f_k. \tag{D.15}$$

Comparison of Eq. (D.14) and Eq. (D.15) shows that $A_{ki}^{\dagger *} = A_{ik} = (\mathbf{A}^T)_{ki}$ or

$$\boxed{\mathbf{A}^{\dagger} = \mathbf{A}^{T*}.}$$

If $\mathbf{A} = \mathbf{A}^{T*}$, the matrix is self-adjoint. Such matrices are called *Hermitian* matrices. For example, a real symmetric matrix is self-adjoint.

D.4 Uses of the Adjoint Operator

D.4.1 The Forward and Adjoint Problems

Consider the problem of finding the function $f(\boldsymbol{\xi})$ from

$$\boxed{\mathcal{O}f(\boldsymbol{\xi}) = S(\boldsymbol{\xi}), \qquad \boldsymbol{\xi} \in \mathcal{R},} \tag{D.16}$$

where $S(\boldsymbol{\xi})$ is some specified function. If $\mathcal{O}$ is a differential operator, the solution $f(\boldsymbol{\xi})$ is also subject to appropriate boundary conditions.

Associated with this *direct* or *forward* problem is an *adjoint problem*

$$\boxed{\mathcal{O}^{\dagger}g^{\dagger}(\boldsymbol{\xi}) = S^{\dagger}(\boldsymbol{\xi}), \qquad \boldsymbol{\xi} \in \mathcal{R},} \tag{D.17}$$

where $S^{\dagger}(\mathbf{r})$ is some specified adjoint source. Again, if $\mathcal{O}^{\dagger}$ is a differential operator, the solution $g^{\dagger}(\boldsymbol{\xi})$ is constrained by *adjoint boundary conditions* which may or may not be different from the boundary conditions used in the forward problem.

Many engineering problems have the form of Eq. (D.16). For example, the steady-state temperature distribution $T(\mathbf{r})$ in some volume V of a thermal conducting medium, which has a thermal conductivity k and an internal heat source $S(\mathbf{r})$, is given by the *heat conduction equation*

$$k\nabla^2 T(\mathbf{r}) = S(\mathbf{r}), \quad \mathbf{r} \in V. \tag{D.18}$$

To uniquely specify a solution, appropriate boundary conditions are also needed. Often this equation is *discretized* by using, for example, a finite difference approximation. With such a discretization, the continuous operator becomes a matrix operator and the functions become vectors, i.e., $k\nabla^2 \to \mathbf{A}$, $T(\mathbf{r}) \to \mathbf{T}$, and $S(\mathbf{r}) \to \mathbf{S}$. Thus, the temperature at discrete nodes in $\mathcal{R}$ is given by

$$\mathbf{AT} = \mathbf{S}.$$

In the discretization process, the boundary conditions on $T(\mathbf{r})$ are explicitly factored into the matrix elements of $\mathbf{A}$, and thus for this matrix operator problem, no extra conditions need be imposed when solving this set of linear algebraic equations.

D.4.2   Solution Using the Green's Function

The Green's function $G(\boldsymbol{\xi}, \boldsymbol{\xi}')$ is the solution of Eq. (D.16) with $S(\boldsymbol{\xi}) = \delta(\boldsymbol{\xi} - \boldsymbol{\xi}')$, i.e., the solution of

$$\mathcal{O}G(\boldsymbol{\xi}, \boldsymbol{\xi}') = \delta(\boldsymbol{\xi} - \boldsymbol{\xi}'), \quad \boldsymbol{\xi}, \boldsymbol{\xi}' \in \mathcal{R}, \tag{D.19}$$

subject to the same boundary conditions used in Eq. (D.16). Here $\boldsymbol{\xi}'$ is a parameter of the problem, since $\mathcal{O}$ operates only on the $\boldsymbol{\xi}$ variables.

If one can find $G(\boldsymbol{\xi}, \boldsymbol{\xi}')$, the solution $f(\boldsymbol{\xi})$ of Eq. (D.16) is easily found. To see this start with the identity

$$S(\boldsymbol{\xi}) = \int_{\mathcal{R}} S(\boldsymbol{\xi}')\delta(\boldsymbol{\xi} - \boldsymbol{\xi}') \, d\boldsymbol{\xi}'.$$

Then multiply Eq. (D.16) from the left by the inverse operator $\mathcal{O}^{-1}$, defined by $\mathcal{O}^{-1}\mathcal{O} = 1$, to obtain

$$f(\boldsymbol{\xi}) = \mathcal{O}^{-1}S(\boldsymbol{\xi}) = \int_{\mathcal{R}} S(\boldsymbol{\xi}')[\mathcal{O}^{-1}\delta(\boldsymbol{\xi} - \boldsymbol{\xi}')] \, d\boldsymbol{\xi}'.$$

But from Eq. (D.19), $\mathcal{O}^{-1}\delta(\boldsymbol{\xi} - \boldsymbol{\xi}') = G(\boldsymbol{\xi}, \boldsymbol{\xi}')$. Hence the desired solution is given by

$$f(\boldsymbol{\xi}) = \int_{\mathcal{R}} S(\boldsymbol{\xi}')G(\boldsymbol{\xi}, \boldsymbol{\xi}') \, d\boldsymbol{\xi}'. \tag{D.20}$$

To show explicitly the form of an inverse operator $\mathcal{O}^{-1}$, consider the following first-order differential equation:

$$\mathcal{O}y(x) \equiv \left[\frac{d}{dx} + \alpha\right] y(x) = S(x),$$

whose solution is easily found to be

$$y(x) = A e^{-\alpha x} + e^{-\alpha x} \int e^{\alpha x} S(x)\, dx.$$

The first term is the most general solution of the corresponding homogeneous differential equation and the second term is a particular solution $y_p(x)$. This particular solution can be expressed in terms of the inverse operator $\mathcal{O}^{-1}$ as

$$y_p(x) = \mathcal{O}^{-1} S(x) = e^{-\alpha x} \int e^{\alpha x} [S(x)]\, dx.$$

As expected, the inverse of a differential operator is an integral operator, i.e.,

$$\mathcal{O}^{-1}[\cdot] = e^{-\alpha x} \int e^{\alpha x} [\cdot]\, dx.$$

Green's Function Solution for Heat Conduction

Consider the heat conduction problem of Eq. (D.18). The Green's function for this problem is the solution of

$$k\nabla^2 G(\mathbf{r}, \mathbf{r}') = \delta(\mathbf{r} - \mathbf{r}'), \quad \mathbf{r}, \mathbf{r}' \in \mathcal{R},$$

with the same boundary conditions used for Eq. (D.18). Physically, $G(\mathbf{r}, \mathbf{r}')$ is the temperature at $\mathbf{r}$ caused by a unit strength heat source at $\mathbf{r}'$. Thus, the contribution to the temperature at $\mathbf{r}$ from the actual heat source in dV' about $\mathbf{r}'$, namely, $S(\mathbf{r}')dV'$, is

$$dT(\mathbf{r}) = G(\mathbf{r}, \mathbf{r}')[S(\mathbf{r}')dV'].$$

Finally, integration over all dV' in V gives the desired temperature at any point $\mathbf{r}$ in V,

$$T(\mathbf{r}) = \int_V G(\mathbf{r}, \mathbf{r}') S(\mathbf{r}')\, dV'. \tag{D.21}$$

Note that because $G(\mathbf{r}, \mathbf{r}')$ satisfies the specified boundary conditions, the right side of Eq. (D.21), and hence $T(\mathbf{r})$, satisfies these conditions.

D.4.3   Relation Between Forward and Adjoint Green's Functions

Associated with the Green's function defined by Eq. (D.19) is an adjoint Green's function defined by

$$\mathcal{O}^\dagger G^\dagger(\boldsymbol{\xi}, \boldsymbol{\xi}'') = \delta(\boldsymbol{\xi} - \boldsymbol{\xi}''), \quad \boldsymbol{\xi}, \boldsymbol{\xi}'' \in \mathcal{R}, \tag{D.22}$$

subject to the same adjoint boundary conditions used in Eq. (D.17). Here $\boldsymbol{\xi}''$ is a parameter of the problem, since $\mathcal{O}\dagger$ operates only on the $\boldsymbol{\xi}$ variables. This adjoint Green's function and the forward Green's function of Eq. (D.19) are related by the *reciprocity theorem*

$$G^*(\boldsymbol{\xi}'', \boldsymbol{\xi}') = G^\dagger(\boldsymbol{\xi}', \boldsymbol{\xi}''). \tag{D.23}$$

Proof: By definition $(\mathcal{O}^\dagger G^\dagger, G) = (G^\dagger, \mathcal{O}G)$ or

$$\int_\mathcal{R} [\mathcal{O}^\dagger G^\dagger(\boldsymbol{\xi}, \boldsymbol{\xi}'')]^* G(\boldsymbol{\xi}, \boldsymbol{\xi}') \, d\boldsymbol{\xi} = \int_\mathcal{R} G^{\dagger*}(\boldsymbol{\xi}, \boldsymbol{\xi}'')[\mathcal{O}G(\boldsymbol{\xi}, \boldsymbol{\xi}')] \, d\boldsymbol{\xi}.$$

Now use Eqs. (D.19) and (D.22) to express $\mathcal{O}G$ and $\mathcal{O}^\dagger G^\dagger$ in terms of the delta function. Thus the above result can be rewritten as

$$\int_\mathcal{R} [\delta(\boldsymbol{\xi} - \boldsymbol{\xi}'')]^* G(\boldsymbol{\xi}, \boldsymbol{\xi}') \, d\boldsymbol{\xi} = \int_\mathcal{R} G^{\dagger*}(\boldsymbol{\xi}, \boldsymbol{\xi}'')[\delta(\boldsymbol{\xi} - \boldsymbol{\xi}')] \, d\boldsymbol{\xi},$$

or

$$\left[\int_\mathcal{R} \delta(\boldsymbol{\xi} - \boldsymbol{\xi}'') G^*(\boldsymbol{\xi}, \boldsymbol{\xi}') \, d\boldsymbol{\xi} \right]^* = \int_\mathcal{R} G^{\dagger*}(\boldsymbol{\xi}, \boldsymbol{\xi}'')[\delta(\boldsymbol{\xi} - \boldsymbol{\xi}')] \, d\boldsymbol{\xi}.$$

The delta functions in the integrand thus allow the integrals to be performed to give

$$\left[G^*(\boldsymbol{\xi}'', \boldsymbol{\xi}') \right]^* = G^{\dagger*}(\boldsymbol{\xi}', \boldsymbol{\xi}''),$$

which, after taking the complex conjugate of both sides, yields the desired result.

D.4.4   Time-Dependent Green's Functions

To describe transient behavior, a derivative with respect to time t is often encountered in a linear differential equation, i.e.,

$$\frac{\partial f(\boldsymbol{\xi}, t)}{\partial t} + \mathcal{O}f(\boldsymbol{\xi}) = S(\boldsymbol{\xi}, t), \quad \boldsymbol{\xi} \in \mathcal{R}, \tag{D.24}$$

where $\boldsymbol{\xi}$ are typically the spatial independent variables x, y, and z. Equation (D.24) is subject to an initial condition in addition to a boundary condition.

The Green's function associated with this equation is now also a function of t and t' (the time an impulse source is applied). Three common ways that the time-dependent Green's function is written are $G(\boldsymbol{\xi}, t | \boldsymbol{\xi}', t') = G(\boldsymbol{\xi}', t' \rightarrow \boldsymbol{\xi}, t) = G(\boldsymbol{\xi}, \boldsymbol{\xi}', t - t')$ and this is the solution of

$$\frac{\partial G}{\partial t} + \mathcal{O}G = \delta(\boldsymbol{\xi} - \boldsymbol{\xi}')\delta(t - t') \quad \boldsymbol{\xi}, \boldsymbol{\xi}' \in \mathcal{R} \quad \text{and} \quad t > t', \tag{D.25}$$

subject to zero boundary and initial conditions, i.e., $G(\boldsymbol{\xi}, t | \boldsymbol{\xi}', t') = 0$ for $\boldsymbol{\xi}$ on the surface of $\mathcal{R}$. Because of *causality*, namely, that in any realistic system the effect cannot precede the cause, $G(\boldsymbol{\xi}, t | \boldsymbol{\xi}', t') = 0$ for $t < t'$.

Recall that $(d/dx)^\dagger = -d/dx$ so that the adjoint Green's function $G^\dagger(\boldsymbol{\xi}, t | \boldsymbol{\xi}', t')$ is the solution of

$$-\frac{\partial G^\dagger}{\partial t} + \mathcal{O}^\dagger G^\dagger = \delta(\boldsymbol{\xi} - \boldsymbol{\xi}'')\delta(t - t'') \quad \boldsymbol{\xi}, \boldsymbol{\xi}'' \in \mathcal{R} \quad \text{and} \quad t > t'', \tag{D.26}$$

again with homogeneous boundary and initial conditions. Note that by reversing the direction of time, i.e., letting $t \rightarrow -t$, this equation can be written as

$$\frac{\partial \overline{G}}{\partial t} + \mathcal{O}^\dagger \overline{G} = \delta(\boldsymbol{\xi} - \boldsymbol{\xi}'')\delta(t'' - t), \tag{D.27}$$

whose solution $\overline{G}(\boldsymbol{\xi}, t | \boldsymbol{\xi}'', t'') = G^\dagger(\boldsymbol{\xi}, -t | \boldsymbol{\xi}'', -t'')$.

Following the proof of Eq. (D.23), it is found that

$$G^*(\boldsymbol{\xi}'', t'' | \boldsymbol{\xi}', t') = G^\dagger(\boldsymbol{\xi}', -t' | \boldsymbol{\xi}'', -t''). \tag{D.28}$$

This time-dependent reciprocity relation can be understood from Eq. (D.28). Interchange of $\boldsymbol{\xi}'$ and $\boldsymbol{\xi}$ transforms the spatial derivatives ($\mathcal{O}$) into the complex conjugate of the spatial derivatives of $\mathcal{O}^\dagger$. By contrast, the interchange introduces a negative sign into the transformed Green's function because of the time derivative. Simple put, spatial orientation has no preferred direction for energy diffusion whereas time does indeed have a preferred direction of flow.

D.4.5   Finding Averages of the Solution

Often the details provided by the solution $f(\boldsymbol{\xi})$ of Eq. (D.16) are not needed; rather some weighted average of the solution is sought, namely,

$$I \equiv \int_\mathcal{R} w(\boldsymbol{\xi}) f(\boldsymbol{\xi}) \, d\boldsymbol{\xi}, \tag{D.29}$$

where $w(\boldsymbol{\xi})$ is some prescribed *weight* function. For example, if $w(\boldsymbol{\xi}) = 1/V$, where V is the "volume" of $\mathcal{R}$, then I gives the average value of $f(\boldsymbol{\xi})$ in V. By contrast, if $w(\xi) = \delta(\xi - \xi_o)$, then $I = f(\xi_o)$, the solution at a particular point of interest ξ_o.

What is remarkable is that the same value of I can be obtained from the solution $g^\dagger(\xi)$ of the associated adjoint problem in which the adjoint source is taken as the weight function, i.e., $S^\dagger(\xi) = w(\xi)$. In particular, it can be shown for the special case that $\mathcal{O}$, f, and $g^\dagger$ are all real,

$$I = \int_{\mathcal{R}} w(\xi) f(\xi)\, d\xi = \int_{\mathcal{R}} S(\xi) g^\dagger(\xi)\, d\xi. \tag{D.30}$$

In this and the following proof, it is assumed that all operators and functions are real. The extension to the complex case is left as an exercise.

Proof: In terms of the Green's functions, the forward and adjoint solutions are

$$f(\xi) = \int_{\mathcal{R}} G(\xi, \xi') S(\xi')\, d\xi', \tag{D.31}$$

$$g^\dagger(\xi) = \int_{\mathcal{R}} G^\dagger(\xi, \xi'') w(\xi'')\, d\xi''. \tag{D.32}$$

With Eq. (D.32), the right side of Eq. (D.30) can be written as

$$I = \int_{\mathcal{R}} S(\xi) g^\dagger(\xi)\, d\xi = \int_{\mathcal{R}} S(\xi) \int_{\mathcal{R}} w(\xi'') G^\dagger(\xi, \xi'')\, d\xi''\, d\xi.$$

But from the reciprocity theorem of Eq. (D.23), $G^\dagger(\xi, \xi'') = G(\xi'', \xi)$. Using this result and then interchanging the order of integration yields

$$I = \int_{\mathcal{R}} S(\xi) \int_{\mathcal{R}} w(\xi'') G(\xi'', \xi)\, d\xi''\, d\xi = \int_{\mathcal{R}} w(\xi'') \int_{\mathcal{R}} S(\xi) G(\xi'', \xi)\, d\xi\, d\xi''.$$

From Eq. (D.31), the inner integral of this last result equals $f(\xi'')$, so that after replacing the dummy variable of integration ξ'' with ξ,

$$I = \int_{\mathcal{R}} w(\xi) f(\xi)\, d\xi,$$

which proves Eq. (D.30).

Although the result of Eq. (D.30) was derived for real $\mathcal{O}$, f, and $g^\dagger$, it can be generalized for the case of complex $\mathcal{O}$, f, and $g^\dagger$. However, in most engineering applications, one deals with real linear operators and Eq. (D.30) is sufficient.

Application to Heat Conduction

For the steady-state heat conduction problem of Eq. (D.18), $\mathcal{O} \equiv k\nabla^2$ or, for the finite difference approximation, $\mathcal{O} \equiv \mathbf{A}$, in which the matrix $\mathbf{A}$ is symmetric.

These real operators are self-adjoint. Suppose that the temperature at only one point $\mathbf{r}_o$ (the "hot spot") is of interest. Then one could solve Eq. (D.18) for $T(\mathbf{r})$ and then evaluate $T(\mathbf{r}_o)$. This is equivalent to calculating

$$I = T(\mathbf{r}_o) = \int_V \delta(\mathbf{r} - \mathbf{r}_o) T(\mathbf{r}) \, dV.$$

Alternatively, one could solve the adjoint problem with $S^\dagger(\mathbf{r}) = \delta(\mathbf{r} - \mathbf{r}_o)$, i.e., solve (for this self-adjoint operator)

$$\mathcal{O}^\dagger T^\dagger(\mathbf{r}) = \mathcal{O} T^\dagger(\mathbf{r}) = \delta(\mathbf{r} - \mathbf{r}_o).$$

Then from Eq. (D.30), $T(\mathbf{r}_o)$ can be obtained from

$$T(\mathbf{r}_o) = \int_V S(\mathbf{r}) T^\dagger(\mathbf{r}) \, dV.$$

For complicated heat sources $S(\mathbf{r})$, it is often preferable to solve a heat conduction problem in which a heat source is only at $\mathbf{r}_o$ and then evaluate the above integral (perhaps by numerical integration).

For the discretized heat conduction model, the forward problem requires the solution of the linear algebraic equations

$$\mathbf{AT} = \mathbf{S},$$

in which $\mathbf{S}$ is the vector corresponding to the complex heat source $S(\mathbf{r})$. The temperature of interest $T(\mathbf{r}_o)$ is (say) the ith component of the solution vector $\mathbf{T}$.

The temperature of interest can be more simply obtained (especially, for Monte Carlo solutions of the algebraic equations) by using the solution of

$$\mathbf{A}^\dagger \mathbf{T}^\dagger = \mathbf{AT}^\dagger = \mathbf{e_i},$$

where $\mathbf{e_i}$ is a vector whose components are zero, except for the ith component, which equals 1. Then, from Eq. (D.30)

$$T_i \, (\simeq T(\mathbf{r}_o)) = (\mathbf{S}, \mathbf{T}^\dagger) = \sum_j S_j T_j^\dagger.$$

D.5 Eigenfunctions and Eigenvalues of an Operator

The nontrivial (i.e., nonzero) functions $y(x)$ that satisfy the following equation (plus possibly boundary conditions if $\mathcal{O}$ is a differential operator)

$$\boxed{\mathcal{O} y(x) = \lambda y(x)} \tag{D.33}$$

are called the *eigenfunctions* of $\mathcal{O}$ and the corresponding constants λ are the *eigenvalues*. This eigenvalue problem is exactly analogous to a matrix eigenvalue problem. Further, if $w(x)$ is some specified positive function, this eigenvalue problem can be generalized to

$$\mathcal{O}y(x) = \lambda w(x)y(x). \tag{D.34}$$

Example D.6 Eigenfunctions of a Differential Operator

To demonstrate explicitly how to find the eigenfunctions and eigenvalues of a differential operator, consider the eigenvalue problem for $\mathcal{O} = d^2/dx^2$ on the interval $[-\pi, \pi]$, namely,

$$\frac{d^2 y(x)}{dx^2} = \lambda y(x) \quad \text{with boundary conditions} \quad y(-\pi) = y(\pi) \text{ and } y'(-\pi) = y'(\pi). \tag{D.35}$$

The solution of this ordinary differential equation (ODE) depends on whether λ is positive or negative. Consider each case separately.

Case $\lambda > 0$: Let $\lambda = v^2$, where $v > 0$. The most general solution of Eq. (D.35) is

$$y(x) = Ae^{vx} + Be^{-vx}.$$

To determine the arbitrary constants A and B, apply the two boundary conditions:

$$y(\pi) = y(-\pi) \quad \text{or} \quad Ae^{v\pi} + Be^{-v\pi} = Ae^{-v\pi} + Be^{v\pi},$$
$$y'(\pi) = y'(-\pi) \quad \text{or} \quad Ave^{v\pi} - Bve^{-v\pi} = Ave^{-v\pi} - Bve^{v\pi}.$$

These two equations are two homogeneous algebraic equations, namely,

$$A[e^{v\pi} - e^{-v\pi}] + B[e^{-v\pi} - e^{v\pi}] = 0,$$
$$A[e^{v\pi} - e^{-v\pi}] + B[e^{v\pi} - e^{-v\pi}] = 0.$$

A nontrivial solution for A and B can be obtained only if the determinant vanishes. The determinant is

$$\det = [e^{v\pi} - e^{-v\pi}]^2 + [e^{v\pi} - e^{-v\pi}]^2 = 2[e^{v\pi} - e^{-v\pi}]^2.$$

But the right-hand term is clearly > 0. Hence, there is only the trivial null solution and there are no eigenfunctions with positive λ.

Case $\lambda < 0$: Let $\lambda = -\beta^2$, where $\beta > 0$. The most general solution of Eq. (D.35) is now

$$y(x) = A\cos\beta x + B\sin\beta x.$$

To determine the arbitrary constants A and B, apply the two boundary conditions

$$y(\pi) = y(-\pi) \quad \text{or} \quad A\cos(\beta\pi) + B\sin(\beta\pi) = A\cos(-\beta\pi) + B\sin(-\beta\pi),$$

$y'(\pi) = y'(-\pi)$ or $-A\beta \sin(\beta\pi) + B\beta \cos(\beta\pi) = -A\beta \sin(-\beta\pi) + B\beta \cos(-\beta\pi)$.

With the properties $\cos(x) = \cos(-x)$ and $\sin(x) = -\sin(-x)$, these two equations reduce to

$$A[\ 0\ ] + B[2\sin(\beta\pi)] = 0,$$
$$-A[2\sin(\beta\pi)] + B[\ 0\ ] = 0.$$

The determinant in this case is $\det = 4AB\sin^2\beta\pi$, which vanishes only if $\beta = 0, \pm 1, \pm 2, \ldots$ since A and B are nonzero (otherwise one would obtain the trivial null solution). Hence the eigenvalues are $\lambda_n = -\beta^2 = -n^2$, $n = 0, 1, 2, \ldots$, and the eigenfunctions are

$$y_n(x) = \cos nx, \quad n = 0, 1, 2, \ldots, \quad \text{and} \quad y_n(x) = \sin nx, \quad n = 1, 2, 3, \ldots.$$

Note the negative values of β are not included since $\beta_n = -n$ gives a multiple (± 1) of the same function with positive n. Also $\beta = 0$ is not allowed for the sine function since this gives the trivial null solution.

Example D.7 Eigenfunctions of an Integral Operator

Consider the integral operator $\mathcal{O}[\cdot] \equiv \int_0^\infty e^{-(x+x')}[\cdot]dx'$. The eigenvalue problem for this operator is

$$\int_0^\infty e^{-(x+x')}y(x')\,dx' = \lambda y(x). \tag{D.36}$$

Define $C = \int_0^\infty e^{-x'}y(x')\,dx'$ so that Eq. (D.36) becomes

$$e^{-x}C = \lambda y(x). \tag{D.37}$$

Multiply this equation by e^{-x} and integrate to obtain

$$\int_0^\infty e^{-2x}C\,dx = \lambda \int_0^\infty e^{-x}y(x)\,dx = \lambda C, \tag{D.38}$$

which reduces to

$$C\left[\lambda - \int_0^\infty e^{-2x}\,dx\right] = 0. \tag{D.39}$$

If $C \neq 0$, then from this result it follows that $\lambda = \int_0^\infty e^{-2x}dx = 1/2$. Thus $\lambda = 1/2$ is an eigenvalue. The corresponding eigenfunction is found from Eq. (D.37) as

$$y(x) = (1/\lambda)Ce^{-x} = 2Ce^{-x} = C'e^{-x},$$

where C' is any nonzero constant.

However, for the case $C \equiv \int_0^\infty e^{-x}y(x)\,dx = 0$, it is seen from Eq. (D.37) that the eigenvalue $\lambda = 0$. For this case, there is an infinite degeneracy in the eigenvalue spectrum because there are an infinite number of linearly independent eigenfunctions $y(x)$ for which

$$\int_0^\infty e^{-x} y(x)\, dx = 0. \tag{D.40}$$

For example, suppose one tries $y(x) = a + bx$. Substitute this into Eq. (D.40) to find that $a = -b$. Thus one eigenfunction, for which $\lambda = 0$, is $y(x) = 1 - x$. In fact any function of the form $y(x) = ax^m + x^n$, $n \neq m$, where m and n are integers, is an eigenfunction with a proper choice of the constant a. Actually, the eigenfunctions with $\lambda = 0$ are well-known functions. Recall the Laguerre polynomials $L_n(x)$ have the orthogonality property

$$\int_0^\infty e^{-x} L_n(x) L_m(x) = N_n \delta_{mn}.$$

Since $L_0(x)$ is a constant, it follows that $\int_0^\infty e^{-x} L_m(x)\, dx = 0$, i.e., the Laguerre polynomials $L_m(x)$, $m = 1, 2, 3 \ldots$, are eigenfunctions of Eq. (D.36), all with an eigenvalue $\lambda = 0$.

D.5.1   Properties of Eigenfunctions

The forward and adjoint eigenvalue problems for any linear operator $\mathcal{O}$ are defined by

$$\text{Direct:} \qquad \mathcal{O}\phi_n(\boldsymbol{\xi}) = \lambda_n w(\boldsymbol{\xi})\phi_n(\boldsymbol{\xi}), \quad \boldsymbol{\xi} \in \mathcal{R}, \tag{D.41}$$

$$\text{Adjoint:} \qquad \mathcal{O}^\dagger \psi_m^\dagger(\boldsymbol{\xi}) = \eta_m w(\boldsymbol{\xi})\psi_m^\dagger(\boldsymbol{\xi}), \quad \boldsymbol{\xi} \in \mathcal{R}, \tag{D.42}$$

where, if $\mathcal{O}$ is a differential operator, the direct eigenfunctions $\{\phi_n\}$ satisfy regular boundary conditions and the adjoint eigenfunctions $\{\psi_m^\dagger\}$ satisfy adjoint boundary conditions. Here $w(\boldsymbol{\xi})$ is a real, positive, nonzero weight function.

The eigenfunctions of $\mathcal{O}$ and $\mathcal{O}^\dagger$ are related to each other. In certain function spaces and for certain classes of operators, these eigenfunctions have several important properties, which are stated below without proof.

Eigenvalue Spectrum: The eigenvalue spectrum of $\mathcal{O}$ and $\mathcal{O}^\dagger$ are complex conjugates. In other words, for any λ_n there is a corresponding η_n, such that $\eta_n = \lambda_n^*$.

Orthogonality: The eigenfunctions $\{\phi_n\}$ and $\{\psi_m^\dagger\}$ form a bi-orthogonal set, i.e.,

$$(\psi_m^\dagger, w\phi_n) = N_n \delta_{nm} \qquad \text{or} \qquad \int_{\mathcal{R}} \psi_m^{\dagger *}(\boldsymbol{\xi})\phi_n(\boldsymbol{\xi})w(\boldsymbol{\xi})\, d\boldsymbol{\xi} = N_n \delta_{nm}. \tag{D.43}$$

Completeness: If $\mathcal{O}$ is a compact self-adjoint operator or a normal compact operator ($\mathcal{O}\mathcal{O}^\dagger = \mathcal{O}^\dagger\mathcal{O}$), then the eigenfunctions $\{\phi_n\}$ and $\{\psi_m^\dagger\}$ form complete basis sets, i.e., "any well-behaved" function $F(\boldsymbol{\xi})$ can be uniquely expanded as[1]

$$F(\xi) = \sum_n a_n \phi_n(\xi), \quad \xi \in \mathcal{R}. \tag{D.44}$$

From the bi-orthogonality property, the expansion coefficients $\{a_n\}$ can be computed as

$$a_n = \frac{1}{N_n}(w\psi_n^\dagger, F) = \frac{1}{N_n}\int_{\mathcal{R}} w(\xi)\psi_n^{\dagger*}(\xi)F(\xi)\,d\xi. \tag{D.45}$$

In particular, the Dirac delta function can be expanded:

$$\delta(\xi - \xi') = \sum_n \frac{1}{N_n}w(\xi')\psi_n^{\dagger*}(\xi')\phi_n(\xi). \tag{D.46}$$

This expansion is sometimes called the *closure relation*.

Example D.8 Demonstration of Eigenfunction Properties

To demonstrate the eigenfunction properties, consider the simple differential operator $\mathcal{O} = d/dx$. Earlier it was found that the adjoint of this operator was $\mathcal{O}^\dagger = -d/dx$. This is a *normal* operator since $\mathcal{O}\mathcal{O}^\dagger = \mathcal{O}^\dagger\mathcal{O} = -d^2/dx^2$. Thus, their eigenfunctions should exhibit the above properties.

Direct Eigenfunction Problem:

$$\frac{dy(x)}{dx} = \lambda y(x), \qquad \text{with boundary conditions } y(0) = y(2\pi). \tag{D.47}$$

The most general solution is $y(x) = Ae^{\lambda x}$. Let $\lambda \equiv \alpha + i\beta$. The boundary conditions then require

$$1 = e^{\alpha 2\pi + i\beta 2\pi},$$

which is true only when $\alpha = 0$ and $\beta = \pm n$, $n = 0, 1, 2, \ldots$. Thus, the direct eigenfunctions are (choosing the normalization $A = 1$ for convenience)

$$y_n(x) = e^{inx} \qquad \text{with eigenvalue } \lambda_n = in, \ n = 0, \pm 1, \pm 2, \ldots. \tag{D.48}$$

Adjoint Eigenfunction Problem:

$$-\frac{dy^\dagger(x)}{dx} = \eta y^\dagger(x), \qquad \text{with boundary conditions } y^\dagger(0) = y^\dagger(2\pi). \tag{D.49}$$

---

[1] $F(\xi)$ must satisfy the Dirichlet conditions:

(1) $F(\xi)$ must be single-valued and continuous except at a finite number of points in $\mathcal{R}$,

(2) $F(\xi)$ has a finite number of extrema in $\mathcal{R}$,

(3) the integral $\int_{\mathcal{R}} F(\xi)w(\xi)\,d\xi$ exists.

The most general solution is $y^\dagger(x) = Ae^{-\eta x}$. Let $\eta \equiv \delta + i\epsilon$. The adjoint boundary conditions then require

$$1 = e^{-\delta 2\pi - i\epsilon 2\pi},$$

which is true only when $\delta = 0$ and $\epsilon = \mp m$, $m = 0, 1, 2, \dots$. Thus, the adjoint eigenfunctions are (again, choosing $A = 1$ for convenience)

$$y_m^\dagger(x) = e^{imx} \qquad \text{with eigenvalue } \eta_m = im, \ m = 0, \mp 1, \mp 2, \dots. \qquad \text{(D.50)}$$

Note that the eigenvalue spectrum is the complex conjugate of that for the forward operator.

Orthogonality: It is easily seen that these eigenfunctions form a bi-orthogonal set:

$$\int_0^{2\pi} y_m^{\dagger *}(x) y_n(x)\, dx = \int_0^{2\pi} e^{-imx} e^{inx}\, dx$$

$$= \int_0^{2\pi} \cos(n-m)x\, dx + i \int_0^{2\pi} \sin(n-m)x\, dx. \qquad \text{(D.51)}$$

Clearly, if $m \neq n$ both integrals vanish, and if $m = n$, one obtains $\int_0^{2\pi} y_n^{\dagger *} y_n dx = \int_0^{2\pi} e^{-inx} e^{inx}\, dx = 2\pi$.

Completeness: One can expand any function $F(x)$ as

$$F(x) = \sum_{n=-\infty}^{\infty} a_n e^{inx} = \sum_{n=-\infty}^{\infty} a_n [\cos nx + i \sin nx]$$

$$= a_0 + \sum_{n=1}^{\infty} [a_n + a_{-n}] \cos nx + \sum_{n=1}^{\infty} i[a_n - a_{-n}] \sin nx$$

$$= a_0 + \sum_{n=1}^{\infty} A_n \cos(nx) + \sum_{n=1}^{\infty} B_n \sin(nx)], \qquad \text{(D.52)}$$

which is seen to be the well-known Fourier series expansion of $F(x)$.

D.6 Eigenfunctions of Real, Linear, Self-Adjoint Operators

Now consider the special case of a real, Hermitian (self-adjoint), linear operator $\mathcal{O}$. Its eigenfunctions are the nontrivial solutions of

$$\boxed{\mathcal{O}[y_n(\xi)] = \lambda_n w(\xi) y_n(\xi), \qquad \xi \in \mathcal{R},} \qquad \text{(D.53)}$$

where $w(\xi)$ is a known positive *weight* function. The $y_n(\xi)$ are also subject to boundary conditions if $\mathcal{O}$ is a differential operator. For a Hermitian operator, $(g, \mathcal{O}f) = (\mathcal{O}g, f)$.

The eigenfunctions and eigenvalues of such a Hermitian operator have some remarkable properties which allow one to solve easily many problems involving the operator $\mathcal{O}$, such as inhomogeneous differential equations with variable coefficients or finding a Green's function. Such applications are discussed in later sections.

D.6.1 Eigenvalues Are Real

To prove the eigenvalues of a Hermitian operator are real, consider two eigenfunctions with different eigenvalues, i.e.,

$$\mathcal{O}[y_n] = \lambda_n w y_n, \tag{D.54}$$

$$\mathcal{O}[y_m] = \lambda_m w y_m. \tag{D.55}$$

Multiply Eq. (D.54) by y_m^* and Eq. (D.55) by y_n^* and integrate the results over the region $\mathcal{R}$ to obtain

$$\int_{\mathcal{R}} y_m^* \mathcal{O} y_n \, d\xi = \lambda_n \int_{\mathcal{R}} w y_m^* y_n \, d\xi, \tag{D.56}$$

$$\int_{\mathcal{R}} y_n^* \mathcal{O} y_m \, d\xi = \lambda_m \int_{\mathcal{R}} w y_n^* y_m \, d\xi. \tag{D.57}$$

From the definition of the adjoint operator $\mathcal{O}^\dagger$, the left-hand side of Eq. (D.57) can be written as $\int_{\mathcal{R}} [\mathcal{O}^\dagger y_n]^* y_m \, d\xi$, which, because $\mathcal{O}$ is Hermitian, can be written as $\int_{\mathcal{R}} [\mathcal{O} y_n]^* y_m \, d\xi$. Thus, Eq. (D.57) can be written as

$$\int_{\mathcal{R}} [\mathcal{O} y_n]^* y_m \, d\xi = \lambda_m \int_{\mathcal{R}} w y_n^* y_m \, d\xi.$$

Now take the complex conjugate of this result to obtain (remember w is real)

$$\int_{\mathcal{R}} y_m^* \mathcal{O} y_n \, d\xi = \lambda_m^* \int_{\mathcal{R}} w y_m^* y_n \, d\xi. \tag{D.58}$$

Note that the left-hand sides of Eqs. (D.56) and (D.58) are the same. Subtract these two equations to obtain

$$0 = (\lambda_n - \lambda_m^*) \int_{\mathcal{R}} w y_m^* y_n \, d\xi. \tag{D.59}$$

When $m = n$, the integrand in this result is real and positive, so the integral is > 0. Hence $(\lambda_n - \lambda_n^*)$ must be zero, which can only happen if λ_n is real. This completes the proof that all the eigenvalues of self-adjoint or Hermitian operators are real.

It turns out that most of the differential equations used in science and engineering have self-adjoint operators and, consequently, real eigenvalues. The eigenvalues, for example, determine the energy levels of electrons around nuclei or the forces required for columns to buckle, which have real values. Indeed much of the physics of our real universe is described by Hermitian operators!

D.6.2 Eigenfunctions Are Orthogonal

There are always an infinite number of eigenfunctions and eigenvalues, although the eigenvalue spectrum may be degenerate, i.e., there are multiple eigenfunctions with the same eigenvalue. However, it is always possible to find a set of eigenfunctions that are orthogonal as shown below.

No Eigenvalue Degeneracy

From Eq. (D.59), it is immediately seen that, for $m \neq n$, $(\lambda_n - \lambda_m) \neq 0$ so that

$$\int_{\mathcal{R}} w y_m^* y_n \, d\boldsymbol{\xi} = 0 \quad \text{for any different } m \text{ and } n. \tag{D.60}$$

For the case $m = n$

$$\int_{\mathcal{R}} w y_n^* y_n \, d\boldsymbol{\xi} = N_n, \tag{D.61}$$

where N_n is a real positive number. Often the eigenfunctions are divided by $\sqrt{N_n}$ to normalize the eigenfunctions to unity.

Case of Spectral Degeneracy

If there are different eigenfunctions with the same eigenvalue (λ_1 say), they often are not orthogonal. Suppose λ_1 is k-fold degenerate, i.e., there are k linearly independent functions y_i such that

$$\mathcal{O} y_i(\boldsymbol{\xi}) = \lambda_1 w(\boldsymbol{\xi}) y_i(\boldsymbol{\xi}), \quad i = 1, 2, \ldots, k,$$

and λ_1 is different from $\lambda_{k+1}, \lambda_{k+2}, \ldots$. Any linear combination of these eigenfunctions, $\psi(\boldsymbol{\xi}) = \sum_{i=1}^{k} c_i y_i(\boldsymbol{\xi})$, is also an eigenfunction with eigenvalue λ_1, as shown by the following manipulations:

$$\mathcal{O}\psi(\boldsymbol{\xi}) = \mathcal{O} \sum_{i=1}^{k} c_i y_i(\boldsymbol{\xi}) = \sum_{i=1}^{k} c_i \mathcal{O} y_i(\boldsymbol{\xi}) = \sum_{i=1}^{k} c_i \lambda_1 w(\boldsymbol{\xi}) y_i(\boldsymbol{\xi}) = \lambda_1 w(\boldsymbol{\xi}) \psi(\boldsymbol{\xi}).$$

Although y_i are generally not orthogonal, it is possible to find linear combinations ψ_i that are orthogonal. This is accomplished by the following *Gram–Schmidt orthogonalization* procedure:

$$\psi_1 = y_1,$$

$$\psi_2 = y_2 - \psi_1 \frac{(\psi_1, wy_2)}{(\psi_1, wy_1)},$$

$$\psi_3 = y_3 - \psi_2 \frac{(\psi_2, wy_3)}{(\psi_2, w\psi_2)} - \psi_1 \frac{(\psi_1, wy_3)}{(\psi_1, w\psi_1)},$$

$$\vdots$$

$$\psi_n = y_n - \sum_{i=1}^{n-1} \psi_i \frac{(\psi_i, wy_n)}{(\psi_i, w\psi_i)}, \quad n = 1, 2, \ldots k. \tag{D.62}$$

The ratios of the inner products are just numbers, so that the ψ_i are just linear combinations of functions that all have the same eigenvalue λ_1. It is easy to demonstrate that the ψ_i, constructed as above, are orthogonal, i.e., $(\psi_i, w\psi_j) = 0$.

D.6.3   Real Eigenfunctions

For the case that all the real eigenvalues are distinct, the corresponding eigenfunctions are also real. This property is easily shown by taking the complex conjugate of Eq. (D.54), to obtain $\mathcal{O}y_n^* = \lambda_n wy_n^*$. Thus y_n^* and y_n are both eigenfunctions with the same real eigenvalue λ_n. If the eigenvalue spectrum is not degenerate, then y_n^* must equal y_n, i.e., the eigenfunctions are all real.

However, for an eigenvalue degeneracy, i.e., two or more eigenfunctions have the same (real) eigenvalue, the eigenfunctions may not be real. But linear combinations of the degenerate eigenfunctions are also eigenfunctions with the same eigenvalue. For example, the two real functions $y_n + y_n^*$ and $i(y_n - y_n^*)$ are also eigenfunctions of $\mathcal{O}$ with the eigenvalue λ_n. At least one of these functions is nonzero. Thus, the eigenfunctions can be made real by forming suitable linear combinations. However, this effort to create real eigenfunctions is seldom necessary. In what follows, it is assumed, for generality, that the eigenfunctions may be complex.

D.6.4   Eigenfunctions Form a Complete Basis Set

Just as the eigenvectors of a matrix form a basis set of vectors, so do the eigenfunctions of a Hermitian operator form a basis set of functions. The reader is referred to Morse and Feshbach (1953) for a proof of this very important eigenfunction property. For any function $F(\boldsymbol{\xi})$ that satisfies the Dirichlet conditions (see the footnote on page 521), there exists a set of coefficients $\{a_n\}$ such that $F(\boldsymbol{\xi})$ can be approximated by

$$F(\boldsymbol{\xi}) \simeq \sum_{n=0}^{\infty} a_n y_n(\boldsymbol{\xi}), \quad \boldsymbol{\xi} \in \mathcal{R}, \tag{D.63}$$

such that

$$\lim_{N \to \infty} \left\{ \int_{\mathcal{R}} \left[F(\xi) - \sum_{n=0}^{N} a_n y_n(\xi) \right] w(\xi) \, d\xi \right\} = 0. \qquad \text{(D.64)}$$

In practice, $\sum_{n=0}^{\infty} a_n y_n(\xi) = F(\xi)$ at almost every point in $\mathcal{R}$.

If $F(\xi)$ is known, the expansion coefficients a_n can be evaluated by using the orthogonality property of the eigenfunctions. Multiply Eq. (D.63) by $w(\xi) y_m(\xi)$ and integrate over the range $\mathcal{R}$ to obtain

$$\sum_{n=0}^{\infty} a_n \int_{\mathcal{R}} w(\xi) y_m(\xi) y_n(\xi) \, d\xi = \int_{\mathcal{R}} w(\xi) y_m(\xi) F(\xi) \, d\xi.$$

But from the orthogonality property of $\{y_n\}$, the integral in the above sum equals $N_m \delta_{mn}$, i.e., only the $n = m$ term in the sum is nonzero. Thus

$$a_m N_m = \int_{\mathcal{R}} w(\xi) y_m(\xi) F(\xi) \, d\xi$$

or

$$\boxed{a_m = \frac{1}{N_m} \int_{\mathcal{R}} w(\xi) y_m(\xi) F(\xi) \, d\xi.} \qquad \text{(D.65)}$$

It is this completeness property of the eigenfunctions of Hermitian operators that, as will be seen, makes the eigenfunctions so useful in many applications such as solving inhomogeneous, second-order, differential equations.

Example D.9 Fourier Series Expansions

To demonstrate how eigenfunctions can be used to expand a well-behaved function $F(x)$, consider the eigenfunctions found in Example D.6 for the operator $\mathcal{O} = d^2/dx^2$ on the interval $[-\pi, \pi]$ and with the boundary conditions $y(\pi) = y(-\pi)$ and $y'(\pi) = y'(-\pi)$. The eigenfunctions are

$$y_n(x) = \begin{cases} \cos \lambda_n x, & \lambda_n = 0, 1, 2, \ldots, \\ \sin \lambda_n x, & \lambda_n = 1, 2, 3, \ldots. \end{cases} \qquad \text{(D.66)}$$

Since d^2/dx^2 is self-adjoint, the eigenfunctions $\{y_n\}$ are a complete basis set for the interval $[-\pi, \pi]$. Thus "any" function $F(x)$ can be expanded as

$$F(x) = \frac{a_o}{2} + \sum_{n=1}^{\infty} a_n \cos nx + \sum_{n=1}^{\infty} b_n \sin nx. \qquad \text{(D.67)}$$

This expansion is just the well-known *Fourier series* of $F(x)$. For these eigenfunctions, the orthogonality property is

$$\int_{-\pi}^{\pi} y_n(x) y_m(x) \, dx = \pi \delta_{mn}$$

and the Fourier expansion coefficients are

$$a_n = \frac{1}{\pi} \int_{-\pi}^{\pi} F(x) \cos nx \, dx \qquad \text{and} \qquad b_n = \frac{1}{\pi} \int_{-\pi}^{\pi} F(x) \sin nx \, dx.$$

This Fourier series can be extended to an arbitrary interval $a \leq x \leq b$. With $L = b - a$ and the same boundary conditions as above, the eigenfunctions are given by Eq. (D.66) but the eigenvalues are now $\lambda_n = 2\pi n / L$. Then any function $F(x)$ on this interval can be expanded as

$$F(x) = \frac{a_o}{2} + \sum_{n=1}^{\infty} a_n \cos\left(\frac{2n\pi}{L} x\right) + \sum_{n=1}^{\infty} b_n \sin\left(\frac{2n\pi}{L} x\right). \tag{D.68}$$

The orthogonality property of $\{y_n\}$ gives the *Euler formulas* for the expansion coefficients, namely,

$$a_n = \frac{2}{L} \int_a^b F(x) \cos\left(\frac{2\pi n}{L} x\right) dx, \quad n = 0, 1, 2, \ldots, \tag{D.69}$$

$$b_n = \frac{2}{L} \int_a^b F(x) \sin\left(\frac{2\pi n}{L} x\right) dx, \quad n = 1, 2, 3, \ldots. \tag{D.70}$$

This same Fourier expansion technique can be applied to the eigenfunctions of other differential operators. For example, there are Fourier–Legendre series, Fourier–Bessel series, and so on. Such series are examined in a later section.

D.7 The Sturm–Liouville Operator

Most second-order, linear, homogeneous ODEs can be written as

$$\mathcal{L}[y(x)] \equiv \frac{d}{dx}\left[p(x)\frac{dy(x)}{dx}\right] + q(x)y(x) = 0, \tag{D.71}$$

where $\mathcal{L}$ is the Sturm–Liouville operator. Equivalently, this equation can be written as

$$\mathcal{L}[y(x)] = p(x)\frac{d^2 y(x)}{dx^2} + r(x)\frac{dy(x)}{dx} + q(x)y(x) = 0, \tag{D.72}$$

where $r(x) \equiv dp(x)/dx$ and p and q are real functions.

To see that most second-order differential operators can be written in the form of Eq. (D.71), consider the most general homogeneous second-order operator

$$\mathcal{D}[y(x)] \equiv f(x)\frac{d^2 y(x)}{dx^2} + g(x)\frac{dy(x)}{dx} + h(x)y(x). \qquad (D.73)$$

Multiply this operator by the function

$$F(x) \equiv \frac{1}{f(x)} \exp\left[\int^x \frac{g(t)}{f(t)}dt\right]$$

to obtain

$$F(x)\mathcal{D}[y(x)] = \exp\left[\int^x \frac{g(t)}{f(t)}dt\right]\frac{d^2 y}{dx^2} + \frac{g(x)}{f(x)}\exp\left[\int^x \frac{g(t)}{f(t)}dt\right]\frac{dy}{dx}$$

$$+ \frac{h(x)}{f(x)}\exp\left[\int^x \frac{g(t)}{f(t)}dt\right]y(x) = 0$$

$$= \frac{d}{dx}\left\{\exp\left[\int^x \frac{g(t)}{f(t)}dt\right]\frac{dy}{dx}\right\} + \frac{h(x)}{f(x)}\exp\left[\int^x \frac{g(t)}{f(t)}dt\right]y(x)$$

$$\equiv \frac{d}{dx}\left\{p(x)\frac{dy}{dx}\right\} + q(x)y(x). \qquad (D.74)$$

The associated eigenvalue problem for the Sturm–Liouville operator $\mathcal{L}$ is

$$\boxed{\mathcal{L}[y(x)] + \lambda w(x)y(x) = 0,} \qquad (D.75)$$

where $w(x)$ is a real positive function. Most second-order differential eigenvalue problems can be cast into the form of this Sturm–Liouville eigenvalue problem (see Table D.1). Thus, by studying this Sturm–Liouville eigenvalue problem, eigenvalue problems for most other second-order differential operators are also solved. In particular, if the Sturm–Liouville operator $\mathcal{L}$ can be made self-adjoint, then the eigenfunctions will form a complete basis set.

D.7.1 Boundary Conditions to Make $\mathcal{L}$ Hermitian

The eigenvalue problem of Eq. (D.75) is usually subject to boundary value constraints. Many typical boundary values make $\mathcal{L}$ a self-adjoint operator so that the eigenfunctions will have the important properties discussed in Section D.6.

Consider real functions f and g that are square integrable so that the adjoint Sturm–Liouville operator $\mathcal{L}^\dagger$ is defined by Eq. (D.3), i.e., $(g, \mathcal{L}[f]) = (\mathcal{L}^\dagger[g], f)$. To determine $\mathcal{L}^\dagger$, integrate $(g, \mathcal{L}[f])$ by parts twice as follows:

$$(g, \mathcal{L}[f]) \equiv \int_a^b g^* \mathcal{L}[f]dx = \int_a^b g^*[p(x)f'(x)]'dx + \int_a^b g^*(x)q(x)f(x)\,dx$$

Table D.1 The reduction of important second-order eigenfunction problems to the Sturm–Liouville form.

Equation	$F(x)^*$	$p(x)$	$q(x)$	λ	$w(x)$
Harmonic oscillator	1	1	0	ν^2	1
Legendre	1	$1-x^2$	0	$\ell(\ell+1)$	1
Associated Legendre	1	$1-x^2$	$\frac{m^2}{1-x^2}$	$\ell(\ell+1)$	1
Bessel $(x \to x/a)$	1	x	$-\nu^2/x$	a^2	x
Tschebyscheff I	$\frac{1}{\sqrt{1-x^2}}$	$\sqrt{1-x^2}$	0	α^2	$\frac{1}{\sqrt{1-x^2}}$
Laguerre	e^{-x}	xe^{-x}	0	α	e^{-x}
Assoc. Laguerre poly.	$x^k e^{-x}$	$x^{k+1}e^{-x}$	0	$\alpha - k$	$x^k e^{-x}$
Hermite polynomial	e^{-x^2}	e^{-x^2}	0	2α	e^{-x^2}
Hermite orthog. func.	1	1	$x^2 - 1$	2α	1
Jacobi	$-x^{q-1}(1-x)^{p-q}$	$x^q(1-x)^{p-q+1}$	0	$n(n+p)$	$x^{q-1}(1-x)^{p-q}$
Mathieu	1	1	$-16b\cos 2x$	a	1
Hypergeometric	$-x^{\gamma-1}(1-x)^\delta$	$x^\gamma(1-x)^{\delta+1}$	0	$-\alpha\beta$	$x^{\gamma-1}(1-x)^\delta$

* Factor by which the standard form of the equation is multiplied to obtain the self-adjoint form.

$$= g^* pf' \Big|_a^b - \int_a^b (g^*)'(pf')\,dx + \int_a^b g^*(x)q(x)f(x)\,dx$$

$$= g^* pf' \Big|_a^b - (g^*)'pf \Big|_a^b + \int_a^b [p(g^*)']' f\,dx + \int_a^b g^*(x)q(x)f(x)\,dx$$

$$\text{(D.76)}$$

This must equal $(\mathcal{L}^\dagger[g], f)$ and for $\mathcal{L}$ to be self-adjoint

$$(\mathcal{L}^\dagger[g], f) = (\mathcal{L}[g], f) = \int_a^b [p(g^*)']' f\,dx + \int_a^b g^* qf\,dx. \qquad \text{(D.77)}$$

For $\mathcal{L}^\dagger = \mathcal{L}$, i.e., for Eq. (D.77) to equal Eq. (D.76), the boundary values of f and g must be such that

$$g^* pf' \Big|_a^b - (g^*)'pf \Big|_a^b = 0. \qquad \text{(D.78)}$$

This condition for $\mathcal{L}$ to be self-adjoint is not very restrictive. For example, any of the following boundary conditions make $\mathcal{L}$ self-adjoint: (1) $h(a) = h(b) = 0$,

(2) $h'(a) = h'(b) = 0$, (3) $h(a) = h'(a)$ and $h(b) = h'(b)$, and (4) $h(a) = h(b)$ and $h'(a) = h'(b)$, where h equals f or g. In fact, any homogeneous boundary condition of the form

$$\alpha_1 h'(a) + \alpha_2 h(a) = 0 \qquad \text{and} \qquad \beta_1 h'(b) + \beta_2 h(b) = 0 \qquad \text{(D.79)}$$

satisfies Eq. (D.78) and makes $\mathcal{L}$ self-adjoint. Here α_1 and α_2 are constants, not both zero. The same holds for β_1 and β_2.

D.7.2 Some Eigenfunction Properties of $\mathcal{L}$

The eigenfunctions and eigenvalues of the Hermitian Sturm–Liouville operator, defined by

$$\mathcal{L}[y_n(x)] + \lambda_n w(x) y_n(x) = 0, \qquad \text{(D.80)}$$

besides having the important properties of any continuous self-adjoint operator (namely, real eigenvalues, orthogonal eigenfunctions, and a complete set of eigenfunctions), have several other general properties. In this discussion, it is assumed that the homogeneous boundary conditions of Eq. (D.79) apply, that $p(x)$, $p'(x)$, and $w(x)$ are continuous over $[a, b]$, and that $p(x) > 0$ and $w(x) > 0$ on this interval.

Number of Eigenvalues:

There are an infinite number of discrete eigenvalues $\lambda_1 < \lambda_2 < \ldots < \lambda_n < \ldots$ and $\lambda_n \to \infty$ as $n \to \infty$.

Simple Eigenfunctions:

The eigenfunctions $y_n(x)$ are simple, i.e., there is only one linearly independent eigenfunction associated with each eigenvalue.

Nonnegative Eigenfunctions:

If, in addition to the assumptions about $p(x)$ and $w(x)$, (1) $q(x) \leq 0$ for all $x \in [a, b]$ and (2) the constants $\alpha_1, \alpha_2, \beta_1$, and β_2 in the boundary conditions of Eq. (D.79) are all ≥ 0, then the eigenfunctions have the property $y_n(x) \geq 0$ for all $x \in [a, b]$.

Alternating Zeros:

If $\lambda_n > \lambda_m$, then $y_n(x)$ must change sign (have a zero) between any two successive zeros of $y_m(x)$.

D.7.3   Important Applications of Sturm–Liouville Theory

The completeness and orthogonality of the eigenfunctions, defined by

$$\mathcal{L}[y_n(x)] + \lambda_n w(x) y_n(x) = 0,$$
(D.81)

subject to boundary conditions to make $\mathcal{L}$ Hermitian, have the following important applications.

1. The eigenfunctions can be used to approximate any function by a generalized Fourier series.
2. Such generalized Fourier series can, in turn, be used to find solutions of a second-order inhomogeneous differential equation

$$\mathcal{L}[y(x)] = g(x).$$

3. The eigenfunction expansion can be used to find Green's functions $G(x, x')$, so the solution of the above inhomogeneous equation can be written as

$$y(x) = \int_a^b G(x, x') g(x') \, dx'.$$

4. Finally, the eigenfunction expansion can also be used to solve second-order partial differential equations by converting them to ODEs.

Some of these applications are briefly addressed in the following sections.

D.8   Generalized Fourier Series

The eigenfunctions $\{y_n(x)\}$ of a self-adjoint operator form a complete basis set of functions so that "any" function $F(x)$ can be expanded as

$$F(x) = \sum_{n=1}^{\infty} a_n y_n(x).$$
(D.82)

The expansion coefficients $\{a_n\}$ can be found using the orthogonality property of the eigenfunctions, namely,

$$a_n = \frac{1}{N_n} \int_a^b w(x) y_n^*(x) F(x) \, dx.$$
(D.83)

Two specific examples are given below.

D.8.1 Fourier–Legendre Series

Legendre's equation is

$$(1 - x^2)y''(x) - 2xy'(x) + n(n+1)y(x) = 0, \quad -1 < x < 1,$$

or in Sturm–Liouville form,

$$\frac{d}{dx}\left[(1 - x^2)\frac{dy(x)}{dx}\right] + n(n+1)y(x) = 0.$$

Comparison of this eigenvalue problem to that of Eq. (D.75) reveals that $p(x) = (1 - x^2)$, $q(x) = 0$, $w(x) = 1$, and $\lambda = n(n+1)$.

When n is an integer, the two independent solutions are the Legendre polynomials $P_n(x)$ and the Legendre functions of the second kind $Q_n(x)$. Thus any function $F(x)$, $-1 < x < 1$, can be expanded as

$$F(x) = \sum_{n=0}^{\infty} a_n P_n(x) + \sum_{n=0}^{\infty} b_n Q_n(x).$$

Recall that the functions $Q_n(x)$ become infinite as $x \to \pm 1$, so, for functions $F(x)$ that are nonsingular in $[-1, 1]$, $b_n = 0$ and

$$F(x) = \sum_{n=0}^{\infty} a_n P_n(x).$$

The Legendre polynomials are orthogonal with

$$\int_{-1}^{1} P_n(x) P_m(x)\, dx = \frac{2}{2n+1} \delta_{nm},$$

so that, in terms of the function being expanded, the expansion coefficients are given by

$$a_n = \frac{2n+1}{n} \int_{-1}^{1} F(x) P_n(x)\, dx.$$

D.8.2 Fourier–Bessel Series

As another example of a generalized Fourier series, consider Bessel's equation

$$x^2 y''(x) + x y'(x) + (x^2 - n^2)y(x) = 0. \tag{D.84}$$

To convert this into a Sturm–Liouville eigenvalue problem, let $x' = kx$, where k is some positive constant (eigenvalue) to be determined. Now express derivatives with respect to x as derivatives with respect to x', namely,

$$\frac{dy(x)}{dx} = \frac{dy(kx')}{d(kx')} = \frac{dy(kx')}{k\,dx'},$$

$$\frac{d^2 y(x)}{dx^2} = \frac{d^2 y(kx')}{d(kx')^2} = \frac{d^2 y(kx')}{k^2 dx'^2}.$$

Substitute these derivatives into Eq. (D.84) and let $x' \to x$ to obtain

$$x^2 \frac{d^2 y(kx)}{dx^2} + x \frac{dy(kx)}{dx} + (k^2 x^2 - n^2) y(x) = 0.$$

This equation can now be written in the form of the Sturm–Liouville eigenvalue problem, i.e.,

$$\frac{d}{dx}\left[x \frac{dy(kx)}{dx} \right] - \frac{n^2}{x} y(kx) + k^2 x y(kx) = 0. \tag{D.85}$$

Comparison of this result to Eq. (D.75) identifies the Sturm–Liouville functions as $p(x) = x$, $q(x) = -n^2/x$, $w(x) = x$, and k is the eigenvalue.

The two independent solutions of Eq. (D.84) are the Bessel functions of the first and second kinds, i.e., $J_n(x)$ and $Y_n(x)$, and for Eq. (D.85) $J_n(kx)$ and $Y_n(kx)$. Here n is simply a fixed parameter. To convert Eq. (D.85) into an eigenvalue problem, boundary conditions on $y(kx)$ must be specified. Suppose one is interested in the interval $[0, R]$ with a boundary condition that $y(kR) = 0$, i.e., $J_n(kR) = 0$ and $Y_n(kR) = 0$.

The Bessel function of the first kind of order n has an infinite number of zeros α_{1n}, α_{2n}, α_{3n}, ... so that $J_n(kR) = 0$ requires k to assume discrete values, i.e., $kR = k_{mn} R = \alpha_{mn}$. Thus, one set of eigenfunctions is

$$J_n(k_{mn} x) \quad \text{with eigenvalues} \quad k_{mn} = \frac{\alpha_{mn}}{R}, \ m = 1, 2, 3, \ldots. \tag{D.86}$$

Similarly, $Y_n(r)$ has an infinite number of positive zeros β_{1n}, β_{2n}, β_{3n}, ..., so that the boundary condition $Y_n(kR) = 0$ requires kR to assume the discrete values β_{mn}. Thus another set of eigenfunctions is

$$Y_n(\overline{k}_{mn} x) \quad \text{with eigenvalues} \quad \overline{k}_{mn} = \frac{\beta_{mn}}{R}, \ m = 1, 2, 3, \ldots. \tag{D.87}$$

From the properties of Hermitian operators, any function $F(x)$, $0 \le x \le R$, with the boundary value $F(R) = 0$, can be expanded as

$$F(x) = \sum_{m=1}^{\infty} a_m J_n(k_{mn} x) + \sum_{m=1}^{\infty} b_m Y_n(\overline{k}_{mn} x). \tag{D.88}$$

Also recall that $Y_n(x) \to -\infty$ as $x \to 0$. So, if the physics of the problem requires $F(0)$ to be finite, then the b_m coefficients must vanish. For such functions, the Fourier–Bessel series reduces to

$$F(x) = \sum_{m=1}^{\infty} a_m J_n\left(\frac{\alpha_{mn}}{R}x\right).$$

(D.89)

The eigenfunctions must also be orthogonal (with weight function $w(x) = x$). It can be shown that

$$\int_0^R x J_n\left(\frac{\alpha_{mn}}{R}x\right) J_n\left(\frac{\alpha_{in}}{R}x\right) dx = \frac{R^2}{2} J_{n+1}^2(\alpha_{mi})\delta_{mi}.$$

Thus, the expansion coefficients in Eq. (D.89) are given by

$$a_m = \frac{2}{R^2 J_{n+1}^2(\alpha_{mn})} \int_0^R x F(x) J_n\left(\frac{\alpha_{mn}}{R}x\right) dx.$$

(D.90)

D.9   Solving Inhomogeneous Ordinary Differential Equations

The earlier use of an eigenfunction series to solve linear differential equations was restricted to homogeneous equations. Now consider the solution of linear inhomogeneous differential equations. Suppose one seeks the solution of

$$\mathcal{D}[y(x)] = g(x), \quad x \in [a,b],$$

(D.91)

subject to prescribed boundary conditions. Here $\mathcal{D}$ is any linear differential operator. There are three widely used methods for solving such equations.

D.9.1   Method 1: Find a Particular Solution

The most general solution of Eq. (D.91) can be written as

$$y(x) = y_{homo}(x) + y_{part}(x),$$

where $y_{homo}(x)$ is the solution of the corresponding homogeneous equation (with $g(x) = 0$) and $y_{part}(x)$ is any solution that satisfies Eq. (D.91). In general, if $\mathcal{D}$ is an nth-order differential, y_{homo} contains n arbitrary constants, but y_{part} has no arbitrary constants.

The trick with this method is to find $y_{part}(x)$. If $\mathcal{D}[g(x)]$ returns a function that is in the same family of functions to which $g(x)$ belongs, one seeks a function in the family as a particular solution. Often, however, finding a $y_{part}(x)$ is a very difficult task.

D.9.2 Method 2: Use the Green's Function

This approach has two steps. First one finds the solution of the so-called Green's function equation

$$\mathcal{D}[y(x)] = \delta(x - x'), \quad x, x' \in [a, b], \tag{D.92}$$

subject to the same boundary conditions as Eq. (D.91). One denotes the solution as $y(x) \equiv G(x, x')$, where x' is a parameter. Formally, if $\mathcal{D}^{-1}$ is the inverse operator to $\mathcal{D}$, the Green's function may be written as

$$y(x) \equiv G(x, x') = \mathcal{D}^{-1}[\delta(x - x')]. \tag{D.93}$$

The second step uses the definition of the Dirac delta function $\delta(x - x')$ to yield the identity

$$g(x) = \int_a^b \delta(x - x')g(x') \, dx'. \tag{D.94}$$

Now operate on Eq. (D.91) with $\mathcal{D}^{-1}$, which operates only on the variable x, and use Eq. (D.94) as follows:

$$y(x) = \mathcal{D}^{-1}[g(x)] = \mathcal{D}^{-1}\left[\int_a^b \delta(x - x')g(x') \, dx'\right]$$
$$= \int_a^b \mathcal{D}^{-1}[\delta(x - x')]g(x') \, dx'.$$

Finally, with Eq. (D.93), this reduces to

$$y(x) = \int_a^b G(x, x')g(x') \, dx'. \tag{D.95}$$

The hard part of this method is solving Eq. (D.92) for $G(x, x')$.

D.9.3 Method 3: The Eigenfunction Expansion Technique

This method is one of the principal reasons that this appendix on linear operator theory has been included in this book. This technique allows one to solve easily *any* inhomogeneous differential equation provided $\mathcal{D}$ is a self-adjoint differential operator. Suppose that $\mathcal{D}$ is Hermitian and that its eigenfunctions and eigenvalues are known, i.e., the nontrivial solutions of

$$\mathcal{D}[y_n(x)] = \lambda_n w(x) y_n(x), \tag{D.96}$$

subject to the same boundary conditions of Eq. (D.91), are known. These eigenfunctions must be orthogonal, i.e.,

$$\int_a^b w(x) y_n^*(x) y_m(x)\,dx = N_m \delta_{mn}, \tag{D.97}$$

and they are also complete so that any function, including the desired solution $y(x)$, can be expanded as

$$y(x) = \sum_{n=1}^{\infty} a_n y_n(x). \tag{D.98}$$

Note that, because $y_n(x)$ satisfy the same boundary conditions as Eq. (D.91),
this expansion forces $y(x)$ to also satisfy the boundary conditions.

The basic idea of the eigenfunction expansion method is, rather than find
$y(x)$ directly from Eq. (D.91), to find first the expansion coefficients $\{a_n\}$. Once
$\{a_n\}$ are known, Eq. (D.98) can be used to reconstruct $y(x)$.

Finding the expansion coefficients *for any Hermitian operator* $\mathcal{D}$ is quite
easy. First combine Eq. (D.91) and Eq. (D.98) as

$$g(x) = \mathcal{D}[y(x)] = \mathcal{D}\left[\sum_{n=1}^{\infty} a_n y_n(x)\right] = \sum_{n=1}^{\infty} a_n \mathcal{D}[y_n(x)].$$

But from Eq. (D.96), this reduces to

$$g(x) = \sum_{n=1}^{\infty} a_n \lambda_n w(x) y_n(x). \tag{D.99}$$

Now multiply Eq. (D.99) by $y_m^*(x)$, integrate over $[a, b]$, and use the orthogonality relations of Eq. (D.97) to obtain

$$\int_a^b g(x) y_m^*(x)\,dx = \sum_{n=1}^{\infty} a_n \lambda_n \int_a^b w(x) y_m^*(x) y_n(x)\,dx = a_m \lambda_m N_m.$$

From this, the needed expansion coefficients are found to be

$$a_m = \frac{1}{\lambda_m N_m} \int_a^b g(x) y_m^*(x)\,dx. \tag{D.100}$$

Finally, substitution of this result into Eq. (D.98) gives the desired solution as

$$y(x) = \sum_{n=1}^{\infty} \left[\frac{1}{\lambda_n N_n} \int_a^b g(x') y_n^*(x')\,dx'\right] y_n(x). \tag{D.101}$$

Note this is a solution to Eq. (D.91) for *any linear self-adjoint differential operator* $\mathcal{D}$! This method is a powerful technique indeed.

But wait, there is more. Write Eq. (D.101) as Eq. (D.101) to obtain

$$y(x) = \int_a^b \left[\sum_{n=1}^{\infty} \frac{y_n^*(x')y_n(x)}{\lambda_n N_n} \right] g(x')\,dx'.$$

Comparison of this result to the Green's function solution of Eq. (D.95) shows that the Green's function is given by

$$G(x,x') = \sum_{n=1}^{\infty} \frac{y_n(x)y_n^*(x')}{\lambda_n N_n}. \tag{D.102}$$

As a final bonus, the eigenfunction expansion technique shows, from Eq. (D.102), that the Green's function $G(x,x') = G^*(x',x)$, i.e., it is symmetric with respect to the interchange of its arguments, a result obtained earlier in Section D.4.3. This is known as the *reciprocity theorem* for Green's functions of self-adjoint operators. This theorem often lets one use problem symmetry to interchange a point source and the point at which the solution $y(x)$ is sought.

And that is why the eigenfunctions of Hermitian operators are so important!

References

Morse, P., Feshbach, H., 1953. Methods of Theoretical Physics, vols. I and II. Intern. Series in Pure and Appl. Phys. McGraw-Hill, Boston.

Appendix E

Some Popular Monte Carlo Codes for Particle Transport

There are many general-purpose Monte Carlo codes for radiation transport simulation. Some of these are well documented and widely used, others less well so, and still others proprietary. Nine popular nonproprietary codes are briefly described in this Appendix. The code descriptions are based, often verbatim, on material provided by representatives of the code collaborations for the first edition of this book. In some cases, these descriptions are supplemented by information found on the codes' web pages. The intent of the summaries presented here is to give the reader an understanding of the problems addressed and capabilities of each code. The extensive bibliographies and past applications of the codes, given in the first edition, have been omitted here. The codes are considered in alphabetical order.

E.1 COG

COG (currently, version 11.1) is a modern, full-featured, high-resolution code for the Monte Carlo simulation of coupled neutron, proton, gamma ray, and electron transport in arbitrary 3-D geometry. It provides accurate answers to complex shielding, criticality, and activation problems. COG was written to be state-of-the-art and free of physics approximations and compromises found in earlier codes. COG uses pointwise cross sections and exact angular scattering, and allows a full range of biasing options to speed up solutions for deep penetration problems. Additionally, a criticality option is available for computing k_{eff} for assemblies of fissile materials. COG can also compute gamma-ray doses due to neutron-activated materials, starting with just a neutron source. Source and random walk biasing techniques may be selected to improve solution statistics. These include source angular biasing, importance weighting, particle splitting and Russian roulette, path-length stretching, point detectors, scattered direction biasing, and forced collisions.

COG will transport neutrons with energies in the range of 5–10 eV to 150 MeV, protons with energies up to hundreds of GeV, and photons with energies in the range of 10 eV to 100 GeV (COG's energy ranges are limited by the available cross-section sets and physics models). Via the EGS4 electron transport kernel, electrons in the range of 10 keV to a few thousand GeV can also be transported. The COG code is a significant upgrade from earlier Monte Carlo

Exploring Monte Carlo Methods. https://doi.org/10.1016/B978-0-12-819739-4.00023-8

transport codes and has been written specifically to make it more versatile, more accurate, and easier to use. COG has provisions for calculating deep penetration (shielding) problems, criticality problems, and neutron activation problems, and retains all of the standard capabilities found in other Monte Carlo transport codes. COG uses high-resolution pointwise cross-section databases and makes no compromises in the transport physics, so that the results of a COG run are limited only by the accuracy of the databases used.

COG can use either the LLNL ENDL-90 cross section set or the ENDFB/VI set. Analytic surfaces are used to describe geometric boundaries. Parts (volumes) are described by a method of Constructive Solid Geometry. Surface types include surfaces of up to fourth order and pseudosurfaces such as boxes, finite cylinders, and figures of revolution. Repeated assemblies need to be defined only once. Parts are visualized in cross-section and perspective picture views. A lattice feature simplifies the specification of regular arrays of parts. Parallel processing under MPI is supported for multi-CPU systems.

Useful websites for COG include:

http://cog.llnl.gov
https://rsicc.ornl.gov/codes/ccc/ccc7/ccc-777.html
http://www.oecd-nea.org/tools/abstract/detail/ccc-0829/

E.2 EGSnrc

EGSnrc is a general-purpose software toolkit that can be applied to build Monte Carlo simulations of coupled electron–photon transport, for particle energies ranging from 1 keV to several tens of GeV. It is widely used internationally in a variety of radiation-related fields, particularly in medical physics.

A notable feature of EGSnrc is its very accurate implementation of the condensed history technique for charged particle transport, which is step size independent and artifact-free at the 0.1% level. Such accuracy results from an exact multiple scattering theory, an improved electron step algorithm, an exact boundary crossing algorithm based on a single scattering simulation in the vicinity of region interfaces, and various other details involved with electron transport. Compared to its predecessor EGS4, EGSnrc includes numerous improvements in the modeling of photon and electron interactions such as (1) binding effects and Doppler broadening in the relativistic impulse approximation for Compton scattering, (2) a full relaxation cascade, which includes both radiative and nonradiative transitions, following an inner-shell ionization event, (3) an ability to utilize user-supplied molecular form factors for Rayleigh scattering, (4) an ability to use total photon cross-section tabulations from XCOM, EPDL97, Storm and Israel, or any other user-supplied tabulation, (5) use of radiative corrections for Compton scattering, (6) use of exact differential pair production cross sections for photon energies to 85 MeV, (7) an explicit simulation of pair production in photon–electron interactions (i.e., *triplet production*), (8) explicit simulation of inner shell ionizations by electron impact, (9) improved bremsstrahlung differential cross-section tabulations from NIST that are the basis for ICRU-recommended radiative stopping powers, and (10) inclusion of spin and relativistic effects in electron and positron elastic scattering. EGSnrc, thus, improves the accuracy and precision of the charged particle transport. Moreover, the charged-particle multiple-scattering algorithm allows large step sizes without sacrificing accuracy, a feature that leads to fast simulation speeds.

The EGSnrc package is not a monolithic application. Instead, it provides a set of functions and subroutines for the simulation of coupled electron and photon transport. To obtain a complete application, the user must supply a so-called "user code" that provides a description of the simulation geometry, the particle source, and a function for extracting information from the simulation. User codes can be written in Mortran, FORTRAN, C, or C++. EGSnrc includes a C++ class library, called egs++, which can be used to model elaborate geometries and particle sources. The C++ class library has permitted the development of several C++ user codes for EGSnrc that are included in the general code distribution. Included in the distribution are applications for computing dose distributions, ionization chamber correction factors, particle fluences, pulse height distributions, and various simple tutorial codes. Also included are the two user codes BEAMnrc, used to model medical linear accelerators, and DOSXYZnrc, used for computing dose distributions in rectilinear geometries.

Useful websites for EGSnrc include:

https://nrc.canada.ca/en/research-development/products-services/software-applications/egsnrc-software-tool-model-radiation-transport
https://nrc-cnrc.github.io/EGSnrc/
https://rsicc.ornl.gov/codes/ccc/ccc3/ccc-331.html
http://www.irs.inms.nrc.ca/papers/egs.biblio/egs_papers.html
http://www.irs.inms.nrc.ca/BEAM/bibliog/omega_pubs.html

E.3   GEANT4

GEANT4 (GEometry ANd Tracking) is a toolkit for Monte Carlo simulation of the interaction and transport of particles in matter. The code provides comprehensive geometry and physics modeling capabilities embedded in a flexible structure base using the Object-Oriented technology and C++. Because of the abstract interfaces, users can choose existing components, tailor and adapt them, or create their own plug-in components.

Models of the interaction of particles with matter are implemented in the GEANT4 toolkit to describe electron, ion, muon, gamma ray, electromagnetic, hadronic, and optical photon processes. For many physics processes, a choice of physics models is offered, with different trade-offs between accuracy and CPU cost. Tools for optimization of configurations of geometry and physics are also provided. The combinations of physical processes for typical applications are recommended in Physics Lists. Users can thus choose an existing physics list/configuration to meet their needs, revise or customize an existing one, or create a new one.

The GEANT4 geometry module enables users to describe their setup, and then locates and moves tracks inside it. It contains a variety of shapes and allows Boolean operations of these shapes, as well as the ability to create a hierarchy of volumes. In complex setups, volumes can be repeated, sharing key characteristics. Volumes can also be grouped into regions. A particular geometry can be described in C++ code, or read in from files in the GDML markup language. Other tools can convert the geometry from a Computer-Aided Design (CAD) system into a GDML file readable by GEANT4. The toolkit also enables the user to simulate motion of volumes in the geometrical setup.

Production thresholds (cuts) used for some electromagnetic processes can be varied by geometrical region to allow different precisions for different parts of model. Other event biasing techniques offered in GEANT4 include geometrical importance biasing (splitting and Russian roulette), cross-section enhancement, options for leading particle biasing, and weight window methods. Parallel geometry descriptions, typically simpler, can be used simultaneously for scoring, importance biasing, and/or shower parameterization.

Electromagnetic fields described by GEANT4 affect the transport of charged particles. A global field can be used, and can be overridden locally for any volume.

Typical scoring, including radiation dose deposition, is a part of the toolkit. Specific "hooks," such as the optional user action classes, provide the user with complete control of the production of hits or scoring of the results. A powerful stacking mechanism enables users, at almost no CPU cost, to control the prioritization of the processing of tracks in an event, to abort particular tracks or events early in the simulation, and to postpone the propagation of certain parts of tracks.

Visualization options range from the fast OpenGL to high-quality rendered pictures. A comprehensive system is available for the visualization of geometries, tracks, and detector hits through a choice of drivers and a variety of user interface options. Time-dependent visualization of particle propagation is also possible.

The GEANT4 toolkit can be used for stand-alone simulations or can be integrated in larger software frameworks, such as discussed above for large HEP experiments. Users can also customize and further extend the GEANT4 software based on requirements for a particular application. For example, changing the run configuration allows for the user to vary the geometry or materials composition during execution.

Useful websites for GEANT4 include:

https://geant4.web.cern.ch/
http://cern.ch/GDML/
http://hep.fi.infn.it/geant.pdf

E.4   MCSHAPE

MCSHAPE is a general-purpose Monte Carlo code developed at the University of Bologna to simulate the transport of unpolarized or polarized X-ray and gamma-ray photons with energies between 1 and 1000 keV in one, two, or three dimensions. This code can describe the full evolution of the photon polarization state as the photons undergo Compton and Rayleigh scattering and photoelectric interactions in the target. All the parameters that characterize the photon transport can be suitably defined: (1) the source intensity, (2) its full polarization state as a function of energy, (3) the number of collisions, and (4) the energy interval and resolution of the simulation. It is possible to visualize the results for selected groups of interactions. MCSHAPE simulates the propagation in heterogeneous media of polarized photons (from synchrotron sources) or of partially polarized sources (from X-ray tubes). MCSHAPE can determine the influence of the detector on the predicted measured spectrum. It also contains a library of all interaction data needed for the transport of photons in any combination of materials.

To correctly represent the polarization of photons, MCSHAPE uses the four Stokes intensity parameters, which contain all the physical information about the polarization state of the photons, i.e., the intensity, the degree of polarization, the orientation, and the ellipticity of polarization, at each point in the geometry and in any given direction.

The Boltzmann–Chandrasekhar vector transport equation describes formally the vector flux $\mathbf{f}(\mathbf{r}, \boldsymbol{\Omega}, \lambda)$ of polarized photons in the Stokes system (having components f_I, f_Q, f_U, and f_V) and can be written as

$$\boldsymbol{\Omega}\cdot\nabla\mathbf{f}^{(S)}(\mathbf{r}, \boldsymbol{\Omega}, \lambda) + \mu(\mathbf{r}, \lambda)\mathbf{f}^{(S)}(\mathbf{r}, \boldsymbol{\Omega}, \lambda) =$$

$$\int_0^\infty d\lambda' \int_{4\pi} d\boldsymbol{\Omega}' H^{(S)}(\mathbf{r}, \boldsymbol{\Omega}, \lambda, \boldsymbol{\Omega}', \lambda')\mathbf{f}^{(S)}(\mathbf{r}, \boldsymbol{\Omega}', \lambda') + \mathbf{S}^{(S)}(\mathbf{r}, \boldsymbol{\Omega}, \lambda),$$

where

$$H^{(S)}(\mathbf{r}, \boldsymbol{\Omega}, \lambda, \boldsymbol{\Omega}', \lambda') = L^{(S)}(\pi - \psi)K^{(S)}(\mathbf{r}, \boldsymbol{\Omega}, \lambda, \boldsymbol{\Omega}', \lambda')L^{(S)}(-\psi')$$

is the kernel matrix in the meridian plane of reference, $K^{(S)}(\mathbf{r}, \boldsymbol{\Omega}, \lambda, \boldsymbol{\Omega}', \lambda')$ is the scattering matrix in the scattering plane of reference, $L^{(S)}$ is the four-by-four rotation matrix which transforms the scattered flux from the scattering plane to the meridian plane of the reference, $\mu(\mathbf{r}, \lambda)$ is the narrow-beam attenuation coefficient, which is independent of the state of polarization of the photons (assuming the matter is isotropic), and $\mathbf{S}^{(S)}(\mathbf{r}, \boldsymbol{\Omega}, \lambda)$ is the source vector flux with the four Stokes parameters as components.

The total vector intensity is obtained (component by component) by adding the contributions $f_i^{(k)}(\mathbf{r}, \boldsymbol{\Omega}, \lambda)$ from the different number of collisions, as is done for the scalar model, namely,

$$f_i(\mathbf{r}, \mathbf{\Omega}, \lambda) = \sum_{k=0}^{\infty} f_i^{(k)}(\mathbf{r}, \mathbf{\Omega}, \lambda).$$

Rayleigh scattering, Compton scattering, the photoelectric effect, and the prevailing interactions in the X-ray regime (1–1000 keV), are included in the simulation using state-of-the-art models. Because MCSHAPE uses the Stokes representation, it is the unique Monte Carlo code which can follow the complete evolution of the polarization state of the radiation. The current version of MCSHAPE uses a new algorithm for the simulation of Compton scattering that includes Doppler broadening effects. This new algorithm corrects the error in the representation of the Compton peak in the standard approach adopted for generating the Compton profile in other Monte Carlo codes.

Some of the features of MCSHAPE include: (1) atomic Rayleigh scattering, (2) atomic Compton scattering with a correct Compton profile, (3) the photoelectric effect with about 1000 edge energies and line widths, (4) infinite or finite target thicknesses, (5) heterogeneous three-dimensional targets, (6) monochromatic or polychromatic sources with arbitrary polarization states, (7) use of reflection or transmission geometries, (8) external detectors with detector responses, and (9) an open database. In future releases of MCSHAPE, electron bremsstrahlung and user-defined geometrical elements are planned.

Useful websites for MCSHAPE include:

https://inis.iaea.org/search/search.aspx?orig_q=RN:47055745
http://shape.ing.unibo.it/html/overview.html
https://inis.iaea.org/search/search.aspx?orig_q=RN:47055745

E.5 MCNP6

The MCNP series of codes has perhaps the longest history of any Monte Carlo radiation transport code. In the 1950s and 1960s, a number of special-purpose Monte Carlo codes, such as MCS, MCN, MCP, and MCG, were developed at the Los Alamos National Laboratory (LANL). They were merged in 1973 to create MCNG, which was merged with the photon code MCP in 1977 to create the first version of MCNP. In 1983, MCNP3 was released for public distribution. Between 1986 and 2011, 19 versions of MCNP have been released, each adding additional capability to the code. Finally, in 2012, MCNP6 Beta2 was released. This version merged MCNP5 with MCNPX (a variant of MCNP eXtended for high-energy applications) and took 12 man-years of effort. MCNP5 with its roughly 100k lines of code increased to almost 500k lines of code in the merged MCNP6. MCNP6.2, the current version, was released in 2018. It is estimated that worldwide there are about 10,000 users, likely making it the most popular Monte Carlo radiation transport code.

The code treats an arbitrary three-dimensional configuration of materials in geometric cells bounded by first- and second-degree surfaces and fourth-degree elliptical tori. Macrobodies can be used to define the surfaces of simple geometric volumes. Lattices and nested universes can also be used to easily define repetitive portions of the geometry. A graphical capability is included to aid in the development of a problem's geometry, often the most challenging part of preparing an MCNP input file.

MCNP6 can transport 37 different types of particles: nine elementary particles (γ, e^-, e^+, μ^-, μ^+, v_e, v_m, $\bar{v}_e$, $\bar{v}_m$), 16 composite particles (n, p, Λ^0, Σ^+, Σ^-, Ξ^0, Ξ^-, Ξ^+, Ω^-, π^+, π^-, π^0, K^+, K^-, K^0_S, K^0_L), seven composite antiparticles ($\bar{n}$, $\bar{p}$, $\overline{\Lambda}^0$, $\overline{\Sigma}^+$, $\overline{\Sigma}^-$, $\overline{\Xi}^0$, $\overline{\Omega}^-$), and five complex particles (d, t, ^{3}He, α, heavy ion [$Z = 3$ to 92]). The nuclear data needed to describe the interactions of these particles as they move through phase space are provided by (1) pointwise cross-section data and other extensive data tables, although some older multigroup cross-section data are also provided, and/or (2) various nuclear physics models, used primarily for high-energy particles (> 150 MeV). The comprehensive ENDF/B-VII data files consist of pointwise (continuous-energy) interaction data and secondary particle production data for many particles, with energies up to 150 MeV, and for a few hundred isotopes. $S(\alpha, \beta)$ data are included for thermal-neutron scattering in 48 materials. Proton interacton data for 48 isotopes are provided for energies from 1 to 150 MeV. Photon and electron/positron data are included for energies of 1 eV to 100 GeV and from 10 eV to 1 GeV, respectively. Photonuclear data for 157 isotopes are included for photon energies up to 150 MeV. Above the maximum energies of these data tabulations, and for all hadrons, interaction physics are based on theoretical models with empirical corrections. These models include Bertini's intranuclear cascade (INC) model, ISABEL INC, the cascade-exciton model (CEM), the intranuclear cascade model developed at Liège (INCL), and the LANL version of the quark-gluon string model (LAQGSM).

The source definition capabilities of MCNP6 allow users to specify almost any source. Source variables may be specified discretely or as distributions. Additionally, source variables may be defined to be functions of other source variables, providing the user a great versatility in defining sources. In addition to fixed-source definitions, the user can specify criticality sources and MCNP6 can write surface sources of particles crossing surfaces for use in subsequent calculations.

MCNP6 offers a significant quantity of built-in Monte Carlo variance reduction techniques. Importance splitting/rouletting can be performed if the user specifies splitting parameters for the geometry. Additionally, space-energy- or space-time-dependent splitting can be performed using the weight-window technique, a weight-dependent splitting/rouletting variance reduction technique. MCNP's weight-window generator assists the user in generating the weight-window parameters. Semideterministic variance reduction techniques are also available through the DXTRAN technique and point detectors (next-event estimators). Time, energy, and weight cutoffs are also included in MCNP, as well as many others.

The tallying capabilities of MCNP6 are extensive. Some of the basic tallies include surface flux and current tallies, volume flux tallies, point and ring detectors, heating tallies, and pulse-height tallies to simulate radiation detector responses. The tallies can be performed over multiple individual cells or surfaces as well as the combinations of cells or surfaces and can be binned by energy, time, direction, special multipliers such as cross sections, surface and cell flagging, and user-specified bins. MCNP also has the capability to perform mesh tallies over a user-specified mesh independent of the geometry which can also be binned by energy. MCNP6 has so many features and options, such as treating activation, transmutation, and burnup of isotopes, that space limitations preclude their discussion in this synopsis.

Finally, MCNP6 is simply and accurately described as the merger of MCNP5 and MCNPX capabilities, but it is much more than the sum of those two computer codes. The initial release of MCNP6 contains 16 new features not previously found in either code. Among these new features are the abilities (1) to import unstructured mesh geometries from the finite element code Abaqus, (2) to transport photons down to 1.0 eV and electrons down to 10.0 eV, (3) to model complete atomic relaxation emissions for any element, (4) to generate or read mesh geometries for use with the LANL discrete ordinates code Partisn, (5) to use adjoint-weighted tallies for perturbation studies, (6) to track all charged particles in magnetic fields, (7) to nest multiple DXTRAN spheres for better control of particle populations, (8) to add time bins for mesh tallies, (9) to use enhanced form factors for coherent and incoherent photon scattering, and (10) to provide a precollision next-event estimator to augment the postcollision next-event estimator used to estimate point fluxes.

Useful websites for MCNP6 include:

https://mcnp.lanl.gov/
https://laws.lanl.gov/vhosts/mcnp.lanl.gov/references.shtml
https://laws.lanl.gov/vhosts/mcnp.lanl.gov/pdf_files/la-ur-17-29981.pdf
https://rsicc.ornl.gov/codes/ccc/ccc8/ccc-850.html

E.6   PENELOPE

PENELOPE (**P**enetration and **EN**Ergy **LO**ss of **P**ositrons and **E**lectrons), whose current version is PENELOPE2016, performs Monte Carlo simulation of coupled electron–photon transport in material structures consisting of homogeneous bodies limited by quadratic surfaces. A detailed and entirely self-contained description of the physical models, tracking algorithms, and code structure can be found in PDF format on the web (https://www.oecd-nea.org/science/docs/2015/nsc-doc2015-3.pdf). It consists of the FORTRAN source files, the interaction database, and a set of auxiliary tools (geometry and shower viewers, examples of main programs, etc.). The interaction database covers the energy interval between 50 eV and 1 GeV. Interaction data for energies higher than 1 GeV are estimated by natural cubic spline extrapolation. Although PENELOPE can transport particles with energies as low as 50 eV, the adopted interaction models (and also the particle trajectory model underlying the Monte Carlo simulation) are reliable only for energies higher than about 1 keV.

PENELOPE simulates the main interactions of photons (Rayleigh scattering, Compton scattering, photoelectric absorption, and electron–positron pair production) and of electrons/positrons (elastic and inelastic collisions, bremsstrahlung emission, and, in the case of positrons, two-photon annihilation). The ionization of K, L, and M electronic shells of atoms caused by Compton scattering, photoelectric absorption, and electron/positron impact and the subsequent emission of X rays and Auger electrons are also simulated. The energies of the X rays emitted from primary vacancies are set equal to their experimental values. The information in the database refers mostly to interactions of radiation with single, isolated atoms. Interactions with compound materials are described using Bragg's additivity rule, i.e., cross sections for molecules are approximated by the sum of cross sections of the atoms in the molecule. The adopted cross sections are the most accurate available to date, limited only by the required generality of the code.

Random photon histories are generated using the conventional detailed simulation scheme. Electrons and positrons are simulated by means of a mixed (class II) scheme in which hard interactions (i.e., interactions involving angular deflections and energy losses larger than preselected cutoff values) are simulated individually from the corresponding differential cross sections. The collective effect of the soft interactions (with energy loss or angular deflection less than the cutoff values) that occur between each pair of consecutive hard interactions is described using multiple scattering approximations. The effect of soft scattering on the spatial displacements of the transported electron or positron is simulated using the random-hinge method. For each considered material, the actual values of the energy loss and angular cutoffs are determined by a set of simulation parameters that are set by the user. These parameters also determine the amount of detail (energy and angular resolutions) and the speed of the simulation.

The most characteristic feature of PENELOPE is the consistent use of class II schemes for electrons and positrons. When the energy loss and angular cutoff values are set to zero, the simulation becomes purely detailed, i.e., free from any possible artifact introduced by multiple-scattering approximations. Because only the effect of soft interactions needs to be accounted for by the random-hinge method, spatial displacements are accurately described. In many practical cases, the simulation results are very stable under variations of the simulation parameters. The geometry package PENGEOM is tailored to minimize the numerical work needed to track particles using either detailed or mixed simulation. The geometry viewers generate images of the geometry by using strictly the same tracking routines as in the simulation. PENELOPE is particularly well adapted to simulations of electron transport at low and intermediate energies.

The cross sections for elastic scattering were obtained from elaborate first-principles calculations (Dirac partial-wave analysis). Inelastic collisions are described by means of the Sternheimer–Liljequist generalized oscillator strength model (the target atom or molecule is represented as a set of delta oscillators), which yields a realistic description of inelastic collisions in the energy range of interest. In particular, it gives collision stopping powers consistent with ICRU-recommended values. The ionization of inner shells by electron and positron impact is described as an independent mechanism.

PENELOPE is structured as a family of subroutine packages. The FORTRAN source files of the basic packages are `penelope.f`—analog simulation of coupled electron–photon transport in unbounded media; `pengeom.f`—tracking of particles through modular quadratic geometries (can operate material structures with up to 10,000 surfaces and 5000 bodies); and `penvared.f`—variance reduction routines (particle splitting, Russian roulette, and interaction forcing). The `penelope.f` and `pengeom.f` subroutines generate particle histories automatically, that is, they track particles through the material structure, simulate particle interactions and the generation of secondary particles, and manage the secondary stack. To take full advantage of the code capabilities, the user should write a steering main program, which only has to define the initial states of primary particles according to the characteristics of the radiation source, control the evolution of the generated showers, and keep score of the relevant quantities. To spare occasional users from having to write their main program, the distribution package includes a generic main program, named `penmain`, that simulates electron–photon transport in quadratic geometries, with different sources and a variety of scoring options, which are defined through an input file.

An alternative to writing a main program is to use pyPENELOPE, which is an open-source software to facilitate the use of the Monte Carlo code PENELOPE and its main program PENEPMA in the field of microanalysis. It consists in a python graphical user interface (GUI) to set up materials, geometry, simulation parameters, and position of the detectors as well as to display the simulation's results. Details about pyPENELOPE can be found at http://pypenelope.sourceforge.net/index.html.

Useful websites for PENELOPE include:

https://www.oecd-nea.org/jcms/pl_46441/penelope-2018-a-code-system-for-monte-carlo-simulation-of-electron-and-photon-transport
http://www.nea.fr/html/dbprog/
https://www.oecd-nea.org/science/docs/2015/nsc-doc2015-3.pdf
https://www.oecd-nea.org/tools/abstract/detail/nea-1525
https://rsicc.ornl.gov/codes/ccc/ccc7/ccc-782.html

E.7 SCALE

The SCALE system (currently version 6.2.3) was developed to satisfy a need for a standardized method of analysis for the evaluation of nuclear fuel facility and package designs. At the heart of the SCALE code package are radiation transport methods that utilize both Monte Carlo and discrete-ordinate methods. In its present form, SCALE has the capability to perform criticality, shielding, radiation source term, spent fuel depletion/decay, and reactor physics analyses using well-established codes.

The following discussion is focused on the Monte Carlo capabilities within SCALE. The SCALE package has the KENO and MONACO Monte Carlo codes that are used to solve eigenvalue and fixed-source 3-D radiation transport analyses, respectively.

KENO consists of two distinct 3-D Monte Carlo criticality safety codes in SCALE (KENO-V.a and KENO-VI). These codes continue to be enhanced to provide users with greater capabilities. In SCALE 6, these codes have been extended to allow use of either multigroup or continuous-energy cross sections. KENO-V.a, which has been part of SCALE for 30 years, has a simplified geometry system that makes it very easy to use and is much more computationally efficient than other Monte Carlo criticality safety codes. The geometry models consist of units that are constructed by nesting partial or whole spheres, cylinders, and rectangular cuboids oriented along x-, y-, and z-axes. These units can be combined in rectangular arrays or inserted in other units. The restricted orientation of the units allows very fast particle tracking, but still permits users to construct very complex geometry models. KENO-VI, which was incorporated into SCALE in 1995, provides users with a more general geometry system known as the SCALE Generalized Geometry Package (SGGP) for constructing geometry models. KENO-VI has a much larger assortment of bodies (e.g., cone, cuboid, cylinder, dodecahedron, elliptical cylinder, hexprism, plane, etc.). All KENO-VI bodies may intersect and may be arbitrarily rotated and translated. In addition to rectangular arrays, KENO-VI provides the following array types: dodecahedral (a 3-D stack of dodecahedrons) and three hexagonal array types.

MONACO is a 3-D Monte Carlo code developed within SCALE for shielding calculations and is the result of a modernization effort combining the multigroup neutron and photon physics of the well-known MORSE Monte Carlo shielding code with the flexibility of the SGGP. MONACO uses the same cross-section package as other SCALE modules. Available tallies in MONACO include point detectors, region-based flux tallies, and mesh tallies. Any MONACO tally can be folded with a response function, either user-entered or from a standard list available with each SCALE cross-section library. Mesh tally values and uncertainties can be viewed with a special Java viewer on all computer platforms that run SCALE.

For fixed-source problems, an advanced variance reduction scheme is used. The MAVRIC sequence in SCALE combines the results of an adjoint calculation from the 3-D deterministic code TORT with the Monte Carlo MONACO. Both an importance map for weight windows and a biased source are automatically generated from the adjoint flux using the CADIS methodology. MAVRIC is completely automated—from a simple user input, it creates the cross sections (forward and adjoint), calculates the first collision source, computes the adjoint fluxes, creates the importance map and biased source, and then executes MONACO. Users can start and stop the calculation at various points so that progress can be monitored and importance maps can be reused for similar problems. For simple problems that do not require the advanced variance reduction, the MAVRIC sequence is an easy way to compute problem-dependent cross sections and execute MONACO with a common user input. SCALE includes the latest nuclear data libraries for continuous-energy and multigroup radiation transport as well as activation, depletion, and decay calculations. One of the significant and powerful capabilities in SCALE is the automated problem-dependent cross-section processing capability for multigroup radiation transport analyses. SCALE includes both continuous-energy and multigroup ENDF/B-VI cross-section libraries. Continuous-energy and multigroup ENDF/B-VII.0 libraries may be released with SCALE 6 in 2008. In terms of visualization capabilities, SCALE has the GeeWiz interface module that serves as a user control center for executing many SCALE sequences. With the GeeWiz tool, the user can develop input, perform the calculation, and view the output within a single graphical user interface package. In addition, GeeWiz provides a direct link with the KENO3D visualization package that allows users to interactively view their KENO geometry models as it is constructed. Additionally, SCALE provides the Javapeño interactive plotting program that permits users to plot TSUNAMI sensitivity data, flux profiles, reaction rate data, cross-section data, and covariance data. The KENO code also offers the option of generating its output in HTML format that can be viewed through a web browser, including interactive 2-D Java applet plots. In summary, SCALE provides numerous Monte Carlo-based calculation sequences that can be used to perform various criticality safety and shielding analyses. These features enhance SCALE's ability to provide a powerful but easy-to-use set of nuclear fuel cycle safety analysis tools.

Useful websites for SCALE include:

https://www.ornl.gov/onramp/scale-code-system
http://www.oecd-nea.org/tools/abstract/detail/CCC-0834
https://www.ornl.gov/sites/default/files/SCALE%20Code%20System.pdf
http://www.oecd-nea.org/tools/abstract/detail/CCC-0834

E.8 SRIM

The SRIM code (**S**topping and **R**ange of **I**ons in **M**atter) simulates the transport of heavy ions of less than 2 GeV/amu in matter. The models are quantum mechanical and statistical, in the sense that the ions make macroscopic moves between collisions but then the collision results are averaged. This procedure, common to most charged-particle transport algorithms, leads to increased efficiency. An ion treated in SRIM is any charged particle as heavy as a proton or greater. Matter is any collection of atoms of a gas, a liquid, or a solid up to atomic number $Z = 92$. The TRIM code was initially based on light ion transport, but SRIM now treats ion transport in matter where both the ions and the atoms in the matter include all elements up to uranium.

SRIM is actually a collection of programs, the main one being the DOS-based TRIM code (for **TR**ansport of **I**ons in **M**atter), which originated in 1983 and is a comprehensive ion transport module. TRIM treats objects that can be quite complex. The object can be made of up to eight layers, each layer being composed of a compound material. TRIM estimates the final distribution of the ions (in three dimensions). TRIM also is able to estimate kinetic phenomena that are induced by an ion's energy loss. These phenomena include sputtering, target damage, the production of phonons, and ionization. SRIM models the cascades that result from ion impact on target atoms. The SRIM package can be used to generate tables of stopping powers, ranges of ions in matter, and straggling distributions. SRIM also has been used to study ion implantation, ion sputtering, and ion beam therapy.

The SRIM code allows interruption and resumption. This feature means that intermediate results are not necessarily lost but the simulation can be resumed following some intermediate inquiry on the part of the user. The SRIM code allows 2-D ion transport plots to be viewed and saved. Plots of ion tracks can be viewed during calculation but the primary output of SRIM are files that summarize the detailed ion transport results. The latest SRIM-2013 version has added, in addition, 3-D graphics display of the calculations. Computers have reached speeds to allow this to be done in real-time.

The principal authors of the SRIM series of codes are James F. Ziegler and Jochen P. Biersack, although others have contributed. This summary is based on material taken from the SRIM web page (http://www.srim.org/). The latest SRIM-2013 code can be downloaded from this web page. Although a Tutorial Manual is included with the code package, the most recent and complete description of SRIM and the models used is provided by the book *The Stopping and Range of Ions in Matter* (Ziegler, J.F., Biersack, J.P., Ziegler, M.D. 2008. Lulu Press, Morrisville, NC).

Useful websites for SRIM include:

http://www.srim.org/
http://www.srim.org/SRIM/SRIMLEGL.htm
http://www.srim.org/SRIM/SRIM%2008.pdf

E.9 TRIPOLI

TRIPOLI (**TRI**dimensionnel **POLI**cinétique) is a 3-D continuous-energy Monte Carlo code. TRIPOLI-4 is the fourth generation of the code and the current version is 8.3.

TRIPOLI is mainly used for radiation protection and shielding, core physics, and criticality calculations. The code is designed for neutron and photon transport in the energy range from 0 to 20 MeV, and it uses continuous-energy cross sections. TRIPOLI-4 produces classical Monte Carlo output such as flux and derived quantities (e.g., reaction rates), current, and k_{eff} for criticality and core physics problems. Time-dependent problems may be addressed, as well as perturbation calculations by employing the correlated sampling method.

The code has a flexible parallel mode, enabling it to easily run on heterogeneous networks of processors as well as on massively parallel machines. The code directly reads ENDF and PENDF files produced by the NJOY nuclear data processing system and is compatible with all nuclear data evaluations in the ENDF/B format such as JEF2, JEFF3, ENDF/B-VI, ENDF/B-VII, JENDL3.3, etc. The user may easily mix materials from different nuclear data evaluations, a feature that also makes TRIPOLI-4 a valuable tool for the nuclear data community.

For radiation protection and shielding, one needs efficient variance reduction tools, among which the geometry and energy domain attractors are necessary to address deep penetration problems. TRIPOLI, thus, offers users an importance map, automatically estimated by the code either analytically or from a simplified adjoint calculation using the Djikstra algorithm. For radiation protection applications, Green's functions may be computed to perform source parametric calculations, i.e., different sources can be multiplied by the precomputed Green's functions, thereby providing high-quality results in a short time. For core physics applications, entropy diagnostics are provided to assess the source convergence (with both Shannon and Boltzmann entropies).

A surface restart feature is another approach for parametric problems. With this capability, an initial run is made and surface sources are tallied and then used for several subsequent runs, each with possibly different variance reduction options. Reactor dismantling studies are a good example of such application; a neutron surface source is computed at the reactor boundary and then used to compute tallies at different locations in the concrete outside the reactor.

The problem geometry may be input in either a combinatorial mode or as surface-bounded volumes, and the particle trajectories are tracked with the native TRIPOLI-4 geometry module. Beginning with version 5, TRIPOLI-4 is compatible with the ROOT geometry module developed by the high-energy physics community and maintained by CERN. All ROOT compatible pre- and postprocessors are therefore available for a TRIPOLI simulation.

Developed separately from TRIPOLI-4 is the SALOME TRIPOLI (www.salome-platform.org) preprocessor, offering a full 3-D graphical environment to make a complete TRIPOLI-4 input deck. This preprocessor is also available from the NEA Data Bank and the Radiation Safety Information Computational Center (RSICC).

Useful websites for TRIPOLI include:

http://www.oecd-nea.org/tools/abstract/detail/nea-1716/
https://rsicc.ornl.gov/codes/ccc/ccc8/ccc-806.html
https://inis.iaea.org/search/search.aspx?orig_q=RN:36093031

Appendix F

Minimal Standard Pseudorandom Number Generator

Numerous pseudorandom number generators are in use and there are many reasons given for preferring one over another. Chapter 5 of this text discusses various ways of generating pseudorandom numbers, by which is meant numbers uniformly distributed on the unit interval [0, 1]. Many of the examples and problems at the ends of the chapters in this text require a method of generating pseudorandom numbers. Different pseudorandom number generators will yield different results and the same pseudorandom number generator will yield a string of pseudorandom numbers that depends on the "seed" that is used to start the sequence. Thus, the answer one person may obtain to an example or a problem that requires pseudorandom numbers may differ from the answer another reader obtains to the same problem. This should not be viewed with alarm, as almost all Monte Carlo answers are approximate and small differences among solutions using different pseudorandom number sequences are inevitable. Nevertheless, the authors feel that there may be some merit to providing a basic pseudorandom number generator that readers can use if they wish.

As discussed in Section 3.5, Park and Miller (1988) proposed a *minimal standard* pseudorandom number generator to produce pseudorandom numbers uniformly over the interval [0, 1]. It is this generator, with an initial seed of 73,907, that is used for the many examples throughout this book. In this appendix, this generator is coded in four computer languages to allow you, the reader, to duplicate the example results. These routines will work correctly on any machine whose maximum integer is $2^{31} - 1$ or larger.

F.1 FORTRAN77

In FORTRAN77, variables beginning with the letters i–n are by default integers and all others are real. Thus, the type declarations in the routine below are not necessary, but are included for good programming practice.

```
      FUNCTION random(iseed)
c -------------------------------------------------
c Park & Miller minimal standard for a RN generator
```

Exploring Monte Carlo Methods. https://doi.org/10.1016/B978-0-12-819739-4.00024-X

```
C  -------------------------------------------------
      INTEGER iseed,ia,im,iq,ir,k
      REAL random,am
      PARAMETER (ia=16807, im=2147483647, am=1./im,
     &           iq=127773, ir=2836)
        k = iseed/iq
        iseed = ia*(iseed-k*iq)-ir*k
        IF(iseed.LT.0) iseed = iseed+im
        random = am*iseed
      RETURN
      END
```

F.2 FORTRAN90

Although the above FORTRAN77 routine also works in FORTRAN90, here it is using the new constructs allowed by FORTRAN90.

```
REAL FUNCTION random(seed)
! -------------------------------------------------
! Park & Miller minimal standard for a RN generator
! -------------------------------------------------
  IMPLICIT NONE
  INTEGER :: temp,seed
  INTEGER, PARAMETER :: a=16807,  m=2147483647,&
                   & q=127773, r=2836
  REAL,PARAMETER :: minv=1./m
    temp = seed/q
    seed = a*(seed-temp*q)-r*temp
    if(seed < 0) seed = seed + m
    random = minv*seed
  return
END FUNCTION random
```

F.3 Pascal

Although Pascal is not used much anymore, it is with this language that Park and Miller (1988) first implemented their algorithm.

```
function random : real;
(* integer version of the minimal standard *)
const
  a = 16807;
  m = 2147483647;
  q = 127773;
  r = 2836;
```

```
var
   lo, hi, test : integer
begin
   hi := seed div q;
   lo := seed mod q;
   test := a * lo - r * hi;
   if test > 0 then
     seed := test
   else
     seed := test + m;
   random := seed / m
end;
```

F.4 C and C++

The following routine works in both C and C++. The initial value of `seed` is set outside this routine (and never altered subsequently except by this routine). A random number is then generated with `random(&seed)`.

```
float random(int *seed)
// --------------------------------------------------
// Park & Miller minimal standard for a RN generator
// --------------------------------------------------
{
  int temp;
  const int A=16807, M=2147483647, Q=127773, R=2836;
  const float minv=1./M;

  temp = *seed/Q;
  *seed = A*(*seed-temp*Q)-R*temp;
  if(*seed < 0)
  {
    *seed = *seed + M;
  }

  return minv * (float)*seed;
}
```

F.5 Programming Considerations

The above routines have the initial seed x_0 set outside the routine and subsequent values of x_i are stored in this variable by the routine. The value of seed must only be changed from its initial value by the pseudorandom number routine. The returned value `random` is $\rho_i = x_i/m$.

Although these function routines work well, they can be modified to use special features of the various languages so that no function argument is needed. For example, in FORTRAN77 a labeled `COMMON` block could be used to store the value of `iseed`. In FORTRAN90, this function would better be written as a module so that the seed need not be passed as an argument. In C++, the function could be written as a new class.

References

Park, S.K., Miller, K.W., 1988. Random number generators: good ones are hard to find. Commun. ACM 31, 1192–1201.

Index

If a topic runs over to a second page, only the first page is referenced. For a topic discussed over several pages the first and last pages are given; such references often indicate a major treatment of the listed topic.

A

Absorbing boundary, 326–328
Absorbing media, 403, 414, 444
Absorption states, 298–300, 355
Accuracy of tally, 137–139
   random number generator, 138
   user errors, 138
Adjoint
   eigenvalue problem, 355
   finite difference equations, 318–321
   flux density, 397–400
   importance, 398–401
   linear equations, 297, 318–321
   operators, 398, 507–512
   transport equation
     see Transport equation
   utility, 400
Amphichiral cellular automata (CA), 92
AND binary operator, 83
Antithetic variates, 172–175
Arithmetic, modulo-2, 83, 85
Assessing random number generators, 101–103
Autoregressive sequence, 80

B

Bayes, Thomas, 220
Bayes' theorem, 39, 123, 222–225
   Bayesian analysis, examples, 223–227
   classical analysis, examples, 223–227
   decision theory, 242
   general form, 223

   making inferences, 241–248
     examples with data, 245–248
Bayesian statistical analysis
   conjugate distributions, 226, 231
   conjugate priors, 225–236
   estimating the prior, 238–240
   examples, 223, 225–227
   importance of conjugate priors, 230–232
   importance of the Gibbs sampler, 232
   influence of PCs, 232
   nonconjugate priors, 236–241
   posterior normal distribution, 234
   posterior probability intervals, 225–236
   use of Metropolis–Hasting algorithm, 232–238
   with Markov chain Monte Carlo, 225–248
Bernoulli, 3, 449, 497
Bernoulli distribution, 449
Beta distribution, 472–474
   as prior distribution, 230
Beyer quotient, 65
Biased estimators, 181, 191
Bienaymé–Chebyshev inequality, 498
Binomial distribution, 37, 450–454
   as likelihood function, 226
Bivariate normal distribution, 488
Boltzmann, Ludwig, 7, 388
Boltzmann equation
   see Transport equation
Brownian motion, 7, 361

557